Affect Engineering

Affect Engineering

A Unified Field Theory of Emotion

Marcus Woods

Squared by Woods Books

Hollister, California

Woods, Marcus C.

Affect Engineering: A Unified Field Theory of Emotion

ISBN: 978-0-9960493-1-3

Editor: Marcus Woods

Cover Design: Marcus Woods

Illustrator: Marcus Woods

Published by Squared by Woods Books, a sole proprietorship,

100 Maple Street # 2762, Hollister, CA, 95023

This book is dedicated

to my family,

for I would not be where I am today without them,

to my friends and colleagues,

for the inspiration and advise they have freely given me,

to every professor or coach I have ever had,

for showing me how to teach,

to all the scholars who came before me,

for doing most of the really hard work,

and to anyone brave enough to read this book.

Seriously, if you have made it this far past the cover then

you must believe I have something to say that is worth reading.

Thank you.

TABLE OF CONTENTS

Familiarity with or knowledge of the math concepts bracketed and in italics is recommended.

Preface

Affect Engineering, like many great projects, began on a tangent. While enrolled in a culture and emotions course during my junior year of college, I began mixing ideas from it with concepts from other classes I was taking that semester. In brief fits of inspiration, such as the kind arising from impending deadlines for term papers, the idea to use math to model emotions surfaced. Eventually, it would lead to the beginnings of a drive theory that I described in my final assignment for the class. Not surprisingly, the assignment entailed the creation of a theory of emotion. Affect engineering still had a very long way to go as far as becoming a new branch of knowledge was concerned, but the seed had taken root. Years of on and off refinement followed, until eventually the only question I had left was how to make it happen. This book is the start of it happening.

Since the actual writing and illustrating of *Affect Engineering* began, I have learned that once people find out someone is writing a book they want to know what it is about. I have also discovered that if the author tells them he is inventing a discipline then they will think him either frighteningly smart or completely crazy, with very little middle ground in-between. I make no claim to be a math or psychology prodigy, so the latter can be assumed for now. I also began this journey armed with little more than a calculator, a few dictionaries, and my sense of humor. While I do not expect everyone to understand all of the math in it right away, I wrote and illustrated *Affect Engineering* with the hope that everyone would be able to get something out of it, regardless of ability beyond basic math. It will not be easy, but many things worth doing in life rarely are.

"I don't want to be at the mercy of my emotions. I want to use them, to enjoy them, and to dominate them." - Oscar Wilde [1]

CHAPTER ONE

Body Praxis: Why a Math-Based Theory of Emotion?

It may be said that four kinds of people exist. There are those who know what they want, some who want what they know, others who want what they do not know, and the rest who do not know what they want. It is well known that emotions play a role in shaping behavior. Emotions can direct individuals to acquire or to avoid what is wanted or unwanted, though, in the courtroom of emotions, if one does not master a tool then the tool becomes the master. Not surprisingly, before any tool can be mastered, one has to know what it is and what it is not. Herein lies the problem.

The daunting challenge in crafting a genuine theory of emotion has not gone unnoticed by psychologists. Emotions mean different things to people or sometimes nothing at all. They have been described as bodily sensations by actors and dancers, metaphors by poets and songwriters, chemicals and hormones by reductionists, intuitive, a body praxis, or ineffable by those who feel no language can capture their essence. However, emotions cannot possibly be all of these and none at the same time. Either something has to give, or something better must be found to explain them.

In education, sports, the arts, and most endeavors, the difference between success and failure often depends on a person's level of preparation. One would be hard pressed to find someone capable of finishing a marathon who had never practiced the art of running, or someone capable of walking a tightrope who had never practiced maintaining balance. Most able-bodied individuals have the capacity to sing, perform gymnastics, dance, or swim by virtue of possessing a body, but this alone does not mean arms, legs, and vocal chords will translate into motor skills.

While natural aptitude enhances praxises, nothing substitutes for the timing, hand-eye-legear coordination, body awareness, endurance, and muscular strengthening that practice, training, and learning from experience provide. A well-endowed athlete may lose in a footrace to an otherwise average person for lack of dexterity or being a klutz. A talented musician who has never played a violin may not play it as successfully as someone who has practiced under tutelage or through experimentation. As the matter at hand concerns achieving mastery over one's emotions, uncovering other tools people have that might help them develop mastery over emotions will also be crucial.

Reasoning skills seem to be one of these tools. Many educators go to great lengths to ensure that pupils utilize their reasoning skills efficiently and productively. In most countries, the use of deductive and inductive reasoning skills, along with their fine tuning, occupies a large portion of an individual's life. Yet, a push is still being made for the need to develop emotional skills and bring their refinement up to par. Emotional intelligence is now being touted as a step toward curing some of society's ailments, such as crime, violence, bullying, and it is being adopted by many businesses to improve relationships with both employees and customers. But are reasoning skills separate from emotional skills, or is this a false dichotomy? If separate, one must show how they interact. If they are linked, then one must show what their common bond is. Most important, a consensus on what constitutes an emotion must first be reached.

Before insights can be made into why some people seem to manage their emotions better than others, emotion must cease to be an ambiguous term. Its equivocal nature is no secret in the scientific community. *Affect Engineering* will address the need to unite different traditions in the study of emotion, such as the Jamesian, the Cognitive, the Social Constructivist, and the Darwinian perspectives.[2] *Affect Engineering* presents itself as the foundation for a grand theory of emotion and outlines the framework necessary for answering questions such as the following:

1) What are emotions and how can they be modeled?

2) How can the understanding of one's own emotions be enhanced?

3) How can the understanding of another's emotions be enhanced?

4) Will training in understanding emotions enable someone to live more productively?

5) What relationship exists between emotion and reasoning? Are they diametrically opposed or separate rooms in the same building?

Until now, defining emotions has felt like a barehanded attempt to grab a bar of soap in a vat of oil. Slipperiness of language, culturally specific anomalies, dissimilarities in usage, innuendo, losses in meaning through translation, and the sheer number of emotions posited throughout history make it seem as if coming to an agreement on what an emotion is, or is not, would require a herculean effort. However, one language is least susceptible to differences in semantics, meaning, and interpretation. That language is math. Figures 1.1 and 1.2 depict the calculus functions that will will be used to direct the explanation of affect engineering throughout this book.

KEY
Anxiety Invested in an Entity (Measured in Emotional Units) = ***y***
Self-Distinction = ***SD***
Appraisal = ***AP***
Existence = ***EX***
Uniqueness = ***UN***
Sufficiency = ***SU***
Sentiment = ***SN***
Perceived Threat Severity = ***TSE***
Perceived Threat Susceptibility = ***TSU***
Attention to Threat = ***ATT***
Time = ***TI***
Reasoning to Threat = ***RET***
Half-Life of Attention = ***HA***
Perceived Response-Efficacy = ***REF***
Perceived Self-Efficacy = ***SEF***
Attention to Efficacy = ***ATE***
Reasoning to Efficacy = ***REE***

PAIN AVOIDANCE FUNCTION

$$y = \sqrt{SD} \otimes AP \otimes EX \otimes (UN \otimes SU \otimes SN \oplus 1) \,\widehat{\ }\, \left| \frac{(TSE \otimes TSU) \otimes (ATT) \otimes (.5) \,\widehat{\ }\, \dfrac{TI \otimes (1 \ominus RET)}{HA}}{(REF \otimes SEF) \otimes (ATE) \otimes (.5) \,\widehat{\ }\, \dfrac{TI \otimes (1 \ominus REE)}{HA}} \right|$$

Figure 1.1 The above is the function for emotions concerning the avoidance of pain. The underlying premise holds that if an entity is positively appraised with respect to restoring equilibrium between a purpose and its complementary purpose, and a threat of harm to the entity is both severe and probable, and the belief in one's ability to acquire or protect the entity is low, then a high amount of anxiety will be invested into valuing the entity until it is acquired, protected from harm, or anxiety investment is directed away by reasoning and attentional processes.

KEY
Anxiety Invested in an Entity (Measured in Emotional Units) = ***y***
Self-Distinction = ***SD***
Appraisal = ***AP***
Existence = ***EX***
Uniqueness = ***UN***
Sufficiency = ***SU***
Sentiment = ***SN***
Perceived Threat Severity = ***TSE***
Perceived Threat Susceptibility = ***TSU***
Inattention to Threat = ***ITT***
Time = ***TI***
Reasoning to Threat = ***RET***
Doubling Time of Inattention = ***DTI***
Perceived Response-Efficacy = ***REF***
Perceived Self-Efficacy = ***SEF***
Inattention to Efficacy = ***ITE***
Reasoning to Efficacy = ***REE***

PLEASURE PURSUIT FUNCTION

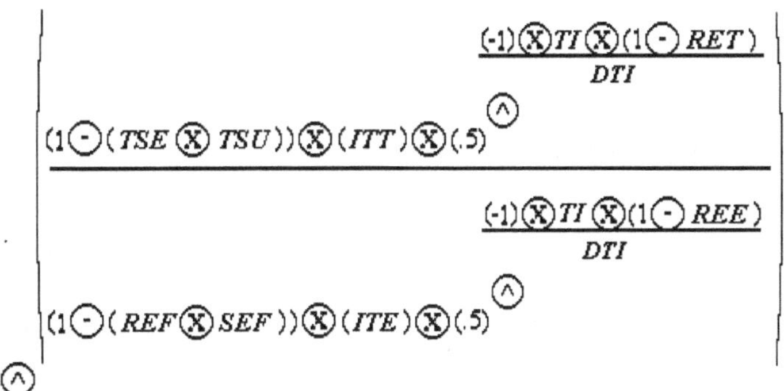

$$y = \sqrt{SD} \otimes AP \otimes EX \otimes (UN \otimes SU \otimes SN \oplus 1) \, \textcircled{\wedge}$$

Figure 1.2 The above is the function for emotions concerning the pursuit of pleasure. The underlying premise holds that if an entity is negatively appraised with respect to restoring equilibrium between a purpose and its complementary purpose, and a threat of harm to the entity has both a low severity and a low probability of occurrence, and the belief in one's ability to acquire or protect the entity is high, then a high amount of negative anxiety will continue to be invested into valuing the entity for as long as it is acquired, protected from harm, or until the negative anxiety invested in it is directed away by reasoning and attentional processes.

Affect engineering does not aim to make concepts more complex by disguising them as variables and numbers. Instead, it is intended to be a novel technique for classifying emotions. Using the language of math and logic, it distinguishes instances of one emotion from another, can be transformed into a neurological model to depict the arousal of emotions at the level of neurotransmitters and substrates, and may be transformed into a model of drives depicting decision making when multiple purposes are in consideration.

The functions will be explained in a step by step manner proceeding from the base and expanding outwards. Logical positivism, being one of the most difficult philosophies to appease, will be used as a guide in the explanation of the equation to ensure everything remains subject to empiricism. Although the exploration of affect engineering here is theory-driven, all aspects of the functions, along with the models derived from them, are structured on the rules of mathematics, logic, principles of psychology, biology, physics, and lend themselves to empirical testing.

Finally, the structure of affect engineering is dynamic. As new scientific discoveries are uncovered, if expected relationships between variables change, and if concepts need modifying, then they can be implemented directly in the equation long after an architect has come and gone. It is hoped that the reader, regardless of ability in math, psychology, biology, physics, computer science, or any of the social sciences, will be able to use affect engineering in order to:

1) Identify, understand, and manage emotions in oneself and others!
2) Become the master of his or her own tools!
3) Translate emotions into the language of math!
4) Conceptualize the structure for a grand theory of emotions!
5) Visualize the many applications of a new, radical discipline, affect engineering!

Notes

1. Wilde, Oscar (2003). *The Picture of Dorian Gray.* New York: Barnes and Nobles Books. Print, p. 112. (Original work published 1890).

2. Cornelius, R. R. *The Science of Emotion: Research and Tradition in the Psychology of Emotion*. Upper Saddle River, NJ. Prentice-Hall, Inc. 1996. Print, p. 12.

"To define is to limit." - Oscar Wilde [1]

CHAPTER TWO

Front-loading

Like other fields of knowledge, psychology is rich with technical language, clumsy phrases, obscure and sometimes archaic jargon. To prevent the tripping over these linguistic hurdles, at the start of each chapter important terminology will be listed in a section entitled Front-loading. Specialized vocabulary is identified here and again at the start of each chapter so it can be realized that each is an instance of a particular meaning that is being used. In most cases their meanings can be inferred from reading the text. At the end of each chapter is a concept check. There, terminology essential for the reader to know before proceeding will be identified. In a few cases, terms have had their meanings invented by the author to describe a specific idea. In such instances, their meanings can be deduced from the basic concepts that make up the components of the compound term. One such concept is that of negative anxiety, which was created by combining the definition of negative with the definition of anxiety. Negative anxiety is conceived as the opposite of anxiety. Other terms that are invented by the author have their meanings explained within the text in a similar fashion.

Recommended sources to obtain definitions of most terms include the *Penguin Dictionary of Psychology*, by Arthur Reber, Rhianon Allen, and Emily Reber, or the *Merriam-Webster Dictionary*, published by Merriam-Webster Incorporated. Terminology unique to specific theories and scholars will be noted when introduced.

Chapter III: Introduction to Affect

Affect, Cognition, Conation, Emotion, Feeling, Negative Affect, Positive Affect

Chapter IV: The Origins of Psychoanalytic Theory

Death Instinct, Defense Mechanism, Drive, Ego, Ego-Ideal, Eros, Id, Life Instinct, Pleasure-Pain Principle, Primary Process, Psychoanalysis, Psychodynamic, Reality Principle, Super Ego, Thanatos

Chapter V: Value and Anxiety: Laying the Foundation for a Numeric Analysis of Affect

Action Potential, Anxiety, Appraisal, Biological Clock, Circadian Rhythm, Cognitive Marker, Emotional Unit, Entity, Existence, Goals, Object, Purpose, Sentiment, Sufficiency, Time, Psychological Time, Uniqueness, Utility, Value Judgement

Chapter VI: A One to One to One to One Ratio 1:1:1:1 and Appraisals

Appraised Value Judgment, Bound Purpose, Complementary Purpose , Model, Double Bind, Drive, Drive Reduction Theory, Homeostasis, One to One to One to One Ratio, Pain, Pleasure, Ray, Vector

Chapter VII: A Roll Cage Model of Drive Reduction Theory, and Identifying Emotions

Front-loading: Anger, Breaking Point, Cone and Disc Apparatus, Content, Courage, Disgust, Drive to Maintain All Drives, Drive to Not Maintain All Drives, Entity-mass, Euphoria, Fear, Grief, Guilt, Happiness, Id-mass, Negative Anxiety, Negative Anxiety-mass,

Front-loading

Positive Anxiety-mass, Roll Cage, Roll Cage Theory of Drives, Sadness, Superego-mass, Utility-mass

Chapter VIII: Stimuli, Modifiers of an Entity's Appraisal, and Category I Emotions, the Intra-Personal Emotions

Avoidance of Pain Equation, Benefits, Bottom-up Processing, Category I Emotions: Intra-Personal, Classical Conditioning, Cognitive-Motivational-Relational Theory, Contingency Theory, Conditioned Stimuli, Cortisol, Distal Stimulus, Excitatory Synapse, Fight-or-Flight Response, Harm, Hormones, Inhibitory Synapse, Negative Reinforcement, Neurotransmitter, Operant Conditioning, Perceived Benefit Intensity, Perceived Benefit Susceptibility, Perceived Response Efficacy, Perceived Self-Efficacy, Perceived Severity of Threat, Perceived Vulnerability to the Threat, Positive Reinforcement, Protection Motivation Theory, Primary Appraisals, Proximal Stimulus, Punishment, Pursuit of Pleasure Equation, Reinforcement, Reward, Secondary Appraisals, Social Constructionism, Top-down Processing, Unconditioned Stimuli

Chapter IX: Neurons, Empathy, and Category II Emotions (Inter-personal)

Antipathetic Mercy, Antipathy, Category II Emotions: Inter-Personal, Companionate Love, Empathy, Four Degrees of Empathy, Gamma-aminobutyric acid, Glutamate, Hateful Humiliation, Hatred, Imaginary Numbers, Indifference, Limerence, Loneliness, Love, Loving Pride, Mercy, Mirror Neurons, Neutrality, Other, Pride, Romantic Love, Self, Self-Distinction, Shame, Sympathetic Shame, Sympathy, Vicarious Humiliation, Vicarious

Mercy, Vicarious Pride, Vicarious Shame, Vicarious Loneliness

Chapter X: Motivation, Perception, Attention, and Reasoning

Associative Percept, Attenuation Theory, Attention, Attentional Decay, Blindsight, Blind Spot, Decay Theory, Deductive Reasoning, Desensitization, Executive Functioning, Extrinsic Motivation, Filter Theory, Forgetting, Gestalt Laws of Organization, Half-life of Attention with Error, Half-life of Attention without error, Inattentional Blindness, Inductive Reasoning, Inferred Percept, Intrinsic Motivation, Macrocosmic, Microcosmic, Motive, Motivation, Multi-tasking, Neglect, Organization, Percept, Perception, Positivism, Realism, Reasoning, Scotoma, Solipsism, Tabula Rasa, Template Matching, Volition, Will

Chapter XI: Category III Emotions: Compound Interactive Emotions

Benevolence, Category III Emotions: Compound Interactive Emotions, Envy, Indulgent Type, Jealousy, Malevolence, Primary Emotions, Protective Type, Secondary Emotions

Chapter XII: Emotive States (Category IV Emotions) and Time

Biological Rhythm, Category IV Emotions: Emotive States, Confusion, Delirium, Entropy, Fatigue, Greed, Joyfulness, Helplessness, Infradian Rhythm, Learned Helplessness, Mood, Post-Traumatic Stress Disorder (PTSD), Restlessness, Surprise, Ultradian Rhythm

Chapter XIII: Multidimensional Models for Multiple Entities, Multiple Purposes, Multiple People, and Ethics

Hypercube, Nature, Non-regulatory Drives, Nurture, Primary Drives, Regulatory Drives, Secondary Drives

Chapter XIV: Application to Defense Mechanisms

Acting Out, Altruism, Anticipation, Conversion, Delusion, Denial, Displacement, Dissociation, Distortion, Fantasy, Humor, Hypochondriasis, Idealization, Identification, Intellectualization, Introjection, Isolation, Passive Aggressive, Projection, Rationalization, Reaction Formation, Regression, Repression, Somatization, Splitting, Sublimation, Suppression, Undoing, Withdrawal

Chapter XV: Conclusion and Future

Cognitive Tradition, Darwinian Tradition, Jamesian Tradition, Natural Selection, Social Constructivist Tradition

Notes

1. Wilde, Oscar. *The Picture of Dorian Gray.* New York: Barnes and Nobles Books. Print, p. 200. (Original work published 1890).

"The aim of life is self-development. To realize one's nature perfectly - that is what each of us is here for." - Oscar Wilde [1]

CHAPTER THREE

Aspirations and an Introduction to Affect

Front-loading: Affect, Cognition, Conation, Emotion, Feeling, Negative Affect, Positive Affect

It is hard to manage something if one is not measuring it. A budget cannot be balanced if one neither knows what is being balanced nor knows how much of it is possessed and being spent. Therefore, the first course of action in establishing a mathematical theory of emotion would to be to start with a tenuous definition of what an emotion is or is not. The second course concerns determining all of the emotions to be considered and how they might be measured or quantified. Etymology, the study of the history of words, will be the starting point.

The root of emotion, from the Latin *emovere*, translates as "to move, to excite, to stir up or agitate" and suggests something about its nature.[2] One look at the major news outlets, tabloids, or the sensationalism of the film industry would confirm that emotions are regularly exploited for their ability to move, excite, or agitate people. Tentatively defining emotion as a force is perhaps the least erroneous description available for now. Indeed, professionals in psychology have observed that coming to an agreement on what an emotion is has proved elusive, noting the term has historically "proven utterly refractory to definitional efforts" and that likely "no other term in psychology shares its combination of non-definability and frequency of use."[3] Contemporary usage has included its employment as an "umbrella term" for a multitude of "subjectively experienced, affect-laden states" whose labels and meanings are arrived at through consensus, or as a label for the scientific fields that investigate and explore "environmental, physiological and cognitive factors that underlie these subjective experiences."[4] The task at hand easily resembles an attempt to put together a puzzle where none of the pieces fit together perfectly with one another. In order to build a grand theory, gaps will need filling in and new pieces will have to be forged or thrown out altogether.

Throughout this book, the term *affect* will indicate feeling or emotion. Feeling is generally defined as "an affective state, as in a sense of well-being, depression, desire, etc."[5] However, before continuing on this endeavor, some basics about the mind need addressing. Given the preponderance of mental techniques available for regulating emotion (e.g., meditation, hypnosis, reflection, introspection), a serious inquiry into the nature of affect must account for all that the mind is capable of doing. Historically, conation, cognition, and affect were the basic mental faculties in psychology.[6]

Cognition is a broad term that "has been traditionally used to refer to activities such as thinking, conceiving, and reasoning."[7] Its scope, being as broad as it is, permits a number of ways that it can be examined or approached. Affect is a "term used more or less interchangeably with various others, such as emotion, emotionality, feeling, mood, etc."[8] It follows that positive affect generally refers to good feelings, such as those arising from the successful completion of a task, while negative affect refers to bad feelings, such as those arising from the failure to achieve a goal. Finally, conation, the last of the three parts, is concerned with volition, striving, and willing.[9] Conation can be thought of as the executive or C.E.O. of the mind in the sense that it transforms cognition and affect into action. If affect, cognition, and conation, are organized into a tentative Venn diagram, then the broad considerations for emotion will be more manageable.

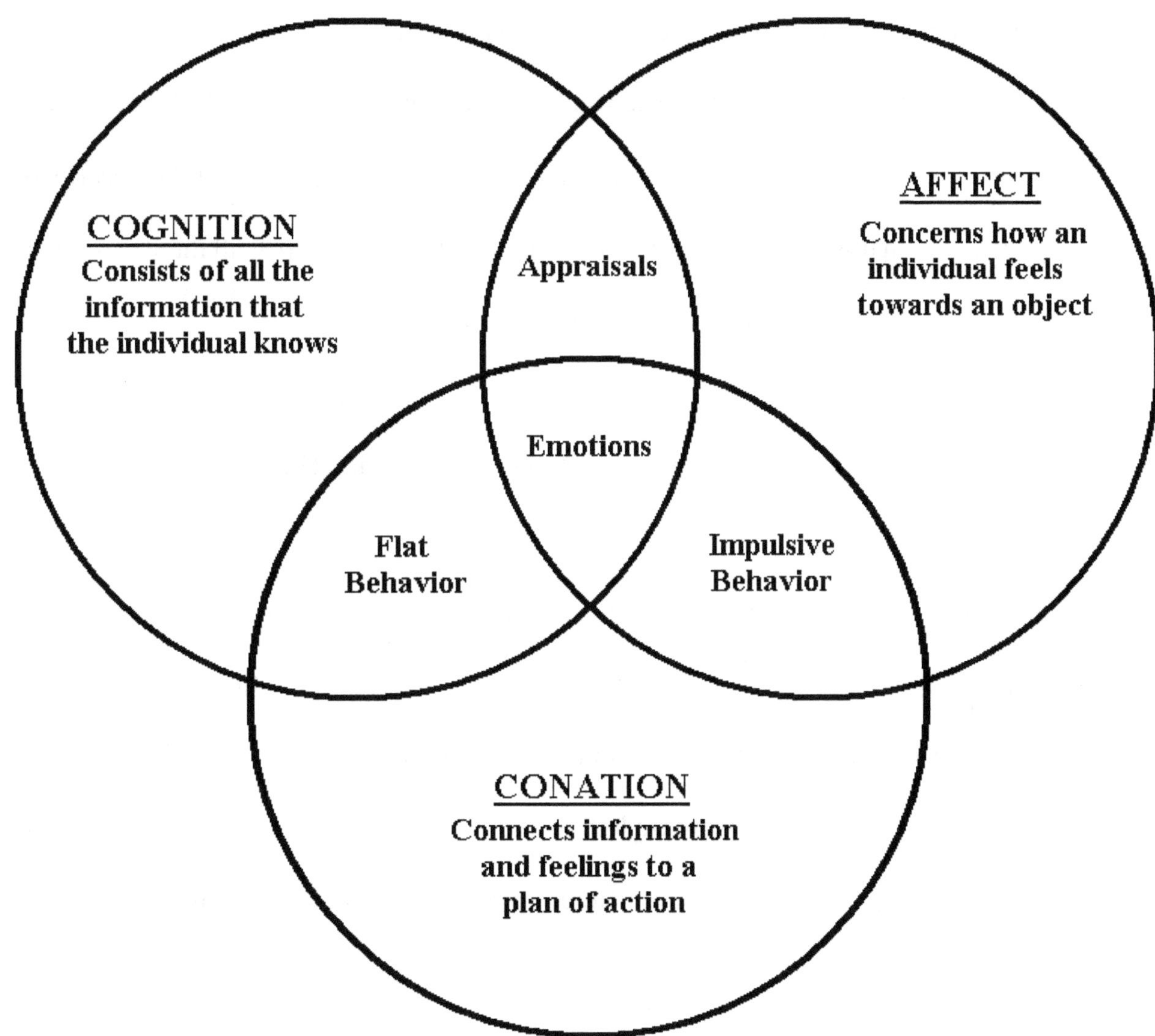

Figure 3.1 The Venn diagram above organizes cognition, affect, and conation.

There are three main circles. One contains affect, which concerns how an individual feels toward an object, another contains cognition, which consists of all the information the individual knows, and the final contains conation, which connects information and feelings to a plan of action. A partial overlap of cognition with conation but not affect would imply flat behavior. Flat behavior, as its name suggests, would be behavior that is devoid of an appropriate emotional response. Impulsive behavior is suggested where conation overlaps with affect but not with cognition. Impulsive behavior would consist of "acts carried out without reflection."[10] Appraisals are suggested where affect overlaps with cognition but before conation enters the mix, and will be described in detail in chapter seven. In this theoretical Venn diagram, a complete overlap of all three components would be an emotion designed for navigating situations in the world.

Plutchik's Emotion Solid and Psychoevolutionary Theory

A Ferris wheel depiction of Plutchik's Psychoevolutionary Theory is shown in figure 3.2 to provide a glimpse of previous theories of emotion that will be considered in this work. The letters A, B, C, D, E, F, G, and H would be basic emotions with their polar opposite being directly across, such as A and E, or B and F. Their lowercase letters a, b, c, d, e, f, g, and h would be less intense manifestations of the basic emotions, while different pieces connecting the spokes of the wheel together would represent emotions that are the product of one or more basic emotion, hence, mixed emotions.[11]

A look at the number of basic emotions listed in such an array offers both an idea of the tremendous work already done by emotional scholars and the enormity of the task that lies ahead. This is but one theory of emotions among several dozen others. Scores of emotions and feelings,

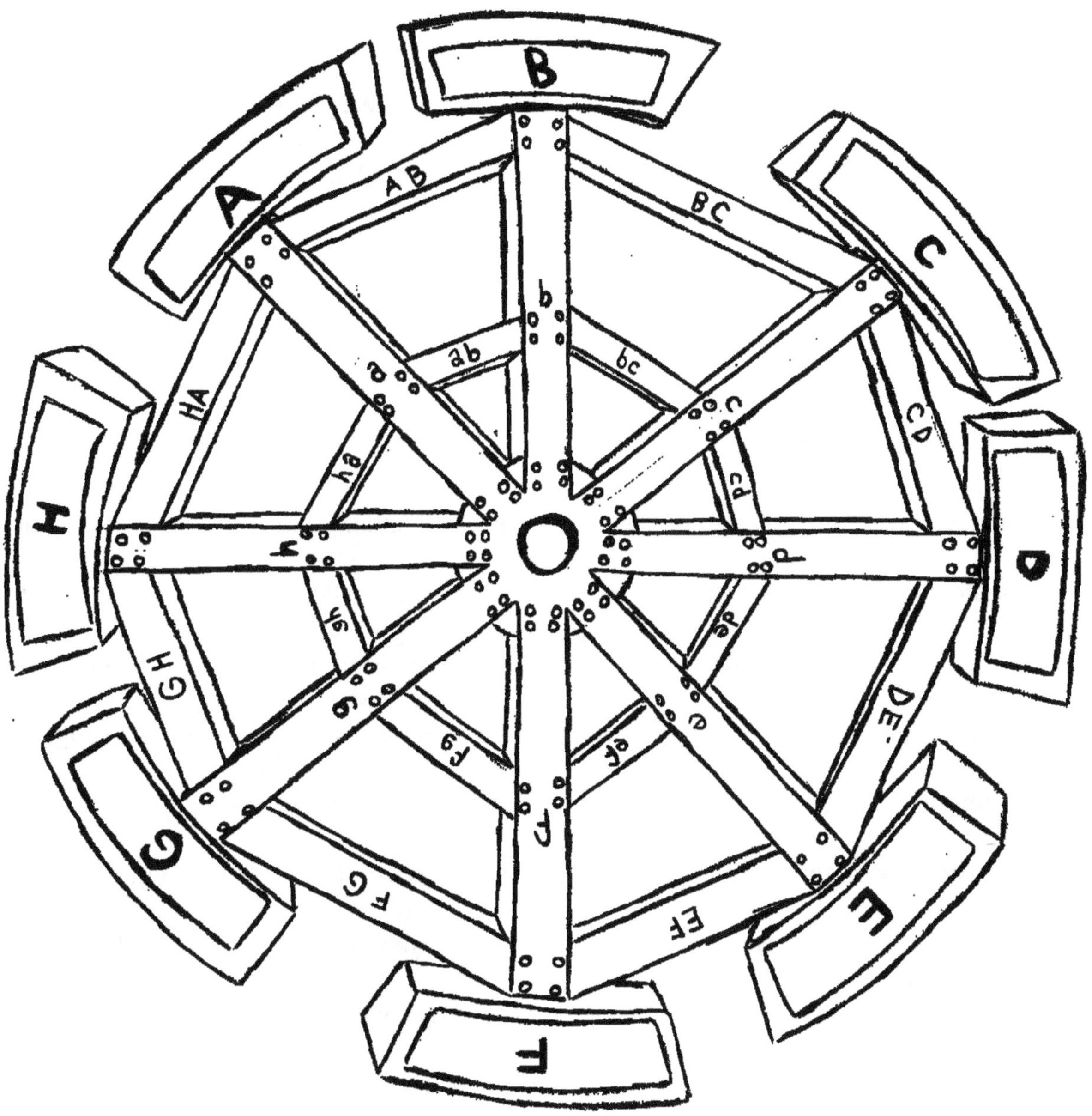

Figure 3.2 A Ferris wheel depiction of Plutchik's organization scheme.

each with different characteristics, intensities, and descriptions, have been suggested throughout the course of history by individual theorists, groups of people, cultures, nations, and languages. If a grand theory of emotions is to be created, whatever approach taken must ultimately be able to account for the empirical findings of Plutchik's model and more. While it might be tempting to give up and simply claim there are as many theories of emotion as there are people in the world who experience them, nothing of the sort will be happening here. There is no need to reinvent the wheel, so to speak, but if it rolls on jilted edges, the spokes are distorted, and the axles do not fit properly, then by all means an attempt should be made to fix it.

The Road Ahead

On a positive note, it is also fortunate that ideas abound. A number of theories are already present on the topic of emotions; from these, inferences can be made and disagreements dissolved. Whenever possible and appropriate, existing models, terms, or definitions will be used, incorporated, or modified to explain a new concept in the equation. Most theories of emotion recognize the following four classes of factors: “cognitive appraisals,” “instigating stimuli,” “physiological and neurological correlates,” and “motivational properties.” [12] Each of these factors will be considered in the explanation of affect engineering. In chapter six cognitive appraisals will be addressed. Instigating stimuli will be addressed in chapter eight. Physiological and neurological correlates will be given attention in chapter nine and motivational properties in the tenth chapter.

Many theories of emotion, personality, and the mind have come before the ideas presented in this book. Some ideas will be borrowed, some built upon, and some transformed in the attempt to create an integrated whole. It is hoped that the conclusions drawn will inspire others to look at their own emotions in the manner suggested and to build on the views and theories presented here.

Before diving into this journey headlong, some consideration will be given to the origins of psychoanalytic and psychodynamic theories, specifically the contributions made by Sigmund Freud to the study of the mind and its components. From there, various concepts concerning cognition, affect, and conation, will be brought together to explain the logic of the mathematical functions from chapter one, how emotions are assessed and interpreted with the functions, how their physiological arousal can be mathematically expressed, managed, and how their influence on decision-making might be modeled. The ultimate goal is to devise an approach to the study of affect that will yield a manner by which to operationally define any emotion or feeling under any context. The practical uses for such a model would be manifold in any number of fields where human choice can influence the outcome of a situation, such as in the explanation of past behavior or the prediction of future behavior. Toward the end of this work, the application of the equation will be set to psychodynamic psychology, where the equation will be employed to demonstrate its explicative prowess by modeling a number of defense mechanisms from psychodynamic theories.

YOU
SHOULD
UNDERSTAND
THIS.
CHAPTER THREE
CONCEPT CHECK
VOCABULARY
Affect
Cognition
Conation
Emotion
Feeling
Negative Affect
Positive Affect

Notes

1. Wilde, Oscar (2003). *The Picture of Dorian Gray.* New York: Barnes and Nobles Books. Print, p. 20. Original work published 1890.

2. Reber, Arthur S., Rhianon Allen, and Emily S. Reber. *Penguin Dictionary of Psychology*. London. Penguin Books, 2009. Print, p. 256.

3. Reber, Arthur S., Rhianon Allen, and Emily S. Reber. *Penguin Dictionary of Psychology*. London. Penguin Books, 2009. Print, p. 256.

4. Reber, Arthur S., Rhianon Allen, and Emily S. Reber. *Penguin Dictionary of Psychology*. London. Penguin Books, 2009. Print, p. 256.

5. Reber, Arthur S., Rhianon Allen, and Emily S. Reber. *Penguin Dictionary of Psychology*. London. Penguin Books, 2009. Print, p. 294.

6. Reber, Arthur S., Rhianon Allen, and Emily S. Reber. *Penguin Dictionary of Psychology*. London. Penguin Books, 2009. Print, p. 152.

7. Reber, Arthur S., Rhianon Allen, and Emily S. Reber. *Penguin Dictionary of Psychology*. London. Penguin Books, 2009. Print, p. 139.

8. Reber, Arthur S., Rhianon Allen, and Emily S. Reber. *Penguin Dictionary of Psychology*. London. Penguin Books, 2009. Print, p. 17.

9. Reber, Arthur S., Rhianon Allen, and Emily S. Reber. *Penguin Dictionary of Psychology*. London. Penguin Books, 2009. Print, p. 152.

10. Reber, Arthur S., Rhianon Allen, and Emily S. Reber. *Penguin Dictionary of Psychology*. London. Penguin Books, 2009. Print, p. 374.

11. Cornelius, R. R.. The Science of Emotion: Research and Tradition in the Psychology of Emotion. Upper Saddle River, NJ. Prentice-Hall, Inc. 1996. Print, p. 46-48

12. Reber, Arthur S., Rhianon Allen, and Emily S. Reber. *Penguin Dictionary of Psychology*. London. Penguin Books, 2009. Print, p. 256.

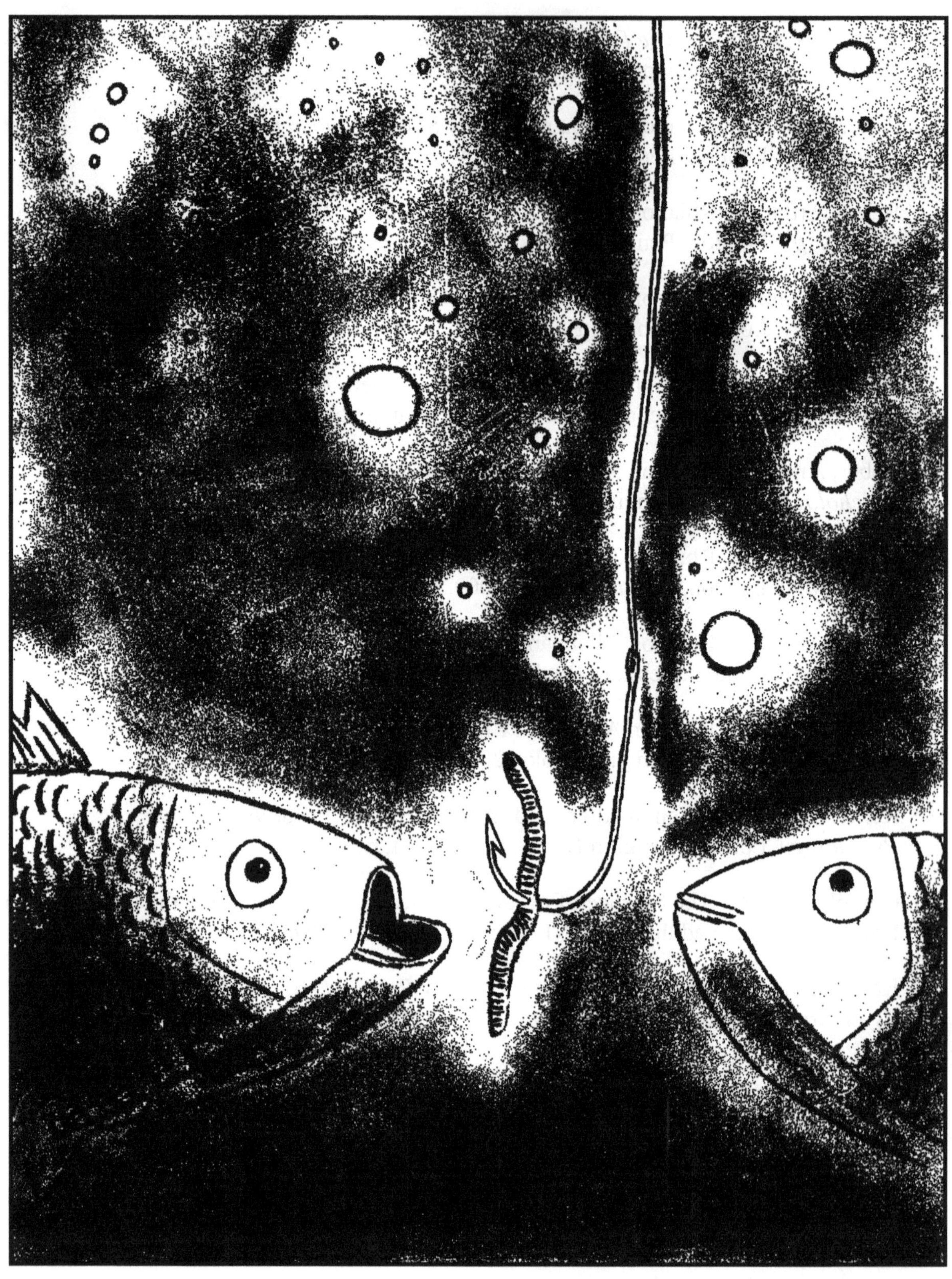

"The only way to get rid of temptation is to yield to it." - Oscar Wilde [1]

CHAPTER FOUR

The Origins of Psychoanalysis

Front-loading: Death Instinct, Defense Mechanism, Drive, Ego, Ego-Ideal, Eros, Id, Life Instinct, Pleasure-Pain Principle, Primary Process, Psychoanalysis, Psychodynamic, Reality Principle, Super Ego, Thanatos

Developed by the Austrian psychiatrist Sigmund Freud (1856-1939), psychoanalysis is a "set of techniques for exploring the underlying motivations of human behavior."[2] As it was the first of its kind, psychoanalysis set the foundation for other psychodynamic theories that followed, with psychodynamic being a label for those theories that "make motivation and drive central concepts."[3] The concepts considered here from psychodynamic theories will be introduced from a psychoanalytic standpoint, as different theorists tend to have their own specialized subsets of meanings for often times the same terminology.

Of the fight for life against death, much has already been written by storytellers, and for good reason. Its ubiquity among life forms suggests that any method for assessing affect will ultimately have to account for the struggle to survive. An instinct is "an unlearned response characteristic of the members of a given species."[4] In the Freudian school of psychology theory, Freud identified a "life instinct" and a "death instinct."[5] The life instinct was designated by the name Eros, the Greek god of love, and referred to the "whole complex of life-preservative instincts."[6] The death instinct was designated by Thanatos, the Greek god of death, and referred to the"theoretical generalized instinct for death and destruction."[7] The life and death instincts may also be thought of as drives, or "motivational states produced by (a) deprivation of a needed substance such as food, a drug or a hormone, or (b) presence of a noxious stimulus such as a loud noise."[8]

Additionally, the mind was viewed as a "tripartite model," consisting of the id, ego and superego, with the id being regarded as the "deepest component of the psyche, the true unconscious," and described as a "primitive, animalistic, instinctual element, the pit of roiling, libidinous energy demanding immediate satisfaction."[9] Moreover, in the classical theory of psychoanalysis, operations of the id consisted of both "strivings for gratification and withdrawals from or avoidances of the unbalanced tension and excessive affect of pain and/or unpleasure," and adhered to the pleasure-pain principle.[10]

Secondly, the ego in classical psychoanalytic theory represented a cluster of perceptual processes that served to "mediate between the primitive instinctual demands of the id, the internalized social, parental inhibitions and prohibitions of the superego, and the knowledge of reality."[11] The ego was thought to be governed by the reality principle, a "recognition of the real environment by a child, the growing awareness of its demands and the need to accommodate to them."[12] Although the ego is sometimes used to refer to the self or person as a whole in many theories, to avoid ambiguity the term *ego* will only be utilized in the psychoanalytic sense throughout this work, while the term *individual* will generally be used to indicate the entirety of a person.

Thirdly, in Freud's model the superego was the part of the psyche "associated with ethical and moral conduct and conceptualized as responsible for self-imposed standards of behavior."[13] The ego ideal, in classical psychoanalytic theory, is distinguished from the superego in that "the former represents prescriptions for life, is modified through growth and experience, and behavior that violates it produces shame," whereas "the latter represents proscriptions, is fixed at a young age, and behavior in conflict with it evokes guilt."[14]

One way to conceptualize these three components of the mind is to think of them through the metaphor of an anthropomorphized sports car. The id would be represented by the physical parts of the automobile (e.g., engine, chassis, fuel, tires). It has places to go and things it wants to do right now! The driver of the vehicle would represent the ego, as the vehicle requires an operator to ensure that it finds the best route to get where it is going safely. Some of the ego's responsibilities would include the decision to drive on a paved road as opposed to off-road, or to determine whether or not a particular path is treacherous (e.g., icy street vs. a detour). A healthy respect for the rules of the road (e.g., instilled by one's parents), such as speed limits, stopping at red lights, or driving on the left or right side of the road, can be thought of as the superego. The driver or ego must always be aware of traffic laws and what an ideal driver does (superego and ego-ideal) to avoid the unthinkable, such as the car being impounded, totaled in an accident, or the id being denied any access to fulfillment whatsoever, as in license suspension.

Figure 4.1 A car, driver, and traffic laws are representing the id, ego, and superego, respectively.

Finally, in psychoanalytic theory the flow of psychic energy was theorized to operate much like heat does in the laws of thermal dynamics, and the term primary process described "mental functioning operative in the id" that was "conceptualized as unconscious, irrational, ignorant of time and space, and governed by the pleasure-pain principle."[15] In figure 4.1 the primary process may be thought of as the uninhibited flow of fuel and energy to the car's engine, radio, stereo system, or air conditioning. An undisciplined adolescent might drive a new car in a reckless, thrill-seeking manner with the engine, stereo and air-conditioning on full blast whereas an adult with more restraint is likely to obey the rules of the road and generally exhibit decorum.

In classical psychoanalytic theory, the ego may employ defense mechanisms to protect itself from tension between the id and superego. For instance, in the car analogy, if the driver put the car in cruise control to avoid the temptation to speed on the highway, then it might be said that the defense mechanism of suppression was used to contain an unacceptable impulse of the car to speed down the road. Necessarily, the inherent complexity of defense mechanisms requires that they be considered after the mathematical function has already been established and described to assess affect. With these items being addressed, it is time to begin.

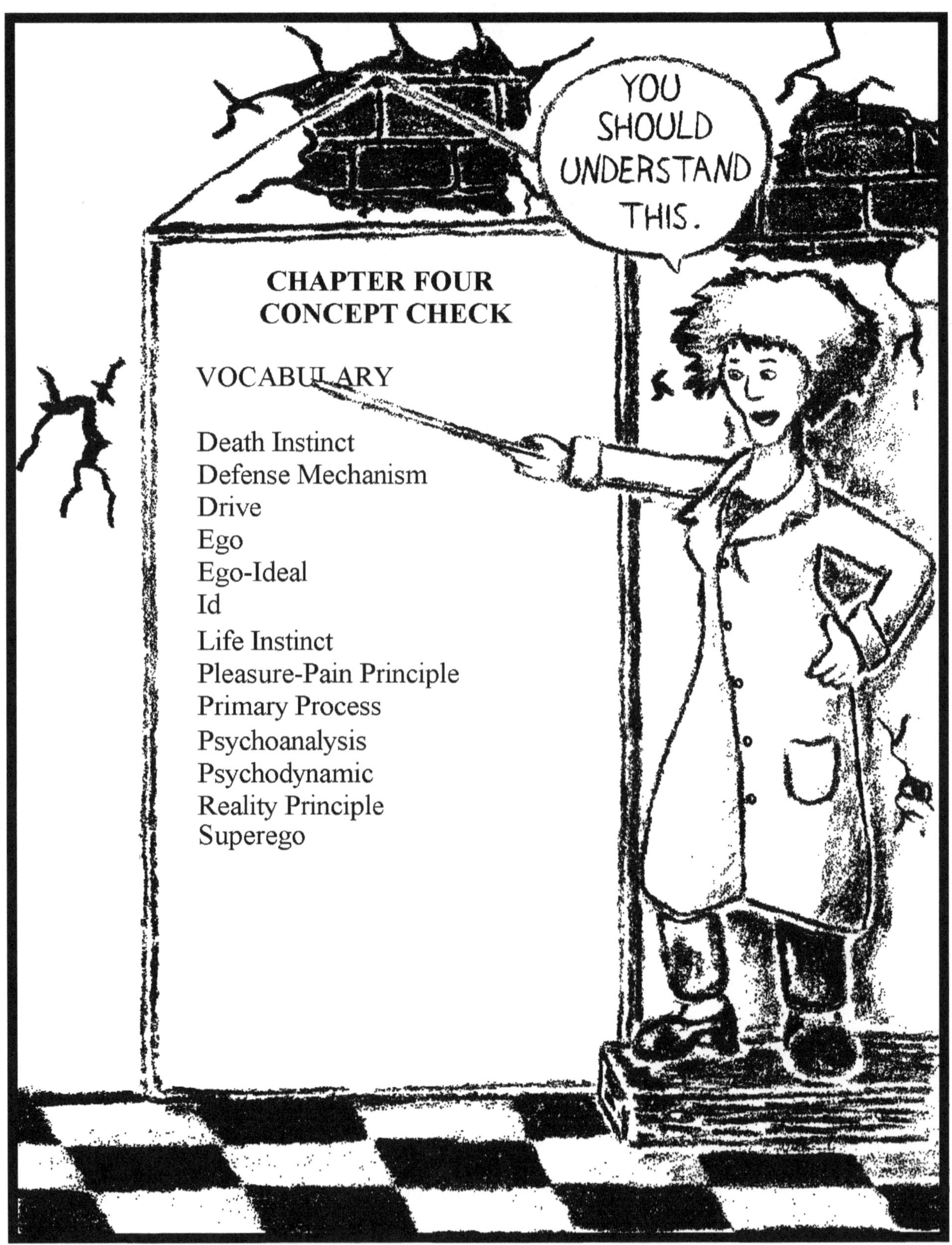
YOU SHOULD UNDERSTAND THIS.
CHAPTER FOUR
CONCEPT CHECK
VOCABULARY
Death Instinct
Defense Mechanism
Drive
Ego
Ego-Ideal
Id
Life Instinct
Pleasure-Pain Principle
Primary Process
Psychoanalysis
Psychodynamic
Reality Principle
Superego

Notes

1. Wilde, Oscar (2003). *The Picture of Dorian Gray.* New York: Barnes and Nobles Books. Print, p. 21. (Original work published 1890).

2. Reber, Arthur S., Rhianon Allen, and Emily S. Reber. *Penguin Dictionary of Psychology*. London. Penguin Books, 2009. Print, p. 631.

3. Reber, Arthur S., Rhianon Allen, and Emily S. Reber. *Penguin Dictionary of Psychology*. London. Penguin Books, 2009. Print, p. 632.

4. Reber, Arthur S., Rhianon Allen, and Emily S. Reber. *Penguin Dictionary of Psychology*. London. Penguin Books, 2009. Print, p. 387.

5. Reber, Arthur S., Rhianon Allen, and Emily S. Reber. *Penguin Dictionary of Psychology*. London. Penguin Books, 2009. Print, p. 387.

6. Reber, Arthur S., Rhianon Allen, and Emily S. Reber. *Penguin Dictionary of Psychology*. London. Penguin Books, 2009. Print, p. 269.

7. Reber, Arthur S., Rhianon Allen, and Emily S. Reber. *Penguin Dictionary of Psychology*. London. Penguin Books, 2009. Print, p. 810.

8. Reber, Arthur S., Rhianon Allen, and Emily S. Reber. *Penguin Dictionary of Psychology*. London. Penguin Books, 2009. Print, p. 235.

9. Reber, Arthur S., Rhianon Allen, and Emily S. Reber. *Penguin Dictionary of Psychology*. London. Penguin Books, 2009. Print, p. 366.

10. Reber, Arthur S., Rhianon Allen, and Emily S. Reber. *Penguin Dictionary of Psychology*. London. Penguin Books, 2009. Print, p. 592-593.

11. Reber, Arthur S., Rhianon Allen, and Emily S. Reber. *Penguin Dictionary of Psychology*. London. Penguin Books, 2009. Print, p. 248.

12. Reber, Arthur S., Rhianon Allen, and Emily S. Reber. *Penguin Dictionary of Psychology*. London. Penguin Books, 2009. Print, p. 657.

13. Reber, Arthur S., Rhianon Allen, and Emily S. Reber. *Penguin Dictionary of Psychology*. London. Penguin Books, 2009. Print, p. 289.

14. Reber, Arthur S., Rhianon Allen, and Emily S. Reber. *Penguin Dictionary of Psychology*. London. Penguin Books, 2009. Print, p. 249.

15. Reber, Arthur S., Rhianon Allen, and Emily S. Reber. *Penguin Dictionary of Psychology*. London. Penguin Books, 2009. Print, p. 614.

"Nowadays people know the price of everything and the value of nothing." - Oscar Wilde[1]

CHAPTER FIVE

Value and Anxiety: Laying the Foundation for A Numeric Analysis of Affect

Front-loading: Action Potential, Anxiety, Appraisal, Biological Clock, Circadian Rhythm, Cognitive Marker, Emotional Unit, Entity, Existence, Goals, Object, Purpose, Sentiment, Sufficiency, Time, Psychological Time, Uniqueness, Utility, Value Judgement

A life without aim or goals is sometimes considered the hallmark of depression, angst, and a host of other maladaptive states and symptoms. What though, might a world without goals look like?

Figure 5.1 A goalkeeper without a goal.

In earnestness, goals are objectives external to individuals that enable tests to be run, such as an end zone in a maze with a reward.[2] Purposes, contrarily, are internal mental aims "set by an individual" that guide behavior.[3] For instance, a hungry mouse might set eating food as its purpose and traverse a maze to acquire a valued object, such as food in a goal box. However, if sated, eating food might not be a purpose for the mouse, and there would be no reason to presume the mouse would move toward the end zone. It may simply go explore. Purposes can be thought of as internalized goals with stipulations specific to an individual.

Figure 5.2 Tom, Dick, and Harry

A foot race between Tom, Dick, and Harry will be considered. The external goal in the study is the area designated by the finish line. Each runner has a different internalized purpose for running the race. Tom's purpose is to finish the race in first, before both Dick and Harry. Dick's purpose is to not finish the race in last place. Harry's purpose is to improve upon his previous race time by cutting a minute off the duration of his run.

Talk of goals and purposes tends to involve value judgments. A value judgment is "a perspective toward a person, object, principle, etc. based on how one values the properties or characteristics thereof."[4] Much like purposes are to goals, a value judgment would be specific to an observer. For example, in the foot race, a pair of size-12 running shoes would have their own intrinsic features useful for running. Tom already owns a pair of size-14 running shoes and the size-12 pair would not fit him. Dick has no running shoes but normally wears a size-10. While Dick could wear them with some discomfort, he would prefer a size-10 pair. Harry also has no running shoes and normally wears a size-12. The pair would fit him perfectly. Objectively speaking, the size-12 shoes' intrinsic characteristics make them most valuable to Harry, then Dick, and least valuable to Tom for running. A general evaluation can be made of the shoes in the context of the

Figure 5.3 The value of the running shoes towards facilitating running to the goal zone (finish line) for Tom, Dick, and Harry.

situation.

Although the size-12 shoes have intrinsic characteristics that give them a value toward running, Tom, Dick, and Harry may each offer different subjective value judgments with consideration to their own demands for the race. For example, Tom cannot use the shoes to run himself, but in the context of his purpose they would be high if he knows that the size-12 pair is the only game in town. Tom already has the advantage of owning a pair of shoes and would have a better chance of finishing first if he prevents anyone else from acquiring the size-12 pair by obtaining them himself. Dick, on the other hand, knows that Tom already has the advantage of possessing a pair of running shoes. If Harry were to acquire the shoes then Dick would be at a disadvantage to both Tom and Harry and in danger of finishing last, but less likely than if Tom acquires them. Meanwhile, Harry, who has been practicing hard and already shaven off a minute from his run time barefoot, merely feels the shoes would be an added measure of security. Whether or not Tom or Dick acquires the shoes will have little to no effect on Harry's purpose. Assessing their individual purposes with this new information, and not merely relying on the external goal of reaching the finish line, the pair of size-12 running shoes would elicit three different value judgments from Tom, Dick, and Harry, and likely three different emotional responses than from before.

Figure 5.4 The new value of the running shoes to Tom, Dick, and Harry if someone else acquires them and it hurts his chance to achieve a purpose.

Purposes and judgments of value are major aspects of affect engineering. Affect engineering is primarily informed by cognitive appraisal theories of emotion. A key component of cognitive appraisal theories of emotion holds that the emotion that arises from a situation depends on "how the individual interprets the situation."[5] It follows that before affect engineering can fully address the concerns of traditions emphasizing bodily responses, such as the Jamesian, or those that emphasize social constructions, such as the social constructivist, its own foundations must be established.[6]

Additionally, whatever approach is used to formulate a grand theory of emotion must adequately consider purposes, value judgments, and appraisals of objects at the level of the individual in order to be worthy of the distinction. Given that objects are the most ubiquitous of the aforementioned terms, they will be the starting point for building an equation. Objects are broadly defined as "anything."[7] For clarity, an entity is "something that has separate and distinct existence and objective or conceptual reality."[1] This will prevent confusion with the verb form and alternate definitions of *object*, such as "goal or end state."[8]

Establishing What *x* and *y* Represent

In a Cartesian plane, the values for one variable are dependent on the values of another variable. For instance, in the equation

y = x,

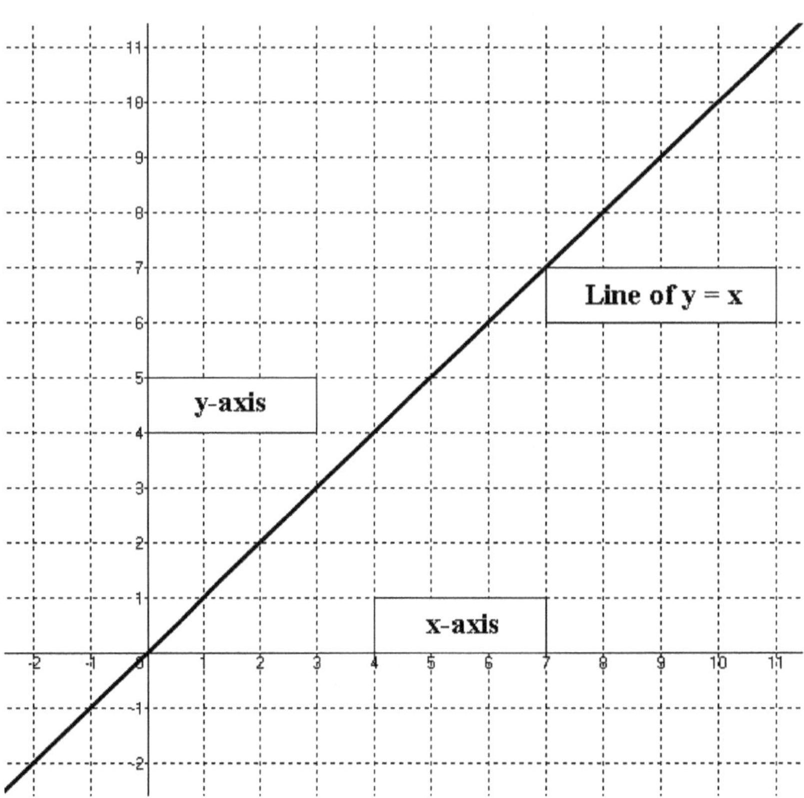

Figure 5.5 The graph of the equation: y = x

[1] "By permission. From Merriam-Webster's Collegiate® Dictionary, 11th Edition ©2014 by Merriam-Webster, Inc. (www.Merriam-Webster.com)."

the y-value is dependent on the x-value. Whenever ***x*** equals zero, ***y*** equals zero. If ***x*** should equal ten, then ***y*** will equal ten. This relationship is a direct one. Whenever ***x*** changes, so does the y-value. The classic way to write an equation is to write it as a function. For example, the equation

$$y = x,$$

would be written as the function

$$f(x) = x.$$

The x-value and y-value, being variables, have to represent something in the world for them to be of any practical application in the study of emotions. In affect engineering, the ***y*** variable will usually gauge the value of an entity as judged by an individual with respect to a single purpose. The unit of measure will be the hypothetical emotional unit (E.U.). Generally speaking, the emotional unit is conceived as the minimal number of neurological resources required to acknowledge or cognize an entity exists over a given time frame. Given that the function is primarily informed by cognitive appraisal theories, emotional units will ultimately have a neurological component for its base. More specifically, a single emotional unit would be the minimum number of activated neurons (i.e., action potentials) in a time frame required to acknowledge that an entity exists. From this, emotional units could alternatively be converted into an energy form, such as the joules of energy in a given time frame required to activate the minimum number of neurons required to acknowledge an entity exists. Two types of value are employed by affect engineering: existential value and utility value. Both are measured in emotional units. Generally, affect engineering holds existential value, or the cognizance of an entity's existence, to be a necessary condition before utility can be assessed. Secondly, the ***x*** variable will have to represent something with enough scope to enable everything to have any practical application. Time is frequently used on the x-axis and appears able to live up to the task.

Figure 5.6 X Represents time.

The progression of time, hence, the time elapsed, will be measured along the x-axis. However, because the relationship between time and the judged value of an entity is not yet established, the starting point will be the equation:

$y = 0$, or the function $f(x) = 0$.

The Base of the Equation: Utility (Sufficiency, Uniqueness, Sentiment) and Existence

In assessing an entity's value with respect to a purpose, a practical definition of purpose must be set up. Thereafter, other questions arise that beg to be answered. How useful is the entity toward helping an individual fulfill a purpose? Does it help reduce tension from a biological drive? Is anything else required? Are there alternatives or groups of items that are more readily available? How important is the purpose anyway? Most of these questions can be generally reduced to the following two characteristics:

1) Is it sufficient or "enough to meet the needs of a situation or a proposed end"?[2]

2) Is it unique, which means "being the only one."[3]

Though cognizance of an entity's existence is essential before sufficiency and uniqueness can be assessed, existence will be explained after the other variables in the base of the equation. Fractions will be used to express an entity's sufficiency and uniqueness toward purpose fulfillment. A purpose, in affect engineering, always concerns the acquisition of an entity or it concerns not acquiring the entity. The purpose need not be related to reducing a biological drive and can be understood in the context of any activity, such as exercising (i.e., acquiring the entity of physical exertion) or a drive to eat food (i.e., acquiring the entity of an apple). To avoid complications, this entity will be called the original entity. For example, an original entity that fulfills a purpose if acquired and requires no assistance from another entity would have a sufficiency of one. If another entity must also be acquired, then the original entity's sufficiency becomes one half. If two more are required then its sufficiency becomes one third and so on. To express sufficiency as a fraction, the number of other entities required to facilitate the fulfillment of a purpose must be known.

Sufficiency = 1 ÷ (1 + Number of subsequent entities required to fulfill a purpose).

Assigning the letter ***S*** for Sufficiency makes this less cumbersome:

Sufficiency = 1 ÷ (1 + S).

To illustrate, Bob wants to see a framed picture on a wall and this is his purpose. He already has a picture but must acquire additional entities in order to successfully acquire the original entity,

---

[2] By permission. From Merriam-Webster's Collegiate® Dictionary, 11th Edition ©2014 by Merriam-Webster, Inc. (www.Merriam-Webster.com).

[3]"By permission. From Merriam-Webster's Collegiate® Dictionary, 11th Edition ©2014 by Merriam-Webster, Inc. (www.Merriam-Webster.com)."

namely, the experience of witnessing a framed picture on a wall:

1) Wall;

2) Frame for the picture;

3) Nail to hang the framed picture;

4) Hammer to secure the nail in wall;

5) Someone to hang the picture on the wall.

Only the picture is currently being considered, and its Sufficiency is not equal to one. Bob requires five other items to assist in fulfilling the purpose. Therefore, the Sufficiency of Bob's picture is one sixth, or sixteen and two-thirds percent.

Sufficiency = 1 ÷ (1 + 5) = 1 / 6 or approximately 16.6 %

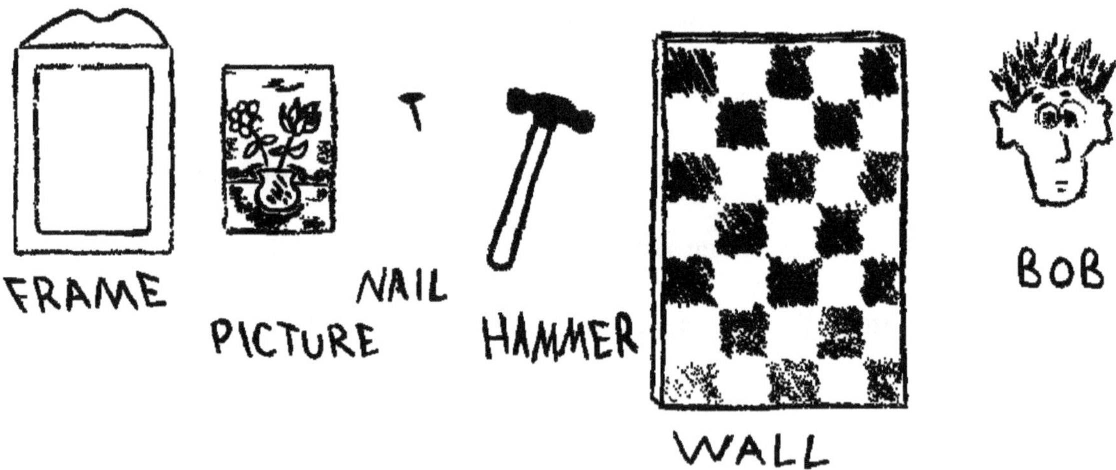

Figure 5.7 The six items for Sufficiency are a frame, Bob's picture, a nail, a hammer, a wall, and Bob.

An entity whose acquisition will satisfy a purpose by itself would appear useful and valuable. However, if a million other entities can serve the same end, then the original entity loses its essentialness. It would not be unique. For instance, when food becomes scarce people are generally willing to do more to acquire it to avoid starvation. Food's judged valuation becomes higher than normal. However, when food is abundant, people are generally less inclined to forage because there is little urgency. Uniqueness can be expressed in a manner similar to Sufficiency. A one of a kind entity would be more unique than an entity that is one of two, or one of three. To express uniqueness as a fraction, the number of alternatives or paired alternative entities that can match the original entity on all levels, concerning the significance of the role it plays for a purpose, must be known.

Uniqueness = 1 ÷ (1 + Alternatives or number of paired alternative entities that can match the original entity on all levels).

Assigning ***A*** to the words above makes this less cumbersome:

Uniqueness = 1 ÷ (1 + A).

If Bob wants to acquire a hammer and it is the only in the world, then it would possess a Uniqueness of one. If another hammer becomes available, then the original hammer's Uniqueness becomes one half or .5.

Uniqueness = 1 ÷ (1 + 1) = 1 / 2 or 50%.

The Uniqueness of Bob's original hammer is halved. Furthermore, if the hammer is one of two hammers, and a pair of alternatives can be found that matches the original hammer on all levels, its Uniqueness drops again. For example, a stick, a stone, and a piece of rope can be fastened together to construct a makeshift hammer, lowering the original hammer's Uniqueness:

Uniqueness = 1 ÷ (1 + 2) = 1 / 3 or 33.3 %

Figure 5.8 Bob has two hammers and a makeshift hammer fashioned from a stone, a stick, and some rope. Therefore, the original hammer's Uniqueness becomes one-third.

Of the ways by which Sufficiency can be combined with Uniqueness, multiplication would appear the most convenient. Neither concept determines the other, as they are separate attributes of an entity, but both play a role in judging an entity's usefulness toward the fulfillment of a purpose. A highly unique and highly sufficient entity would maintain a maximum value of one due to the identity property of multiplication and would not be affected by order of operations. Under multiplication, large swings toward zero of one variable by altering the other's influence without negating it completely, meaning Sufficiency's relationship to Uniqueness would be coupled.

Under multiplication, an entity's utility toward satisfying a purpose would equal:

$[1 \div (1 + S)] \times [1 \div (1 + A)]$.

This simplifies to: Sufficiency × Uniqueness

However, a check of the other basic operations will be done to make certain that this choice is ideal.

First to be considered is subtraction. If an entity is considered to have maximum Uniqueness and maximum Sufficiency, subtraction would yield a value of zero:

Sufficiency - Uniqueness = 1 - 1 = 0 emotional units

It is not logical to assign a value of zero to an entity that will both completely satisfy a purpose if acquired and the entity is one of a kind. Subtraction also runs into a problem with order of operations, yielding different results depending on the arrangement. Subtraction can be eliminated as a candidate operation.

Figure 5.9 Subtraction can be ruled out as it may yield illogical conclusions.

While addition does decouple Sufficiency and Uniqueness, one runs into distortion with the identity property of addition. For instance, if the Sufficiency of an entity is near or approaches zero, but the entity is the only one of its kind, then one can be misled into believing the entity is valuable to the fulfillment of a purpose solely because of its uniqueness. For example:

Sufficiency + Uniqueness = approximately 0 + 1 = approximately one emotional unit.

Sufficiency is not exactly equal to zero here as the value for Sufficiency can only approach zero if the denominator goes to infinity. Notwithstanding, even if Bob possesses the only unicorn in the world it would do him little good toward the purpose at hand of hanging a framed picture on a wall.

Figure 5.10 With addition, a high Uniqueness would inflate the utility of a unicorn for the purpose at hand, and Bob will need a new wall.

Similarly, an entity may be sufficient to facilitate the satisfaction of a purpose, but might have its value overestimated if alternatives are readily available and everywhere at once:

Sufficiency + Uniqueness = one + approximately zero = approximately one emotional unit.

If Bob is in need of a hammer, but everyone else is born with a hammer in place of their left hand, then it is not all that important he invest a lot of effort looking for one over something more difficult to find. It is worth noting that addition could be used if coefficients were affixed beforehand. Sufficiency and Uniqueness would be decoupled, such as where the utility of an entity might equal

(.5) × Sufficiency + (.5) × Uniqueness.

However, one obstacle to using addition in this manner is that one would have to know how many attributes influence utility (e.g., Sufficiency, Uniqueness, Sentiment) in order to determine a proper coefficient or none at all. The choice to use multiplication over addition depends on what one wants to assume about the wiring of a person's brain (i.e., are utility components coupled or decoupled).

Figure 5.11 If everyone else is born with a hammer in place of a left hand, it is not all that important that Bob spend more effort on finding a hammer than something else that may be harder to obtain. Addition, though a viable choice, will be put aside.

Division, too, runs into trouble with the order of operations. No reliable means to determine whether Sufficiency should be divided by Uniqueness, or vice versa, is clear. Nevertheless, an example will be considered. If a hammer is wholly sufficient for satisfying a purpose and it is one of two hammers in the world, then two different values would be possible:

Sufficiency ÷ Uniqueness = 1 ÷ (1 / 2) = 2 emotional units

Uniqueness ÷ Sufficiency = (1 / 2) ÷ 1 = 1 / 2 of an emotional unit

If it was instead revealed that the hammer was one of a kind, then dividing would yield a lower value than in the first case above:

Sufficiency ÷ Uniqueness = 1 ÷ (1 / 1) = 1 emotional unit

It is not logical to value a one of a kind hammer as less useful than a hammer that is one of two with respect to the purpose of hanging a framed picture on a wall and solely based on Uniqueness. While the argument can be made that for some devices a decrease in the Uniqueness makes an entity more valuable, such as a telephone, this is not actually the case. With a telephone, the desired entity is the link or connection between two phones. A single phone, then, has a Sufficiency of one half toward the purpose of creating a connection. A group of three phones, for instance, may have anywhere between zero, one, two, or three connections between them. If there is only one connection available, then that connection is unique and valuable. If, however, another connection is being used as a backup, or there are already hundreds of thousands of connections available, the loss of a single connection is less important because other alternatives are readily accessible. Division, then, can be eliminated as a candidate operation.

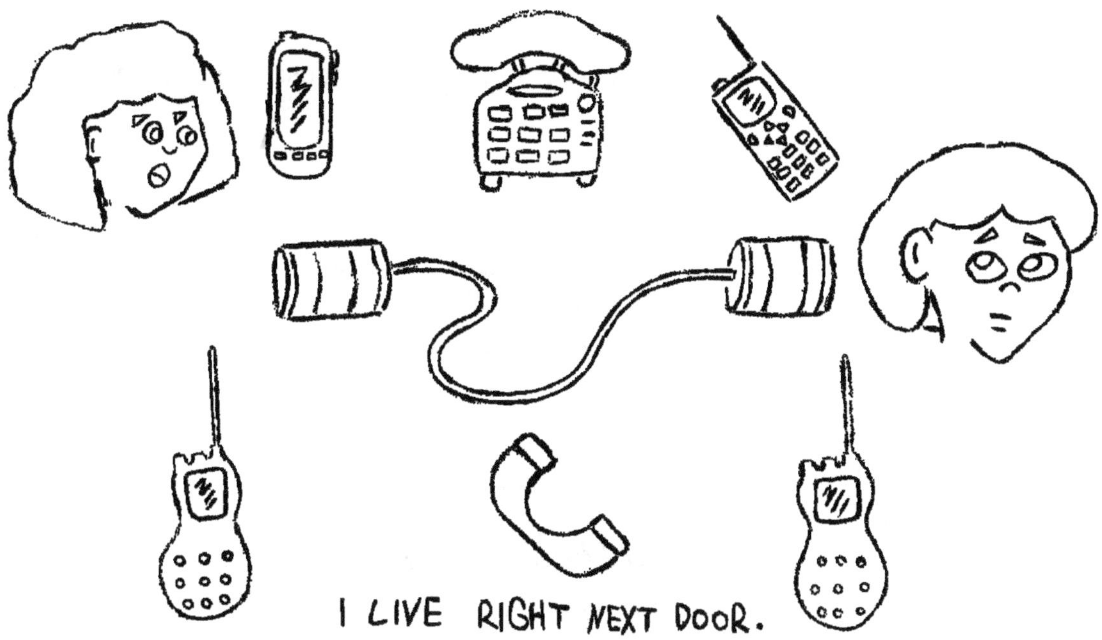

Figure 5.12 Division can be ruled out. The order of operation leads to mixed results and possessing more of an entity makes each one expendable.

Not all purposes, however, are held in the same esteem. Sentiment is proposed as a wildcard variable and measure for a single purpose's ranking of importance against all other purposes the individual possesses. Physiologically, Sentiment would be indicated by the signal strength (i.e., firing rates) of neurons from a hub of neurons that govern arbitrary preferences. Sentiment's makeup would also be influenced by experience, and as more purposes are added into the mix become more fastidious. While it can account for concepts like aesthetic preference, its role would be closer to that of the ego-ideal. Sentiment would be separate from both Sufficiency and Uniqueness, and typically serve as a tiebreaker, for instance, when the same entity is being valued for multiple purposes. It might also be employed to reassess a purpose if different entities, which are being judged for the same purpose, have a value too similar to distinguish based on their Uniqueness and

Sufficiency components alone. Lastly, if two entities being valued for two separate purposes have a similar value based upon their Uniqueness and Sufficiency components for their respective purposes, then Sentiment might also be used to determine which purpose will be given precedence, and which entity acquired first or not at all.

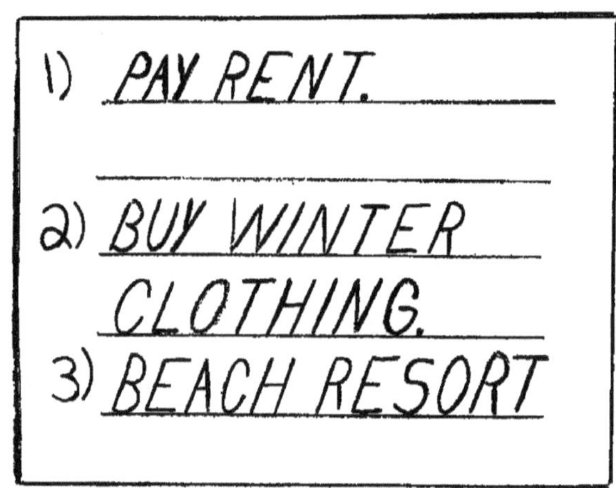

Figure 5.13 To do list: ranking priorities

In the second case above, the original purpose might be broken down into two or more purposes that are more specific and an individual would then value the entities for two different purposes with potentially separate Sentiment rankings based on other, secondary features. If the entities are still judged to possess the same value after this, then the prediction of which entity will be acquired first, solely based upon the variables discussed thus far, would revert to chance. Sentiment, theorized as a ranking of the priority of different purposes, guides but does not dictate behavior.

For the variable of Sentiment, different purposes are ranked along a ratio scale to measure their importance relative to the individual's most important purpose, which lies at the origin. If survival is an individual's most important purpose, then its absolute ranking would be zero because it is zero units away from the origin. If buying a home is the next most important purpose the individual has, then this purpose would be second in priority and might have a distance of one from the origin. If marrying is the third most important purpose for the individual, then it might have a distance from the origin of two, so on and so forth. Although any positive real number would be adequate, for simplicity's sake intervals of whole numbers will be used to gauge the Sentiment ranking of a purpose (e.g., 1, 2, 3, 4). Physiologically, a Sentiment value of one would correspond to a peak signal strength from a Sentiment hub, .5 would be a 50% peak signal strength, and .25 would be 25% peak signal strength. To obtain the Sentiment value, the ranking of the purpose in question and its distance from the most important purpose (top ranked purpose at the origin) must be known. The process for identifying the value of Sentiment is similar to that of Sufficiency and Uniqueness. It would carry an equal weight to the other two in the valuation of an entity's utility.

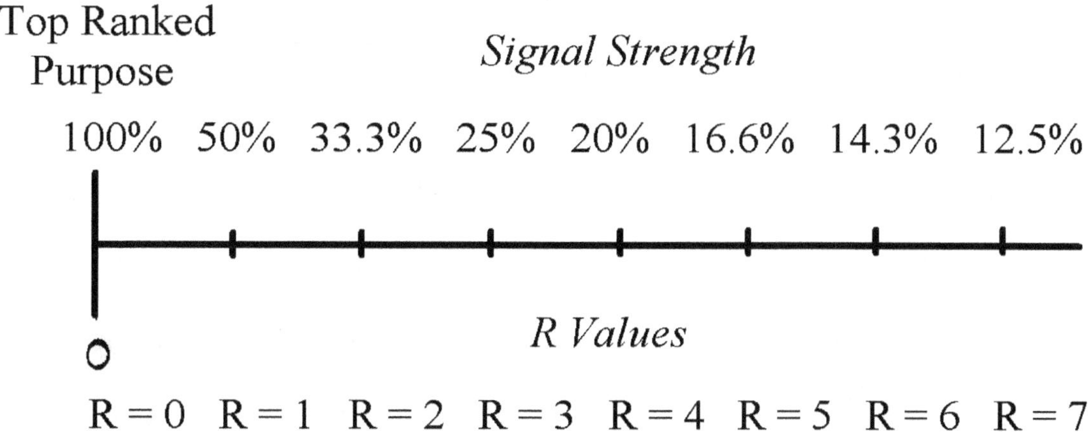

Figure 5.14 Sentiment is a ratio scale ranking of purposes or an absolute ranking with the top ranked purpose as the origin. Signal Strength would be from a Sentiment hub of neurons acting in conjunction with Sufficiency and Uniqueness neuron hubs on neurons that indicate Utility value.

Sentiment = 1 ÷ (1 + Ranking of the purpose in question as measured by its distance from the most important purpose the individual holds)

Another way to conceptualize this rank is through the notion of absolute value. A number's absolute value is its distance from an origin; in most cases, zero is the origin. The numbers 13 and -13 both have absolute values of thirteen because they are thirteen whole numbers away from the origin zero. If the purpose with the highest priority is thought of as the origin, zero, a purpose with the second highest priority might be one unit away from the origin, while the purpose with the third highest priority might be two units away. This can be thought of as the purpose's absolute rank, meaning the distance of the rank of the purpose at hand from the rank of the purpose with the highest priority. Hence:

Sentiment = 1 ÷ (1 + Absolute rank of the purpose)

Assigning ***R*** to the italicized words makes this less cumbersome:

Sentiment = 1 ÷ (1 + R). The ***R*** value may be any real number.

For example, Kate wants five distinctively colored balloons for a party and sets the task of acquiring five distinctively colored balloons as her purpose. The sight of two balloons with the same color would lower her judgement for the Uniqueness of each. However, upon arriving at a party store, she discovers they have ten distinctively colored balloons: white, black, red, green, blue, yellow, orange, violet, brown, and magenta. She now has ten options and value judgments to consider for one purpose. Kate does not want to pick them randomly, but after assessing all ten balloons for their value, is unable to decide on five. All the balloons are one of a kind in color, meaning they possess a Uniqueness of one. They are also distinctively colored balloons, meaning each possesses a Sufficiency of 1 / 5. Each meets her demands to the same extent.

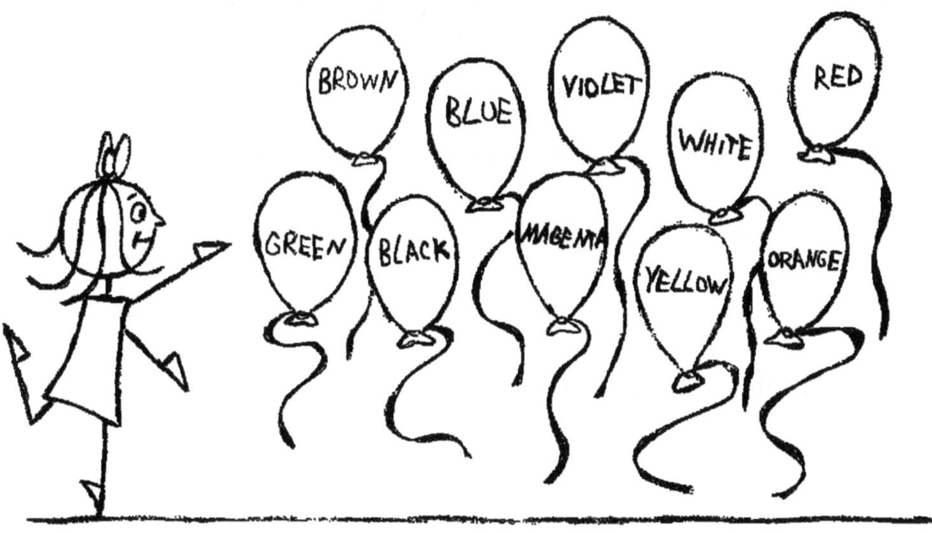

Figure 5.15 Kate initially values the balloons equally.

Kate realizes that her purpose will have to be more specific if a decision is to be made, so she decides to create ten purposes from her original one based off of secondary characteristics of the balloons (i.e., to acquire a distinctly colored green balloon, a distinctly colored red balloon, etc.). Initially, the balloons would all have the same value again, given that the Sufficiency and the Uniqueness for all the balloons now all equal one. However, in Kate's past experience she remembers that she has an affinity for the following colors in descending order of preference: green, blue, violet, red, orange. The other colors she likes equally the same, but not more than any of her five favorites. With Sentiment taken into account progress can be made. The balloons that correspond to the highest ranked purposes will be picked first. They would have the most emotional units or anxiety invested in them and the highest values. Using the variables established thus far,

Value judgment of an entity = Sufficiency × Uniqueness × Sentiment =

1 × 1 × Sentiment.

Applying this to each color will determine Kate's value judgments:

1) The green balloon is the color with the highest Sentiment,

Judged value = $1 \times 1 \times [1 \div (1 + 0)] = 1$ emotional units;

2) The blue balloon is the color with the second highest Sentiment,

Judged value = $1 \times 1 \times [1 \div (1 + 1)] = 1 / 2$ or .5 emotional units;

3) The violet balloon is the color with the third highest Sentiment,

Judged value = $1 \times 1 \times [1 \div (1 + 2)] = 1 / 3$ or .33 emotional units;

4) The red balloon is the color with the fourth highest Sentiment,

Judged value = $1 \times 1 \times [1 \div (1 + 3)] = 1 / 4$ or .25 emotional units;

5) The orange balloon is the color with the fifth highest Sentiment,

Judged value = $1 \times 1 \times [1 \div (1 + 4)] = 1 / 5$ or .2 emotional units;

6) The black, white, brown, magenta, and yellow balloons all possess the same value for Sentiment,

Judged value = $1 \times 1 \times [1 \div (1 + 5)] = 1 / 6$ or .16 emotional units.

Figure 5.16 Green, Blue, Violet, Red, and Orange are Kate's top five color choices for the balloons.

Sentiment, in Kate's case, refers to the importance of acquiring a specific balloon color over another balloon color. Sufficiency would be determined by whether or not each balloon aided the fulfillment of Kate's purpose by being a distinctive and specifically colored balloon. Uniqueness would still refer to whether or not the balloon's color is one of a kind. It can be observed that the value judgments of each balloon descend from highest to lowest despite them all being unique and sufficient balloons for Kate's original purpose. If Kate were to pick five balloons based upon this information, the green, blue, violet, red, and orange balloons are the ones with the highest value judgments, the most anxiety invested in them, and would be picked first. The other five purposes and corresponding balloons she would neglect because the values of the other balloons would be too low in Kate's eyes to merit picking. If she only required three balloons, then Sentiment would model

Kate picking the green, blue, and violet balloons because they have the highest judged values. Conversely, if Kate realized she required six balloons, then Sentiment would model that her five favorite colors would be chosen first and another color from one of the other five would be picked at chance.

For comprehension purposes, typically only one purpose will be considered at a time, and Sentiment's influence would be neutralized by setting it to one. This means ***R*** would equal zero and the purpose being examined would be considered to have the utmost importance. Similarly, Sufficiency and Uniqueness can also be neutralized by setting them both to one in order to examine other variables in the equation while holding these constant. An entity being valued for a purpose and whose utility variables have been neutralized would be wholly sufficient, ***S*** = 0, the entity would be wholly unique, ***A*** = 0, and the purpose in question would have the same priority as the top ranked purpose, ***R*** = 0. This is important to keep in mind when higher level variables and operations are under inspection and it is wished to hold these constant. Alternately, if one wanted to model ethics with affect engineering then Sentiment would play a greater role. Doing so will require expanding the Cartesian plane into a fourth dimension for different purposes. The third dimension would consist of different entities being valued for the same purpose.

Last, and most important, in establishing a baseline for the value of an entity as it relates to a purpose, an acknowledgment of the entity's existence must be present. If a person acknowledges an entity, then that entity possesses a value given to it by the person; it at least occupies space in the person's mind (i.e., neurons fire in response to thought or perception of it), even if it makes no claim to be a tangible object. The concept of the entity has a real presence, and that presence is represented by plus one emotional unit (+1), or the minimum number of neurons required to cognize that an entity exists. Additionally, Existence is all or nothing, and equals plus one or zero. It may be construed as the absolute threshold needed to cognize an entity's existence. Although fractions of a single emotional unit are being used for utility right now, these values must eventually be amplified above the threshold of existence before they can be of consequence (i.e., detected by an individual). Moreover, whatever value is added to an entity's existential value by Sufficiency, Uniqueness, or Sentiment is on top of the value given by cognizance of the entity's Existence. Lastly, it bears repeating that the entity need not physically exist in order to be conceived, as in the case of an imaginary four-headed giraffe.

Figure 5.17 A four-headed giraffe does not exist in the actual world, but one could imagine what it might look like without having seen the image above.

The emotional unit, or base number of activated neurons over a given time frame required to cognize an entity exists, is the existential value of the entity and measuring stick for utility value. Both the utility value of an entity and the existential value of the same entity toward the fulfillment of a purpose are separate and measured in emotional units, with the former being a quality and the latter being a matter of fact. As noted, utility is an entity itself, and would have a value of at least + 1 emotional unit when substantial. By incorporating this new idea of existence into the function being built, the function would now look like the following:

Judged value of an entity = Sufficiency × Uniqueness × Sentiment + Existence

Existence is given more weight than the three utility variables. An individual can think about an object's existence without necessarily thinking of it as useful for the fulfillment of a purpose. The entity, in such a case, simply exists to the individual (e.g., figure 5.18). However, an individual

Figure 5.18 A pair of useless entities with Sufficiency values near zero still exists.

cannot contemplate an entity's utility toward fulfilling a purpose if no acknowledgment of the entity is cognized. Cognizance of an entity's existence is a necessary condition for Sufficiency, Uniqueness, and Sentiment. Existence must be spread across the other variables. Therefore:

Judged value of an entity =

Existence × (Sufficiency × Uniqueness × Sentiment + 1)

To minimize clutter, an entity can simply be understood as existing and the above would be written in distributed form:

Judged value of an entity = Sufficiency × Uniqueness × Sentiment + 1

Temporality (x-axis)

Though the x-axis is being used to represent time, a suitable location to place the ***x*** variable is not yet available. A relationship between time and a value judgment of an entity has not yet been established, so the endeavor of building an equation must continue at least until a suitable location is found. In light of this, a moment will be taken to consider the nature of time on its own. Like the earlier observations made on the difference between goals and purposes, and between values and value judgments, the assessment of time has both an objective quality and an "experienced or subjective" distinction.[9] Psychological time is a subjective meaning of time in which an individual's senses of duration are independent of external cues like clocks, calendars, or sunrise and sunset.[10] Instead, they are dependent on internal events such as biological clocks, circadian rhythms, and cognitive markers.[11] For the purpose of assessing the value of entities as they relate to purposes, standard time, as measured by a clock (e.g., an atomic clock) will be used. Other means by which to measure the passage of time can and eventually will be offered as substitutes (e.g., cognitive markers) or accounted for (e.g., circadian rhythms) once time is officially implemented into the equation with the x-variable. For demonstrative aims, however, it would unnecessarily complicate the constructive process of the equation to introduce them here, so they will be put aside for now. Time will be revisited in greater depth at a later point once a proper relationship between time and the judged value of an entity can be established and the x-value is given a home in the equation (chapter ten).

Figure 5.19 "x" marks the spot

Anxiety and Emotions in Affect Engineering

The principle tenet behind affect engineering upholds the position that anxiety is best thought of as a resource (e.g., energy and neurons) to be invested to value entities. A second tenet holds that an emotion's type and magnitude are derived from the derivative of an entity's valuation. For the purpose of building a math equation to model emotion, the conception of anxiety as value is the most practical. A merger with the concept of anxiety with value means that within the context of affect engineering, value is anxiety and anxiety is value. Anxiety corresponds with value and the emotional unit on a one to one basis. This comes at no cost to the other interpretations of anxiety's definition, or anxiety disorders, which would be understood in affect engineering as a misallotment or overinvestment of one's anxiety resources into particular entities. Within affect engineering, for someone valuing an entity for a purpose, the following positions would be defended:

1) One has invested anxiety toward acknowledging an entity's existence;
2) One has anxiety invested toward securing access to the entity in order to attend to purposes that have a high priority and Sentiment;
3) One frets if the entity is not available in ample supply due to having a high Uniqueness, and so invests more anxiety into it;
4) One ponders over the entity's well-being and integrity if it is has a high Sufficiency value for whatever purpose it is intended, and so invests more anxiety into it.

The ability to value comprises the notion of anxiety, its investment, and is measured in the y-axis of the function.

For instance, if a gunman takes a number of people hostage, the amount of anxiety felt by hostages for fear of their lives would be found with affect engineering by determining how much the hostages value their own lives at any given point. Hostages' anxiety may fluctuate, lowering if the captor demonstrates kindness, or elevate if they are constantly reminded of their own mortality.

Figure 5.20 Would hostages who are constantly threatened with death have more anxiety, and therefore, temporarily value their lives more so than calmed hostages?

Although harm has not been officially introduced into the explanation of the equation, it can be presumed a threat of destruction would cause an individual to value his or her own life more so than before, subsequently elevating anxiety invested in it. This can be tentatively demonstrated here mathematically.

By observing the equation and its variables thus far, it can be seen four have been considered: Sufficiency, Uniqueness, Sentiment, and Existence.

Judged value of an entity =

Existence × (Sufficiency × Uniqueness × Sentiment + 1)

If Uniqueness is considered, then by most accounts an individual's life is the only one possessed. It would be unique; ***A*** equals zero.

1 ÷ (1 + 0) = 1 for Uniqueness, with zero alternatives

The numerator is the life itself, with the denominator corresponding to the Uniqueness of an individual's life for fulfilling a purpose and alternatives. If the denominator were reduced to zero, where the Uniqueness of the individual's life became extinguished, then

1 ÷ (0 + 0) = Undefined for Uniqueness. Dividing by zero in math is not permitted.

However, a check of limits reveals that at infinitesimally small numbers in the denominator the Uniqueness value skyrockets toward positive infinity. For instance, at one billionth:

1 ÷ (.000000001 + 0) = 1,000,000,000 for Uniqueness's value

The same amplification would hold true for the other utility variables of Sufficiency and Sentiment. For most purposes that a person has, the integrity of his or her body is also a Sufficiency requirement, for without it, the purpose usually goes unfulfilled. It follows that the purpose of staying alive, under most conditions, might be given an absolute rank of zero. Reducing its denominator to zero would have a ripple effect on every other purpose harbored by the individual. The possibility that the final variable, Existence, may be reduced to zero, death, would offer little consolation, save as a refuge from the barrage of mounting anxiety being invested.

Some exceptions worth mentioning concern if the individual believes in an afterlife, reincarnation, or some form of recurrence. For instance, if an individual believes in an afterlife, the Uniqueness variable in a hostage situation would be somewhat different:

$1 \div (1 + 1) = 1 / 2$ for the Uniqueness of an individual's life, suggesting that belief in an afterlife would help mitigate an overinvestment of anxiety into a threatened life and serve as a coping mechanism.

Generally, the functions hold that one cannot feel anxiety for an entity without first acknowledging the entity's existence, be the entity tangible or intangible. Although even the unknown occupies a place in one's mind, if one does not know whether something is unknown or not, then attempting to mathematically quantify anxiety invested in the entity that has not been cognized by an individual generally falls outside the realms of what affect engineering is best at modeling. Such occurrences could be paralleled to noise in a frequency.

Figure 5.21 The idea in the statement, "I do not know whether or not I do not know something exists, but whatever it is, or is not, I value it," would be referred to as noise or static by affect engineering. The emotional units, unrelated to an entity and purpose, are unavailable to do work to value entities and fall under entropy.

The primary case where this would be of concern would be if an individual's neurons responsible for valuing entities become hijacked by other physiological or non-cognitive factors (e.g., drug use, physical injury). These exceptions can be accounted for when entropy (energy, emotional units, or neurons not available to do work by valuing entities) and biological rhythms are introduced in affect engineering as a coefficient. Entropy is the primary means by which concerns from the Jamesian tradition in the study of emotion can be accounted for in affect engineering.

As anxiety and value are one and the same within affect engineering, measuring anxiety requires finding the judged value of an entity for a purpose. Typically, value is represented by the y-value and measured in emotional units. In the following example, Jane wants a breakfast bar from a vending machine and this is her top priority. Breakfast bars cost a dollar. Jane's value judgment of the original dollar is as follows:

Jane's initial judged value of the original dollar =

$[\, 1 \div (1 + 0) \,] \times [\, 1 \div (1 + 0) \,] \times [\, 1 \div (1 + 0) \,] + 1 = 2$ emotional units

Figure 5.22 Jane needs a dollar.

Jane's initial valuation of the dollar is two emotional units. Her anxiety invested is also at two emotional units. Another way to conceptualize the y-value is to find the area under the curve and between the x-axis. This would enable one to measure Jane's anxiety over a given amount of time. Jane's judged value of the dollar would equal two emotional units per unit of time.

Though the x-value has yet to be placed in the equation, it may be passively considered for the example. If the first value judgment was taken at 8:00 A.M. when Jane retrieves a dollar to buy a health bar, then she has judged the value of the dollar to be worth two emotional units during the course of a minute. Her anxiety level would be two emotional units per minute. However, at 8:01 A.M., when Jane arrives at her purse she discovers that she actually has two dollars there. This changes the Uniqueness of the original dollar. So now:

Jane's second judged value of the original dollar =

$[1 \div (1 + 0)] \times [1 \div (1 + 1)] \times [1 \div (1 + 0)] + 1 = 1.5$ emotional units.

Figure 5.23 Jane has two dollars.

Jane's anxiety level is now at 1.5 emotional units per minute. Upon arriving at the vending machine at 8:02 A.M., Jane discovers its price has doubled to $2.00. Jane still only wants one breakfast bar, but this new information again changes her judged value of the original dollar. The original dollar now only has half its initial Sufficiency from before, and so another dollar is required to purchase the breakfast bar. Jane's anxiety being invested in the original dollar is now at 1.25 emotional units per minute.

Jane's third judged value of the original dollar =

$[1 \div (1+1)] \times [1 \div (1+1)] \times [1 \div (1+0)] + 1 = 1.25$ emotional units

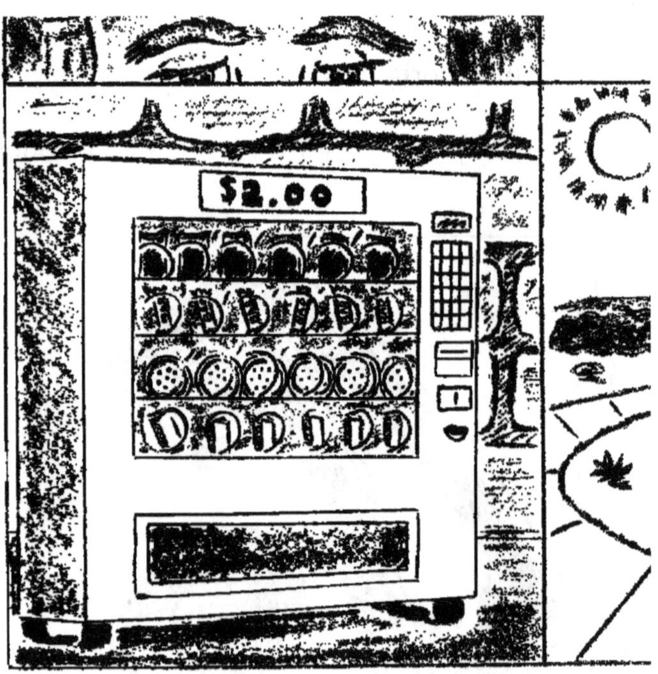

Figure 5.24 Breakfast bars are now $2.00.

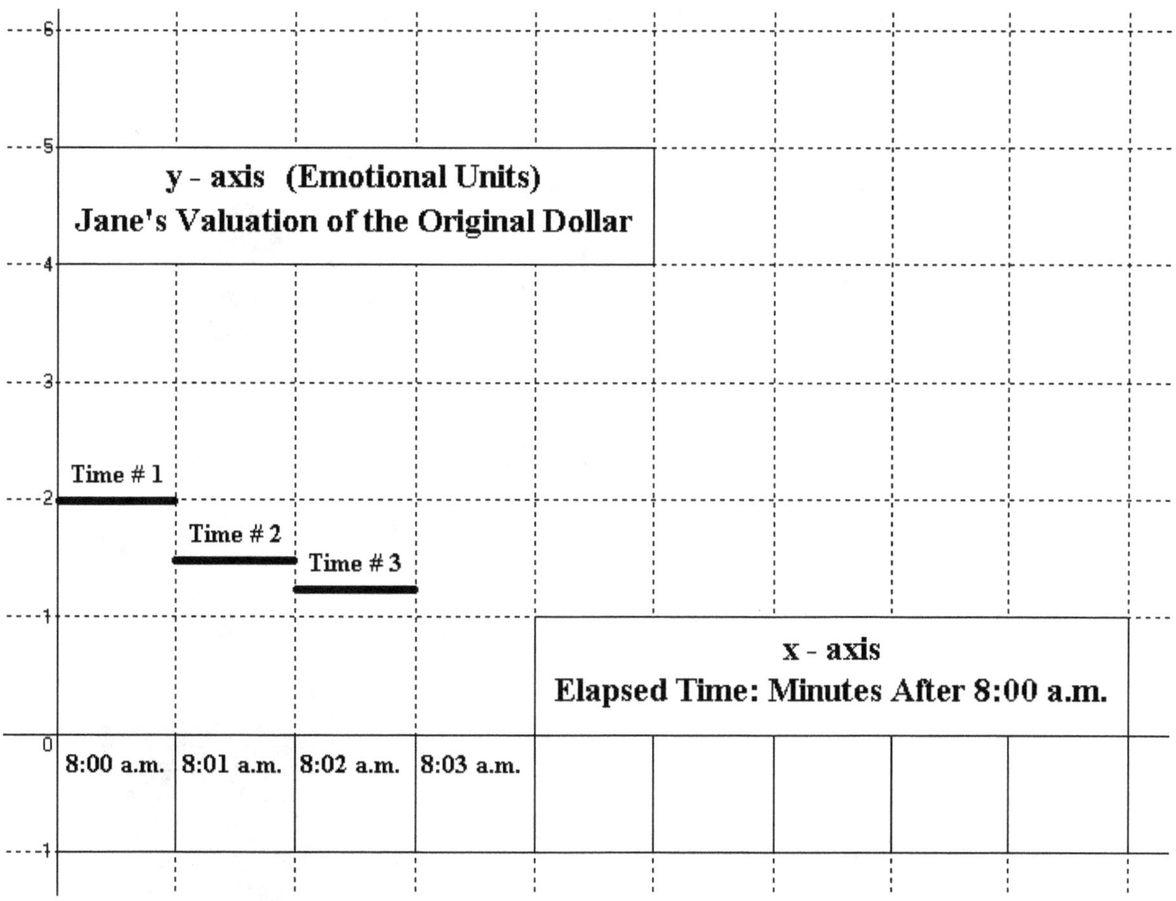

Figure 5.25 Jane's valuation of the dollar drops between time 8:00 a.m., 8:01 a.m., and 8:02 a.m.

Within the span of two minutes, Jane's valuation of the original dollar dropped from two, to one and a half, and then one and a quarter emotional units per minute due to her acquisition of more dollars and its diminished purchasing power resulting from a loss in both Sufficiency and Uniqueness of the original dollar. The utility value of the dollar decreased to one-fourth of its original value, but the dollar's existential value remains intact. If Sentiment were considered and Jane decided to place the purpose of going for a walk as the highest ranked purpose, bumping the purpose of eating a breakfast bar down toward $\boldsymbol{R} = 1$, then the judged value of the dollar with respect to the purpose of buying a breakfast bar falls further, to 1.125 emotional units.

Jane's judged value of the original dollar with respect to the purpose of buying a breakfast bar =

$$[\,1 \div (1+1)\,] \times [\,1 \div (1+1)\,] \times [\,1 \div (1+1)\,] + 1 = 1.125 \text{ emotional units}$$

Figure 5.26 Jane's valuation of the original dollar's utility is reduced to 1 / 8 it's original level. The Uniqueness, Sufficiency, and finally Sentiment variables are halved. It's existential value remains unchanged at + 1.

Although the x-variable's location has not been explained, it may be passively considered like in the example with Jane. Integration will also be a helpful technique for finding the area between a function for an entity's value and the x-axis. With simpler equations, such as f(x) = 2, and for the scenario where the dollar's Sufficiency, Uniqueness, Sentiment, and Existence values do not change:

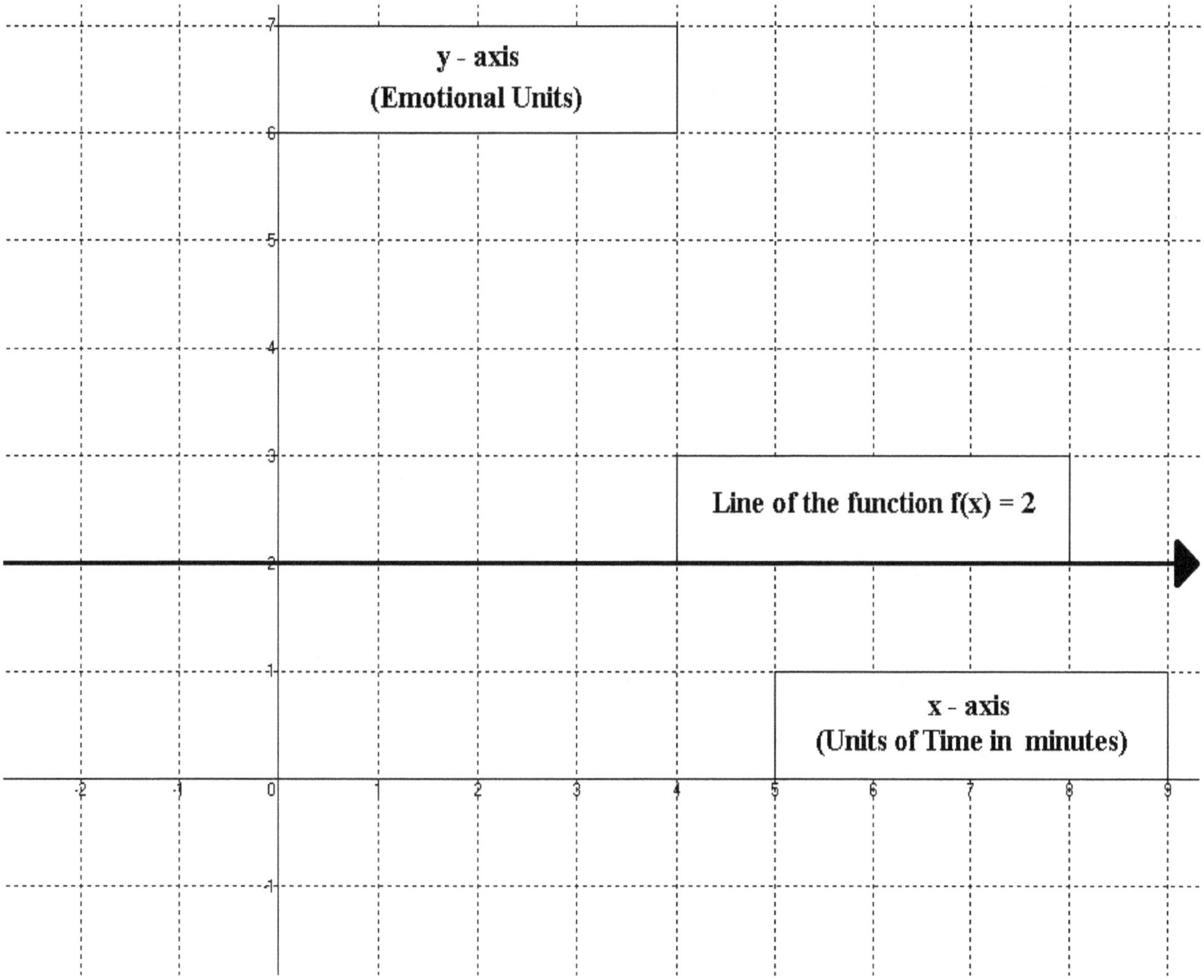

Figure 5.27 f(x) = 2

$\int f(x)$ is the integral of f(x) and is represented by:

$y = 2x + 2$.

The integral reveals anxiety invested over time. For instance, between the x-values of three and zero, the integral function yields:

$[2(3)+2] - [2(0)+2] = 6$ square emotional units over three minutes if the value of the dollar, in emotional units, were to remain unchanged.

This reduces to two emotional units per minute.

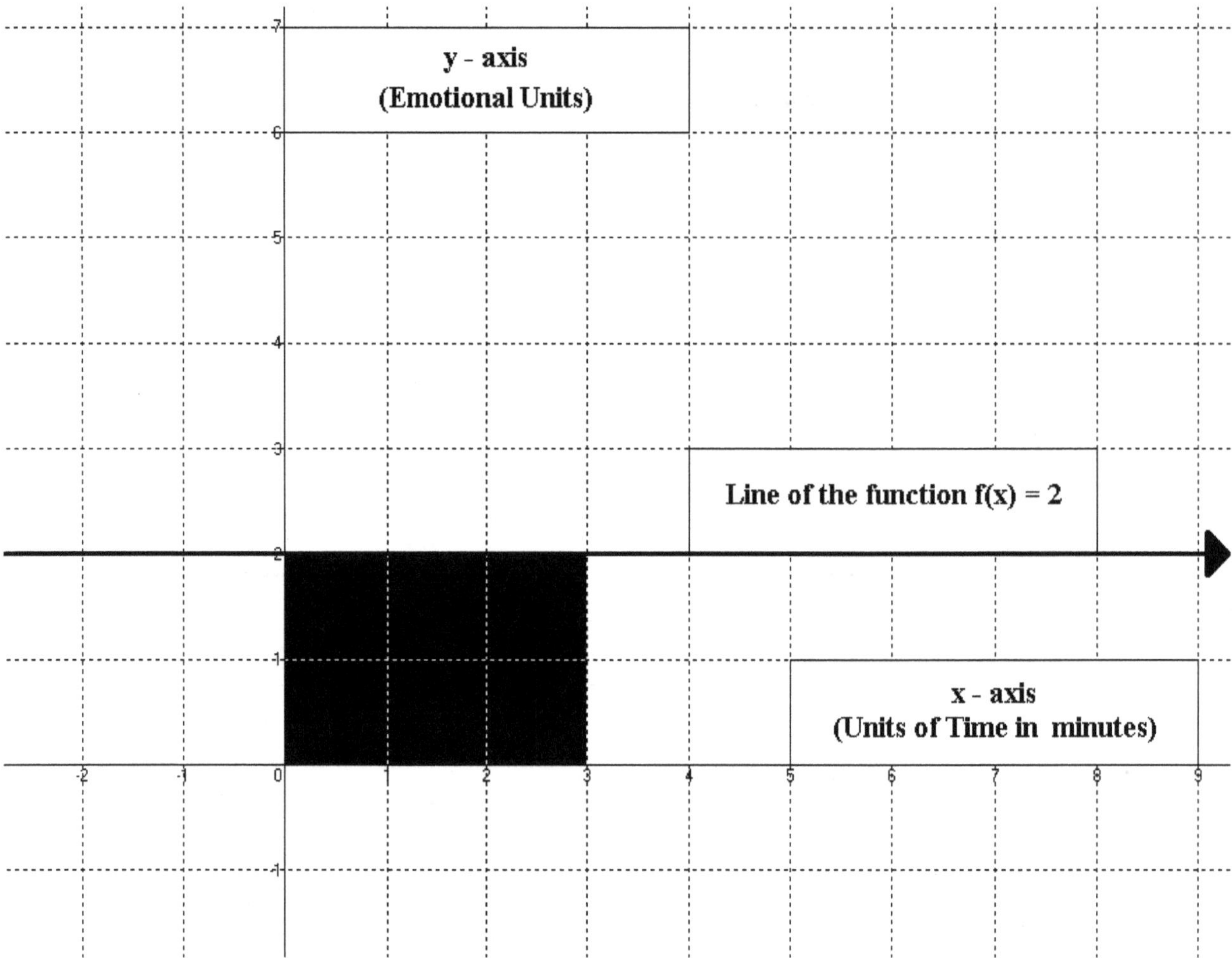

Figure 5.28 Area under the curve of the function f(x) = 6 square emotional units between 0 and 3 minutes, or 2 square emotional units per minute.

The Emotional Unit, the Nervous System, and the Conversion to Joules of Energy

Neurons are recognized as the nervous system's basic unit, and they should be addressed.[12] All neurons consist of a "cell body (soma), a nucleus, its neural processes, an axon, and one or more dendrites."[13] The axon and the dendrites are of the most importance in understanding the neuron's functioning. The axon is a "nerve-fibre projection" that "serves to transmit action potentials from the cell body to other adjacent neurons or to an effector such as a muscle."[14] Neurons can be thought of as the sender of the message and axons the medium, where an action potential is the impulse transmitted when a neuron fires. Starting at the top of the chain, one will find dendrites, "richly branching, tree-like processes attached to the cell body (soma) of a neuron that receive information for the neuron and are stimulated by neurotransmitters."[15] Dendrites, in essence, are recipients of neurotransmitter messages. Neurotransmitters "flow across the synapse from the terminal button" of the axon of one neuron to the dendrites of neighboring neurons and they can be thought of as the actual message.[16]

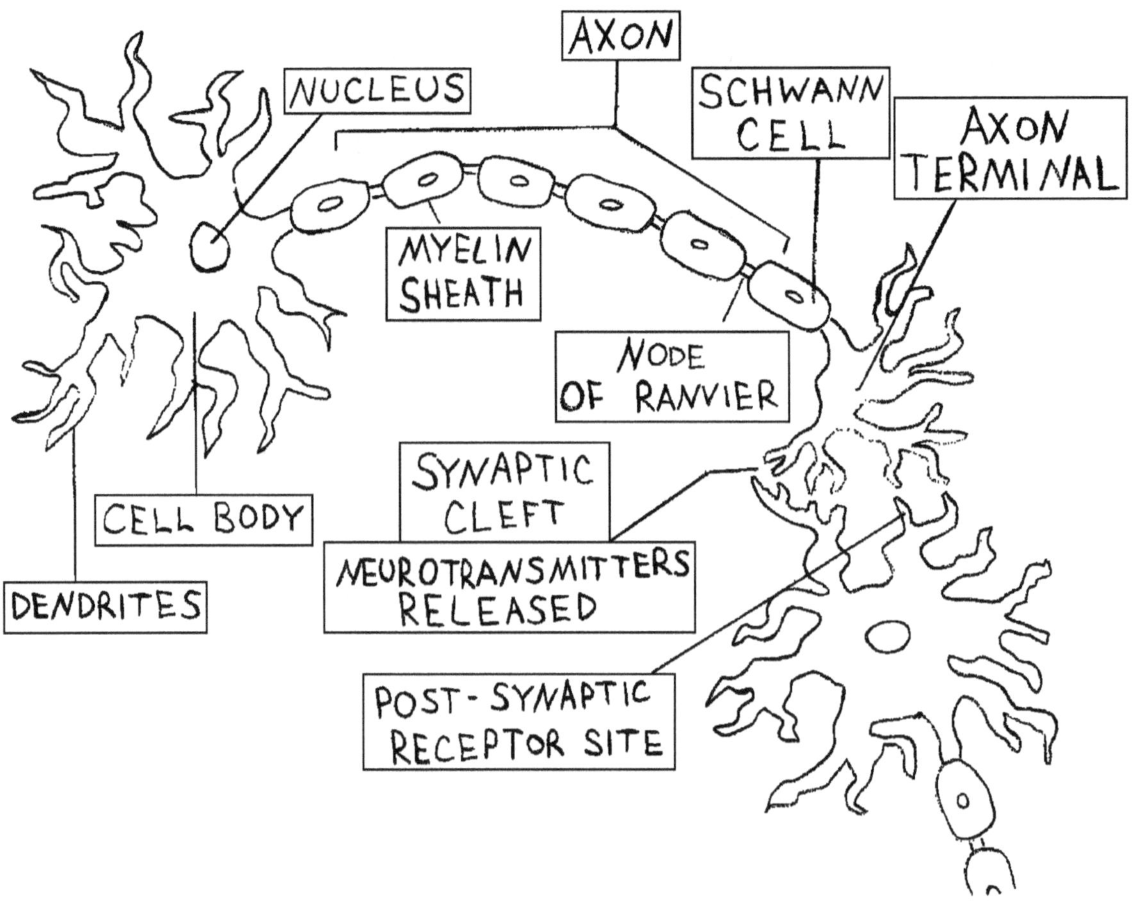

Figure 5.29 Anatomy of a neuron and a neural link. A basic understanding of neurons and how they work will be helpful in the chapters to come.

The intent in bringing neuroscience into consideration is in no way an attempt to try to explain what each specific section of the brain and neuron does. Rather, it is an explanation, offered at the neurological level, for the use of the term emotional unit. As stated earlier, the term emotional unit refers to the minimum number of activated neurons required to cognize that an entity exists and has a value. This may be one, ten, one hundred, or one thousand neurons. For simplicity's sake, it can be presumed that one neuron is equivalent to one emotional unit. This implies the measuring stick would be set at one emotional unit per one neuron's activation as opposed to ten neurons, fifteen, or any other whole, natural number. Given that emotional units are conceived as the minimum number of activated neurons in given time frame that are required to enable an individual to cognize that an entity has existential value, this can be converted into an energy form, which will be vital to keep in mind when biological rhythms and entropy are introduced.

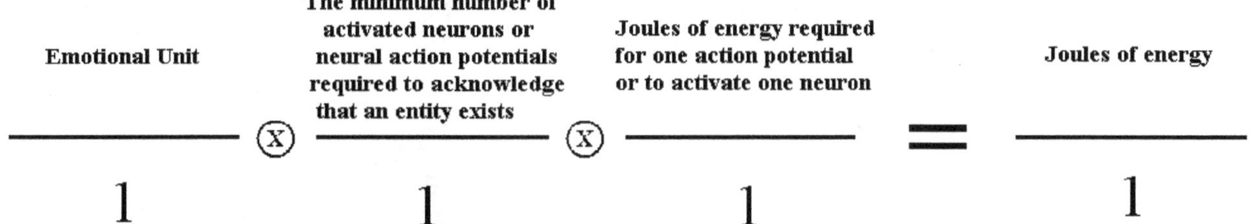

Figure 5.30 Conversion of one emotional unit into joules of energy

Conclusion

Taking into account the possibility that each neuron within a neighborhood can potentially link to hundreds of other neuron in complex networks, operations more powerful than the standard four operations will eventually be required. The natures of some emotions are often explosive, subtle, complex and intricate. Although only the surface of affect engineering has been skimmed, its versatility provides it with a host of means to address the concerns of incorporating drives, appraisals, context in environments, empathy, and whatever else may come its way.

YOU
SHOULD
UNDERSTAND
THIS.
CHAPTER FIVE
CONCEPT CHECK
VOCABULARY
Action Potential
Anxiety
Appraisal
Biological Clock
Circadian Rhythm
Cognitive Marker
Emotional Unit
Entity
Existence
Purpose
Sentiment
Sufficiency
Time
Psychological Time
Uniqueness
Utility
Value Judgement

Notes

1. Wilde, Oscar (2003). *The Picture of Dorian Gray.* New York: Barnes and Nobles Books. Print, p. 50. (Original work published 1890).

2. Reber, Arthur S., Rhianon Allen, and Emily S. Reber. *Penguin Dictionary of Psychology*. London. Penguin Books, 2009. Print, p. 329.

3. Reber, Arthur S., Rhianon Allen, and Emily S. Reber. *Penguin Dictionary of Psychology*. London. Penguin Books, 2009. Print, p. 642.

4. Reber, Arthur S., Rhianon Allen, and Emily S. Reber. *Penguin Dictionary of Psychology*. London. Penguin Books, 2009. Print, p. 854.

5. Reber, Arthur S., Rhianon Allen, and Emily S. Reber. *Penguin Dictionary of Psychology*. London. Penguin Books, 2009. Print, p. 258.

6. Cornelius, R. R.. The Science of Emotion: Research and Tradition in the Psychology of Emotion. Upper Saddle River, NJ. Prentice-Hall, Inc. 1996. Print, p. 12.

7. Reber, Arthur S., Rhianon Allen, and Emily S. Reber. *Penguin Dictionary of Psychology*. London. Penguin Books, 2009. Print, p. 521.

8. Reber, Arthur S., Rhianon Allen, and Emily S. Reber. *Penguin Dictionary of Psychology*. London. Penguin Books, 2009. Print, p. 521.

9. Reber, Arthur S., Rhianon Allen, and Emily S. Reber. *Penguin Dictionary of Psychology*. London. Penguin Books, 2009. Print, p. 819.

10. Reber, Arthur S., Rhianon Allen, and Emily S. Reber. *Penguin Dictionary of Psychology*. London. Penguin Books, 2009. Print, p. 819.

11. Reber, Arthur S., Rhianon Allen, and Emily S. Reber. *Penguin Dictionary of Psychology*. London. Penguin Books, 2009. Print, p. 141.

12. Reber, Arthur S., Rhianon Allen, and Emily S. Reber. *Penguin Dictionary of Psychology*. London. Penguin Books, 2009. Print, p. 506.

13. Reber, Arthur S., Rhianon Allen, and Emily S. Reber. *Penguin Dictionary of Psychology*. London. Penguin Books, 2009. Print, p. 506.

14. Reber, Arthur S., Rhianon Allen, and Emily S. Reber. *Penguin Dictionary of Psychology*. London. Penguin Books, 2009. Print, p. 82

15. Reber, Arthur S., Rhianon Allen, and Emily S. Reber. *Penguin Dictionary of Psychology*. London. Penguin Books, 2009. Print, p. 202.

16. Reber, Arthur S., Rhianon Allen, and Emily S. Reber. *Penguin Dictionary of Psychology*. London. Penguin Books, 2009. Print, p. 202.

NOON

LATER THAT DAY

"In this world there are only two tragedies. One is not getting what one wants, and the other is getting it." - Oscar Wilde[1]

CHAPTER SIX

A One to One to One to One Ratio (1:1:1:1) and Appraisals

Front-loading: Appraised Value Judgment, Bound Purpose, Complementary Purpose Model, Double Bind, Drive, Drive Reduction Theory, Homeostasis, One to One to One to One Ratio, Pain, Pleasure, Ray, Vector

In chapter four, Freud's conception of psychoanalysis was introduced along with the two drives he theorized, a life instinct or libido, and a death instinct. Under classic drive-reduction theory, pioneered by Clark Hull and then E.L. Thorndike, drives interfere with homeostasis (balance), motivate an individual to reduce the drive, and those responses reducing the drive become reinforced.[2] Freud's theory of psychoanalysis possessed two drives in opposition to one another that at first glance might appear difficult to reconcile both with the equation and most contemporary theories. Fortunately, this is not the case. To achieve the aim, the concept of purpose will be considered along with the notion of the double-bind.

A double-bind occurs when an individual is ". . . confronted with a series of contradictory messages from a powerful or socially significant other," where it seems that ". . . no course of action will prove satisfactory and all assumptions about how to act will be disconfirmed."[3] The afflicted individual essentially feels to be caught in an impasse. For example, if Rose's supervisor wants her to go to an important out-of-town conference, but she and her spouse had already made reservations in town for a dinner and show elsewhere, then she is in a double bind as she cannot be in two places at once to satisfy both. She cannot both leave town and stay in it to please both husband and boss.

Figure 6.1 Rose cannot dance at two weddings.

Once a drive is described as a purpose with an entity whose acquisition fulfills it, for instance, the entity of water being acquired for the purpose of drinking water to hydrate, purposes behaving like double-binds arise. For example, the opposite purpose to that of drinking water would be the purpose of not drinking water. These bound-purposes, instead of originating from someone else, would have their source within the individual. They are bound in the sense that they restrict each other and are mutually exclusive. An individual, therefore, would possess both a drive to hydrate and a drive to not hydrate.

For this to work inside the restraints of the equation, a one to one to one to one ratio must be established and upheld between the individual, the purpose in question, the entity being valued, and the value judgment of the entity. With the equation as it stands now, only one valuation can be given to an entity as it relates to one purpose for one individual. For instance, in the case of Rose it will be assumed that she chose to go with her husband to the restaurant. Rose is hungry. She establishes a purpose along the following lines:

1) Individual: Rose

2) Drive or purpose: Eat food

3) Entity: Food

4) Value Judgment of the Entity: High

The above indicates Rose has determined that eating food is a valuable means to deal with the threat of starvation and to reduce the drive of hunger. However, the drive to alleviate hunger can easily be misinterpreted if it is not correctly implemented into the equation and in accordance with the 1:1:1:1 ratio. If Rose ate until she was full, then she would be in possession of two purposes, not a solitary one. The first purpose would be eating food. The second purpose would be not eating food at some point later. Necessarily, each purpose must periodically overtake the other if Rose is to stay alive. The entity of food then, will have two separate valuations for two separate purposes, the main purpose and its complementary purpose. Merging them prematurely, violates the 1:1:1:1 ratio, and the 1:1:1:1 ratio is always upheld in affect engineering.

1) Rose's main purpose is to eat food (i.e., gorge).

2) Rose's complementary purpose is to not eat food (i.e., fast).

For example, if Rose ate without stopping (e.g., binging) this would soon lead to death, either from Rose's intestines rupturing due to a surplus of food or less dramatically from choking, like an individual overeating pie at a pie eating contest. Likewise, if Rose did not eat food indefinitely, then she would starve to death. The same result would be found if Rose engaged in any other activity without being impeded such as drinking water, running, sleeping, painting, or working.

Figure 6.2 Overeating can result in death, as in the case of hyperphagia.

Together, these bound or complementary purposes enable Rose to stay alive. Separately and unrestrained, they would each lead to her untimely death. Henceforth, a traditional drive, such as the drive to eliminate hunger, rather than being thought of as a single purpose, must be construed as its two separate purposes in the equation. This avoids the indiscriminate mixing of two separate purposes into one, a trend that is too prevalent among drive theories. It also enables a clearer line of thought. The math concepts of rays and vectors will help elucidate this.

Independently, a drive may be mathematically represented as a ray. In math, a ray is a line that has a starting point on one side and extends infinitely in the other direction. Without a complementary drive to offset it, if Rose simply ate food then there would be no foreseeable end in sight. Death would be the result.

However, under a main purpose - complementary purpose model of motivation, the influence of each purpose and its corresponding drive can be thought of as a set of diametrically opposed rays that create a vector. A vector is "a quantity that has magnitude and direction and that is commonly represented by a directed line segment whose length represents the magnitude and whose orientation in space represents the direction."[1] While the force of each drive and complementary drive may extend out to infinity like a ray, and because the drives are in opposition to one another, there are bounds that the individual cannot transgress if he or she is to continue living. This gives each ray the qualities of a vector, that is, only if the individual chooses to remain within the boundaries.

Figure 6.3 A ray

[1] By permission. From Merriam-Webster's Collegiate® Dictionary, 11th Edition ©2014 by Merriam-Webster, Inc. (www.Merriam-Webster.com).

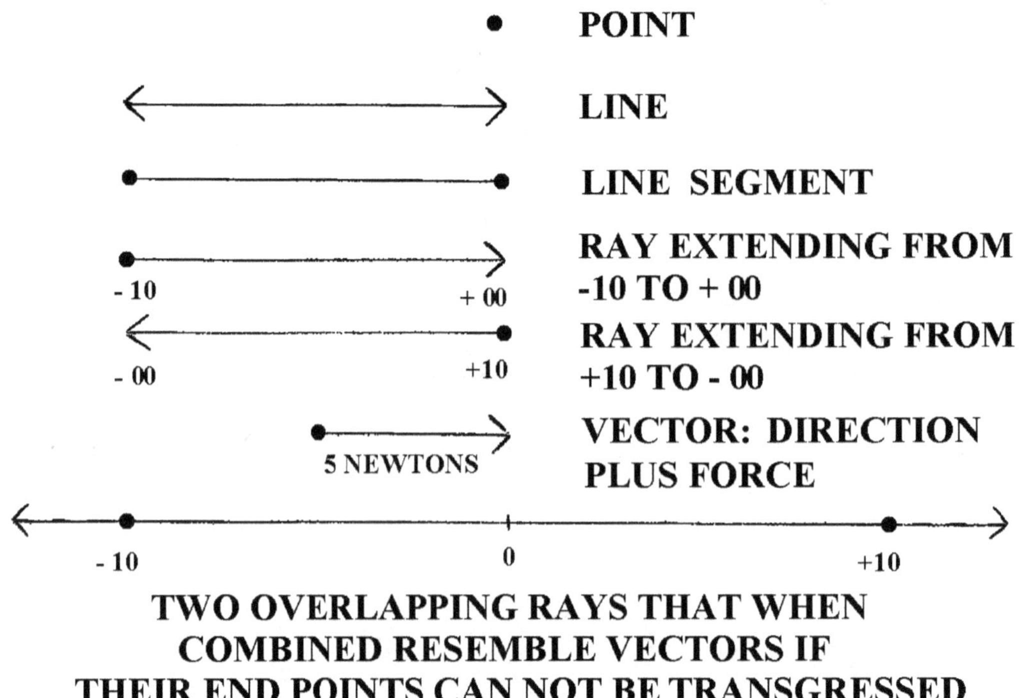

Figure 6.4 Creating a vector model of drives from two opposing rays

If hunger and satisfaction were modeled as rays with their corresponding purposes, then they would look like figure 6.5 and figure 6.6.

Figure 6.5 Ray for the drive to eat

Figure 6.6 Ray for the drive to not eat

If the two Rays intersected each other, then the following results (figure 6.7)

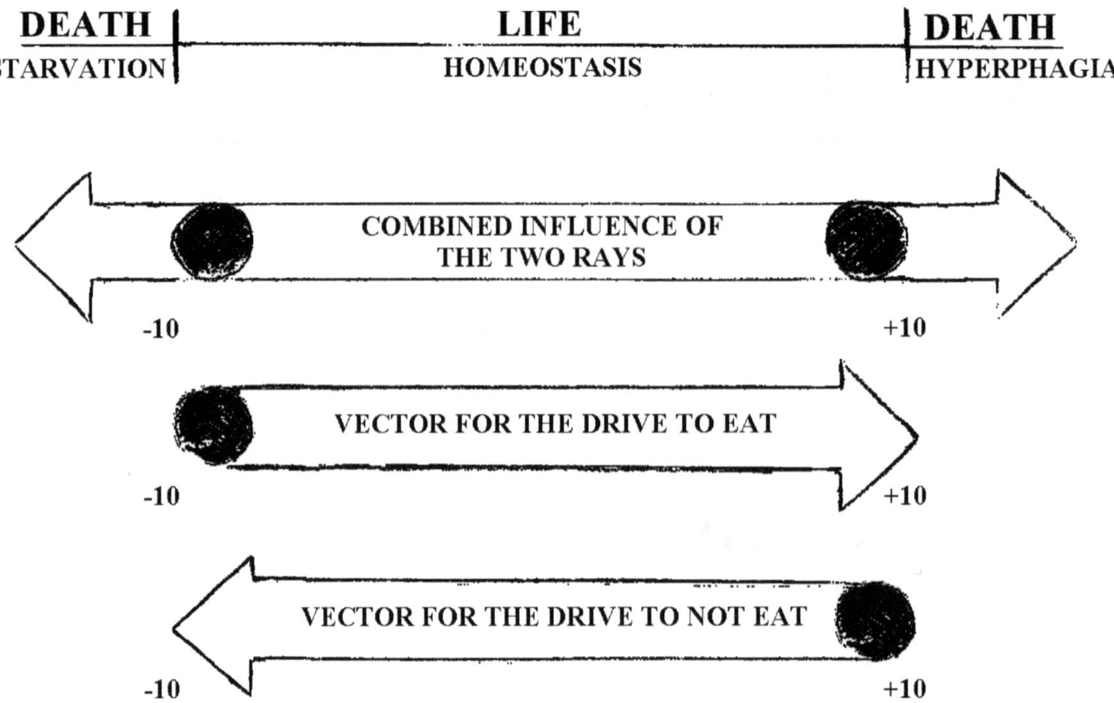

Figure 6.7 The two crossed rays and resulting vectors for the drive to eat and to not eat

From the image it can be seen that life is sustained by not progressing beyond either endpoint for eating or for fasting. However, each drive, left to their own devices, would lead to death. Subsequently, any action, if enacted upon perpetually, would result in self-annihilation. Homeostasis and the desire to maintain a tendency toward balance, however, keep an individual alive. In essence, a "feedback loop" results in which an individual modifies its own behavior to maintain optimal functioning and to balance drives.[4]

A One to One to One to One Ratio 1:1:1:1

One to One to One to One Ratio Explained

Imposing a ratio of values to entities, purposes, and people does not mean that an item can only have one value. On the contrary, the same entity can have multiple values, but only if multiple purposes or people are in question. For example, a knife can have value to someone for the purpose of cutting a banana. The same knife would possess another value for a different purpose, such as opening a package or as a potential murder weapon. An entity can have as many values as there are purposes for which it is intended to be used or people that value it for such purposes.

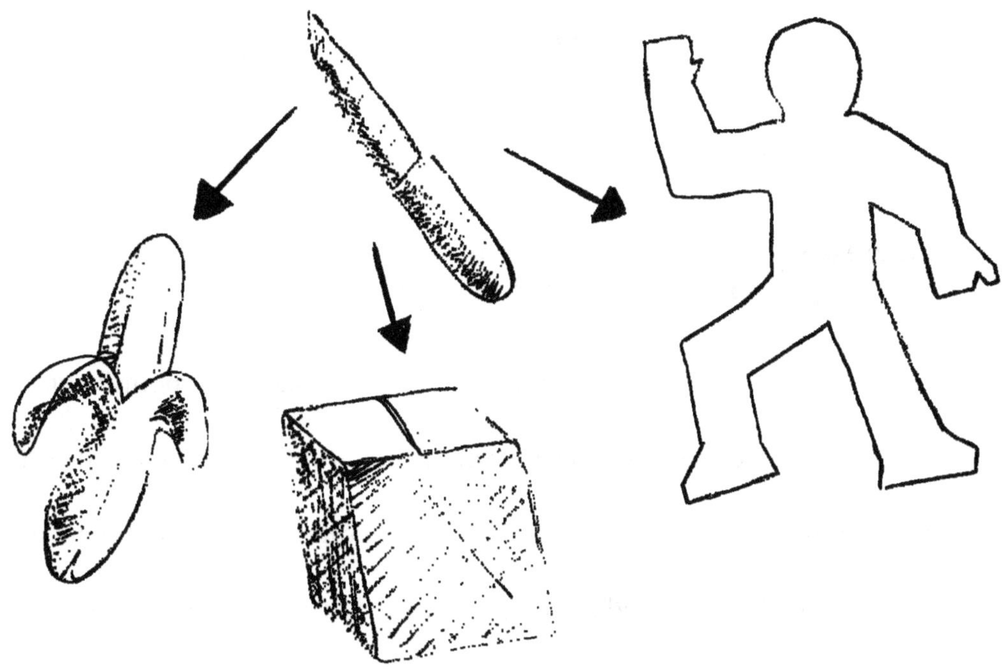

Figure 6.8 The entity of a knife has different values for different purposes.
1) Cutting a banana
2) Opening a box
3) Potential murder weapon

Likewise, multiple entities may receive the same value judgment as they relate to the same purpose or different purposes. A nail and a pin could each be judged to have the same value toward the purpose of popping a balloon, but each of those values belongs to that specific entity even though the nail and pin are equally valuable to popping a balloon.

Figure 6.9 Different entities may have an equivalent value for the same purpose, but each value belongs specifically to the entity.

Similarly, if Ted has two purposes, painting room *A* red and painting room *B* blue, then there are two entities in question: the red paint and the blue paint. Each entity could have the same judged value toward their respective purposes if their Sufficiency, Uniqueness, and Sentiment are equal or they balance out. If both the blue paint and red paint have the Sufficiency to color the room, if they possess the same level of Uniqueness, and if Ted has no preference as to which paint job should be held as more important, then the two separate paints would receive the same value judgment from Ted based on the variables discussed thus far.

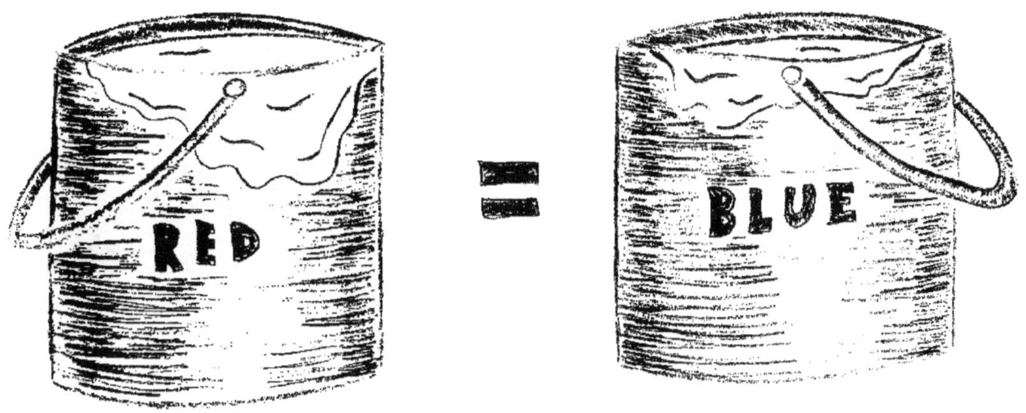

Figure 6.10 If the blue paint and the red paint are being valued for two different purposes, they may still have equivalent values if their Sufficiency, Uniqueness, and Sentiment components balance out.

In another case, one purpose could have multiple entities that are valued differently by an individual. With respect to the purpose of reducing darkness by lighting a room, a lightbulb, a candle, a flashlight, and a black chalkboard could all have different judged values given to them by an individual. The first three entities require additional items that may or may not be available and would alter their Sufficiency: lighting fixtures and electricity, a match, and batteries. The fourth item has the effect of making a room appear darker and is of no practical use for the purpose.

A One to One to One to One Ratio 1:1:1:1

Figure 6.11 The different value judgments of separate items for a common purpose:

1) The lightbulb, candle, and flashlight all have uniqueness values of 1 / 3. The blackboard's uniqueness is no higher than 1 / 4.
2) Lightbulb: Sufficiency value of 1 / 3
3) Candle: Sufficiency value of 1 / 2
4) Flashlight: Sufficiency value of 1 / 2
5) Blackboard: Sufficiency value tending toward 0
   The black chalkboard is not sufficient for the purpose of reducing darkness in a room by lighting it, as it absorbs light and makes the room darker. Its Sufficiency will tend towards zero and its total value becomes reduced to it's base or existential value of + 1. Though the argument can be made that its Sufficiency could potentially be 1 / 2, meaning that it simply needs another entity such as a lit candle in order to be sufficient to light a room, its Sufficiency value would be dependent upon whatever else is available. Moreover, its Uniqueness value, in addition to being judged against the other three entities that light up the room, would be judged against all the other entities that can match the blackboard's ability on all levels to light up the room, and this will likely include everything.

What Value Judgments Do

One way to conceptualize the role played by value judgments of entities is to think of value as a marker or a flag. Entities that possess the most importance toward the achievement of a purpose are given the biggest flags and they divert a proportionally larger share of anxiety resources in their direction because they are the most pressing. It follows that value judgments, in the equation, are

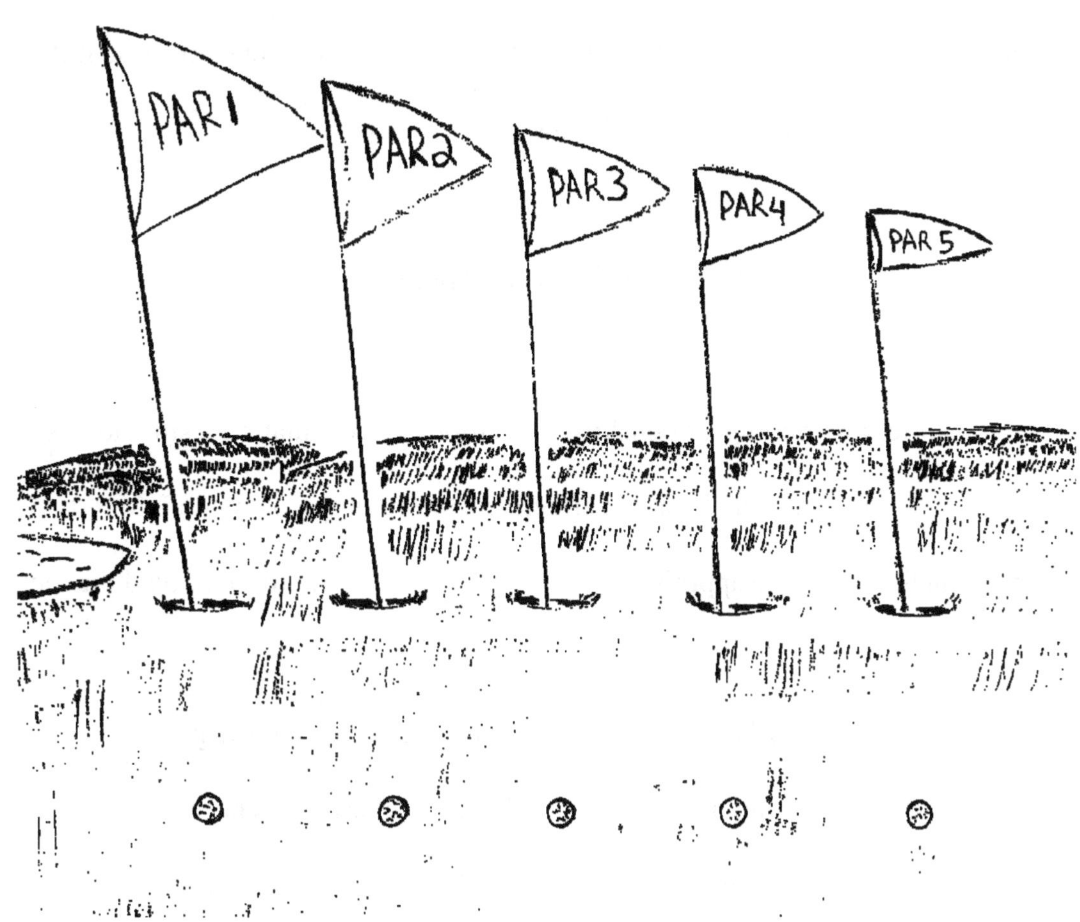

Figure 6.12 Golf flags above are used as a metaphor for anxiety. Although these five putting holes are right next to one another and the distance of the putt is the same for each, their values and subsequently, the amount of anxiety invested in making each shot are not the same. The par 1 flag is the largest because more anxiety is invested in it. An individual would only have one chance to make par on the shot. In descending order, the par 2, 3, and 4 flags would be the next largest flags. The par 5 flag would be the smallest flag of all because the least amount of anxiety is invested into making par for that hole, as there are five chances to make par.

markers for how an individual's anxiety resources have been allocated. They incite a call to action that marshals behavior toward a specific end. It bears mentioning that mismanagement or a poor investment of one's anxiety resources can often prove disastrous.

A key concept in affect engineering is the idea that emotions arise as a result of how value judgments for a particular entity may fluctuate or stay the same with respect to purpose fulfillment. Appraisals and cognitive appraisal theories will now be introduced.

Introduction to Cognitive Appraisals

Appraisal theories are theories of behavior that emphasize evaluating a situations's nature, determining if a situation is controllable and if effective responses can be made.[5] Magda Arnold, one of the pioneers in the cognitive appraisal approach to the study of emotion, held the view that

> "To arouse an emotion, [an] object must be appraised as affecting me in some way, affecting me personally as an individual with my particular experiences and my particular aims. If I see an apple, I know that it is an apple of a particular kind and taste. This knowledge need not touch me in any way. But if the apple is of my favorite kind and I am in a part of the world where it does not grow and cannot be bought, I may want it with a real emotional craving."[6]

For Arnold, the sequence of events leading up to an emotion was the following: perception of an object, an appraisal of the object, and an emotion.[7] Arnold conceptualized an appraisal as the "immediate sense judgment of weal or woe."[8] Her term "'sense judgment'" emphasized the "'direct, immediate, non-reflective, nonintellectual, and automatic'" nature of appraisals, which she exemplified in the example of a baseball player's attempt to catch a fly ball.[9] In her scenario, if a baseball player trying to make a catch stopped to think about his or her own movement, or that of the ball's flight path, then he "'would never stay in the game.'"[10]

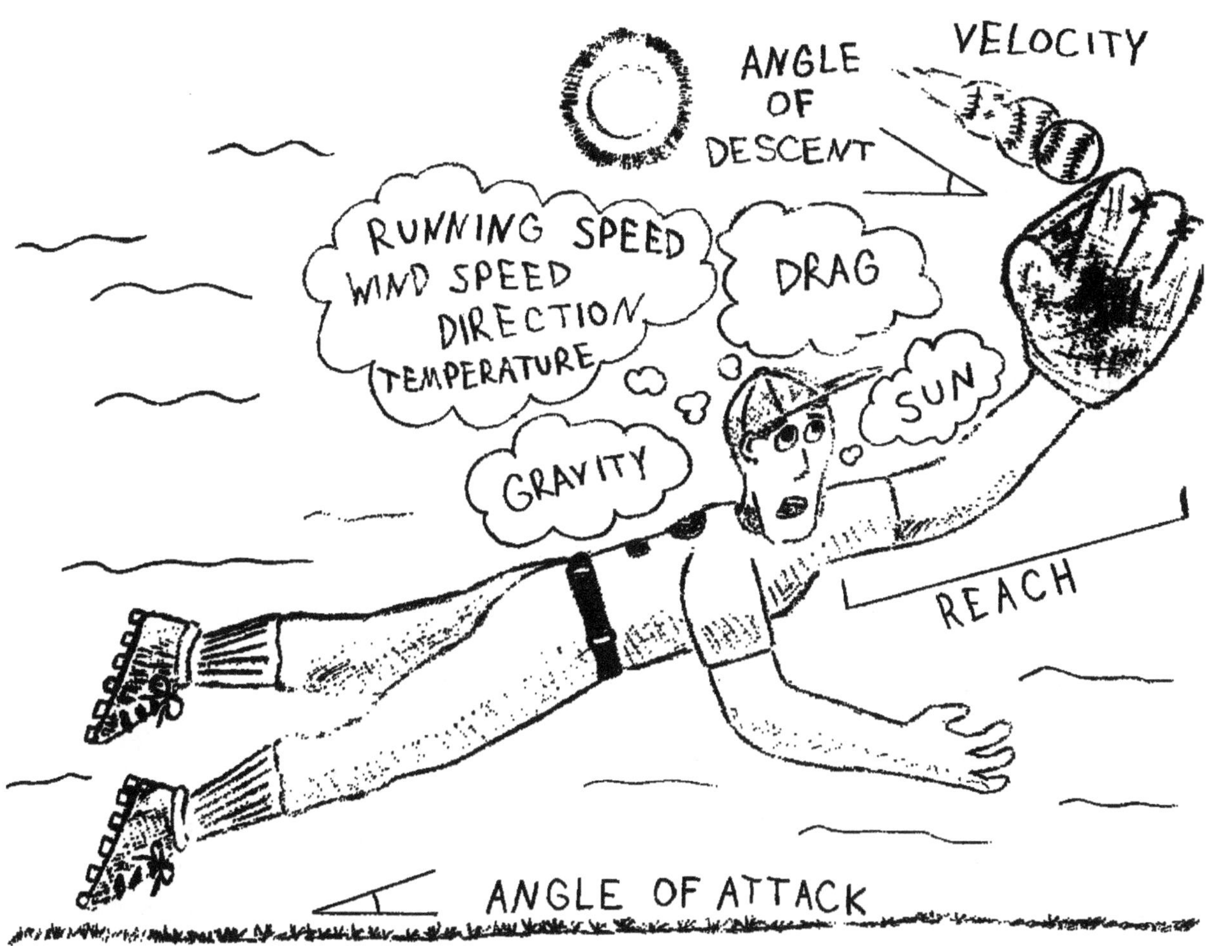

Figure 6.13 The catch

Appraisals, a Roll Cage Model of Drive Reduction Theory, and Identifying Emotions

Arnold defined emotion as,

> "'The felt tendency toward anything intuitively appraised as good (beneficial), or away from anything intuitively appraised as bad (harmful). This attraction or aversion in turn, is accompanied by a pattern of physiological changes organized toward approach or withdrawal. The patterns differ for different emotions.'"[11]

Based on Arnold's description of objects as either good and eliciting approach behavior, or as bad and eliciting avoidance behavior, there would be two valences to consider if Arnold's theory of appraisals were directly applied to the equation. While it may be tempting to assign a positive value to good entities and a negative value to bad entities, a closer inspection would reveal this to be a too simplistic and a wrong decision for the equation. The conception of an appraisal will have to be made more precise before it can be implemented into the equation.

In the equation, instead of labeling an entity as simply good or bad, currently, entities have been thought of as valuable toward both the fulfillment of a particular purpose and valuable toward that purpose's opposite or complementary purpose. This will continue as it also helps to avoid unnecessarily imparting motives or intentions onto inanimate objects.

Figure 6.14 Earth Dude says, "Imparting motives or intentions onto inanimate objects is to be avoided."

Considering an entity to be valuable for two opposing purposes instead of being harmful or beneficial in and of itself has the added bonus of freeing up the term harm to be applied to something else for the equation. In affect engineering, harm will eventually be used to describe relationships, such as between two entities, where one event or phenomenon (i.e., a second entity) threatens the original entity with destruction. Additionally, it can be surmised that the two values of an entity, with respect to a purpose and the purpose's complement, would be oppositely related to each other.

For instance, if an article of clothing that is being worn is sufficient toward the purpose of keeping a person warm, it **cannot** also be sufficient toward the purpose of not keeping the same person warm unless it is being employed for use in a different manner and a different purpose. It would, in the first case, be detrimental to the purpose of not keeping warm. It's Sufficiency then, toward the purpose of not keeping warm, would be significantly less and in opposition to the purpose of keeping warm for that particular use. The uses are mutually exclusive.

Figure 6.15 The different uses of a faux-fur coat include:

1) In a winter wonderland
2) On a hot beach
3) As a fan on a hot beach
4) In the first two cases (left and center) the individual has the purpose of wearing clothing to keep warm vs. its complement to not keep warm. In the third case (right) the individual has the purpose of finding something to serve as a fan or not to serve as a fan, meaning that four purposes are being considered instead of two. The 1:1:1:1 ratio is upheld.

What this means is that the Sufficiency of the entity, a coat, toward the purpose with which it is associated, wearing clothing for warmth, must be found first. Thereafter, its Sufficiency for the complementary purpose of not wearing for warmth could be gauged as being in opposition to this value, as the Sufficiency of the coat for keeping warm plus the Sufficiency of the coat for not keeping warm must equal one. Likewise, a threat of harm to the faux-fur coat, such as a thief, stealing it, would be more advantageous to the individual for the purpose of not keeping warm than to the purpose of keeping warm.

Figure 6.16 The acquisition of a coat to be worn has a specific Sufficiency value for the purpose of keeping warm. It cannot also be sufficient for not keeping the individual warm. The Sufficiency of an entity for a purpose plus its Sufficiency for the complementary purpose must equal one.

Measuring Value for the Fulfillment and Non-fulfillment of Purposes

In order to find out how useful an entity is toward not doing something (i.e., the non-fulfillment of a purpose), its Sufficiency for that specific purpose must first be known. Sufficiency for not doing something is found indirectly. To elaborate, an entity cannot both be wholly sufficient for doing something and sufficient for not doing the same thing. Knowing this, the following may be said.

> The Sufficiency of an entity for the fulfillment of a purpose
> plus the Sufficiency of an entity for not fulfilling the purpose
> equals one.

Stated differently, the Sufficiency of entity *A* for not fulfilling purpose # 1 would be equal to one minus the Sufficiency of entity *A* for the fulfillment of purpose # 1 or (1 - ***S***), where ***S*** is the Sufficiency of entity A for the fulfillment of purpose # 1.

If the Sufficiency of entity *A* for the fulfillment of purpose # 1 is one over one, then the Sufficiency of entity *A* for the non-fulfillment of purpose # 2 would be zero. If its Sufficiency is one-half for the fulfillment of the purpose, then it would be one-half for the nonfulfillment; if it is one-third for the fulfillment, then it would be two-thirds for the nonfulfillment and so on.

However, the same would not hold true for Uniqueness, which is a question of the number of other entities that can match the entity's Sufficiency. The Uniqueness of an entity is determined by analyzing the properties of the entity for a specific purpose against all of the other entities the individual knows. Sentiment is determined separately for every purpose harbored by the individual, but relative parity between a purpose and complementary purpose should be expected if the individual aims to achieve homeostasis between the two.

For most of the examples throughout the book, the main purpose under observation (i.e., doing a specific action) will be considered to have a maximum Sufficiency, a maximum Uniqueness, and often a maximum Sentiment value, meaning its valuation will tend toward + 2 emotional units.

In such cases, the complementary purpose, (i.e., of not doing the above specific action), will always have a valuation that is + 1 emotional unit due to Sufficiency being zero for the complement.

For example, if, as part of an apple boycott, not buying apples at a grocery store is the purpose in question, and a pumpkin is being investigated, then the pumpkin's Sufficiency for the purpose of buying apples would need to be determined first. The pumpkin's value would tend toward zero as something else would be required to buy an apple. It would be one in the numerator divided by one plus the number of other entities required in the denominator. Thereafter, its Sufficiency for the purpose of not buying apples could be determined to be headed close to one, as one minus Approximately zero = Sufficiency of the pumpkin for not buying apples. For instance, if nine other entities are required to make buying the pumpkin sufficient for buying an apple, then the Sufficiency of the pumpkin for not buying apples would be .9, as:

1 - (1 / (1 + 9)) = 9 / 10 = Sufficiency of the pumpkin for not buying apples.

However, if a buy-one-pumpkin-get-one-apple-free special is occurring, where accepting the apple is mandatory, then the Sufficiency of the pumpkin for the purpose of not buying apples might be two-thirds with the added entities of the special, the apple, and where if price is considered no object (i.e., the entity of money is ignored):

1 - (1 / (1 + 2)) = 2 / 3 = Sufficiency of the pumpkin for not buying apples.

Similarly, this value could be one-half if the only way to buy a pumpkin involved first buying an apple.

If one were looking at the entity of an apple, the Sufficiency of the apple for the purpose of buying an apple is found to be one (1 / 1). The apple's Sufficiency for the purpose of not buying an apple would be zero.

Figure 6.17 The Sufficiency of the pumpkin for the purpose of not buying an apple would have a potential minimum value of two-thirds (with the special and the apple being considered) and a potential maximum value heading toward one. The Sufficiency of any single apple for the purpose of not buying an apple would be zero, because they are apples.

In short, trying to find out how sufficient an entity is for not doing something can be gathered by obtaining its Sufficiency for actually doing the purpose, and thereafter by subtracting that total from one. For example, if a tomato is sufficient for eating food, then it cannot also be sufficient for the purpose of not eating. Consequently, a deficit results in the sense that a lesser proportion of anxiety resources would be invested into valuing the tomato for the purpose of not eating. Hence, it is possible to distinguish an entity as valuable for fulfilling a purpose or a drive and thereafter measure its value for the complementary purpose of not fulfilling the purpose or drive. In affect engineering this entails that a purpose is best expressed as doing something or not doing something with respect to the acquisition of an entity. For example, the purpose of losing weight would not be the complementary purpose of gaining weight. The complementary purpose to acquiring the entity of weight loss would be the purpose of not acquiring the entity of weight loss; likewise, the complementary purpose of gaining weight would be not gaining weight. Although losing weight and not gaining weight convey similar ideas, as do gaining weight and not losing weight, they are four, distinct purposes as opposed to two. Eventually, it will become necessary to associate purposes that are related to each other, but for now this concern can be put aside.

Concerning the Uniqueness of an entity for the purpose of not doing something, this value will have to be assessed for each entity and purpose. For instance, in figure 7.5, the Uniqueness of the pumpkin for the purpose of buying an apple was the following:

1 / (1 + 49) = 1 / 50.

The other 49 apples matched the entity of the pumpkin on all levels for this purpose. However, for the purpose of not buying an apple, the pumpkin's Uniqueness was the following:

1 / (1 + 0) = 1.

No other entity matched the pumpkin's Uniqueness for the purpose of not buying an apple. Similarly, the Uniqueness of an apple for the purpose of buying an apple would be 1 / 49, as there are 48 alternatives to any single apple:

1 / (1 + 48) = 1 / 49.

The Uniqueness of an apple for the purpose of not buying an apple would be 1 / 50:

1 / (1 + 49) = 1 / 50.

To illustrate Uniqueness in another example, with respect to the purpose of not eating food, if a direct attempt to measure a nonedible pillow's Uniqueness were made then its Uniqueness would be one at first, and then one-half once the entity of nothingness is considered. The number of alternatives that could match the pillow's non-edibility would be accounted for when finding its Uniqueness value for the purpose of not eating food, and would consist of everything that exists and anything that is not categorized as food:

Uniqueness = [1 ÷ (1 + any other entity that the individual knows is not food or that could be an alternate for not being food, such as the act of eating nothing)]

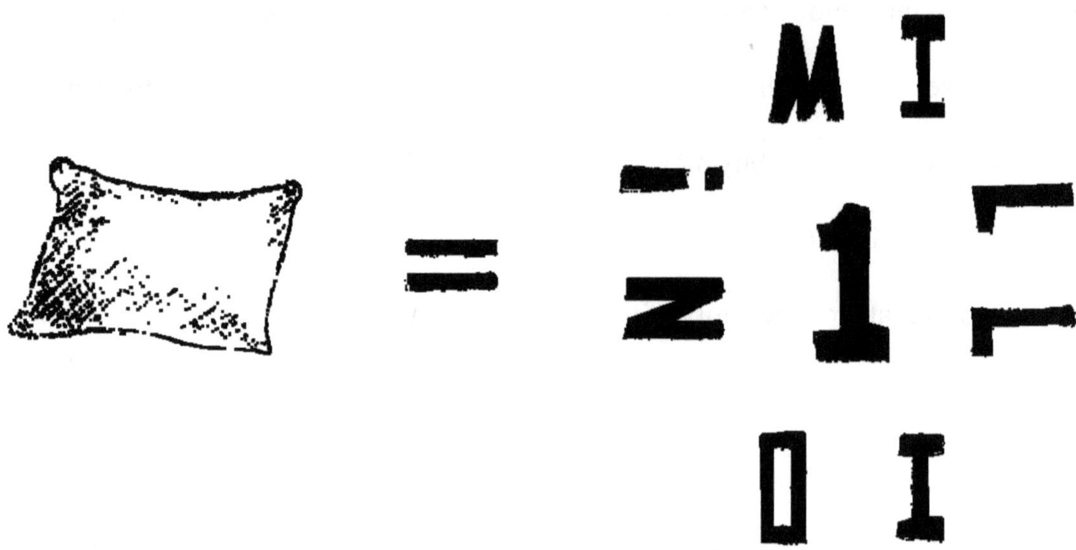

Figure 6.18 A pillow may literally be one in a million objects that an individual considers.

If Peter were only aware of ten objects in the universe (seven of them being food items, the eighth, ninth, and tenth being the pillow, nothingness, and Peter himself), then the valuation of the pillow toward the purpose of not eating food would equal the following:

1 × [1 - (1 ÷ (1 + 1 other entity that must be acquired for the purpose of eating))] ×

[1 ÷ (1 + 2 alternatives, as Peter can eat nothing and he cannot eat himself)] ×

[1 ÷ (1 + 0 displacements from the highest ranked purpose)]

and + 1 for the existence value.

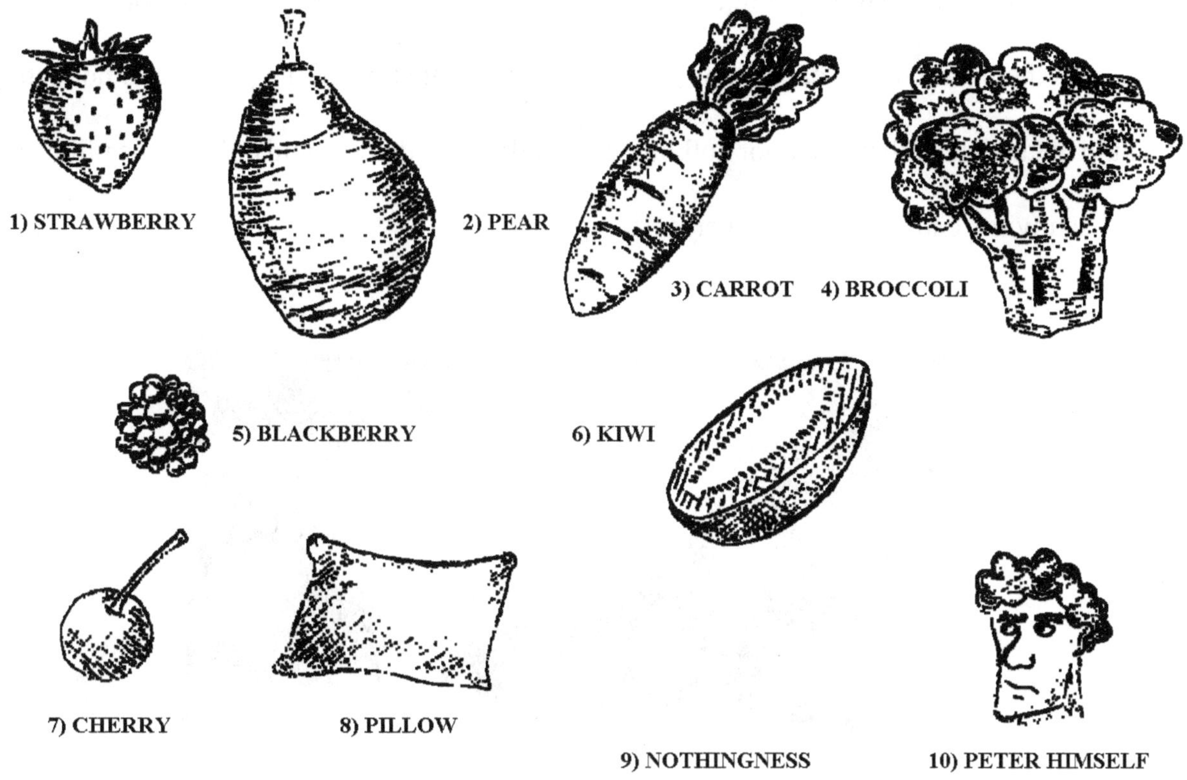

Figure 6.19 Peter and his nine other objects have values both toward the fulfilment of the drive to eat food and toward its complementary drive to not eat food. In this case, Peter is only considering their values for the complementary drive to not eat food.

The value judgment of the pillow would reduce to one and one-sixth emotional units for the purpose of not eating food. As Peter grows more aware of nonedible entities, the value judgment of the pillow will continue to progress infinitesimally close to one emotional unit for the purpose of not eating food. The Uniqueness of the pillow for the purpose of eating food is even lower, initially at one tenth, as the seven food items and two other non-foods match its Sufficiency:

$$1 / (1 + 9) = 1 / 10$$

As noted, the Sufficiency value of the pillow for the purpose of eating food can be no higher than one-half.

For the pillow to be wholly unique for the purpose of not eating food, it would have to be the only entity in the universe and capable of judging its own value toward the purpose of not eating. This would mean that neither Peter nor nothingness would exist, there would only be one pillow and it would occupy the entirety of the universe.

Figure 6.20 If all were pillow...

1 × [1, *because no other entities are required for Sufficiency*] × [1 ÷ (1 + 0, *because no alternatives exist anywhere*)] × [1 ÷ (1 + 0 displacements from the highest ranked purpose, *because the purpose in question has no equal*)] and + 1 for its existence value.

Although such a scenario is improbable, it also bears mentioning that if the pillow is the only entity in the universe and made of cotton candy for instance, then its Sufficiency for the purpose of not eating food would never be greater than one-half. At least one other entity would be required to make the original food inedible, such as something else that makes it inedible. This would cut its Sufficiency value for the purpose of not eating in half before anything else is considered, then to zero as no other entities actually exist in this case.

The utility portion of the equation, namely, the Sufficiency, Uniqueness, and Sentiment variables, can collectively be thought of as the drive reduction or purpose fulfillment component of the equation. To clarify, in the ray model of drives and in affect engineering, drives do **not** reduce on their own, but become balanced against their opposite or nemesis drive. Hence:

The drive reduction or purpose fulfillment component of the equation equals:

Sufficiency × Uniqueness × Sentiment

For an entity strongly associated with reducing a drive by fulfilling a purpose, the value of this drive reduction component will approach + 1 and tend to be well above zero:

zero ≤ Sufficiency × Uniqueness × Sentiment ≤ one

For the same entity, its association with not reducing the above drive or not fulfilling the above purpose, results in the value of the drive reduction component approaching and shrinking infinitesimally close to zero. Hence, if Sufficiency equals one in the above case for a purpose, then Sufficiency and the drive reduction component will equal zero for its complementary purpose:

Sufficiency (if zero) × Uniqueness × Sentiment = zero

Stated differently, if the acquisition of an entity is wholly sufficient for the fulfillment of a purpose, then that same entity's Sufficiency for the fulfilment of the complementary purpose would be zero.

Therefore, if the Sufficiency of an entity is not high for the fulfillment of a purpose, then the object's net value with respect to the main purpose and complement could swing the other way, or be closer to zero depending on what Uniqueness and Sentiment are determined to be. In example, an individual finds a novel entity of an axe that is not yet associated with being valuable to either a purpose of cutting or that purpose's complement of not cutting. He or she may at first simply give the novel entity of an axe an existential value of + 1 emotional unit. Thereafter, if the individual had a predisposition to assume that the axe's Sufficiency for a purpose was zero, as a default, then the initial value of the axe for a purpose and its complementary purposes might look like the following:

1) Purpose # 1 is to cut down a tree. The value of the axe = + 1 emotional unit. The Sufficiency variable equals zero, leaving only the existential value intact.

2) Purpose # 2 is to not cut down a tree. The value of the axe is between + 2 emotional units. The Sufficiency is equal to one, initially, and the entity, being novel, also has a Uniqueness value of one; Sentiment is additionally being held constant at one)

At first inspection, the axe is not associated with fulfilling purpose one; the axe has a maximum Sufficiency value for not cutting down a tree only by default before it is assessed for its utility.

If, after assessing the axe, it becomes associated with facilitating purpose # 1, cutting down a tree, then the axe takes on a higher value for purpose # 1, cutting down a tree, while the value for the other complementary purpose decreases to one due to the change in Sufficiency:

3) Purpose # 1 is to cut down a tree. The value of the axe now becomes + 2 emotional units (Sufficiency, Uniqueness, and Sentiment all equal one).

4) Purpose # 2 is to not cut down a tree. The value of the axe now decreases to its existential value of + 1 emotional unit, as the drive reduction component of the base has become zero due to Sufficiency becoming zero.

Subtracting the value of the axe for purpose # 2 from the value of the axe for purpose # 1 would lead to the following:

2 - 1 = 1 emotional unit.

The axe would be more valuable toward the purpose of cutting down a tree than not cutting down a tree by one emotional unit. If the value of the axe for purpose # 1 were subtracted from the value of the axe for purpose # 2, then it would lead to:

1 - 2 = - 1 emotional unit.

What this would suggest is that the axe is less valuable toward the purpose of not cutting down a tree by one emotional unit against its value for the purpose of cutting down a tree. It could be said that the overall net value of the axe is at + 1 emotional unit for the purpose of cutting down a tree and at - 1 emotional unit for the purpose of not cutting down a tree. This will hold true for so long as the individual associates the axe with being useful for the purpose of cutting down a tree and the Sentiment variable remains constant. If, however, the axe suddenly becomes blunt and ineffective, then it will lose its association with the fulfillment of the purpose to cut down a tree, due to a reduction in Sufficiency, and the disparity between its value for purpose # 1 and the value for purpose # 2 would disappear. In short, by maintaining the one to one to one to one ratio, the muddling of the interpretation of an entity's value is prevented. The premature mixing of a valuation for a purpose with its valuation for a complementary purpose is averted.

As a theory of learning and within the context of the equation, it will be helpful to keep in mind that an entity can at first only be considered valuable with respect to its existence, and thereafter to a purpose or that purpose's complement. Entities here are not being considered harmful to a purpose or to an individual in and of themselves. Rather, they are considered to possess two discrete values, one for a purpose and another for the complement of the purpose. Initially, both of these values might be equal to + 1, the Existential or base value for an entity, if the entity is not being considered for the fulfillment of a purpose nor for its complementary purpose. In such a scenario, neither value would vary from one, hence, the drive-reduction components of Sufficiency,

Uniqueness, and Sentiment variables would not even be under consideration. They would be irrelevant. Until an entity is considered for the reduction of a drive or purpose fulfillment, the model holds that a cognized entity's valuation will stay at its existential value of + 1 emotional unit.

Eventually, when using the equation, only one purpose out of a complementary pair will need to be considered to gain a sense of an entity's valuation. In the examples throughout this book, the complementary purpose will always be described as not doing an action, while the main purpose will always be described as doing an action. This will help to maintain clarity later. Subsequently, if the Sufficiency value is maximized with respect to fulfilling a purpose or a drive, then it will have the potential to approach + 2. Meanwhile, the same entity's valuation for not fulfilling a purpose or not reducing a drive will tend to approach + 1, its existential value.

As demonstrated earlier, if an entity has become strongly associated with fulfilling a purpose, such as the axe with cutting a tree, then the value of the axe for the purpose of not cutting a tree can only approach one due to the drive reduction or purpose fulfillment component of the equation heading toward zero and it garnering only + 1 emotional unit for its existence. However, the negative of the entity's valuation for the original purpose can be loosely understood as an indirect measure for its valuation of the complementary purpose. For example, if the value of the axe toward the original purpose of cutting a tree is the following.

Value Judgment of axe = 1 $(1 \times 1 \times 1 + 1) = 2$ emotional units.

The value of the axe toward the complementary purpose of not cutting a tree can be estimated at - 2 emotional units. As the equation becomes further developed later in the book, the main purpose will come to have valuations that can be substantially greater than two, while the complementary purpose's valuations will only contract toward one. The complementary purpose will still be measured, in accordance with the 1:1:1:1 ratio, but relative to a main purpose its valuation will become trivial at higher values for the main purpose.

Figure 6.21 The above depicts the tendency of an entity's value towards fulfillment of a purpose and the fulfillment of its complementary purpose. If an entity, in this case a strawberry, is associated with facilitating the fulfillment of a purpose (e.g., eating food), then it's value approaches + 2 for that specific purpose and descends infinitesimally close to + 1 for the purpose's complement of not reducing the same drive (not eating).

What this means is that the anxiety invested in the axe for the fulfillment of purpose # 1, cutting down the tree, would be equivalent to negative anxiety being invested in the axe toward the fulfillment of purpose # 2, not cutting down the tree. Although both negative and positive valuations can be given to an entity, these valuations represent different concepts in the equation than what may be expected. In the confines of the equation, positive values given to an entity mean that anxiety and subsequently pain, are being used to judge the entity's value. Conversely, negative values given to an entity mean that negative anxiety, and subsequently pleasure, are being used to judge the entity its value.

With respect to traditional conceptions of positive affect and negative affect, the use of negative and positive anxiety might be confusing initially, but will become less so as one grows familiar with their implementation in the equation. Positive affect in the equation would occur whenever anxiety decreases or negative anxiety increases for a single purpose (figure 6.22). Negative affect, in the equation, would occur whenever anxiety increases or negative anxiety decreases for a single purpose (figure 6.23). These can be seen in the following images.

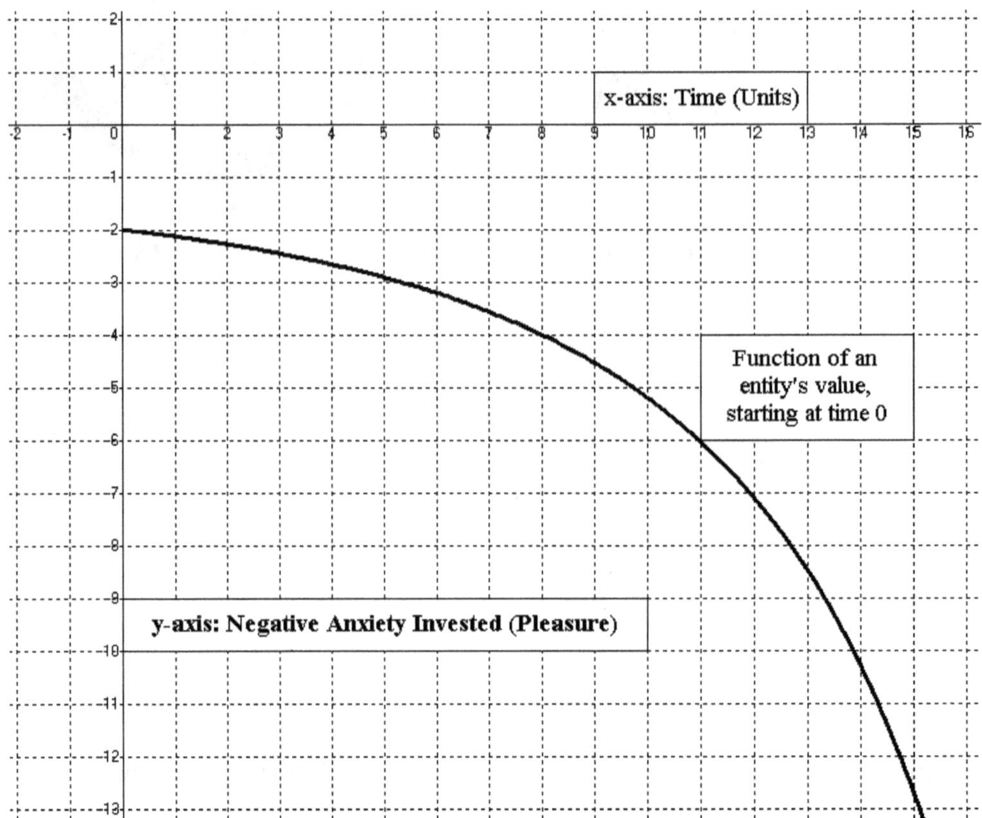

Figure 6.22 A function below the x-axis represents negative anxiety invested in an entity and pleasure felt with respect to a specific purpose. Negative anxiety investment is increasing for the specific purpose, meaning pleasure felt for it is elevating.

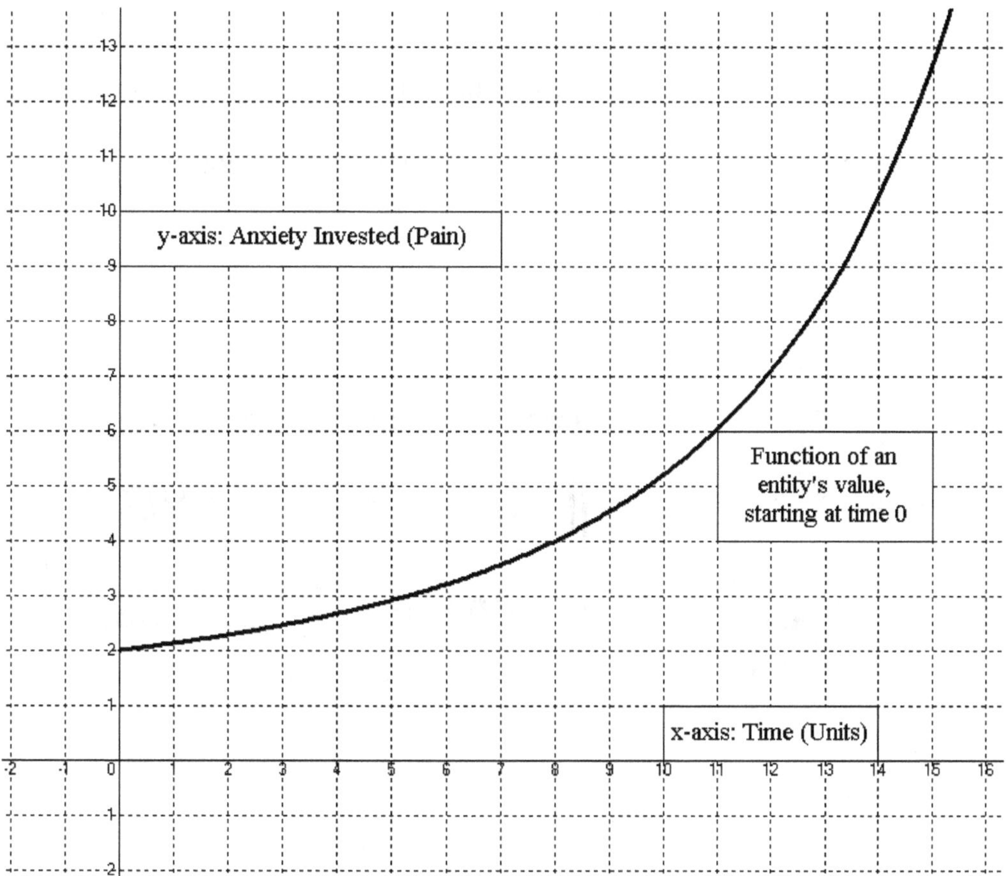

Figure 6.23 A function above the x-axis represents positive anxiety invested in an entity and pain felt with respect to a specific purpose. Anxiety invested in an entity is increasing, meaning pain felt for it is increasing.

Positive and Negative Appraisals in the Context of the Equation

A situation where an entity has its existence threatened by harm, such as Arnold's example of the apple, will now be considered for the person of Adam. If a particular apple that Adam wants is in limited supply and cannot be found or bought elsewhere in a region, then it will have a Uniqueness value close to or equal to one. If only one apple is required for Adam's purpose, then it would also have a Sufficiency value equal to one. Regardless of whether or not the apple is in abundance in Adam's homeland, if only one was brought on the journey then it would be valuable nonetheless in the new location and for Adam's purpose of eating an apple during the trip. For this example, only the purpose of eating the specific type of apple will be considered.

Figure 6.24 Apples far from home

A threat of harm to the apple might come from a thief who wants the apple or from the possibility that worms may burrow into and ruin it. Adam's value judgment of the apple would become elevated by the consideration of danger or harm to the apple. While it is not yet certain how much the judged value of the apple elevates due to a threat of harm, it can be ascertained that the individual's judged value of the apple will elevate toward positive infinity if it is stolen or becomes unusable. There would be no other means for recourse in the purpose of eating that apple if someone else gets to it first. For instance, the denominator in the Uniqueness variable for this entity, if it headed toward the limit of zero, would create a vertical asymptote in the function. Though its Uniqueness would be undefined if the denominator were reduced to zero, the variable can be estimated by using a number incredibly close to zero as was done in chapter five (figure 5.20), such as .000000001 for the denominator, or one billionth. This would result in one being divided by .000000001, which equals one billion. It can be calculated that the Uniqueness of the apple would approach infinity as values smaller than one billionth are used. If no other apples are available as well, then a great deal of anxiety would be invested in trying to acquire an apple that no longer exists. Adam would be forced to abandon the purpose of eating an apple, such as by drastically lowering its Sentiment ranking to compensate. Otherwise he will end up investing the remainder of his anxiety resources trying to acquire an apple that cannot be found. In the traditional sense of the term anxiety, one might say that Adam is experiencing an abnormal amount of anxiety due to the apple's imminent destruction. However, in the context of the equation, it would be more appropriate to say that Adam depleted his available anxiety resources upon discovering news of the apple's imminent destruction and choosing to invest anxiety in the apples substantially above their existential level.

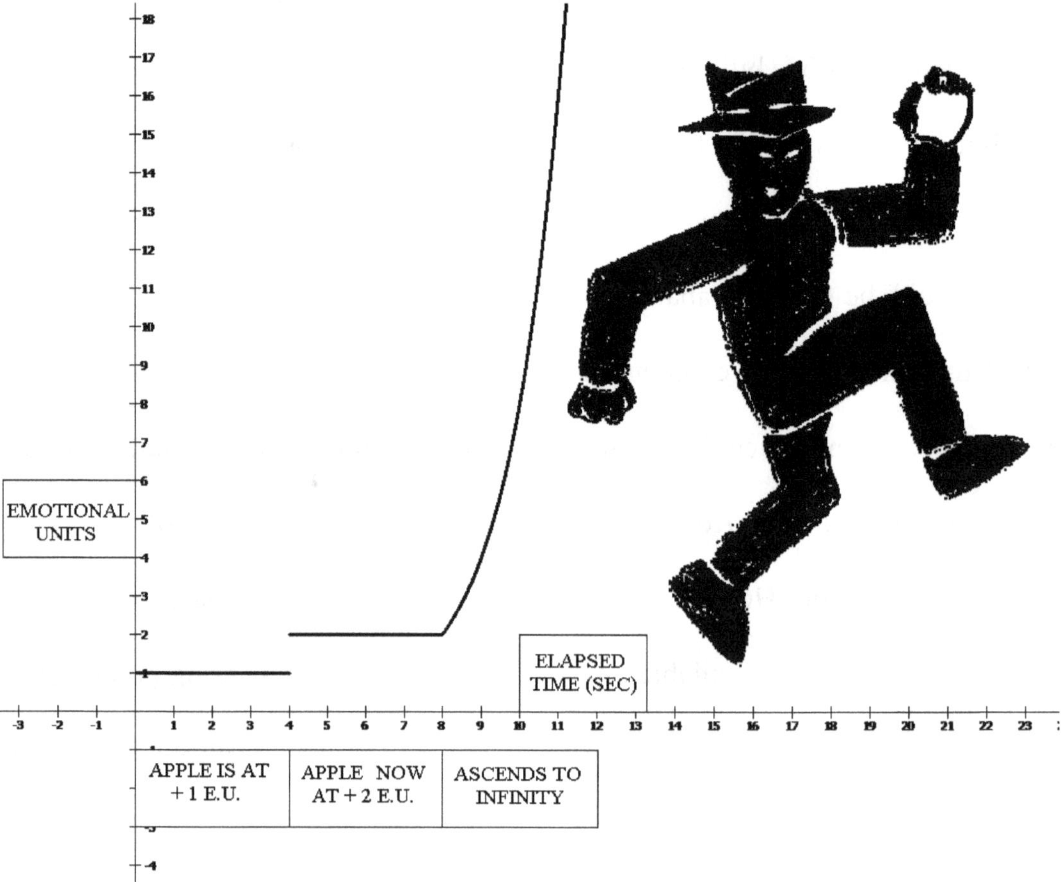

Figure 6.25 Adam's anxiety invested in the apple starts off at + 1 once the apple is discovered to exist. It then becomes + 2 once the purpose of eating it during the journey is set as a purpose. The apple finally approaches infinity after a thief steals it and starts to eat it.

Conversely, if someone were in the same situation as Adam and had brought along an ample supply of apples just in case the one needed for the purpose of eating became stolen or unusable, then this would result in a negative valuation of the extra apples if they were already in one's possession. The following situations will be considered where:

The value of the apple = Existence × [Sufficiency × Uniqueness × Sentiment + 1]

Figure 6.26 Adam, Granny Smith, Bob, Paula Red, and Washington are on a guided tour in Mini-Appleless, the land without apples. Each has the goal of wanting to at least eat one apple during the course of the trip. After witnessing the theft of Adam's apple, none of the travelers revealed to anyone else that they themselves are carrying apples. Each believes that his or her apples are the only ones in Mini-Appleless.

1) Adam has no apples left: Adam's anxiety investment was approaching infinity because there were no apples to be found so he was compelled to abandon the purpose of eating an apple by reducing his Sentiment for the purpose in order to properly allocate his anxiety resources elsewhere. The denominator in the Uniqueness variable for Adam became reduced to zero. However, because it is not possible to divide by zero, the value can only approach the limit of zero. As the Uniqueness variable's denominator approached zero, Adam's anxiety approached infinity and so his only option was to forsake the hope of ever acquiring an apple to eat.

2) Granny Smith brought one apple: Granny Smith's net anxiety is at + 2 emotional units. Her single apple carries a value of + 2 emotional units.

3) Bob brought two apples: Bob's net anxiety investment is at zero emotional units. His original apple carries a value of +1.5 emotional units while the second or back up apple carries a value of -1.5 emotional units.

4) Paula Red brought three apples: Paula Red's net anxiety investment is at -1.33 emotional units. Each of Paula Red's three apples carries an absolute value of 1.33 emotional units. Her two extra ones, however, carry a negative charge.

5) Washington brought four apples: Washington's net anxiety investment is at -2.5 emotional units. Washington's apple he plans to eat carries a value of 1.25 emotional units. The three extra apples each carry a value of -1.25 emotional units.

Because extra of a desired entity are possessed, the additional resources become an added measure of security against threats of harm, in this case harm to the apple required for the purpose. They become insurance apples. If an individual's anxiety is thought of as an electrical grid, then instead of drawing out two volts of power from an energy reserve for instance, three volts might be put back into the system and become available for other purposes. Hence, instead of using up anxiety resources, there is now a surplus of anxiety resources available. What is proposed is that the apples, being in more than ample supply, contribute to a negative anxiety. Rather than using up available anxiety resources, the extra apples impart a sense of carefreeness and act as a buffer against anxiety resources being used up in too great an amount for the purpose of eating an apple. Consequently, in order for individuals to maintain optimum potential, flexibility, and work efficiency, they would want to allocate their anxiety and negative anxiety resources wisely. For instance, deciding to pack 100 apples would reduce the anxiety felt for the purpose of eating one apple well below zero. Unfortunately, one would also have 100 apples and that would make walking around difficult. For the purpose of traveling, for instance, the possession of only a dozen apples would be less of a hindrance than possessing 100. The purpose of traveling and the purpose of eating an apple would then have to be weighed against one another to determine an ideal number of apples to bring. In such a scenario, Sentiment would play a more decisive role in anxiety investment if the two purposes of traveling and eating an apple are not ranked the same.

Figure 6.27 It's Super Sentimental Apple, the sometimes final arbiter!

To apply the concept of appraisals to the equation, positive or negative valences will be used to distinguish between an entity whose acquisition will either lead toward homeostasis or lead away from homeostasis. Anxiety resources will be said to have been invested in an entity that has a positive value and whose acquisition will lead to homeostasis. Alternatively, negative anxiety will be said to have been invested in an entity that has a negative value and whose acquisition will lead away from homeostasis because it has already been acquired in an ample amount for homeostasis.

Henceforth, a coefficient of positive one or negative one will be affixed to the equation to indicate the appraised value judgment of an entity. The expression for an appraised value judgment is:

Appraisal × Existence × (Sufficiency × Uniqueness × Sentiment + 1)

In the context of the equation, Appraisal is defined as the self or ego's valuation of an entity toward the purpose of restoring equilibrium between a purpose and its complement. Equilibrium, or homeostasis in the case of drives, would be the point where the overall anxiety or negative anxiety being invested in an entity for a purpose and its complement is as close to the origin, zero, as possible. Mathematically, Appraisal is represented by a value that equals either the coefficient (+1), leading toward equilibrium, or the coefficient (-1), leading away from equilibrium. If the ego appraises an entity positively (+1), then this means that the entity's acquisition will lead to a restoration of equilibrium between two complementary purposes, in favor of homeostasis. This would also mean anxiety is being invested into the entity and that the entity is a cause of worry, concern, or angst because it is either beneficial to a purpose and not yet in one's possession, or it is not safe from threats of harm despite being in one's possession. It would look like the following:

Appraised Value Judgment =

(+1) × Existence × (Sufficiency × Uniqueness × Sentiment + 1)

Another way to view this is to think of a positive Appraisal value as an indicator of pain. In addition to being an unpleasant sensation caused by nerve tissue damage or nerve stimulation, pain also indicates "psychological or psychic distress."[12] Valuing something too much or too greatly, henceforth, would be akin to experiencing acute anxiety and feeling a great deal of pain.

Alternatively, if the ego gives an entity a negative Appraisal value toward the restoration of equilibrium (-1), this would mean that the entity's acquisition will not lead to a restoration of equilibrium between two complementary purposes, but away from homeostasis. More negative anxiety would be invested into the entity than positive anxiety. The entity would become a source of ease, peace of mind, and reassurance with respect to the purpose at hand, but not toward maintaining equilibrium between the purpose and its complementary purpose if it were acquired. If the entity is threatened or in jeopardy of not being acquired, then the negative anxiety felt invested in it will diminish. In the context of the equation, it would look like the following:

Appraised Value Judgment =

$$(-1) \times \text{Existence} \times (\text{Sufficiency} \times \text{Uniqueness} \times \text{Sentiment} + 1)$$

Another way to view this is to think of a negative Appraisal value as an indicator of pleasure. In affect engineering, valuing something negatively would be akin to experiencing acute negative anxiety and feeling a great deal of pleasure as the entity relates to a specific purpose; pleasure is considered by some to be the opposite of either "unpleasure or pain."[13] In another sense, it may "simply be a fundamental emotional experience characterizable as a desire to have the stimulation that produced it repeated."[14] This interpretation of pleasure is accommodated for with the definition of Appraisal. Both pain and pleasure are represented by values with an opposite valence toward one another. The pursuit of pleasure is a plausible explanation for why an individual would continue to acquire an entity that was no longer required for maintaining homeostasis, such as collecting apples in excess. What remains to be analyzed in depth, however, is the relationship between drives and the rest of the equation. This will be explored in the next section.

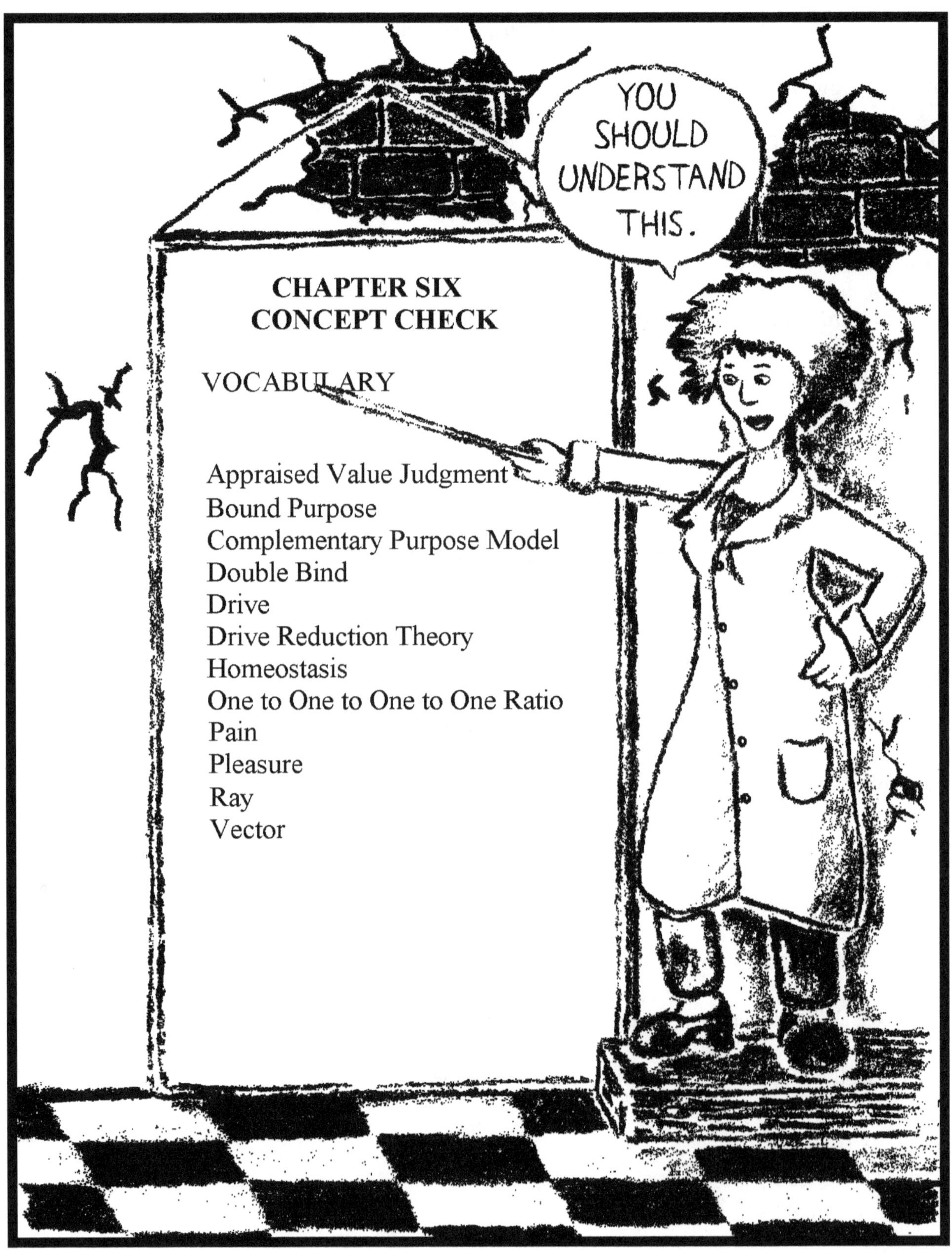
YOU SHOULD UNDERSTAND THIS.
CHAPTER SIX
CONCEPT CHECK
VOCABULARY
Appraised Value Judgment
Bound Purpose
Complementary Purpose Model
Double Bind
Drive
Drive Reduction Theory
Homeostasis
One to One to One to One Ratio
Pain
Pleasure
Ray
Vector

Notes

1. Wilde, Oscar (2003). *The Importance of Being Earnest and Four Other Plays.* New York: Barnes and Nobles Books. Print, p. 143. (Original work published 1892, 1893, 1894, and 1895).

2. Reber, Arthur S., Rhianon Allen, and Emily S. Reber. *Penguin Dictionary of Psychology*. London. Penguin Books, 2009. Print, 236.

3. Reber, Arthur S., Rhianon Allen, and Emily S. Reber. *Penguin Dictionary of Psychology*. London. Penguin Books, 2009. Print, p. 233.

4. Reber, Arthur S., Rhianon Allen, and Emily S. Reber. *Penguin Dictionary of Psychology*. London. Penguin Books, 2009. Print, p. 293.

5. Reber, Arthur S., Rhianon Allen, and Emily S. Reber. *Penguin Dictionary of Psychology*. London. Penguin Books, 2009. Print, p. 54.

6. From *Emotion and Personality: Vol. 1. Psychological Aspects*, by Arnold, M. B. Copyright © 1960. Page 171. Columbia University Press. Reprinted with permission of the publisher.

7. Cornelius, R. R. *The Science of Emotion: Research and Tradition in the Psychology of Emotion.* Upper Saddle River, NJ. Prentice-Hall, Inc. 1996. Print, p. 116.

8. From *Emotion and Personality: Vol. 1. Psychological Aspects*, by Arnold, M. B. Copyright © 1960. Page 175. Columbia University Press. Reprinted with permission of the publisher.

9. From *Emotion and Personality: Vol. 1. Psychological Aspects*, by Arnold, M. B. Copyright © 1960. Page 175. Columbia University Press. Reprinted with permission of the publisher.

10. From *Emotion and Personality: Vol. 1. Psychological Aspects*, by Arnold, M. B. Copyright © 1960. Page 175. Columbia University Press. Reprinted with permission of the publisher.

11. From *Emotion and Personality: Vol. 1. Psychological Aspects*, by Arnold, M. B. Copyright © 1960. Page 182. Columbia University Press. Reprinted with permission of the publisher.

12. Reber, Arthur S., Rhianon Allen, and Emily S. Reber. *Penguin Dictionary of Psychology*. London. Penguin Books, 2009. Print, p. 548.

13. Reber, Arthur S., Rhianon Allen, and Emily S. Reber. *Penguin Dictionary of Psychology*. London. Penguin Books, 2009. Print, p. 592.

14. Reber, Arthur S., Rhianon Allen, and Emily S. Reber. *Penguin Dictionary of Psychology*. London. Penguin Books, 2009. Print, p. 592.

"Difference of object does not alter singleness of passion. It merely intensifies it." - Oscar Wilde[1]

CHAPTER SEVEN

A Roll Cage Theory of Drive Reduction Theory and Identifying Emotions

Front-loading: Anger, Breaking Point, Cone and Disc Apparatus, Content, Courage, Disgust, Drive to Maintain All Drives, Drive to Not Maintain All Drives, Entity-mass, Euphoria, Fear, Grief, Guilt, Happiness, Id-mass, Negative Anxiety, Negative Anxiety-mass, Positive Anxiety-mass, Roll Cage, Roll Cage Theory of Drives, Sadness, Superego-mass, Utility-mass

The Avoidance of Pain, Pursuit of Pleasure, and a Roll Cage Theory of Drives (Supplemental)

From the addition of a new variable into the equation, Appraisal, the concept of negative anxiety has arisen. Along with it, two other items, pain and pleasure, have been brought into the equation. Most life forms seek to avoid or minimize pain and will pursue or maximize pleasure if given the opportunity, so these tendencies will be given the most consideration. What these ideas suggest for the equation will also be expanded upon. Freud's two controversial drives, a creative life instinct or libido and a destructive death instinct, were reconciled earlier by proposing a ray model of opposing drives that defined the physical limitations of an organism. Under a ray model of drives, all drives would lead to death, which, even if only coincidental, lends support to the argument for a death instinct. However, this also implies there would be no opposing life or survival instinct so to speak, because all drives would lead to death. What must emerge instead is a balancing act of the many death drives available to an individual. This balancing act of the death drives will be given the distinction drive to maintain all drives, which would be the equivalent of maintaining homeostasis. A drive to maintain all drives would parallel Freud's conception of the life-instinct, Eros or the libido, and the pursuit of pleasure. Conversely, a drive in and of itself would parallel Freud's conception of the death instinct, Thanatos, or the pursuit of pain.

In the equation, what all of this entails for a life instinct, or drive to maintain all drives, is as follows. The life instinct is a perversion of the death instinct in affect engineering. Given that individual drives lead to death by themselves in the ray model of drives, the only way to prevent death and maintain life would be to subvert singular death drives by setting each of them against their nemesis or opposing death drive. If at any point one drive overwhelms its nemesis, or vice versa, then death would result. If, however, the individual is able to account for the magnitude of and counterbalance opposing drives, then death would be subverted while life and the drive to maintain all drives would be upheld.

In order to conceptualize this, a Roll Cage Theory of Drives will be presented as an illustrative model that incorporates the following psychoanalytic and affect concepts: ego; id; superego; life instinct; death instinct; cognition; conation; affect; value; anxiety; negative anxiety; pain; pleasure; drive to maintain all drives; and a drive to not maintain all drives. If it is imagined that one's self, or life essence, and ego are inside of a roll cage being balanced on top of a disc that is also balanced on top of a cone, then the ray model of drives can be depicted as the Roll Cage Theory of Drives. It is one of two models that will be used to help explain the math functions and their use. Understanding the Roll Cage Theory of Drives is not essential, but all variables in the equation can be incorporated within it. The primary function the Roll Cage Theory of Drives serves in this book is to illustrate homeostasis (equilibrium) between a purpose and its complementary purpose, which it does more vividly than the equation can on a two-dimensional Cartesian plot. A secondary function it serves is to link theories of affect to psychodynamic theories. Hence, it is an application of the equation to model true to life scenarios. It also serves as a more visual representation of the equation. The other model that will be used to explain the variables in the equation is a neurological network model that will be introduced in chapter eight. The neurological network model will present a microscopic approach to understanding the equation at the biological level of action potentials and neurotransmitters. Oppositely, the Roll Cage Theory of Drives presents a macroscopic approach to understanding the equation and is a graphic illustration of how the functions would explain affect and its role in decision making when multiple factors are at play.

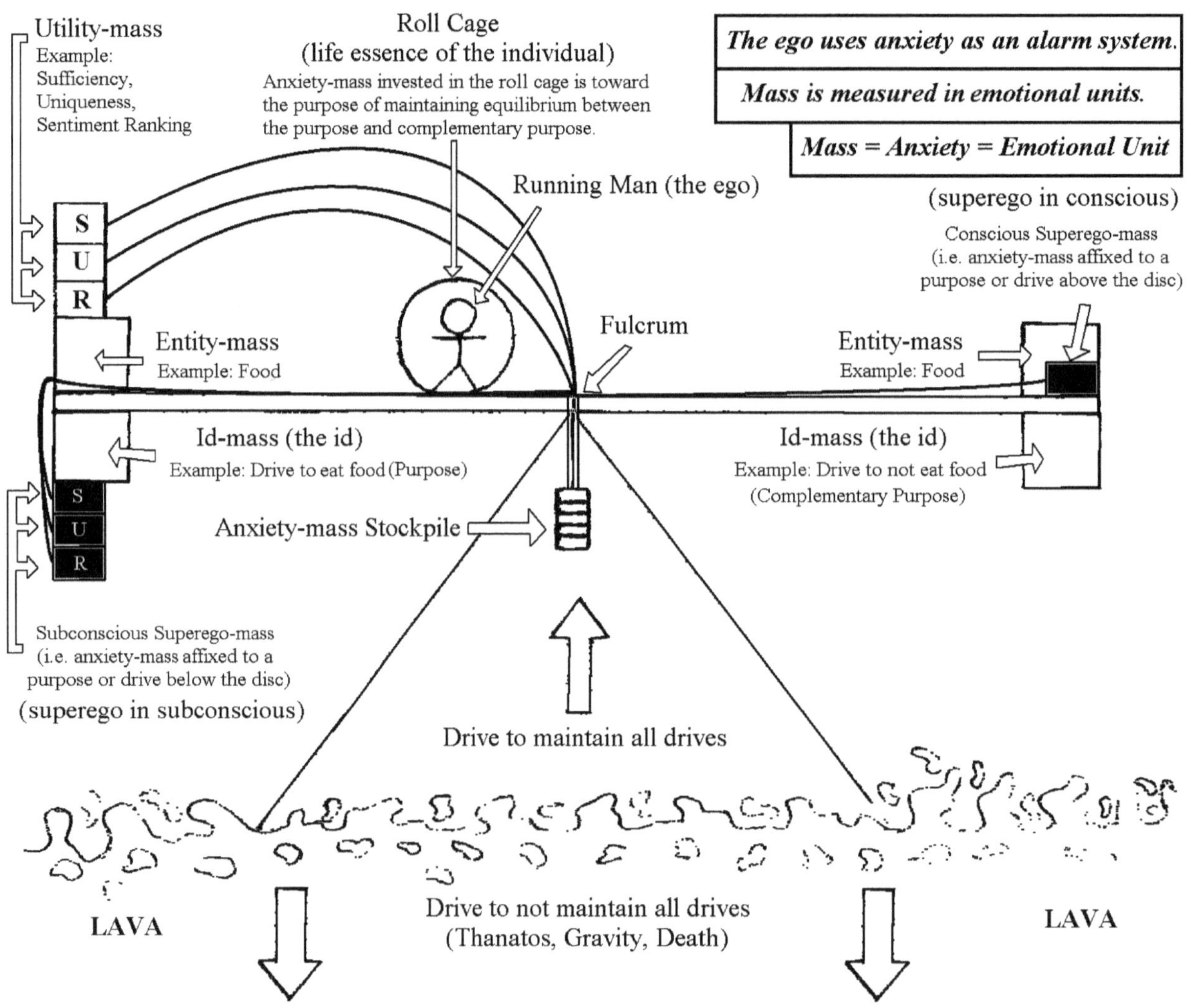

Figure 7.1 Roll Cage Theory of Drives: drives (id-masses) are all shown in the subconscious (or unconscious) in the examples, but may be in the conscious.

Roll Cage

This is where all is felt or can be known by the self, where decisions are made, and where the life essence and ego are contained. The location of the roll cage along the spectrum between two opposing drives (i.e., drinking water and not drinking water) represents how much of an entity, which is related to fulfilling a drive or its complement, has been acquired. The very edge of the disc on both sides represents the individual's physical limitations. The ego or self is represented by a person running inside the roll cage. The ego inside the roll cage runs nonstop away from the direction of whichever drive it aims to reduce or fulfill by counterbalancing it. Equilibrium, or homeostasis between a drive (purpose) and its complement is understood as a ratio of success between the two drives (purposes). For instance, a 1:1 ratio of success for the purpose of hydrating against not hydrating would mean that an individual is hydrating at the same rate of success that he or she is not hydrating, if the mass on each side of the disc is equal.

With respect to affect and the desire to achieve homeostasis between a purpose and its complementary purpose, pleasure would be felt by the ego as the sensation that the disc's equilibrium is being restored as a result of the movement of the roll cage in a particular direction; hence, the disc is becoming balanced with the horizon. Pain would be felt by the ego as the sensation that the disc's equilibrium is not being restored as a result of the movement of the roll cage; hence, the disc is becoming unbalanced with the horizon. If the ego's goal is to maintain homeostasis, then the sensation of backsliding against one's desired direction typically becomes likened to pain while the sensation of progressing in a desired direction typically becomes likened to pleasure. It would follow that the amount of pleasure felt for the purpose of achieving homeostasis between a purpose and its complementary purpose would be greatest whenever the ego manages to elevate a drive (drives are represented by id-mass) upwards toward the horizon. Contrarily, pain felt for the purpose of achieving homeostasis between a purpose and its complement would be greatest whenever an id-mass (e.g., drive or purpose) descends and the ego is not able to move in the desired direction to elevate the id-mass to the horizon and restore balance.

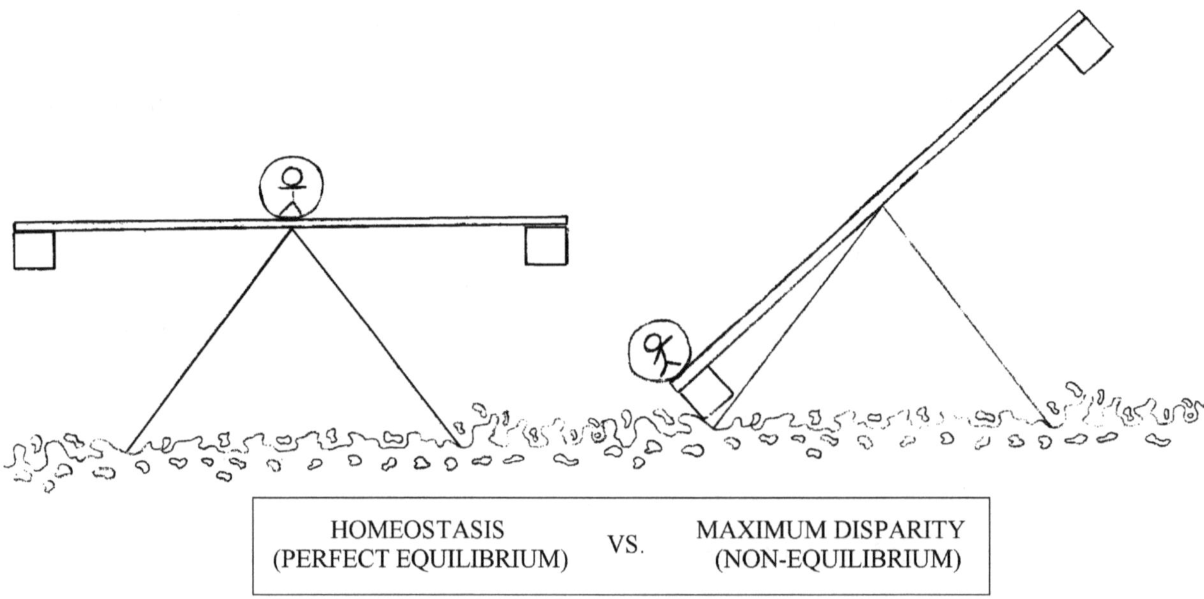

Figure 7.2 Equilibrium vs. Non-Equilibrium

If an individual only considers one purpose from a complementary pair, however, then the maximization of pleasure felt for that one purpose will come at the expense of pain felt for that purpose's complementary purpose and any other purposes associated with it. The ego's ability to feel whether or not it is falling or ascending is influenced by the pressure of the roll cage against the flat of the disc while running; a steeper gradient in the disc versus the horizon leads to a sensation that the disc is more off-balance, hence not at equilibrium. If the gradient of the disc is angled downward in a specific direction, then the ego will feel that movement in that a direction is nearly effortless and leads away from equilibrium between the purpose and complementary purpose.

Cognition and conation are both controlled from the ego inside the roll cage. The roll cage is transparent, enabling the ego to see everything around itself and on top of the disc but not on the underside of the disc. Cognitive ability, therefore, consists in the ego's ability to appropriately survey the disc, id-mass (id components), entity-mass (existential value), utility-mass (utility value), and superego-mass (superego components) to best assess how to maintain its balance and survival. Conation is represented by the ability of the ego to move the roll cage around the disc to either acquire an entity, or to not acquire an entity, which may ultimately change the balance of the disc on top of the cone to restore equilibrium. Affect is represented by all mass on the cone and disc apparatus (anxiety), be it in the form of a purpose, complementary purpose, an entity's existential value, or the utility value of an entity. Affect is also represented by the anxiety invested in the roll cage itself for the purpose of maintaining equilibrium between a complementary pair of purposes.

Because the roll cage is a sphere, it is always in motion; this means that the individual is always engaged in action even when not doing anything. To not drink, for instance, is to be engaged in the action of not drinking. An entity is given an existential value for both the fulfilment of a purpose and for not fulfilling a purpose, so it lies on both sides of the disc. Where an entity is found to be wholly Sufficient and Unique for a main purpose, and has a Sentiment value of + 1, it will have a collective mass of + 2 emotional units from the entity-mass and the utility-mass. The value of the entity for the non-fulfillment of a purpose will tend to its existential value and approach + 1 emotional units and be balanced out or overwhelmed by its value for its nemesis, the fulfillment of the purpose. Its drive reduction component will be reduced to zero for a lack of Sufficiency.

Moreover, in the roll cage model of drives, the collective mass disparity between items on opposing sides of the disc corresponds to the disparity between the distance of each id-mass to the lava bed and death. Excluding the mass of the roll cage, the ratio of anxiety-mass between the items on opposite sides of the disc is also the ratio of success between the two purposes that the ego would feel at equilibrium. If both sides have the same mass, then equilibrium would be felt in the center. If the left side has twice the mass of the right side, then equilibrium would be felt two-thirds of the distance (i.e., the diameter of the disc) from the left to the right side in figure 7.3.

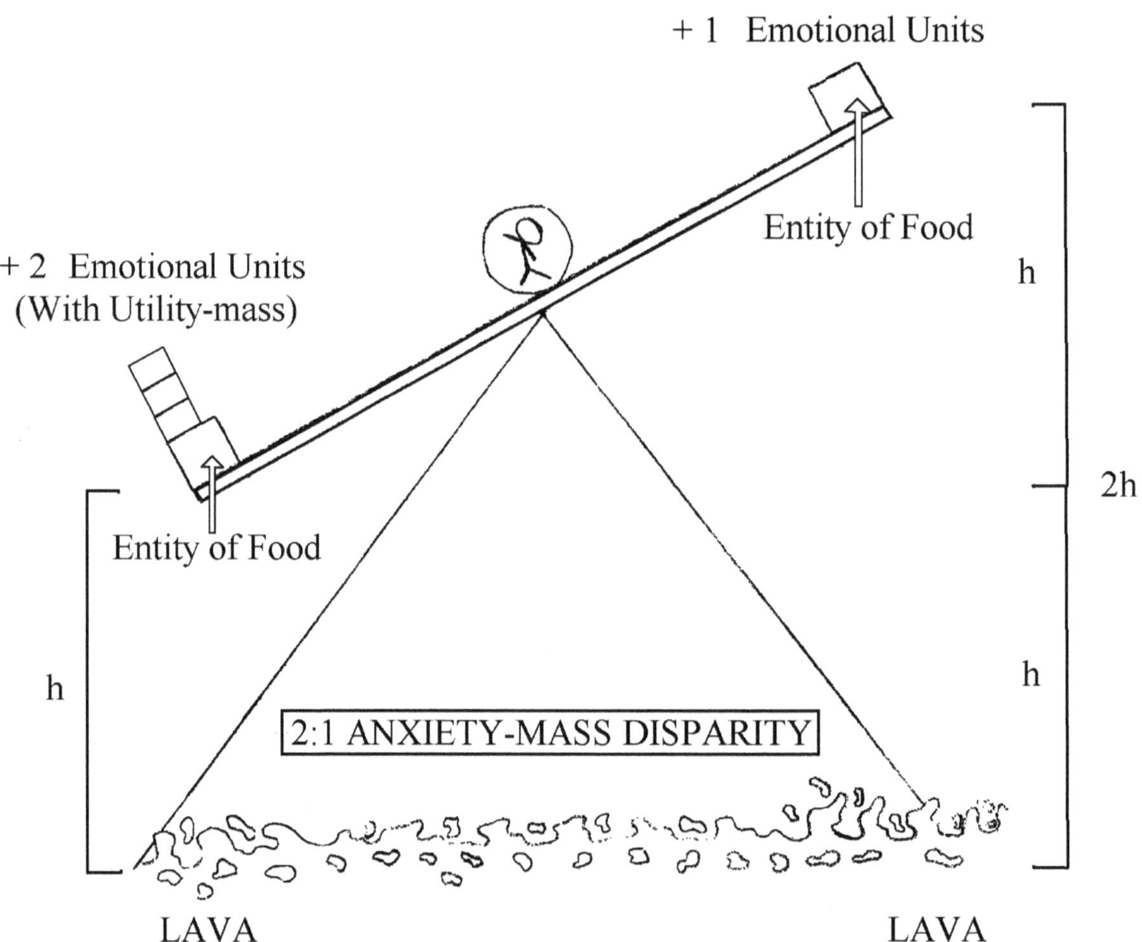

Figure 7.3 A two to one disparity or imbalance of anxiety-mass means that the more massive side for which an entity has a value (left in the image) compels an individual to move towards death at twice the rate as the entity's value for the complementary drive (right in the image). Food, in this example, has twice the anxiety-mass invested in it for a purpose on the left than its complement on the right. The variable "h" on the left side is the distance to the ground where the roll cage will strike lava. Although the id-masses are not shown for simplicity's sake, (e.g., drive to eat and the drive to not eat), for this example they may be assumed to be varying with the entity's themselves.

Along a dimension, the roll cage can move in one of two directions. In the case where an entity has significant Sufficiency, Uniqueness, and Sentiment values for fulfillment of a purpose, if the roll cage is moving away from the heavier entity-mass and/or utility-mass representing the entity's value for fulfilling the drive beneath it, then that entity is being acquired to facilitate that purpose. Oppositely, if the roll cage is moving toward the heavier entity-mass (representing the entity's value for reducing the drive beneath it), then that entity is not being acquired to facilitate the purpose.

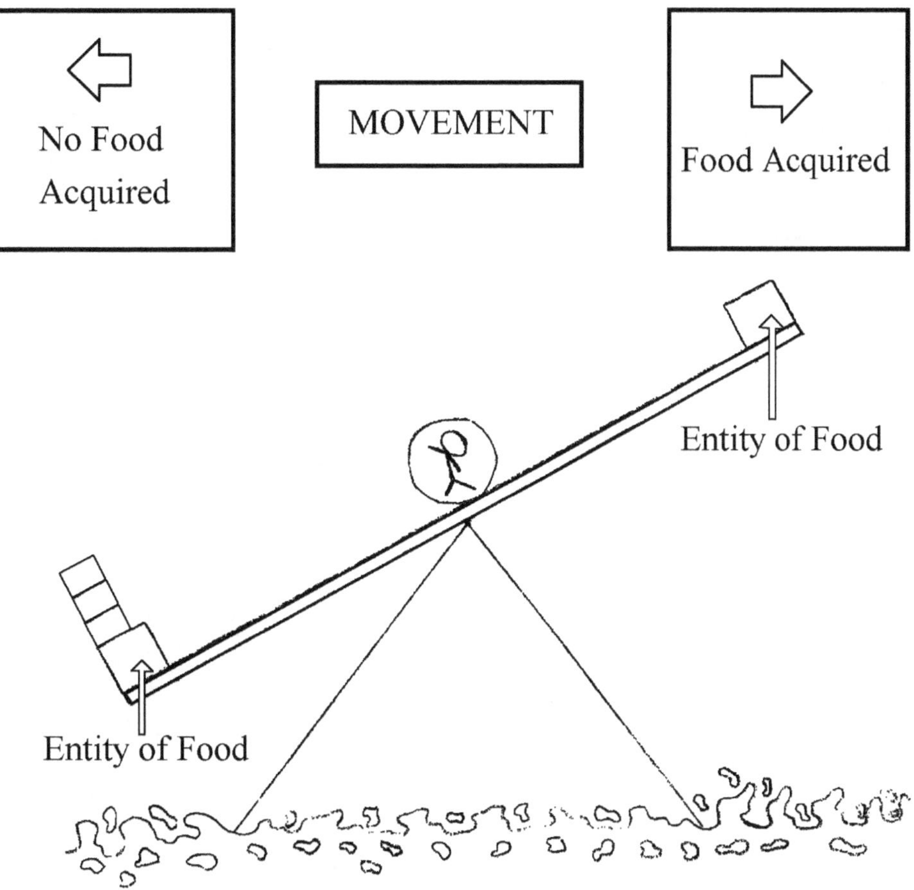

Figure 7.4 Acquisition vs. Non-acquisition: Because the entity of food is associated with alleviating the more massive drive to eat (LEFT) than to not eat (RIGHT), if the roll cage were to move to the right by acquiring food then it could balance out the disc to restore equilibrium.

Because the entity's value for the complement only approaches its existential value of + 1 emotional unit, it does not need to be considered beyond this. Once the ego has acquired enough of the entity to achieve equilibrium, the disc becomes level. The ego would thereafter give the entity a negative valuation for the purpose of restoring equilibrium, as further acquisition of the entity will destabilize the disc. If the roll cage continues to heedlessly acquire the entity, however, it will fall off of the disc in the other direction.

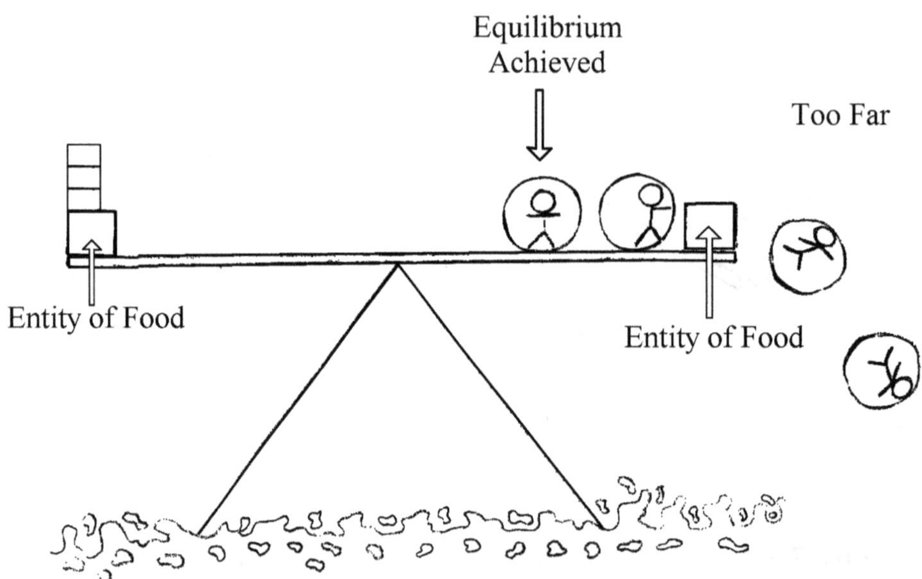

Anxiety: **Figure 7.5** Equilibrium achieved and then transgressed. The ego and roll cage must then change direction to avoid death by over-eating (hyperphagia).

A Roll Cage Theory of Drive Reduction Theory and Identifying Emotions Positive Anxiety Mass, and Negative Anxiety Mass

In the Roll Cage Theory of Drives, mass is measured in emotional units, represents anxiety, and it depicts the psychic energy of the individual. The mass of the roll cage itself represents the total amount of anxiety that an individual has invested toward balancing a complementary pair of purposes and entities valuable to both the purpose and complementary purpose. Anxiety in the Roll Cage Theory is much like an alarm system to the ego. The anxiety-mass on the edges of the disc (e.g., id-mass, entity-mass, utility-mass) represents the anxiety that the self has invested into different entities as they relate to purposes. The anxiety-mass of the roll cage itself is primarily a marking tool to help the ego feel at equilibrium in a given environment, and its location determines the actual ratio of success between fulfilling a purpose versus its complementary purpose. Though anxiety-mass added to the roll cage makes it more difficult to move against gravity, they assist the ego in balancing the disc to achieve and feel at equilibrium. It can be imagined that extra anxiety-mass is normally stockpiled in the center underneath the disc and attached to the roll cage whenever the ego deems that more mass will be needed to balance the disc between the fulfillment of a purpose and its complement.

An entity's Appraisal as positive or negative with respect to a purpose (negative anxiety vs. positive anxiety) is distinguished by whether or not the corresponding id-mass and entity-mass are above or below the fulcrum. Id-mass, entity-mass, and utility-mass that are below the fulcrum (beneath the horizon) signify positive anxiety, pain, or negative affect invested and felt. Id-mass, entity-mass, and utility-mass that are above the fulcrum (higher than the horizon) signify negative anxiety, pleasure, or positive affect invested and felt.

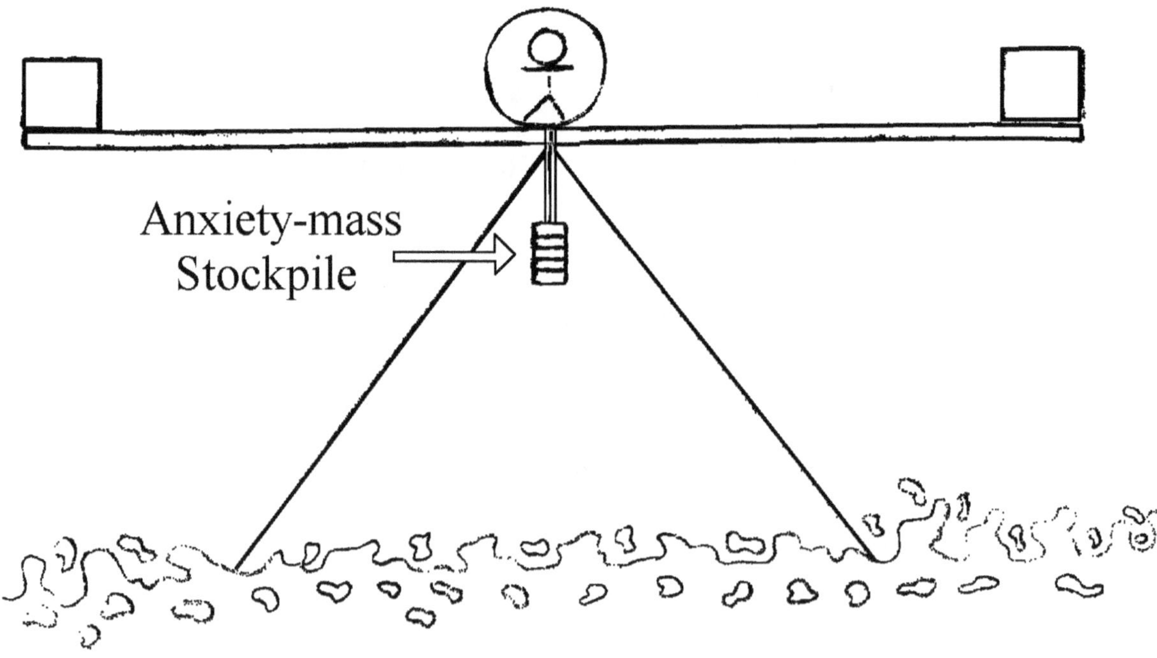

Figure 7.6 Stockpile of anxiety-mass.

Entity-mass

Entities and their representation in the equation require an investment of energy and anxiety (emotional units). Anxiety, in the Roll Cage Theory of Drives, is represented by mass. The chosen unit of measurement, the emotional unit, is the minimum amount of psychological energy required to cognize that an entity exists over a given time frame. Because anxiety, value, and psychological energy all correspond to the emotional unit in the equation (or the minimum joules of energy required to cognize that an entity exists) all entities in the roll cage model of drives subsequently have their mass measured in emotional units. Different types of entity-mass litter the disc top and, depending on their collective mass with other factors along a spectrum, influence the balance of the disc by making a drive they are associated with feel more massive than the complementary drive. An entity-mass initially only carries a value of + 1 emotional unit, but may become more massive as more value, anxiety, or mass from utility components becomes associated with it.

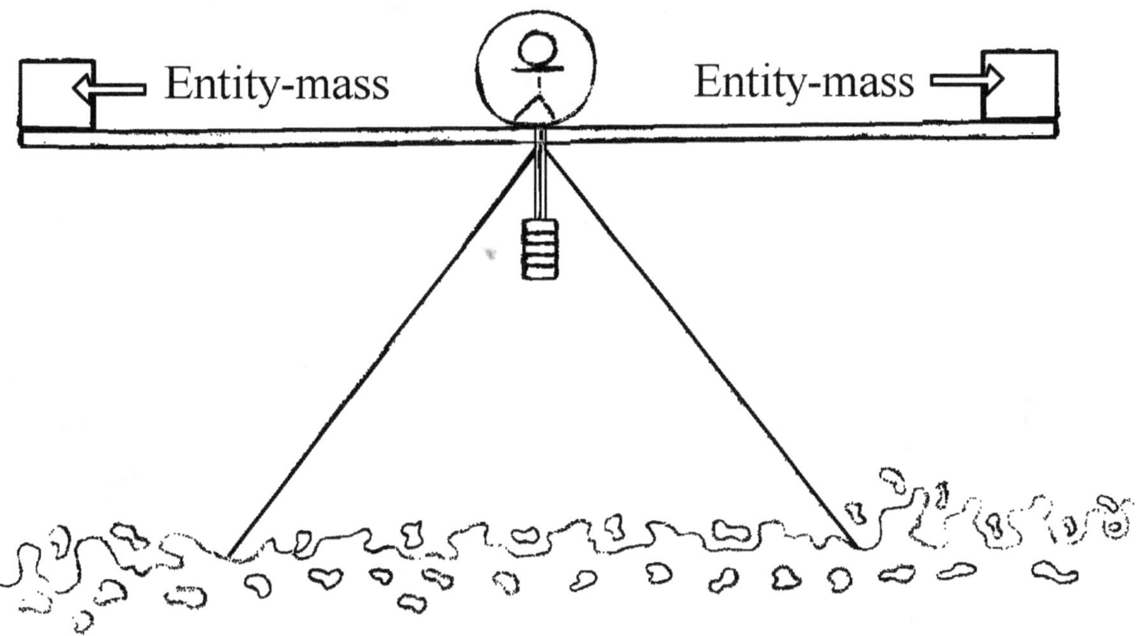

Figure 7.7 Each entity-mass on the disc has a mass of one emotional unit.

Id-mass on the Edge of the Disc

The id is represented in the equation as the source of psychic energy (e.g., anxiety and negative anxiety in the equation), and is represented by the totality of all mass on the disc. The purposes or drives that an individual possesses, the id-masses (boxes underneath the disc), simply organize psychic energy. The top of the disc-cone apparatus is ultimately the playing field where most psychic energy is dispersed in the Roll Cage Theory of Drives. Psychic energy, being measured by the hypothetical emotional unit, is a gauge for both anxiety and negative anxiety in the equation.

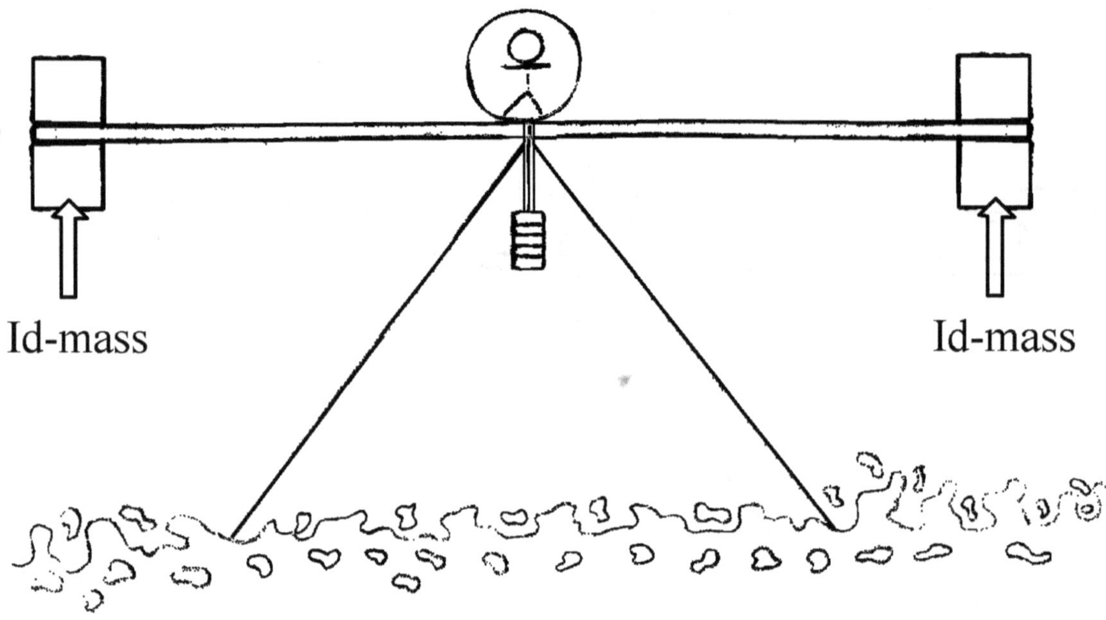

Figure 7.8 Each id-mass on the disc has a mass of one emotional unit and represents a purpose or the complement to that purpose.

Id-mass can be felt by the ego as a force that tips the scale or disc and it represents the id components of psychoanalytic theory. By themselves, id-mass would not alter the balance of the disc, as they are all canceled out by their nemesis. For instance, the drive to eat food would be counterbalanced by the drive to not eat food. The purposes that reduce each of these drives, eating and not eating, are on opposite ends of the disc along its diameter. If food is considered to be an entity-mass that reduces the drive to eat, then the id-mass corresponding to the purpose of eating would be felt to be more pressing than the id-mass corresponding to the purpose of not eating. The ego would be compelled to take action to restore equilibrium by eating food in order to move back toward the center of the disc to balance the disproportionate anxiety invested.

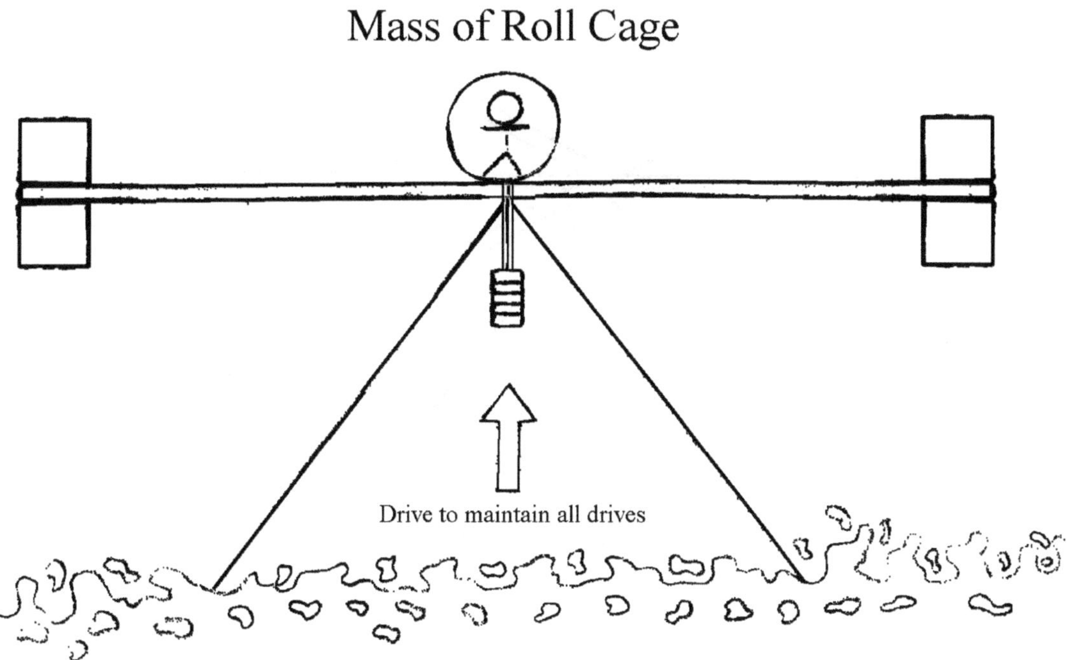

Figure 17.9 The ego and roll cage, entity-mass, and id-mass are depicted on the disc. Anxiety invested in the drive to maintain equilibrium between the purpose and the complementary purpose is represented by the mass of the roll cage. The mass of the roll cage is equivalent to the sum of all other mass on the disc (i.e. id-mass of a drive and its complementary drive, entity-mass, and utility mass). For instance, if the mass on the left side totaled +2 emotional units and the mass on the right side totaled +1 emotional unit, then the anxiety invested in maintaining equilibrium between the two would be + 3 emotional units (mass of the roll cage). The mass of the roll cage above is + 4 emotional units because there are two units of mass on each side.

Once the roll cage and ego travel far enough along the horizontal part of the disc, then the appraised valuation of the entity associated with balancing the more massive of the two id-masses, in this case the one representing food's value toward reducing the drive to eat, becomes negative by elevating and breaking the plane of the horizon at the fulcrum.

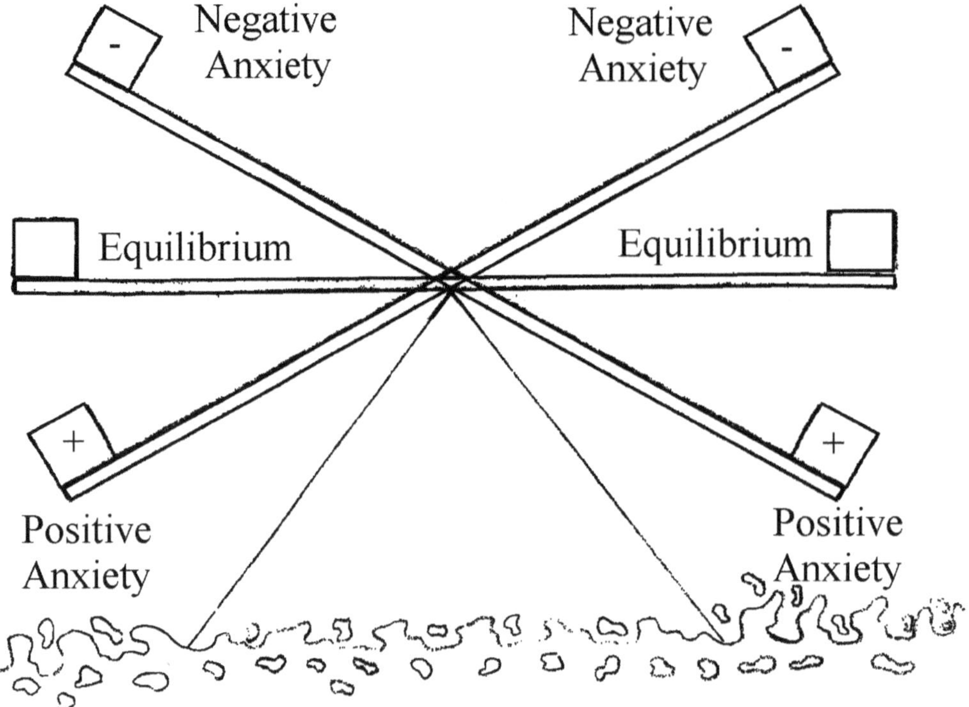

Figure 17.10 Whether an entity's Appraisal starts as positive or negative depends on if the anxiety invested in it is above or below the horizon. If it is below the horizon, then it is positive anxiety-mass (negative affect felt for a specific purpose) because the entity needs to be acquired to restore equilibrium. If it is above the horizon then it is negative anxiety-mass (positive affect felt for a purpose) because the entity has already been acquired in sufficient amount and further acquisition of the entity will lead away from equilibrium.

Thereafter, if the individual ate too much food, the id-mass for not eating would be perceived by the ego as more pressing and tip the disc toward not eating and satiation. At that point, the ego will have to correct itself by not eating if homeostasis between the purpose and the complement is to be maintained. Id-masses, although fixed in their distance across the diameter from their complementary drive or nemesis drive, can rotate around the circumference of the disc to become associated or linked with other drives to reflect the three-dimensional nature of the disc.

Although id-masses themselves are entities that may or may not acquire additional utility-mass (Sufficiency, Uniqueness, Sentiment), their representation underneath the disc in the following scenarios is simply that of a place holder to map out the balance of drives/purposes and complements held by an individual while keeping the explanation neat looking. Although the value of the drive itself may be ignored in some of the examples to follow when an entity above the disc is being considered (e.g., for simplifying an explanation), their value would still be contributed to the overall weight on a particular side because the id-mass itself is an entity that the individual can conceive. The drive typically exists, receives an existential value for this, and is used more or less as a landmark around which to organize the values for other entities. Over time, or as a result of learning from experience, an individual may choose to value some id-masses differently, for instance, based upon their utility components (Sufficiency, Uniqueness, or Sentiment) in promoting a sibling drive. In the Roll Cage Theory of Drives, anxiety-mass that is added to the id-mass itself comprises the superego from psychoanalysis, and is distinguished as superego-mass. If superego-mass corresponding to a specific id-mass is visible to the ego, lying in the conscious, then it is above the disc. However, if the anxiety-mass making up the superego is not visible and lies in the subconscious or unconscious, then it is attached to the id-mass underneath the disc. This would occur if the entity of a purpose is given a different utility value than its complementary purpose.

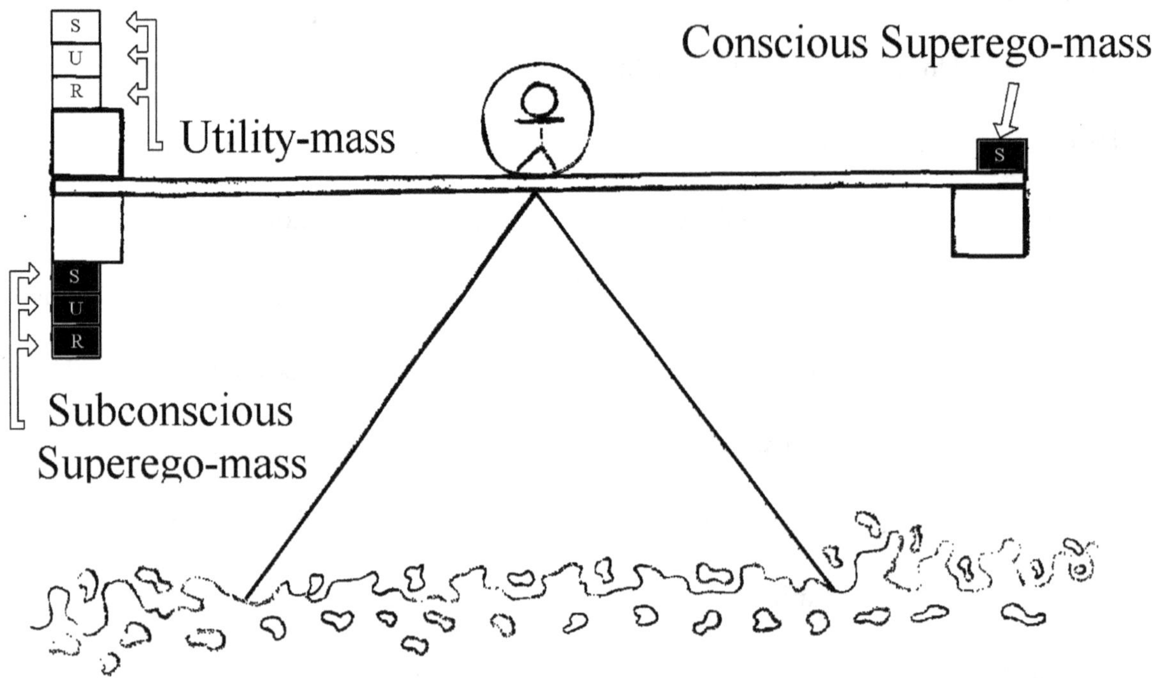

Figure 17.11 Normally, all id-masses would have the same mass, thus cancelling themselves out. Elevating one id-mass over another by investing anxiety beyond its existential value (i.e. Sufficiency or Uniqueness) gives it a heightened mass against its complement. Any utility-mass (Sufficiency, Uniqueness, or Sentiment) applied directly to an id-mass, and not another type of entity-mass, makes up the superego in the Roll Cage Theory of Drives and is distinguished as Superego-mass. Superego-mass that is above the disc lies in the conscious, while superego-mass lying underneath the disc is in the subconscious or unconscious.

If, however, all drives and their complements only carry their existential value of + 1 emotional unit, then an individual has no need to concern himself or herself with investing anxiety into the drive itself; the investment of anxiety may instead be directed to entities related to the balancing of those drives.

Superego-mass on Bottom or Top of the Disc and Other Utility-mass

In the equation, the superego is represented by the variables of Sufficiency, Uniqueness, and

Sentiment when they are directly applied to the entity of a drive/purpose (i.e., id-mass). Therefore, superego-mass (a type of utility-mass) corresponds to the variable of Sufficiency, Uniqueness, and Sentiment in the equation when the entity in question is the purpose or drive itself. Superego-mass may become associated with specific id-masses in order to give them and entities associated with balancing them an enhanced sense of importance. In effect, superego-mass represents the individual's superego (internalized moral code) and distorts the true balance distribution based upon the id-mass alone. Whereas id-masses have an existential value that is equal to one, and whereas the stockpile of anxiety-mass underneath the disc is generally reserved for the ego to attach to the roll cage in order to balance out the disc based on what it sees in the conscious (on top of the disc), the influence and role of superego-mass may sometimes be hidden from the ego. Like utility-mass for other entities, superego-mass in the model is tied to the center of the disc, and may become associated with and freely move alongside an id-mass around the circumference of the disc. However, the superego-mass may also lie either in the conscious or subconscious. If the superego-mass lies in the conscious, hence, on top of the disc, then the ego is aware of it and can readily compensate for it by adding the appropriate amount of anxiety-mass to the roll cage. However, if the superego-mass lies in the subconscious, hence, underneath the disc, then the ego is not able to see it and cannot compensate for it by adding the appropriate amount of mass to the roll cage. Nevertheless, the effect of the superego-mass in the subconscious would still be felt by the roll cage on top of the disc, creating a very real effect on the disc that the ego feels but cannot accurately gauge. The ego must compensate for this in some other manner. This aspect of unconscious anxiety in the Roll Cage Theory of Drives will be vital to keep in mind when defense mechanisms are addressed toward the end of the book.

It can also be imagined that an id-mass and an entity-mass are a pole, while superego-mass and utility-mass are pieces of metal tied to the center of the disc. Superego-mass represents additional anxiety invested into an id-mass (drives) and this additional mass effectively distorts the balance of the disc in favor of the individual's internalized moral code. Although superego-mass is depicted as additional mass imposed on an id-mass (drive) by the ego, it can arise as a result of the influence of parents or authoritative figures in addition to originating from the individual. One such example might be the internalization of a doctor's order to not skip meals, tipping the scale in

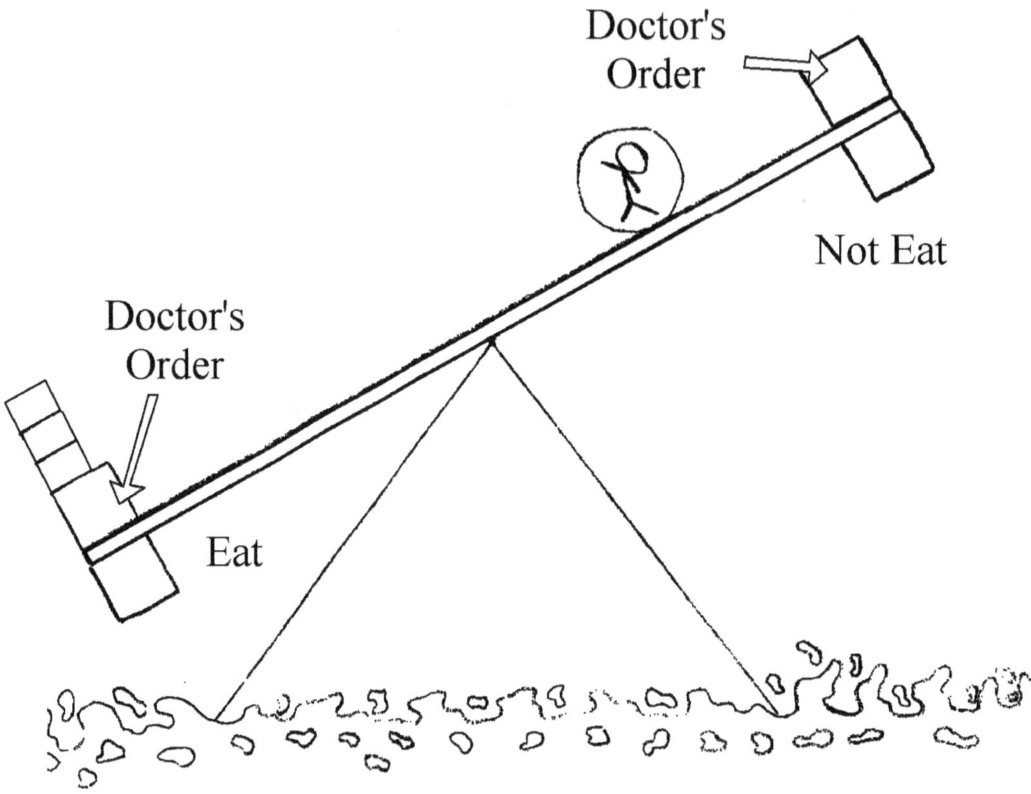

Figure 7.12 Additional anxiety-mass has been invested into the doctor's order to not skip meals and has tipped the scale in favor of the id-mass to eat.

favor of the id-mass to eat, making that drive feel more massive.

The disc would tip toward the id-mass of eating, and the ego, in response, would want to restore equilibrium by moving further away from the id-mass via eating. Because the ego is attempting to maintain homeostasis and keep the disc balanced in this case, it will spend more time eating and moving away from the superego-mass of the doctor's order; it will also respond to the drive to eat more readily because the center of balance on the disc has become further displaced from the fulcrum, toward the opposite side. If, however, on the other side are added two other superego-masses, a diet that an individual has been trying to stick to, humiliation of not being able to fit into a particular pair of jeans, and subconscious anxiety arising from an instance where she failed to stay on a diet (e.g., superego-mass affixed to the drive to not eat), then the balance would change.

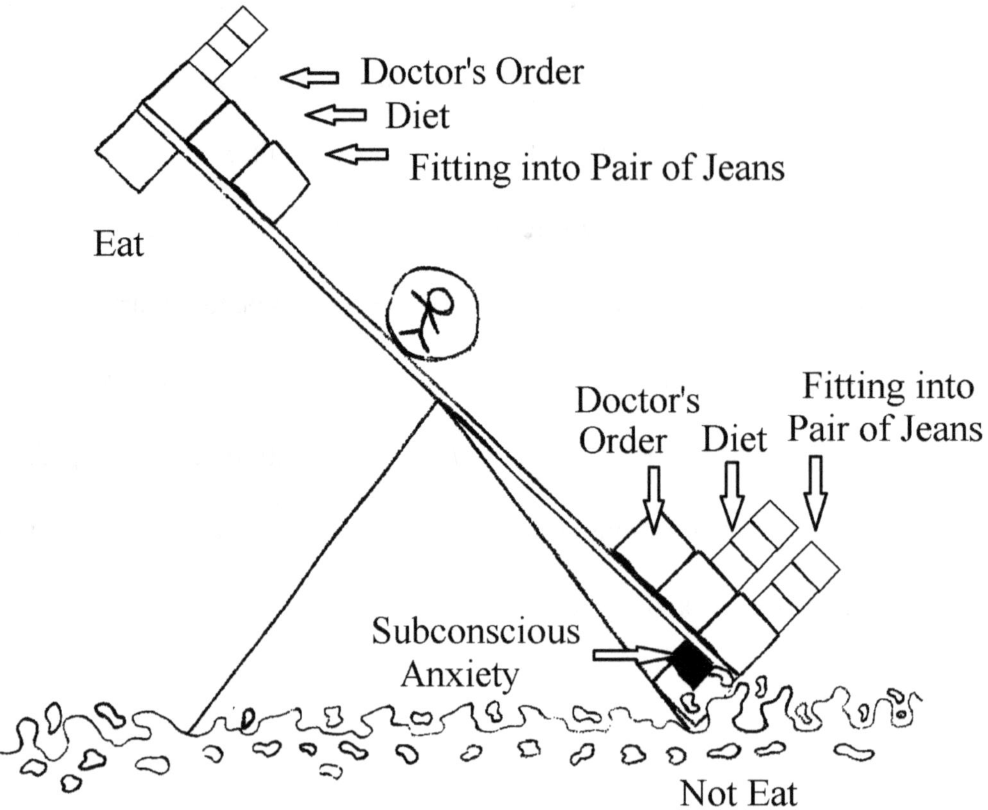

Figure 7.13 The balance swings the other way. Diet and humiliation of not fitting into jeans is in conscious, while anxiety from having failed to stay on a diet in the past is in the subconscious.

The scales would tip the other way in favor of fasting, or the id-mass of not eating. The ego would feel that it needs to move back in the other direction and would spend more time trying to balance out these two drives by not eating.

Drives with superego-mass, arising from the individual's assessment of id-masses, are depicted as entities with additional anxiety-mass affixed to them that is attached to the center of the disc by a rope. With respect to a purpose and a complementary purpose, every entity has at least two values, one for an id-mass and another for its complementary id-mass. In accordance with psychoanalytic theory, superego-mass, thereafter, might distort the anxiety-mass invested in drives to direct the ego to behave in a specific manner in order to balance out competing drives. The Appraisal charge of the superego-mass and the id-mass it is attached to would depend on whether or not these masses, collectively, are above the fulcrum. Superego-mass that is below the horizon is a barometer for positive anxiety, as fulfilling its corresponding purpose or id-weight would lead toward a restoration of equilibrium. Superego-mass that is above the fulcrum is a barometer for negative anxiety, as fulfilling its corresponding purpose or id-mass would lead away from a restoration of equilibrium. While this explains the valuation of purposes/drives in the Roll Cage Theory of Drives, the effects of the utility variables of Sufficiency, Uniqueness, and Sentiment are not totally accounted for on other entities. Their conception would be the same as that of the superego-mass in the model. Additional mass affixed to specific entities above the disc due to Sufficiency, Uniqueness, or Sentiment is referred to as utility-mass.

For instance, an entity, such as a piece of cake, is associated with the purpose of eating and valued more so by an individual than other types of foods. If an individual is in the throes of hunger pains, then the value of the cake's mass (in emotional units) is felt in addition to the hunger drive (e.g., the cake’s mass plus the id-mass corresponding to eating food), making the individual feel more famished and the ego more determined to restore equilibrium by eating. Because the balance of the disc is tipped in favor of the drive to eat, the ego will attempt to restore equilibrium sooner than would be the case if they were equal. If, however, the individual eats five pieces of cake when only one was required and has two more ready to be eaten, the roll cage, having moved to the other side of the disc, will have tipped the scale drastically toward the side of the complementary drive, not eating (figure 7.29). The valuation of the cake toward the purpose of restoring equilibrium became negative but was pursued anyway, so the ego must make an about face to balance the disc.

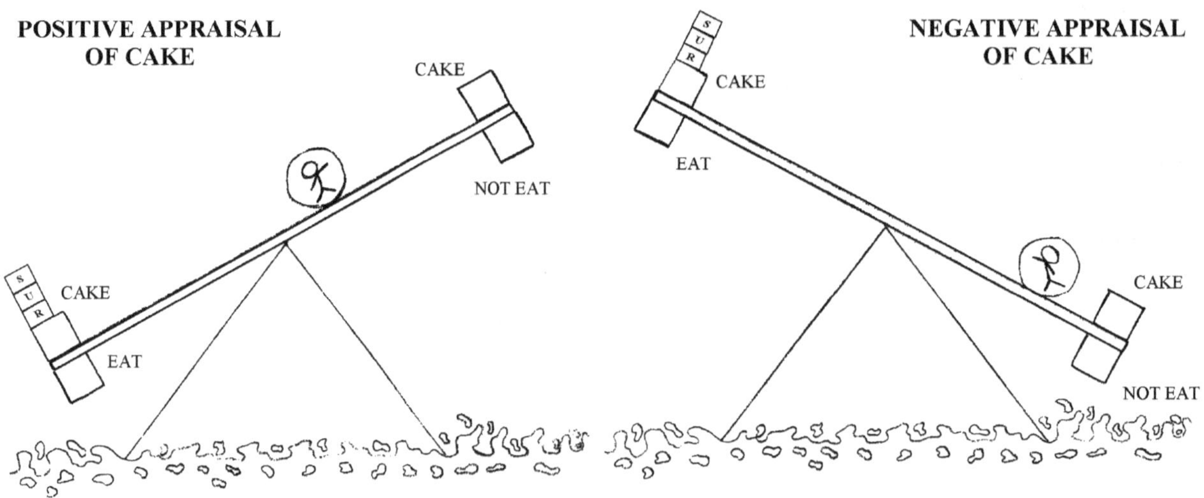

Figure 7.14 A change in the cake's Appraisal towards the restoration of equilibrium from a positive value to a negative value

However, if anxiety invested in the cake changes, such as by lowering, then the id-mass of not eating would be felt even more strongly than before. Despite the influence of the anxiety-mass on the drive to eat, the balance of the disc could be further altered by a drop in the anxiety invested in the cake, such as one precipitated by a loss of either Uniqueness or Sufficiency. For example, a loss of Uniqueness could be due to cake becoming more prevalent, and as such, more easily acquired. This would lower anxiety-mass invested in the Uniqueness of the cake, exaggerating the tilt of the disc even further at an inopportune time (figure 7.15).

It would become more difficult to reduce the drive of not eating because cake would be more plentiful. Moreover, if another entity, such as a brownie, could match the entity of cake on all levels with respect to the purpose of reducing the drive of hunger, then the cake's Uniqueness value for the purpose of eating would diminish further and the weight of the drive to not eat would be felt even more strongly if the individual chooses to remain in what used to feel like equilibrium. Likewise,

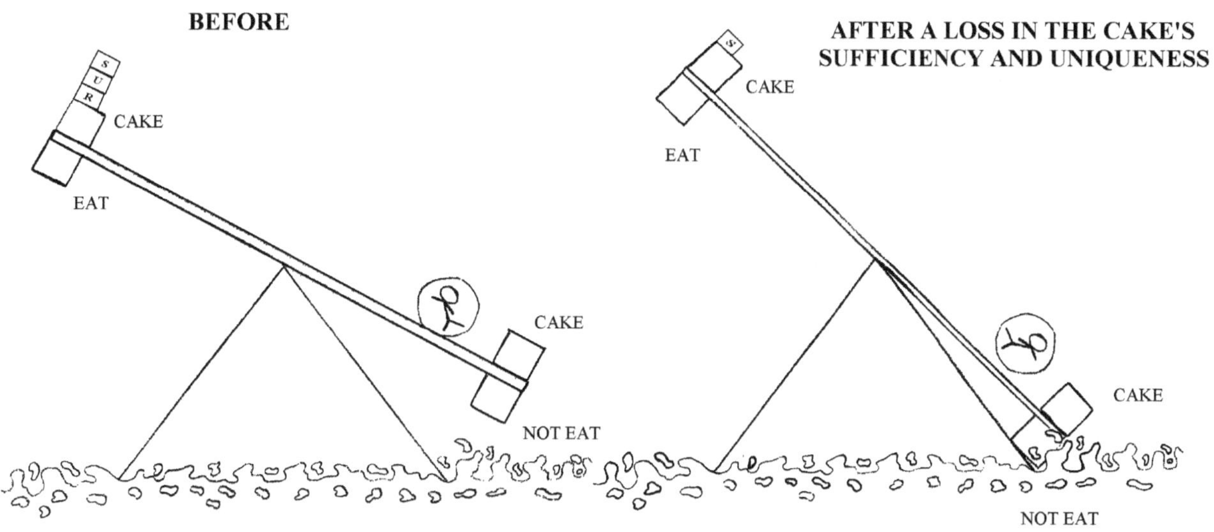

Figure 7.15 A decrease in the cake's Uniqueness and Sufficiency at an inopportune time.

a further drop in the cake's value could be precipitated by a decrease in its Sufficiency for the purpose of eating (e.g., if a fork and plate are also needed to eat) and would further reveal to the ego how far the center of balance has been shifted off the fulcrum.

With respect to the drive of eating food, the cake might normally have a mass approaching + 2 emotional units (Existential value and utility values combined), a positive valuation, and represent anxiety if the individual were famished. Eating it would lead to a restoration of equilibrium between the purpose of eating and its complement, not eating. In the aforementioned instance, however, the cake's valuation toward the purpose of restoring homeostasis became negative because the entity was acquired in excess. If the ego is sensing that the mass of the complementary drive of not eating is tipping the scales more strongly than its opposite, then changes in the cake's value would also feel like a change in pleasure or pain felt with respect to achieving equilibrium between the purpose and its complement. A decrease in the cake's value, if the ego were trying to fulfill the complementary drive of not eating, would feel painful, like someone who is on a diet suddenly finding himself or herself at a free snack food convention. Conversely, an increase in the cake's value would parallel pleasure, as it would make it easier to fulfill the purpose of not eating and staying on a diet. Like superego-mass to an id-mass, Sufficiency, Uniqueness, and Sentiment act as utility-mass in the model when they concern an entity that is not a drive or purpose but is being valued for one.

Figure 7.16 While on a diet at a snack food convention, whether pain or pleasure is felt would depend on which purpose is being considered.

Cone and Disc Apparatus

The cone is directly underneath the disc and the two together resemble a seesaw or lever from a side view. The cone's point strikes the disc in its exact center, so that the fulcrum between the disc and the cone is also in the exact center of the disc. The cone's upward force represents the drive to maintain all drives: life instinct, libido, Eros, and maintaining homeostasis. The drive to maintain all drives is felt as an upward force against the disc by the cone and is strongest at the fulcrum. Pleasure, with respect to the purpose of maintaining equilibrium between a purpose and its complement, is generally ensured for the ego by always striving to ascend in order to keep the disc balanced and by maintaining proximity to the fulcrum to achieve homeostasis. Pain, with respect to the purpose of not maintaining equilibrium between a purpose and its complement, is generally ensured for the ego by always striving to descend to keep the disc unbalanced. Though called a cone, the drive to maintain all drives can be shaped more like a needle, as the disc may tilt up to 90 degrees (one-half pi radians) to either side. A tilt of ninety degrees in the disc always results in death, as it implies the success level for at least one drive in a complementary pair is zero (e.g., two to one). The roll cage will fall into the lava uncontested by the disc.

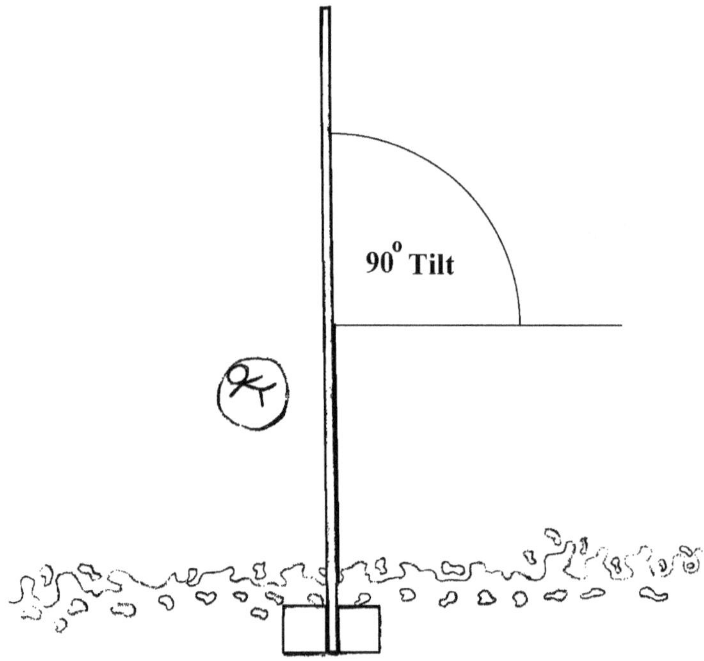

Figure 7.17 A 90 degree tilt of the disc, (if angle p could equal zero) would imply that the ego believes it has no chance at either restoring equilibrium or recovery from an imbalance.

Hence, the angle of the cone's point, angle *p*, to the radius of the cone's base, influences the maximum degree to which the disc may tilt at a given total mass before the ego feels there is no chance at recovery. For a specific total amount of total mass on the disc, angle *p* is called the breaking point angle. For a specific amount of total mass on the disc, the breaking point is reached at a particular success ratio between a purpose and a complementary purpose. The ego cannot transgress beyond the breaking point without losing the integrity it needs to cope with the demands of maintaining equilibrium. If the success ratio of the breaking point is known for a given total mass, then angle *p* can be calculated. Also, for a given total mass, the gradient of the disc may correlate to only one ratio. For instance, at a ratio of 1:1 the disc apparatus is always at equilibrium. At a ratio of 3:1 the disc apparatus is always at a 45-degree angle (one-quarter pi radians) to the

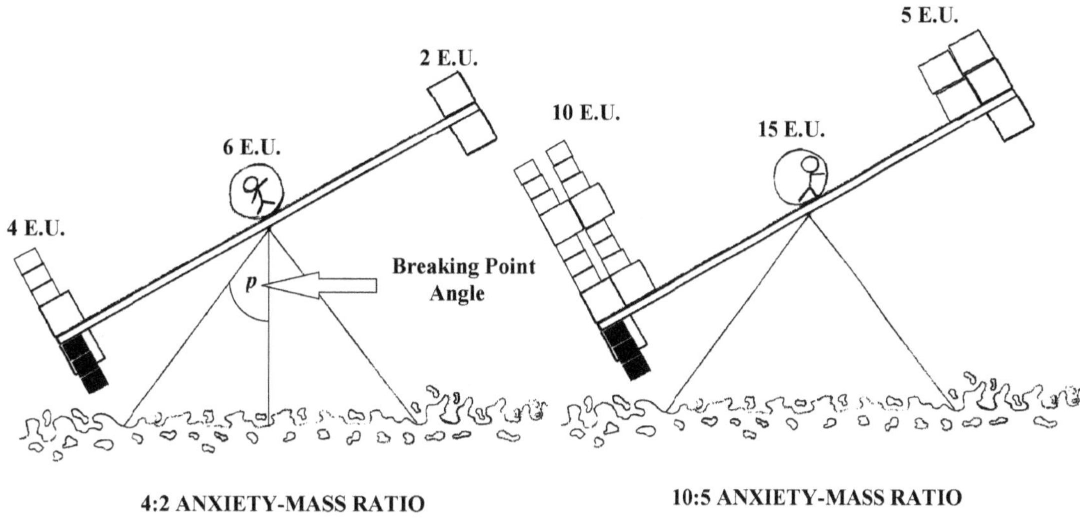

Figure 7.18 The two cone and disc apparatuses have a similar ratio of success but different total masses. The breaking point angle of the image on the left is angle p.

horizon. However, if the total mass on the disc is greater in one case, then the ego will have to work harder in one scenario versus another scenario. For instance, if two disc apparatuses both have a two to one anxiety-mass ratio, but one has a balance of four units of mass against two, while the other has a balance of ten units of mass against five, then the ego will have to work harder to balance the disc in the second case. Subsequently, the breaking point angle for the ego in the second case would be reached sooner as the ego has more mass to handle.

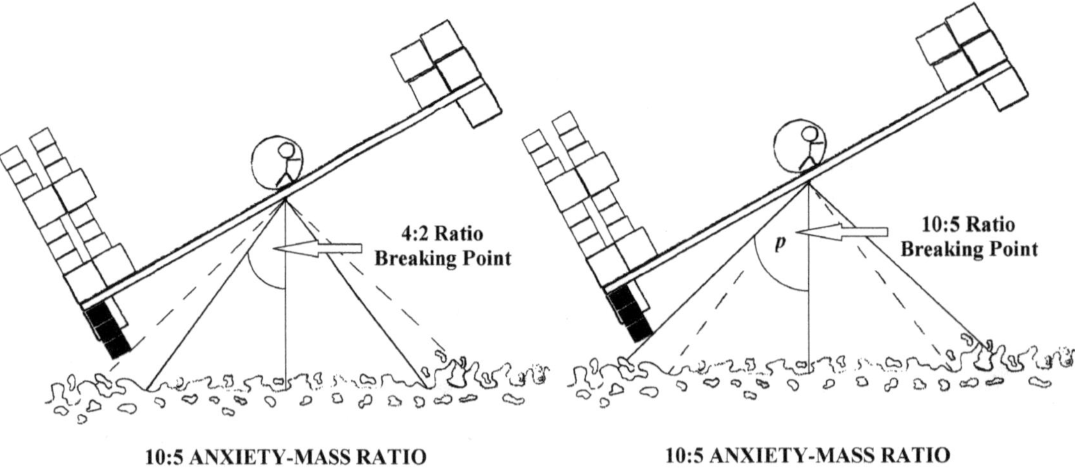

Figure 7.19 Due to the difference in total masses, the disc apparatus with the larger total mass will have a breaking point that is reached sooner, meaning that angle *p* will be larger for the cone and disc apparatus on the right than on the left.

Outside the Cone

Outside the cone and beyond the disc on all sides is death. A sea of boiling lava can be imagined here. If the ego is unable to balance the drives, then the roll cage will fall off of the disc, into the sea of boiling lava, and the individual will perish. Pain, with respect to maintaining equilibrium, is felt by the ego as the sensation that the disc is falling off balance, and the angle of the disc against the fulcrum where the cone strikes the disc's center can increase this sensation if the gradient is a steep one. For example, if the disc is tilted at an extremely sharp angle then the ego will have to work much harder to restore the equilibrium of the disc so as to not slide into death. Pain is generally minimized by keeping the roll cage balanced on the center of the disc. If the roll cage moves to the outskirts of the disc, then the disc will usually tip sharply in that direction with respect to its orientation to the horizon. Subsequently, pain felt is maximized by not keeping the disc level, thus leading to a dynamic state if homeostasis is not reached.

Gravity

Gravity in the model counters the life instinct and the drive to maintain all drives. Namely, it is the death drive and it can be thought of as the drive to **not** maintain all drives. It is represented by the combined effect of the id-masses, superego-masses, entity-masses, utility-masses, the mass of the roll cage, anxiety-mass stockpiled in the cone, and gravitational pressure exerted against the cone toward the ground (lava). Like the drive to maintain all drives, the drive to not maintain all drives has an investment of anxiety. Normally, it tends toward + 1 emotional unit, a place holder value like other complementary purposes, but may become elevated if its utility components change (e.g., Sufficiency). In the examples to come, the breaking point of the ego, concerning the force felt between the roll cage and the lava, will hold the mass of the lava to be at its existential value of one emotional unit.

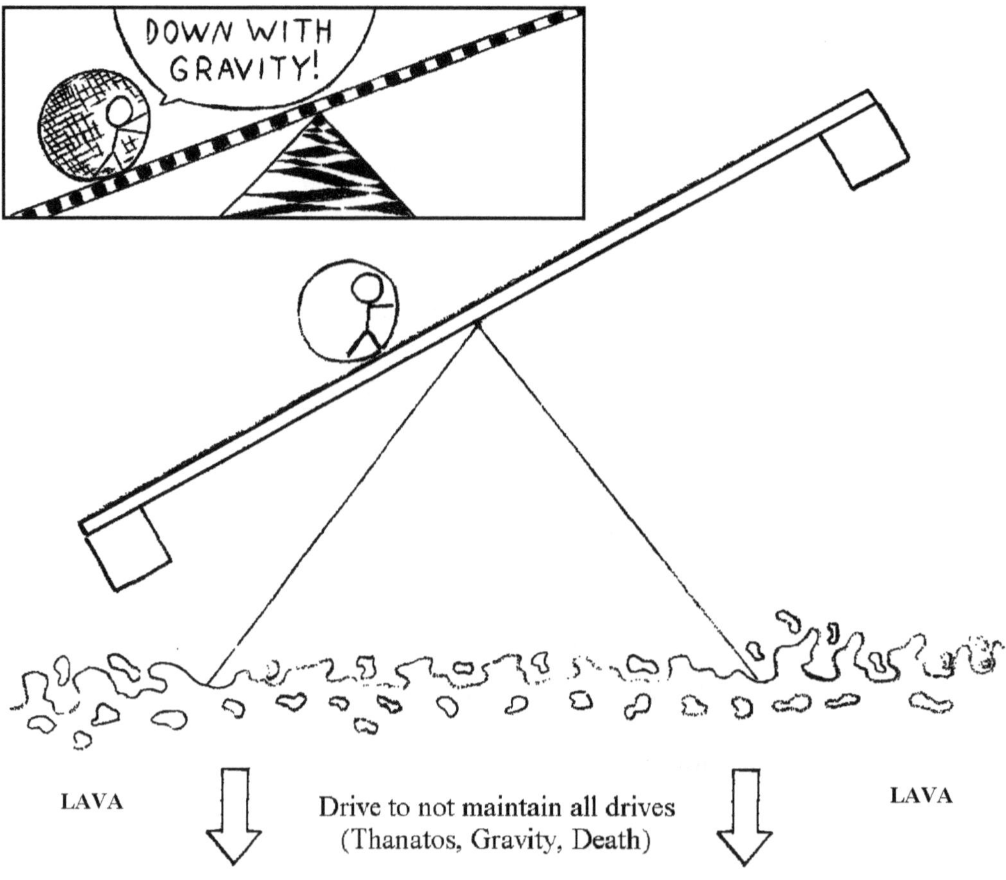

Figure 7.20 Gravity: The drive to not maintain all drives

Id-masses's Location and Entity-masses's Location

The individual's many drives are not fixed at any particular point around the edge and underside of the disc, but they are fixed in their distance from their nemesis drive across the diameter of the disc. Two drives can become associated with each other if their fulfillment is tied to a common entity. One such occurrence would be eating and drinking. If an individual is always eating salty foods, then this will fulfill the drive to eat but at the expense of the drive of drink. In turn, the purpose of eating would become associated with the purpose of not hydrating while the complementary purposes of not eating and hydrating would become associated and closer to one another on the perimeter of the disc. Their respective id-masses would align accordingly.

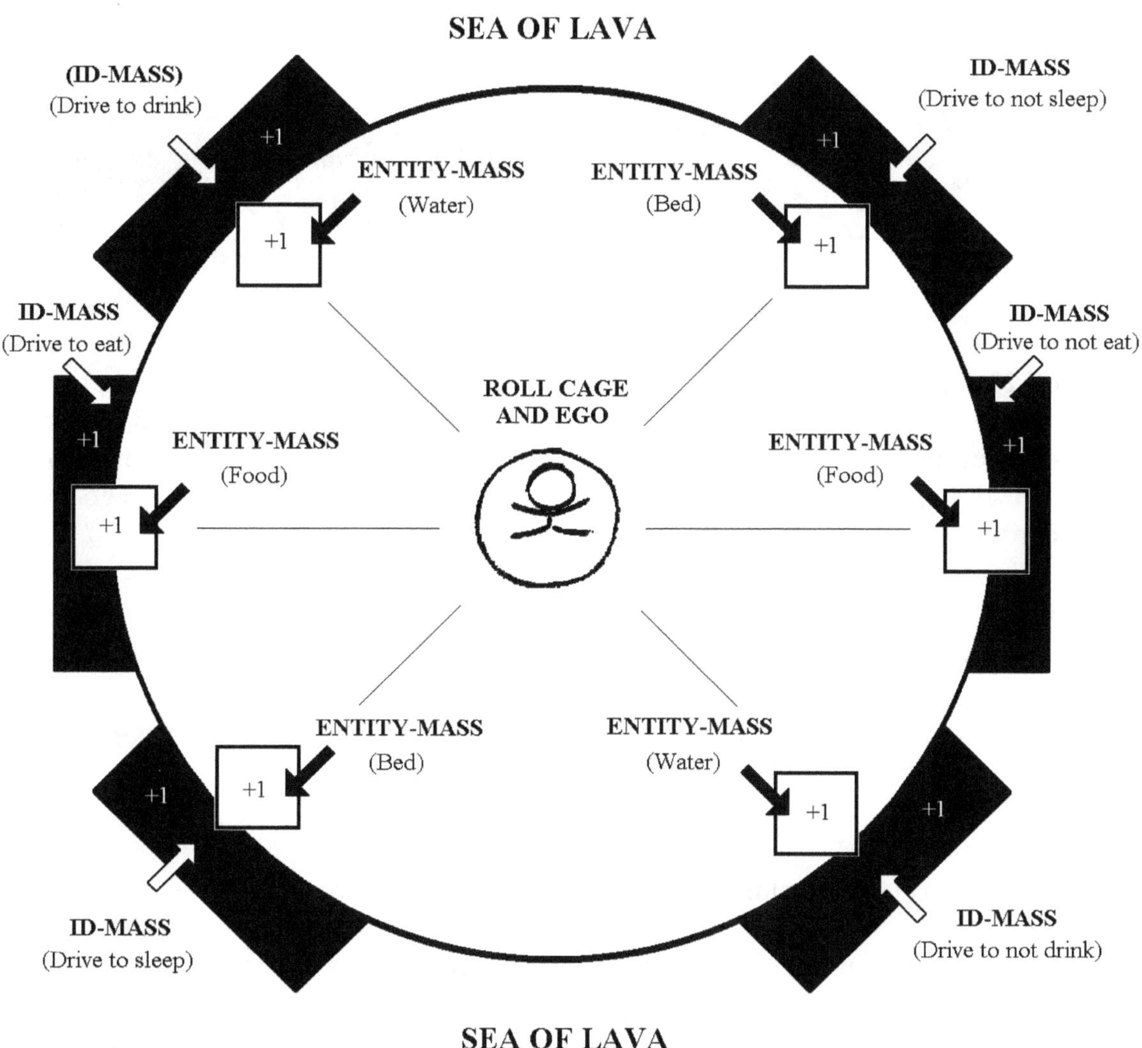

Figure 7.21 Flattened, bird's eye view of the roll cage model with three drives and their complementary drives modeled.

Figure 7.22 All id-masses, acting as magnetic poles, rotate freely and in opposition to their complements along the edge of the disc apparatus, which is three-dimensional in nature. Entity-masses, utility-mass, and superego-mass (all metallic and tied to the center) move in tandem with the id-masses. Id-masses are depicted above with flattened superego-mass. in the conscious.

Superego-mass's Location and Location of Utility-masses concerning Sufficiency, Uniqueness, and Sentiment of Other Entities

It has been explained that the utility-values of drives are represented as superego-mass that have become associated with an id-mass, and that these drives tilt the disc to spur the ego to acquire or not acquire an entity. Over time a multitude of superego-mass and other utility component masses will arise and be organized to correspond to specific id-masses and entity-masses. These may assist or perplex the ego in its efforts to figure out how to counterbalance complementary purposes. Being metallic, these superego-masses and utility-masses are magnetically oriented to an id-mass, nemesis id-mass, and corresponding entity-mass.

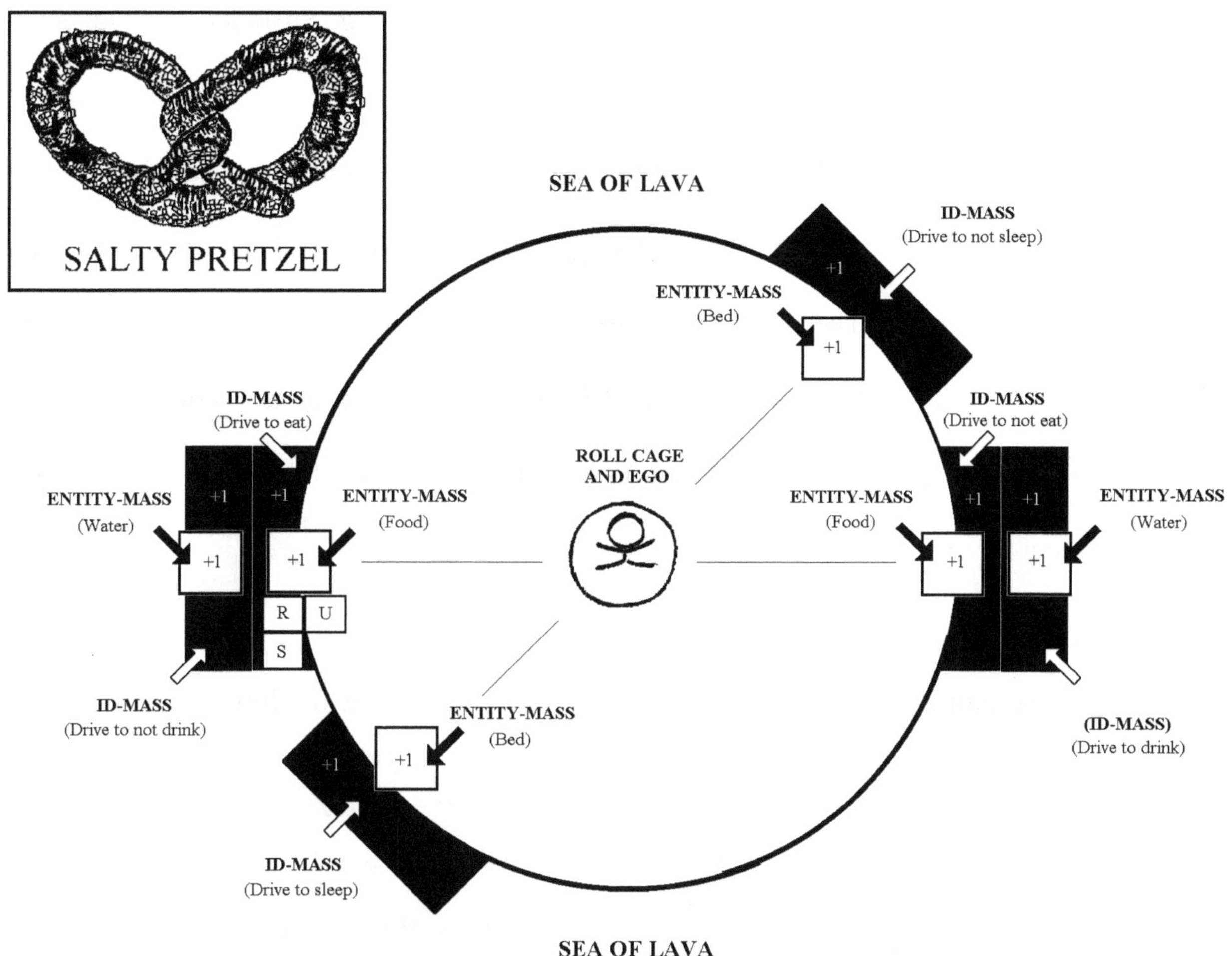

Figure 7.23 Re-orientation of utility-mass to entity-mass and associated id-masses from a flattened, bird's eye perspective is shown. The entity of a salty pretzel, valuable towards the purpose of eating, is also valuable towards the purpose of dehydrating oneself (e.g.,. not drinking). If salty pretzels are the only food item one eats, then over time, these two id-masses, of eating and dehydrating, will become associated with one another. Meanwhile, their complementary drives or id-masses, of not eating and not dehydrating (e.g., drive to drink), will also become associated. Three separate entities are shown above being valued for three separate purposes. Only the pretzel has utility-mass depicted here.

Moreover, the superego-mass, also being attached to the center of the disc with a rope, may move freely with id-masses as they rotate around the disc to become associated with other id-masses. In the case of a salty pretzel for instance, it is given a value judgment of + 1 emotional unit for the fact that it exists. Thereafter, its value toward the fulfillment of the drive to eat and the purpose of eating may be assessed along with any superego-mass or utility-mass (e.g., Sufficiency, Uniqueness, or Sentiment) attached to it.

Such valuations would still hold true even with the addition of other purposes. The pretzel, if being considered for a different purpose than eating, would have additional anxiety-mass affixed to an additional entity mass. The salty pretzel, for instance, is more valuable to the purpose of dehydrating oneself than not dehydrating oneself. Consequently, the pretzel's acquisition would cause the purpose of eating to become more associated with the purpose of dehydrating oneself and it would receive a separate valuation.

Homeostasis (Equilibrium) and Isaac Newton's Inverse Square Law of Gravitational Force

From an isometric 2-D side view of the 3-D Roll Cage Theory cone and disc apparatus, the roll cage itself may be seen to move left or right and up or down. While movement of the roll cage to the left or right represents the acquisition of an entity or lack thereof, hence an actual physical state, vertical movement of the roll cage represents another concept, homeostasis, balance, or equilibrium. The mass of a particular entity or drive, measured in emotional units, signifies an idealized ratio of success between a purpose and its complementary purpose based upon the existential and utility values (Sufficiency, Uniqueness, and Sentiment) of an entity with respect to a purpose. While a complete understanding of force in the Roll Cage Theory of Drives is not essential, it is presented here for those who are interested in a physics model of affect engineering.

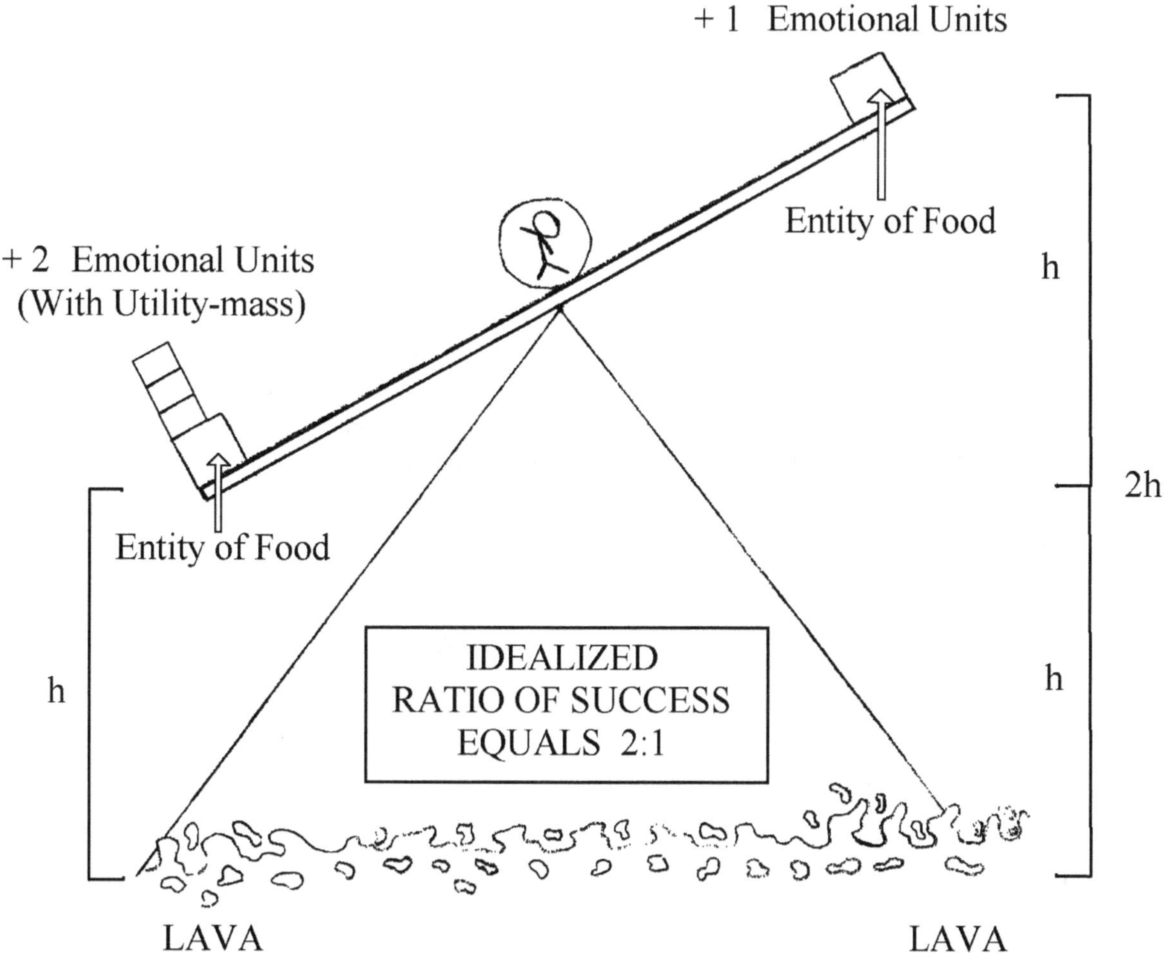

Figure 7.24 The Idealized Ratio of Success vs. True Equilibrium (Fulcrum) is depicted. True equilibrium is always achieved when the roll cage is directly above the fulcrum. Disparities in anxiety, however, always distort where equilibrium is felt by changing the ideal ratio of success between a purpose and its complementary purpose. Although the roll cage is at true equilibrium in the above illustration, anxiety has distorted where equilibrium is felt so that the idealized ratio of success now feels to be 2:1 instead of 1:1. (The influence of the roll cage is nil when it is over the fulcrum.) The id-masses of eating and not eating are not displayed, but may be assumed to be correlating with the entity-mass of food).

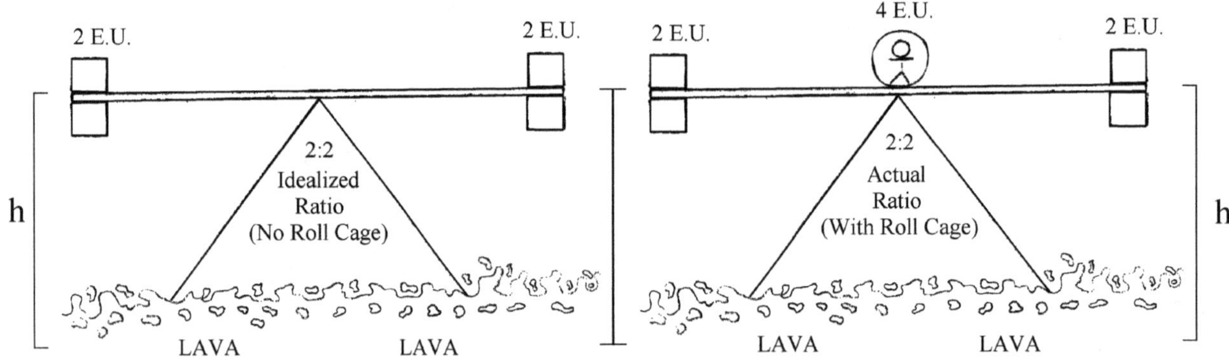

Figure 7.25 Idealized Ratio vs. Actual Ratio: The idealized anxiety-mass disparity is the idealized ratio of success between the purpose and the complementary purpose without any influence from the roll cage (left). The actual ratio of success takes into account the mass of the roll cage (right).

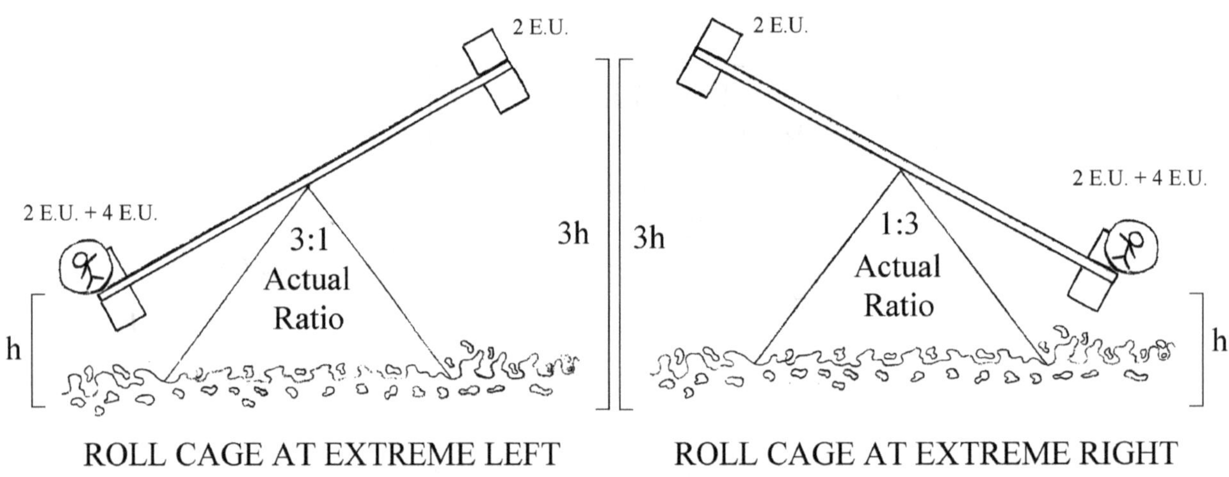

Figure 7.26 The actual success ratio between a purpose and complementary purpose is depicted above when the roll cage is at the extreme left or right and the total mass on the disc apparatus is eight emotional units (E.U.).

However, as noted earlier, one can have a lot of anxiety invested in purposes and entities while at a particular mass disparity, or a little bit of anxiety invested while at a particular mass disparity (See Figure 7.18). Similarly, one can have very little anxiety invested in purposes and entities while not being at equilibrium, or a lot of anxiety invested in purposes while not being at equilibrium. Given that anxiety and negative anxiety are tools serving more or less as an alarm system to warn the ego/self that equilibrium is not being maintained, they can have their sensitivity altered or shut off if the ego is able to successfully withdraw anxiety invested back toward the anxiety-mass stockpile. As for finding the breaking point of the ego, hence, the amount of anxiety it can handle before losing its integrity, Isaac Newton's famous inverse square law of gravitational force will be helpful. It can be used to model the force of gravity that the ego feels between itself and the lava and depict a breaking point in terms of force felt due to anxiety-mass.

Isaac Newton's Universal Law of Gravity holds that "any two objects in the universe attract

$$\text{Force} = \frac{G \otimes \text{"Mass(1)"} \otimes \text{"Mass(2)"}}{\text{Square of the "Distance Between Objects"}}$$

Figure 7.27 Isaac Newton's Universal Law of Gravity

each other" with a force that is proportional to the product of their masses, inversely proportional to the square of the distance between the objects, and where the coefficient G corresponds to the universal gravitational constant.[2]

For a specific total amount of mass on the disc, at a specific point of force felt between the roll cage and lava, there will be an instance at which the ego cannot compensate and it will break. The angle of the disc against the horizon at which this occurs will be called angle *p*.

The lava, which represents the drive to not maintain all drives, and being the complement to the drive to maintain all drives, will be given an anxiety-mass value of + 1 emotional unit like all of the other complements being considered here. However, that is not to say this would always be the case.

If there is a 1:1 ratio and d = 0, then:

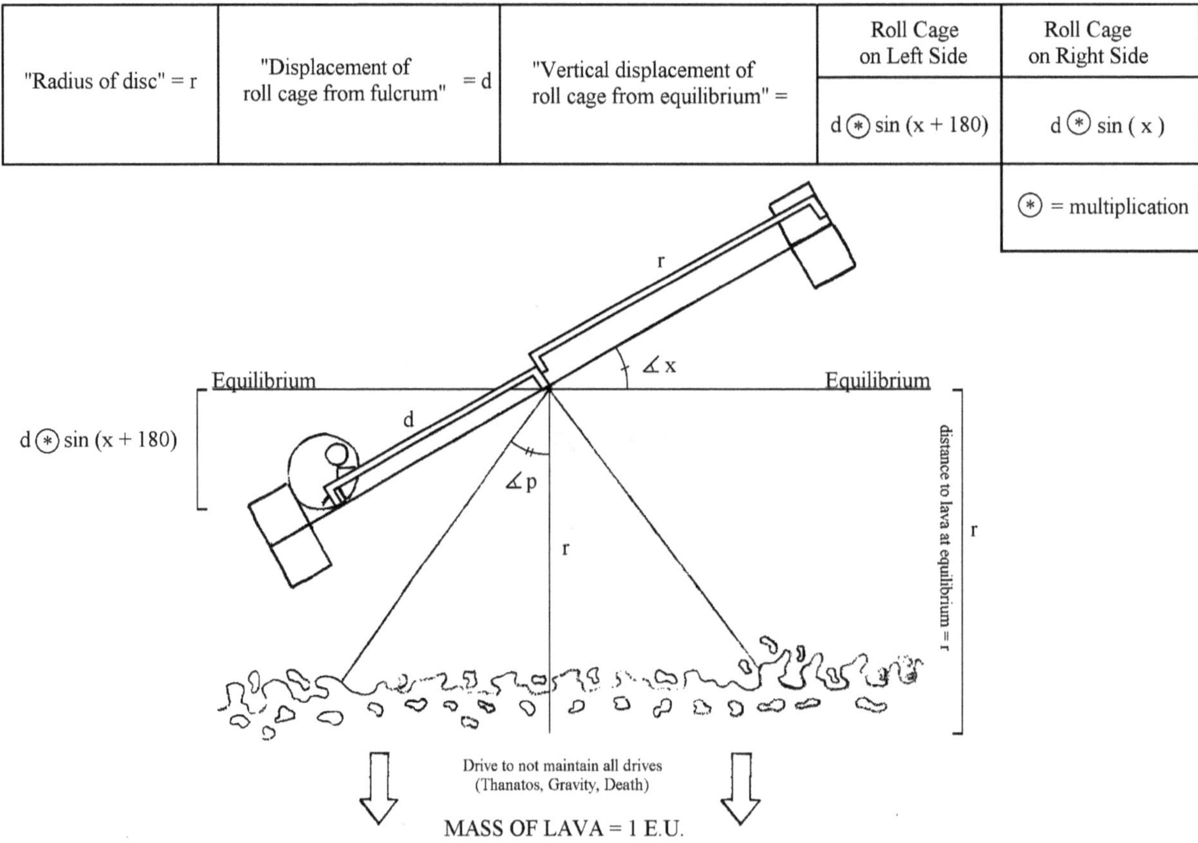

"Radius of disc" = r	"Displacement of roll cage from fulcrum" = d	"Vertical displacement of roll cage from equilibrium" =	Roll Cage on Left Side	Roll Cage on Right Side
			d ⊛ sin (x + 180)	d ⊛ sin (x)
				⊛ = multiplication

Figure 7.28 Radius of Disc, Displacement of Roll Cage from Fulcrum, Vertical Displacement of Roll Cage from Equilibrium, Mass of Lava (1 E.U.), and the Mass of the Roll Cage (4 E.U.) are depicted above.

Equilibrium is felt at sin (x) = 0 or sin (x + 180) = 0

If sin (x) is not equal to zero, then perfect equilibrium is not felt even if the individual is at true equilibrium or directly above the fulcrum. Quadrants one and four concern the right side of the disc, quadrants two and three concern the left side.

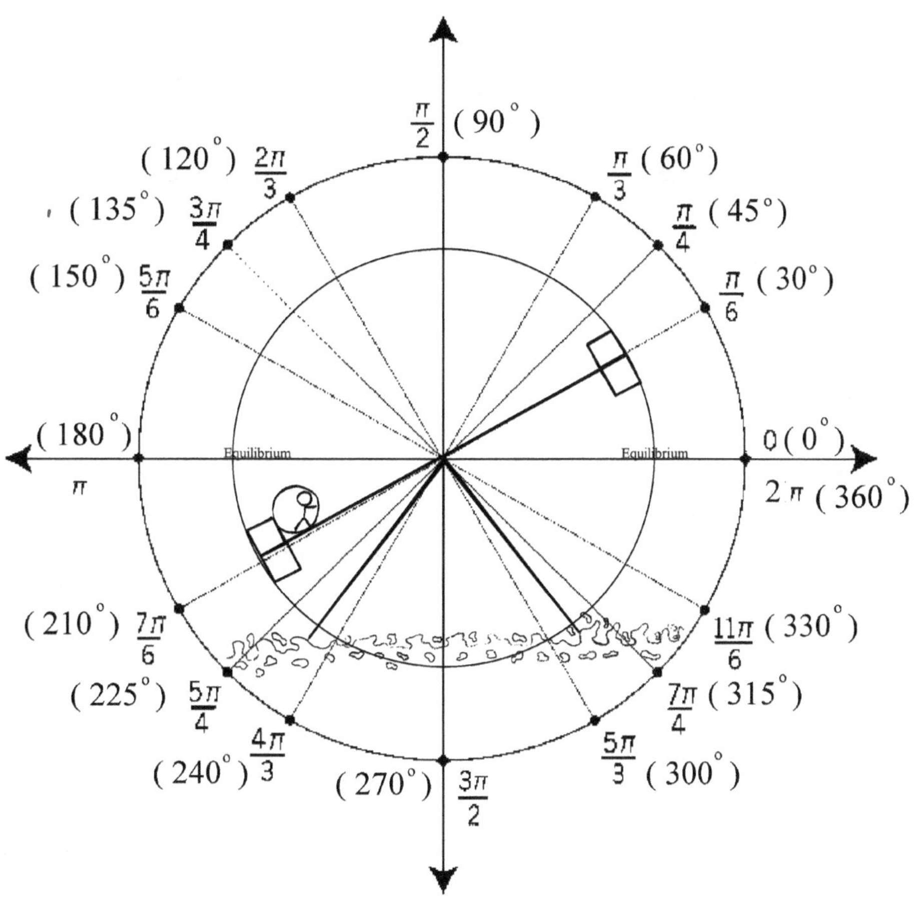

Figure 7.29 The tilting of the disc can be mapped on a unit circle.

Calculating Gravitational Force

At a specific total amount of anxiety-mass on the disc, every anxiety-mass disparity correlates to a single value for angle x. As total mass increases, so does the force between the roll cage and the lava, and subsequently the x value for the breaking point decreases, meaning angle p becomes larger.

If one thought of the model in terms of physics, then the breaking point would be construed as a specific amount of gravitational force (downward) felt between the roll cage and the lava, that the ego is not able to cope with. Force is influenced at first by the initial mass disparity before the roll cage moves (the idealized success ratio), the disparity after the roll cage moves (actual success ratio), and the total anxiety-mass on the disc. Given that the anxiety-mass of the lava is being held at + 1 emotional units like all other complementary purposes, gravitational force can be calculated. If the lava's mass (Drive to not maintain all drives) equals to one, then the gravitational force exerted between the lava and the roll cage would be represented by:

F = *G times mass #1 times mass #2 divided by the square of the distance between the two*

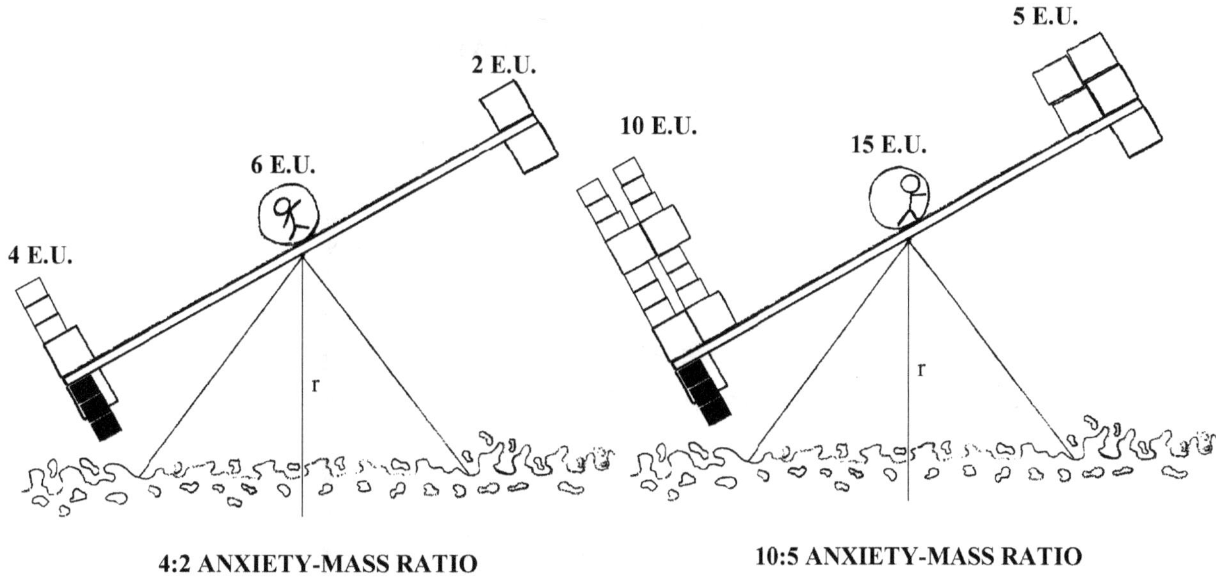

Figure 7.30 The gravitational force that the ego feels in the image on the left is less than the force felt by the ego in the image on the right despite them both being at true equilibrium. Using Newton's Universal Law of Gravity, where G = the gravitational constant, and the distance between the roll cage and the lava = r or the radius of the disc, the force felt by the roll cage on the left would be:

Force (Left) = G * (6 E.U.) * (1 E.U.) ÷ (r ^ 2)

While the force felt by the roll cage on the right would be

Force (Right) = G * (15 E.U.) * (1 E.U.) ÷ (r ^ 2)

The force felt by the roll cage on the right is two and a half times greater than the force felt by the roll cage on the left.

If it is given that for a specific total anxiety-mass (e.g., eight units) the ego's breaking point is reached at an anxiety-mass disparity of 3:1, and the thickness of the disc is inconsequential, then angle *p* can be found from the following information in figure 7.31. The radius of the disc will be considered the height of the fulcrum from the lava.

6 : 2 ANXIETY-MASS DISPARITY

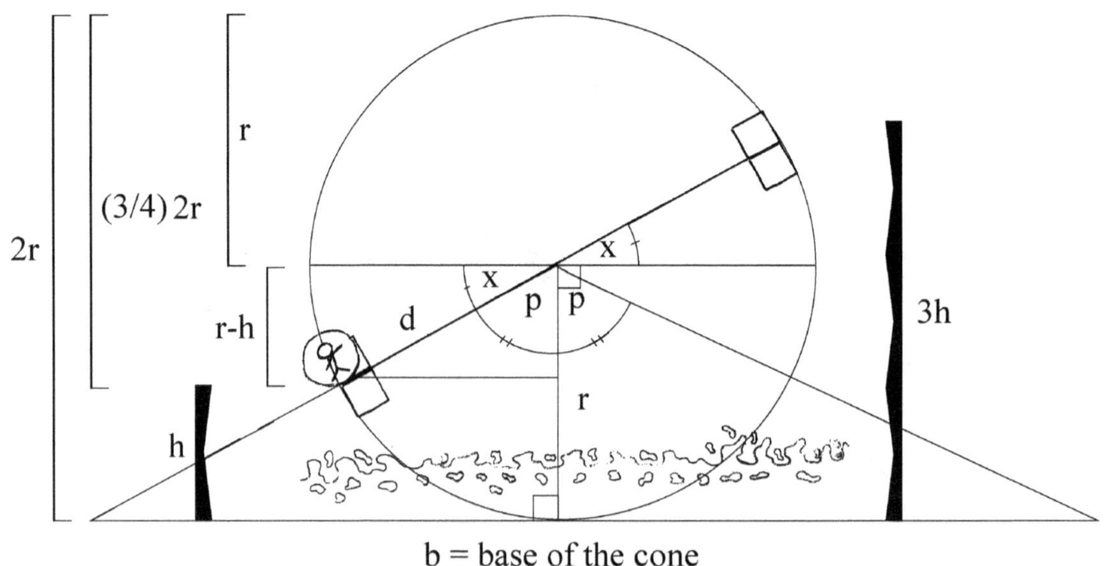

$$\cos p = \frac{(3/4) \circledast 2r - r}{r} \Rightarrow \arccos \frac{(3/4) \circledast 2r - r}{r} = \angle p$$

$(3/4) \circledast 2\,r$ is obtained from:

$$\left(\frac{(\text{"Exterior Mass"} + (d / r) \circledast \text{Mass of Roll Cage})}{\text{Total Mass on the Disc}} \right) \circledast \text{"Diameter of the Disc"}$$

The proportional influence of the roll cage is represented by:

$(d / r) \circledast$ Mass of Roll Cage)

$$\text{Gravitational Force} = \frac{G \circledast \text{"Mass 1"} \circledast \text{"Mass 2"}}{\text{"Distance" squared}}$$

$$= \frac{G \circledast (4 \text{ E.U.}) \circledast (1 \text{ E.U.})}{(r - d \circledast \cos(p))\ \hat{}\ 2}$$

Figure 7.31 Calculating Angle *p* and the Gravitational Force at the Breaking Point.

Expressed in words:

cos p = twice the radius of the disc (times (the mass of extremity on one side plus the proportional mass of the roll cage based on how far it has traveled over the total mass on the disc) minus the radius of the disc) divided by the radius of the disc. The influence of the roll cage on the tilt of the disc is proportional to its mass times its distance from the fulcrum divided by the radius of the disc.

For whatever value of r is chosen, in each case angle p = 60 degrees or pi over 3 radians. The gravitational force can simply be calculated by knowing the mass of the roll cage and its distance from the lava. In the above case, the distance of the roll cage to the lava is conveniently equal to the distance of the disc's endpoint to the lava, h, but it will not always be the case the displacement equals the radius. In such a case, the breaking point angle will need to be known. For instance,

Gravitational force to lava = G * (mass of roll cage) * (mass of lava) divided by the square of the difference between the radius and the displacement of the roll cage from the fulcrum times the cosine of the breaking point angle. In this case it is 16 G.

If the roll cage is at the fulcrum, then the anxiety-mass ratio of the exterior mass is represented by the ratio of their distance toward the lava. The anxiety-mass ratio of just the id-mass, entity mass, and utility-mass (including superego-mass) when the roll cage is at the fulcrum is the idealized ratio of success between a purpose and a complementary purpose. The ratio of anxiety-mass on each side, when the roll cage is at the fulcrum, can be represented by the following expression (figure 7.32).

$$\frac{r - r(\sin(x + 180))}{r - r(\sin(x))} = \frac{\text{Mass Influence on Left Side}}{\text{Mass Influence on Right Side}}$$

Figure 7.32 Idealized ratio, when the roll cage is at the fulcrum

The roll cage's position represents the actual ratio of success between a purpose and a complementary purpose. The roll cage's position also adds mass to a particular side (i.e., left or right). When at the fulcrum, no additional mass is added to one side or the other. The amount of mass added to a side is proportional to the fraction of the roll cage's displacement from the fulcrum over the radius of the disc times the mass of the roll cage (figure 7.33).

4 E.U. 3 E.U. 2 E.U. 1 E.U. 0 E.U. 1 E.U. 2 E.U. 3 E.U. 4 E.U.

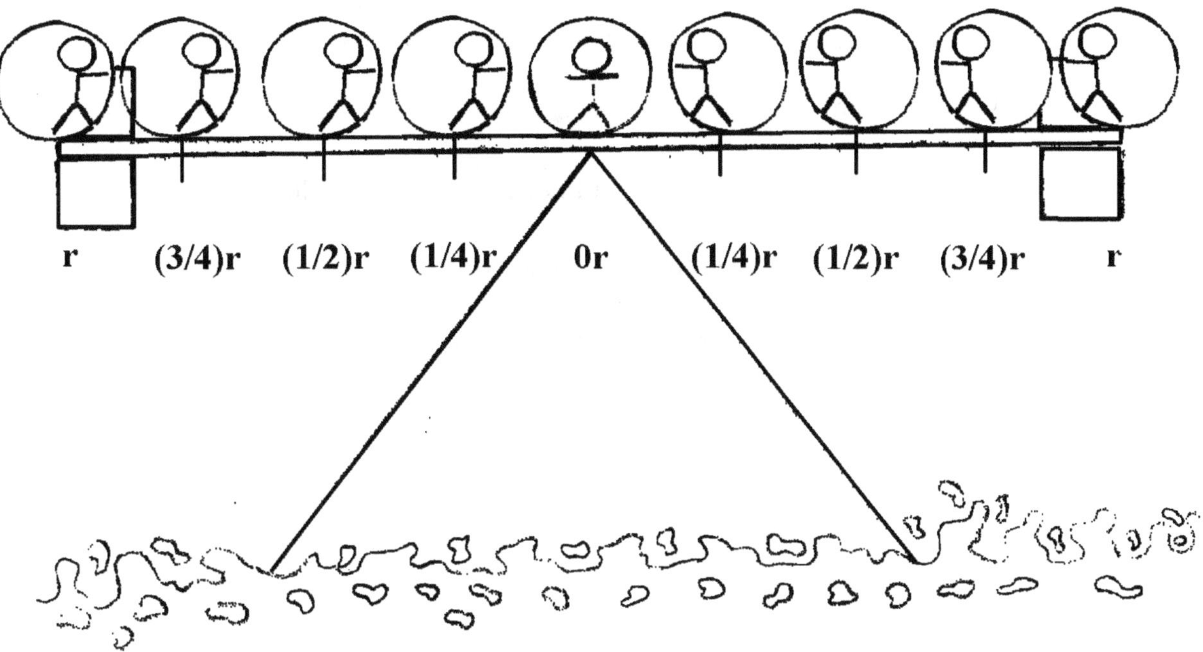

Proportional Influence of the Roll Cage

Total Mass on Disc = 8 E.U.

Figure 7.33 The proportional mass of the roll cage added to a side is based upon its distance away from the fulcrum and towards the edge of the disc.

To find the force between the roll cage and the lava, the following must be known or calculable:

1) The total mass on the disc must be known.

2) The side of the disc that the roll cage is on and subsequently which drive it is closer to (purpose or complementary purpose) must be known.

3) The ratio of mass between the roll cage, purpose, and complementary purpose must be known.

4) Either angle *x* between the right side of the disc and the horizon, or the distance *d* of the roll cage from the fulcrum in order to obtain the roll cage's proportional mass added to a particular side. Angle *p* is the complementary angle to angle x, meaning that (x + p = 90).

5) The radius of the disc, *r*, can be set to one like a unit circle.

6) The mass of the lava (in emotional units), which represents anxiety invested in the drive to not maintain all drives, must be known.

Given:

1) 6 : 2 Anxiety-Mass Disparity with the Roll Cage's Influence
2) Total Mass on Disc = 12 E.U.
3) d = 1 / 3 of r

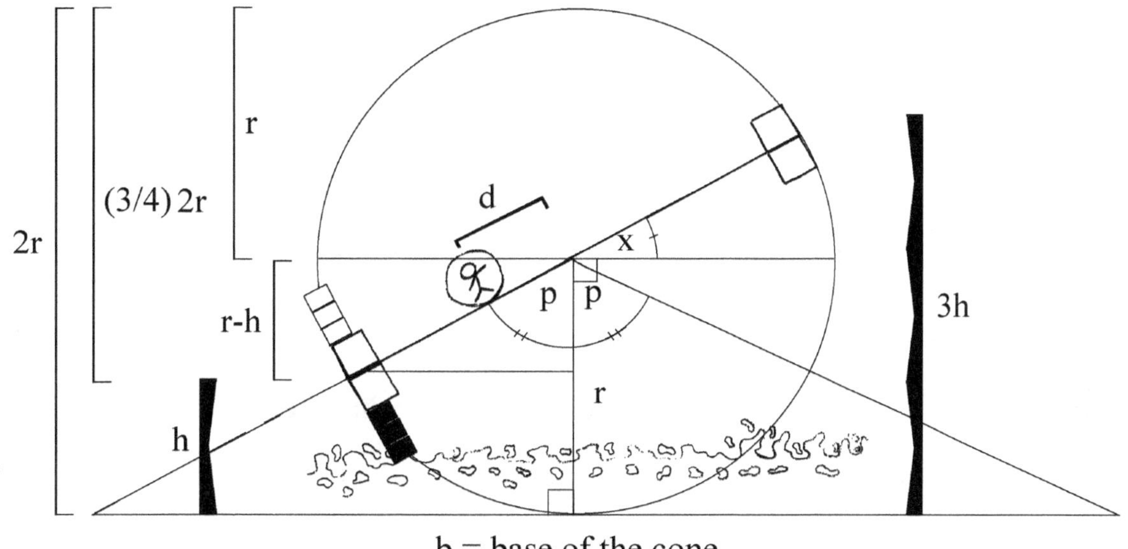

$$\text{Gravitational Force} = \frac{G \circledast (\text{ Mass of Roll Cage }) \circledast (\text{ Mass of Lava })}{(r - d \circledast \cos(p))^{\wedge}2}$$

It was found earlier that angle p = 60 degrees

After the mass values are input:

$$\text{Gravitational Force} = \frac{G \circledast (6) \circledast (1)}{(r - d \circledast \cos(60))^{\wedge}2}$$

If ***r*** is equal to one, then:

$$\text{Gravitational Force} = \frac{G \circledast (6) \circledast (1)}{(1 - (1/3) \circledast \cos(60))^{\wedge}2}$$

Reducing to:

$$\text{Gravitational Force} = \frac{216\ G}{25}$$

Figure 7.34 Sample Gravitational Force Calculation

The x value may equal any number between -90 and + 90 degrees. At x equals zero there is equilibrium. Finally, d must be equal to or less than the radius of the disc, r, if the individual is still alive.

Synopsis

The roll cage, along with all contained within it, represents the self, ego, and life essence of the person. It contains the ego and is the source of cognition, conation, and interpretations of affect. The id is represented by id-masses along the perimeter of the disc while the superego is represented by small, superego-masses on top of or below the disc that become associated with id-masses. The ego is tasked with maintaining homeostasis by keeping the roll cage from rolling off of the disc and falling into the sea of boiling lava that lies on all sides of the apparatus. The ego is able to feel affect through the desperate sensation of falling backwards and down (pain), or the conquest of ascending forward and upwards (pleasure). Affect becomes measured as the felt pressure that the ego feels exerted on it by a combination of the forces from the id-masses, entity-masses, superego-masses, utility-masses, the cone through the disc, lava, and the roll cage itself. Cognition is represented by the ego's knowledge of the layout of the disc along with a sense of the forces acting on the roll cage that may be lying underneath the disc in the subconscious. Last of all, conation is measured as the direction in which the ego decides to move the roll cage, henceforth, the capacity of the ego to act or its intended behavior.

If the ego is unable to identify an entity that will reduce a drive, then it cannot guarantee its own survival and is at the mercy of the elements around it. Superego-mass and utility-mass distort the ego's ability to sense where the true center of balance is on the disc to alert it to act, for instance, due to environmental demands. If the individual were in an area with a lack of water, such as a desert, more anxiety might be invested into the entity of water or the drive to hydrate than if the individual were living near a freshwater lake. It may be said that the role of anxiety, more often than not, is to distort the individual's conception of the world so as to direct it toward specific courses of action. In spite of these distortions, anxiety also provides the ego with a frame of reference for understanding the relationship between entities in the real world and how they may be reconciled with the individual's own drives.

Because the life instinct is a perversion of the death instinct in the Roll Cage Theory of Drives, perpetual life or immortality is theoretically possible but cannot be guaranteed. Neither the drive to maintain all drives, nor the drive to not maintain all drives can extinguish the other one outright. As a result of the life instinct being a drive to maintain all drives, this necessarily includes its nemesis, the drive to not maintain all drives. Conversely, as a result of the death instinct being a drive to not maintain all drives, this necessarily includes the death instinct. The life instinct, in effect, must always stay one step ahead in order to outwit the death instinct. On the other hand, the death instinct inadvertently gets in its own way. This would be much like a navigator plotting a course that keeps a reckless captain of a ship going in circles to prevent the ship's sinking. The navigator is unable to mutiny and throw the captain overboard, however, because the captain is the only one who knows how to pilot the ship. A course, then, is plotted to keep them going in circles. Whereas one seeks to prolong the inevitable sinking of the ship, the other's actions would almost certainly realize it.

Figure 7.35 The Captain and the Navigator

Even though drives are represented as opposing forces along a spectrum, with the center of the spectrum being the fulcrum where the disc and cone meet, their corresponding id-masses would not always be spaced evenly around the disc except against their complement. Drives that are nearby one another may become linked with one another over time to indicate that reducing one influences the other. One id-mass, for instance, might move closer to a separate id-mass and in opposition to its complement. If the three drives of eating, drinking, and sleeping are considered with their complementary drives of not eating, not drinking, and not sleeping, they may all start evenly spaced from one another in no particular order (figure 7.21).

If an individual is always eating salty foods, then eventually the reduction of the drive to eat will become linked with the reduction of the drive to not drink or to dehydrate oneself. This will not change the total mass on the cone and disc apparatus, only the distribution of mass along the edges of the disc and where equilibrium is felt in three dimensions. Eating salty food will subsequently intensify the drive to drink or hydrate. Thereafter, the complementary drive of not eating will become linked to the drive to drink by default, as a drive and its complementary drive must always be on opposite ends of the spectrum if the disc apparatus is to stay balanced (figure 7.23).

For example, it is discovered by an individual that eating a heavy, salty meal induces the urge to sleep. For example, eating a soft, salty pretzel with all the toppings might make the drive to sleep feel more massive while simultaneously making the drive to drink feel massive as well. Additionally, should the individual only drink caffeinated beverages that prevent sleep, then a new dynamic enters the mix. If fulfilling the purpose of drinking (e.g., a caffeinated beverages) makes the individual feel as if the drive to stay awake is being alleviated, then the following associations between id-masses might be made (figure 7.36).

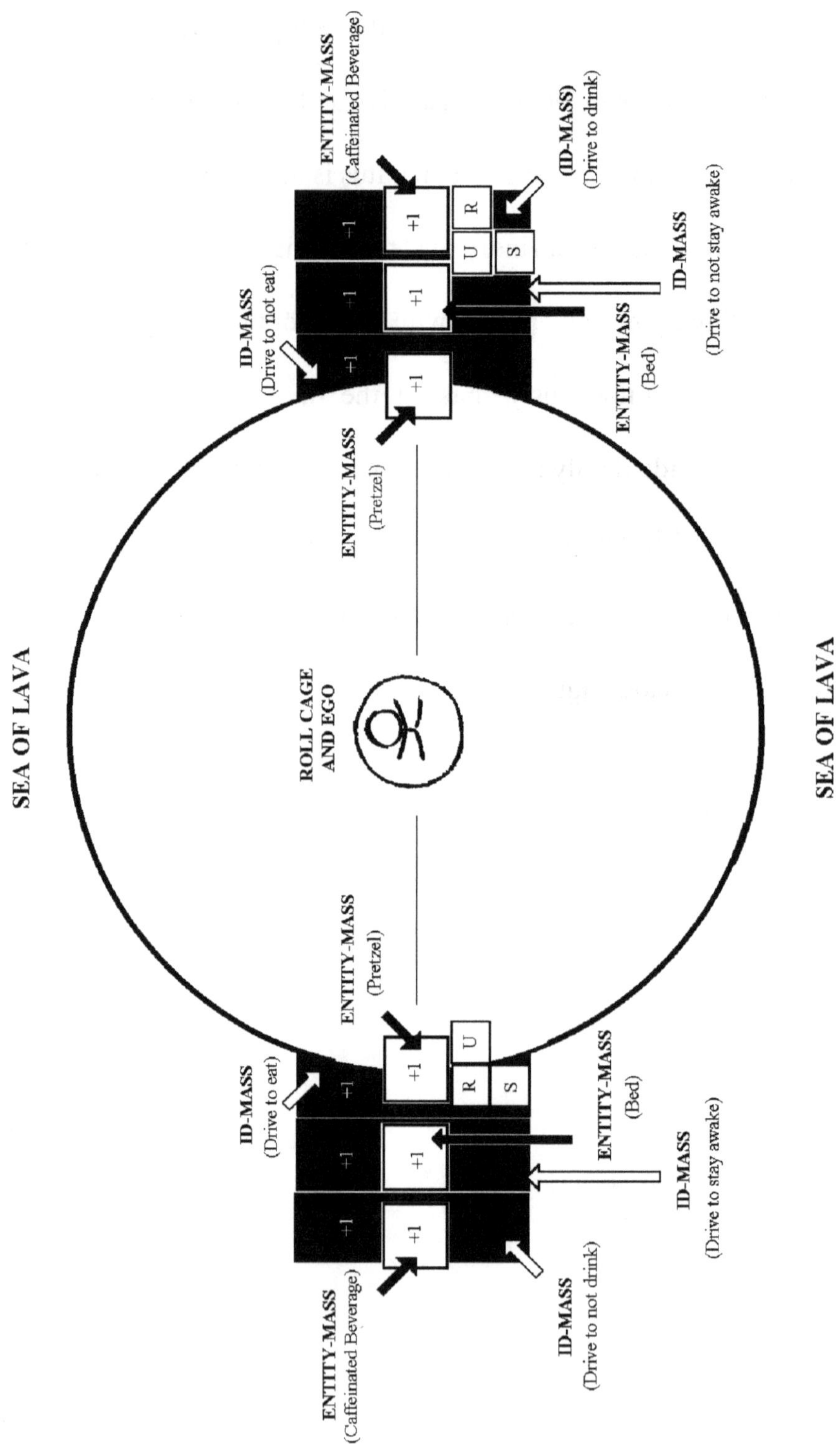

Figure 7.36 The association of different id-masses is depicted.

In figure 7.36, the fulfillment of the following drives becomes linked:

1) Not Staying Awake, Not Eating, and Drinking: Drinking caffeine makes the drive to stay awake feel more pressing, as caffeine is associated with preventing sleep. Not eating and drinking are associated because the individual always eats salty foods.

2) Staying Awake, Not Drinking, and Eating: Eating a heavy, salty meal makes the drive to not stay awake feel more pressing, as the heavy meal is associated with inducing sleep. Eating and not drinking are associated because the individual always eats salty foods.

If, however, instead of drinking a caffeinated beverage, an individual always drank a glass of coconut milk to induce sleep after eating a salty meal, then the balancing of the drive to not stay awake would neither become linked with the drive to not eat nor become linked with the drive to drink. Likewise, balancing of the drive to stay awake would become linked neither with reduction of the drive to eat nor with the drive to not drink. Therefore, staying awake and its complementary drive of not staying awake would end up moving halfway between the other four drives (figure 7.37).

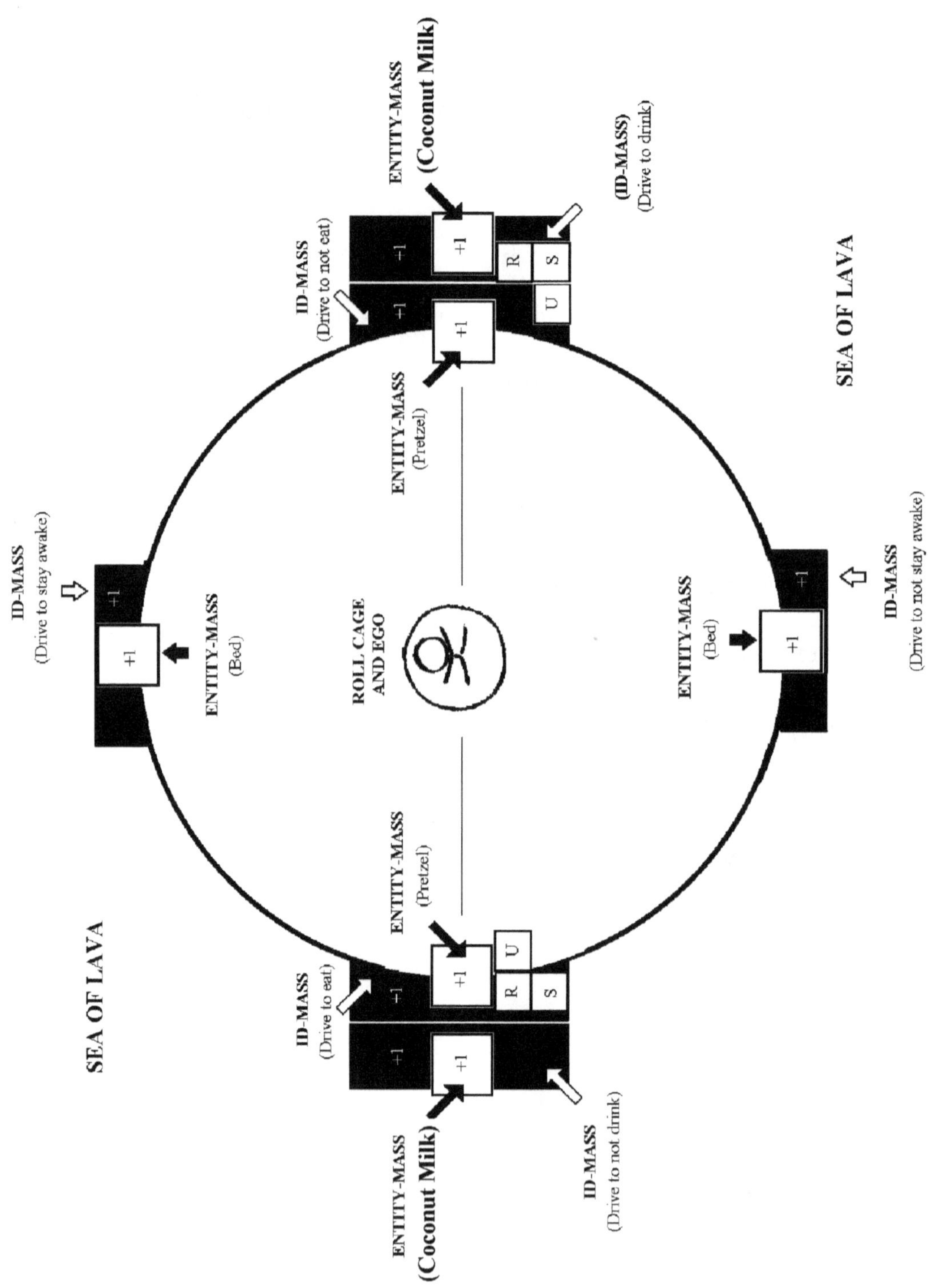

Figure 7.37 The drive to stay awake and not stay awake are isolated.

A Roll Cage Theory of Drive Reduction Theory and Identifying Emotions

In figure 7.37 The reduction of the following drives would be arranged as such:

1) Not Eating and Drinking would be associated

2) Eating and Not Drinking would be associated

3) Staying awake and Not Staying Awake would be halfway between.

The only thing yet to be accounted for in the Roll Cage Theory of Drives is how the mass of the roll cage (i.e., anxiety-mass invested in the drive to maintain all drives) can become larger than two emotional units in some of the examples above. Valuations of entities larger than two emotional units have not been possible up to this point. However, they will become possible once contingencies and relationships between two separate entities are explained in chapter eight. A return to the equation is inevitable in order to achieve this, but not before one final example.

The Case of Flo (Supplemental)

A more concrete demonstration of the Roll Cage Theory of Drives is in order and will be considered for the hypothetical person of Flo, who has just been released from a carefully controlled, bubble environment after a number of years.

Flo is assessing the drive of raising her body's water composition against its complementary drive of not raising her body's water composition and is now valuing water with respect to these two drives. Flo loses water in her body's composition at the same rate that she can raise it by drinking. It is also given that the drive to raise Flo's water composition was originally felt as equal to the drive to not raise her body's water composition, with equilibrium being felt at a success ratio of 1:1 and a 70% water composition. However, after conceptualizing the entity of water and valuing it both toward the purpose of hydrating and toward its complementary purpose of not hydrating, the two idealized success ratios no longer feel equivalent to Flo. Flo's valuation of the entity of water has subsequently redistributed her anxiety resources to the extent that the drive to hydrate and the value of the entity of water for it now feel to have twice the anxiety-mass of the drive to not hydrate and the value of the entity of water the complement.

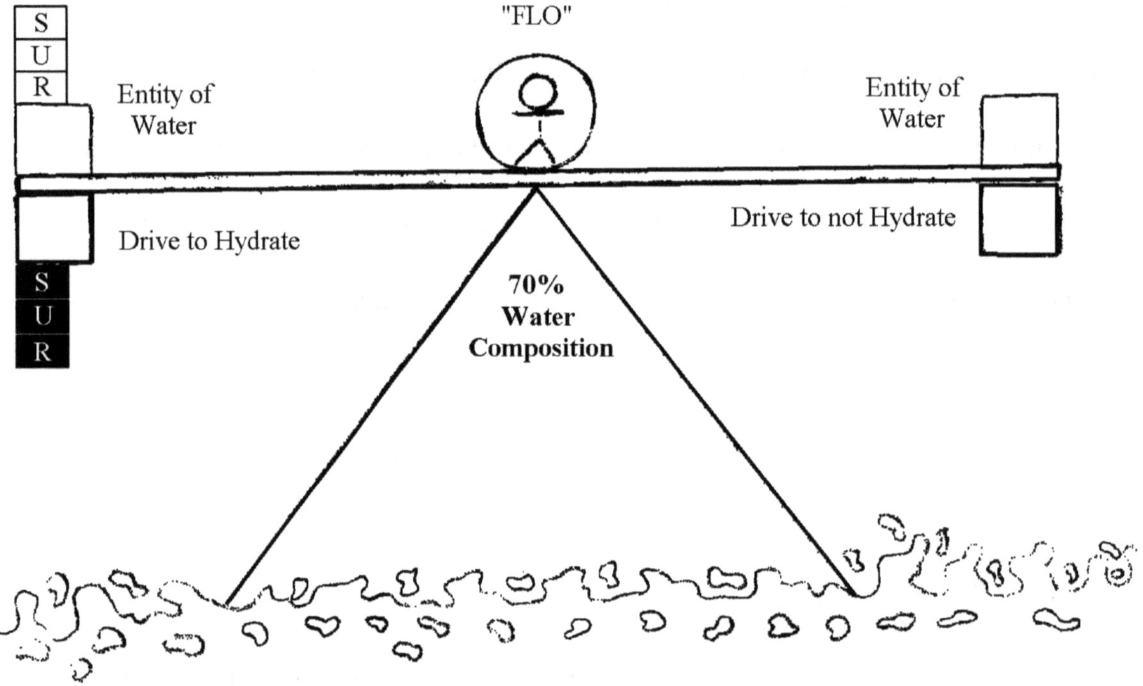

Figure 7.38 The case of Flo

1) The id-mass of hydrating and the entity of water now have twice the magnitude of the id-mass of not hydrating and the entity of water.
2) The id-mass of hydrating and the entity of water (left) have a combined mass of + 4 emotional units with their utility-mass.
3) The id-mass of not hydrating and the entity of water (right) have a combined mass of + 2 emotional units.
4) In the example of Flo, the id-mass of hydrating and the entity-mass for the value of water for this purpose will be said to concur with one another. The id-mass of not hydrating and the entity-mass for the value of water for this purpose will also be said to concur with one another as well. This eliminates the need to do double the work and enables the total mass on each side to be assessed with only the entity-mass and whatever utility-mass (Sufficiency, Uniqueness, Sentiment) is concerning it.

With Sentiment, the Uniqueness of water, and the Sufficiency of water to raise Flo's water composition all being held constant at + 1, Flo's valuation of water for the purpose of raising her body's water composition is + 2 emotional units, compared to + 1 emotional unit for the purpose of not raising her body's water composition. Moreover, in addition to now valuing water differently, Flo also reassessed the drive to hydrate to have twice the anxiety-mass of its complement of not hydrating. Flo intends to avoid both dying from becoming dehydrated and dying from over-hydration (e.g., hyponatremia). She wants to know the following:

A) What is the maximum amount of water she can have in her body and her highest possible body weight in pounds?

B) What is the minimum amount of water she can have in her body and her lowest possible body weight in pounds?

C) Now that her valuations of water, the drive to hydrate, and the drive to not hydrate have changed her estimation of where she feels homeostasis, how much water weight will Flo need to possess in order to feel at equilibrium between the drive to raise her water composition and the drive to not raise her water composition?

D) What will the water composition of Flo's body be when she feels at equilibrium?

It is given:

1) Flo weighs 100 pounds and 70% of her body weight is water. Flo, therefore, initially has 70 pounds of water weight.

2) If Flo's water composition falls below 60% then she will die of dehydration.

3) If Flo's water composition exceeds 80% then she will die of over hydration, hyponatremia.

4) While not hydrating, Flo's water composition decreases at the same rate that it would increase if she were hydrating.

5) Because it is fully sufficient, unique, and has the utmost priority, Flo has valued the entity of water at + 2 emotional units toward the fulfillment of the purpose of raising her body's water composition. Water's value for the purpose of raising her water composition and hydrating will be held constant at + 2 emotional units. Due to water's zero Sufficiency for the complementary purpose of not hydrating, the value of water for the complementary purpose will become + 1 emotional units. Water's value for the purpose of hydrating will be + 2 emotional units.

6) Flo's valuation of the drive to hydrate (id-mass and superego-mass) is concurring with her valuation of water for the purpose to hydrate (entity-mass and utility-mass). Also, Flo's valuation of the drive to not hydrate (id-mass and superego-mass) is concurring with her valuation of water for the purpose to not hydrate (entity-mass and utility-mass). The mass of the former is felt as twice that of the latter, which means that Flo's valuation of the entity of water, in this example, is equivalent to the id-mass representing each drive. Henceforth, neither the value of the id-mass nor superego-mass, on either side of the fulcrum, will need to be shown in this example, and will not be included for the sake of simplicity.

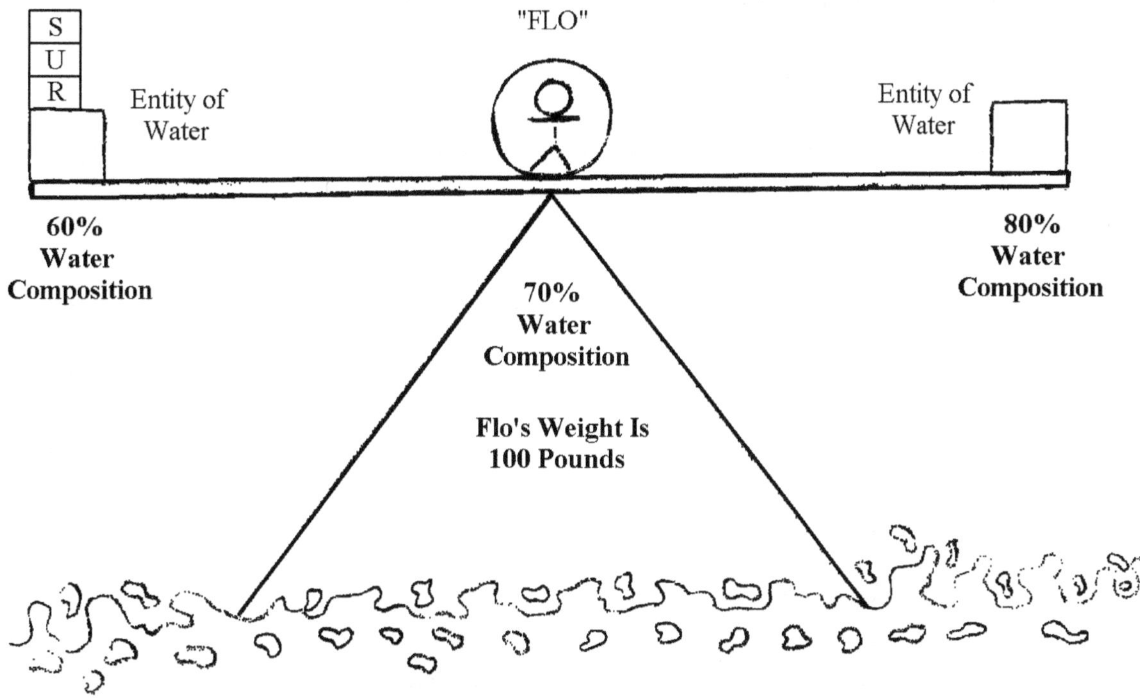

Figure 7.39 Flo's Upper Limit and Lower Limit

Solution

The first task will be to find Flo's upper and lower thresholds for water retention. For the lower threshold, the x-value, measured as pounds of water, will be used for the following equation:

(70 "Flo's original water weight" + x) ÷ (100 "Flo's original weight" + x) = 60 / 100

x = - 25 pounds, meaning that if Flo loses more than 25 pounds of water weight, then her water composition will fall below 60% and she will perish. Hence, her water weight would have to fall below 45 pounds, which means her total weight would have to fall below 75 pounds.

For her upper threshold,

(70 "her original water weight" + x) ÷ (100 "her original weight" + x) = 80 / 100

x = 50 pounds, meaning that if Flo gains more than 50 pounds of water weight, then her water composition will exceed 80% and she will perish. Hence, her water weight would have to exceed 120 pounds and her total weight would have to exceed 150 pounds.

If Flo's current water weight is at 70 pounds, then this means that she can lose no more than 25 pounds of water and gain no more than 50 pounds of water. Her lower threshold for total weight would be 100 - 25, or 75 pounds at 60% water weight while her upper threshold for weight would be 100 + 50, or 150 pounds and 80% water weight. There is a total range of 75 pounds of water that Flo can lose or gain. This means that if Flo's water weight is at its absolute lowest of 45 pounds, and her total weight is at 75 pounds, then she would be comprised of 60% water. If her water weight is at its absolute highest of 120 pounds, and her total weight is at 150 pounds, then she would be comprised of 80% water. A look at the three values also reveals that more and more water is required to change her water composition by percentage as her total weight increases. At a water composition of,

60%, Flo's water weight is 45 pounds

70%, Flo's water weight is 70 pounds, a difference of 25 pounds

80%, Flo's water weight is 120 pounds, a difference of 50 pounds

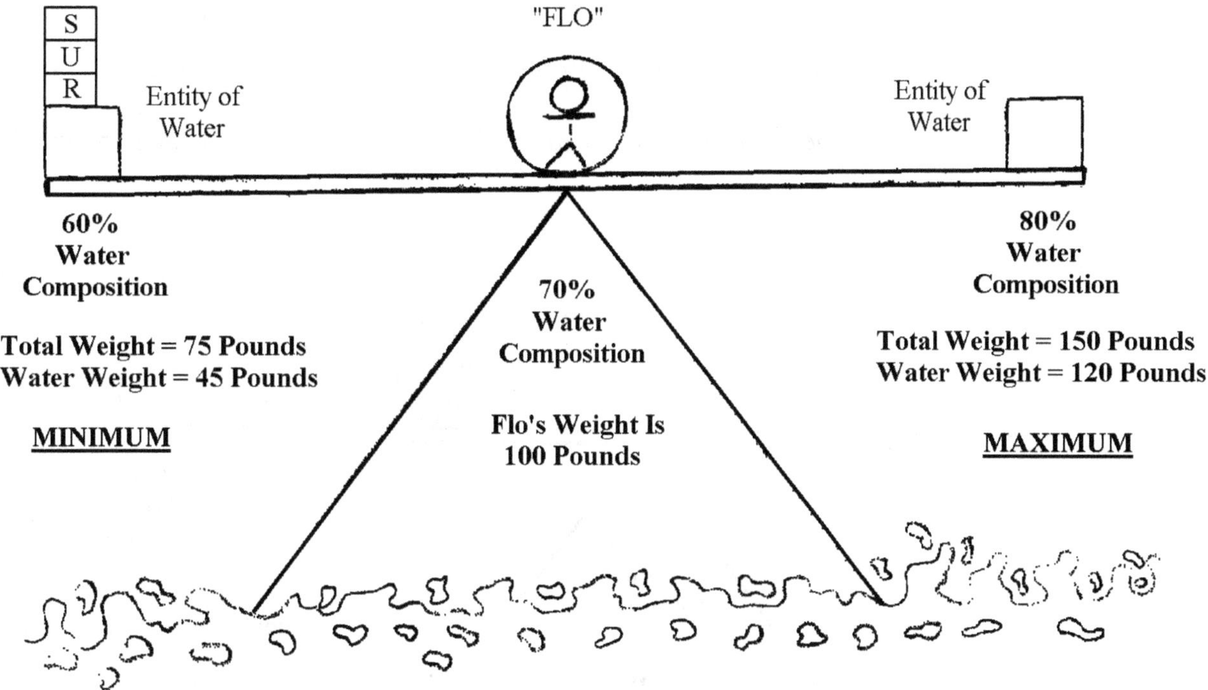

Figure 7.40 Flo's roll cage will have a lower bound of 60% water composition, an upper bound of 80% water composition, and a fulcrum at 70% water composition. It can be seen that there is a disparity in the pounds of water required to raise Flo's body composition on each side: 25 pounds between 60% and 70%, and 50 pounds between 70% and 80%.

However, in this case the entity of water is being valued for each percentage increase in Flo's water composition, and not merely the weight of water. Moreover, the entity-mass for raising Flo's water composition, due to the effect of the entity of water, is felt as twice that of the entity-mass to not raise Flo's water composition. Hence, the ego will spend more time keeping the roll cage on one side than the other to find equilibrium based upon the distribution of anxiety. The ratio can be thought of in terms of the portion of the total percentage points available on each side times the anxiety-mass. Hence, ten percentage points of water composition times one emotional unit is being applied to the drive to not raise Flo's water composition while ten percentage points of water composition times two emotional units is being applied to the drive to raise Flo's water composition.

Although Flo's water composition is initially at 70%, halfway between the lower threshold of 60% and the upper threshold of 80%, the drive to raise Flo's water composition would not feel balanced by its complement if she were at a composition of 70%. The center of balance on the disc

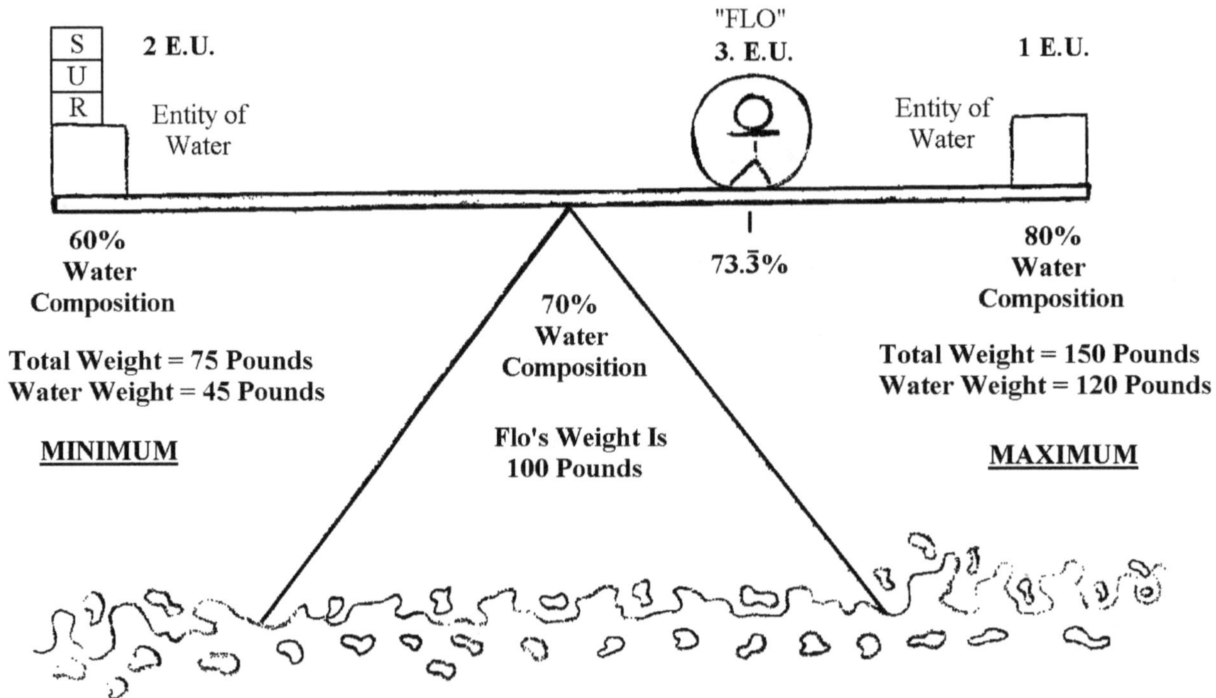

Figure 7.41 Imbalance : if only ten percentage points worth of downward mass are felt on the side closest to the drive to not raise Flo's water composition (near the 80% upper bound), then twenty percentage points worth of downward mass would be felt on the side closest to the drive to raise Flo's water composition (near the 60% lower bound) and tip the scale. Even if Flo keeps the roll cage at 70% water composition, which is true equilibrium, she will not feel at equilibrium because of how her anxiety-mass is distributed. The apparatus will not feel balanced even though she is at the true middle between the extremes. The roll cage takes on additional anxiety-mass in order to help it balance the disc. One might ask at this point, "Where does this anxiety come from? What if there is not enough?" If there is not enough anxiety available in the stockpile, then the ego may have to strip free-floating anxiety-mass (e.g., superego-mass or the utility-mass Sufficiency and Uniqueness) that is associated with other drives around the disc.

is displaced off the top of the fulcrum point due to the entity of water possessing different masses for each purpose. The entity of water, more valuable toward the purpose of raising Flo's water composition than to the purpose of not raising it, amplifies the drive concerning the former by a factor of two emotional units against one emotional unit for the complementary drive:

The value judgment of water for the purpose of raising Flo's water composition

divided by

The value judgment of water for the purpose of not raising Flo's water composition

= 2 / 1, or = 2 : 1.

In order to balance this, the ego must acquire additional anxiety-mass. Additional anxiety, in the roll cage model of drives, is represented by a stockpile of weights underneath the disc in the cone. The first task the ego does, with respect to an entity's value toward a purpose and its complementary purpose, is to acquire an amount of anxiety-mass that is proportional to the total mass being invested in the entity for both the drive to hydrate and the drive to not hydrate. In the current case, the total anxiety being invested into the value of water toward both the purpose and its complementary purpose is + 3 units of mass (i.e., emotional units). The value of water for the purpose of raising Flo's water composition is + 2 units of mass and the value of water for the purpose of not raising Flo's water composition is + 1 unit of mass.

Subsequently, to feel at equilibrium with water amplifying one drive over another at a ratio of two to one, of the 20% percentage points for water composition available within the range between 60% and 80%, two thirds, or 13.3 (thirteen and one-third) would have to go toward raising Flo's water composition and the other 6.6 would go toward not raising her water composition. Hence, Flo would feel balanced with a water composition of 73.3% (seventy-three and one-third percent) slightly off the fulcrum. This would balance the mass on both sides of the disc evenly if the roll cage, with a mass of + 3 units, were to move three and one-third percentage points away from the more massive drive corresponding to the purpose of raising Flo's water composition, thus elevating it from 70% to 73.3%. One third of the roll cage's three units of anxiety-mass would be added to the right side.

This is one-third of the way toward the upper bound of 80% water composition, but because of the roll cage's mass, it can balance out the drive on the other side that corresponds to the value of water for the purpose of raising Flo's water composition.

Originally, before the ego and roll cage move,

The mass of the drive to raise water composition equals $(+ 2) \times 10 = 20$ units of mass.

The mass of the drive to not raise water composition equals $(+ 1) \times 10 = 10$ units of mass.

Afterwards, when the roll cage moves one-third of the way toward the upper bound:

The mass of the drive to raise water composition equals $(+ 2) \times 10 = 20$ units of mass

The mass of the drive to not raise water composition equals

$(+ 1) \times 10 + 3 \times (3.3 / 10) \times 10 = 20$ units of mass

The value judgment for the entity of water, in essence, biases Flo toward staying hydrated and raising her body's water composition. To reach a water composition of 73.3%, Flo would need to add approximately 12.5 pounds of water:

$(70 \text{ "her original water weight"} + x) \div (100 \text{ "her original weight"} + x) = 73.3 / 100$

If taken to three significant figures, x would $=$ 999 / 80 or approximately 12.5 pounds of water. Hence, taking into account the valuation of the entity of water and its effect on the drive to drink and the drive to not drink, in order to balance the drive of hydrating against the drive of not hydrating Flo would have to consume approximately twelve and a half pounds of water to raise her water composition to 73.3%.

If one looks closely, it can be noted that the halfway point of consumed water between the two extremes would be 75 / 2, or 37.5 pounds of water from either end. This would be 82.5 pounds for her water weight, and 112.5 pounds for her total weight: 45 + 37.5 = 82.5 pounds for her water weight. The extra 30 pounds gives 112.5 pounds for her total weight. 82.5 pounds divided by 112.5 pounds = 73.3% or 73 and 1 / 3 percent. 120 - 37.5 also equals 82.5 pounds. This seems to be a short cut to getting the answer.

However, it is not always the case that this strategy would work to solve the problem as the amount of water required to change Flo's water weight percentage increases with each subsequent pound. Moreover, the entity of water was originally being valued with respect to its ability to change Flo's water composition by percentage and this alternate strategy does not adhere to this. If, hypothetically, the value of water for the drive to raise Flo's water composition were somehow made greater and amplified by a factor of 3:1 against its value for the complement, then of the 20% points available between 60% and 80%, three fourths, or 15% would need to go toward the purpose of raising Flo's water composition. Equilibrium would then be felt at a water composition of 75%. This would be achieved by consuming 20 pounds of water. Flo's water weight would become 70 + 20, and her total body weight 100 + 20, which equals 90 / 120, reducing to 75%. The ratio between Flo's water consumed out of the total amount she could consume, if her weight began at 45 pounds, would be 45 / 75, which equals .6, 60%, or 45 of the total 75 pounds of water her water weight might fluctuate.

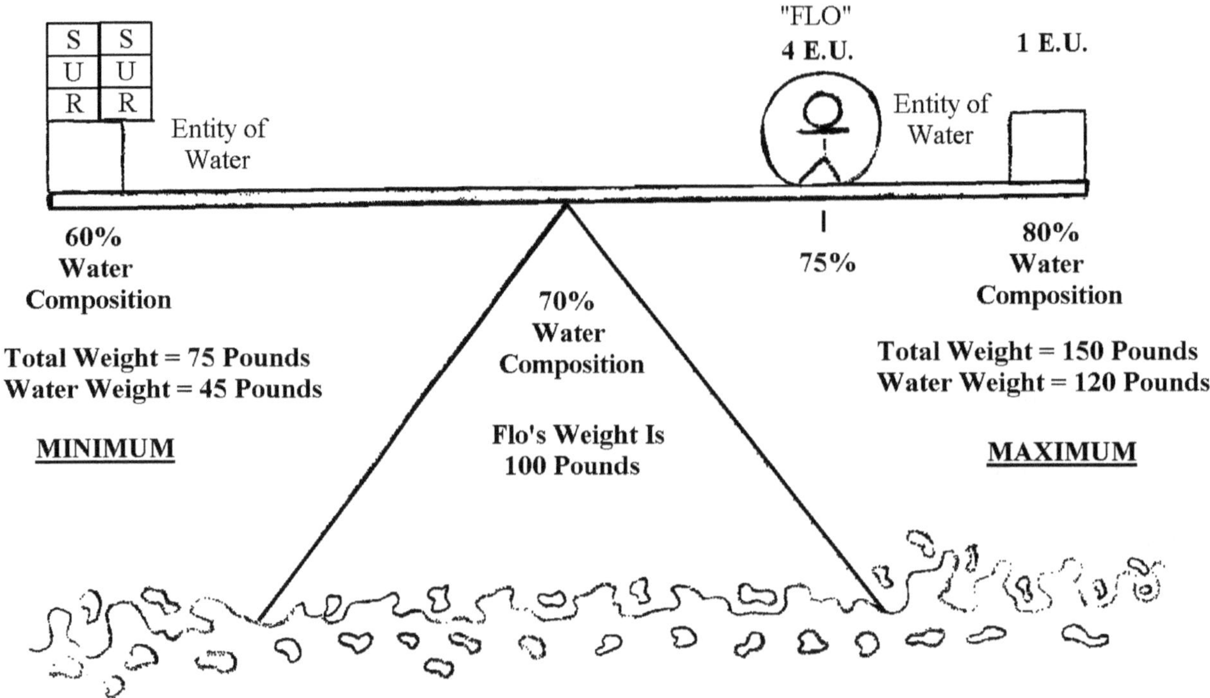

Figure 7.42 At a 3:1 idealized ratio of success between the drive to hydrate against the drive to not hydrate, equilibrium would be felt with a water composition of 75%. Because the roll cage is halfway to the edge of the right side, half of its total mass of four is added to the mass already present on the right side.

Moreover, using the original approach, if the ratio of values were 3:1, then the ego would require additional negative anxiety-mass proportional to the total value for the entity of water. This total would be four units of anxiety-mass. At a 3:1 ratio, the valuation of water for the purpose of raising Flo's water composition would be three times the valuation of water for the purpose of not raising Flo's water composition and a 3:1 distribution of the percentage points in the twenty point spread would be required if the roll cage took on a mass of four emotional units. Before the ego and roll cage take action:

The mass of the drive to raise Flo's water composition equals (+ 3) × 10 = 30 units of mass.

The mass of the drive to not raise Flo's water composition equals (+ 1) × 10 = 10 units of mass.

Afterwards, with the roll cage having moved from the fulcrum to halfway toward the upper bound.

The mass of the drive to raise Flo's water composition equals $(+ 3) \times 10 = 30$ units of mass.

The mass of the drive to not raise Flo's water composition equals

$(+ 1) \times 10 + 4 \times (5 / 10) \times 10 = 30$ units of mass, and the water composition feels balanced at a 75% water composition in Flo’s body.

It is soon discovered that taking three-fourths of the total 75 pounds of water available between 45 and 120 pounds of total water weight yields 56.25 pounds of water, which would translate to Flo having a total water weight of 45 + 56.25 pounds or 101.25 pounds for equilibrium. Flo's total weight would be 131.25, and 101.25 / 131.25 reduces to 81 / 105, or 77.1% water composition. This is a completely different value from the 75% water composition found before.

However, an even closer inspection reveals the following. If the emotional value for the entity of water were being measured from water's usefulness at raising Flo's water weight instead of being valued for its usefulness to raise Flo's water composition, then it would be appropriate to use the aforementioned process. It will be assumed that everything else is held the same as before (3:1 anxiety-mass ratio before the roll cage moves) except that water is being valued for the purpose of raising Flo's water weight against not raising Flo's water weight. Instead of being valued for its usefulness to raise or not raise Flo's water composition by percentage, pounds of water would be the item of concern. The fulcrum would be at 82.5 pounds, the upper bound would be 120 pounds of water, and the lower bound would be 45 pounds of water. Flo would be attempting to balance the drive to raise her water weight against the drive to not raise her water weight, two completely

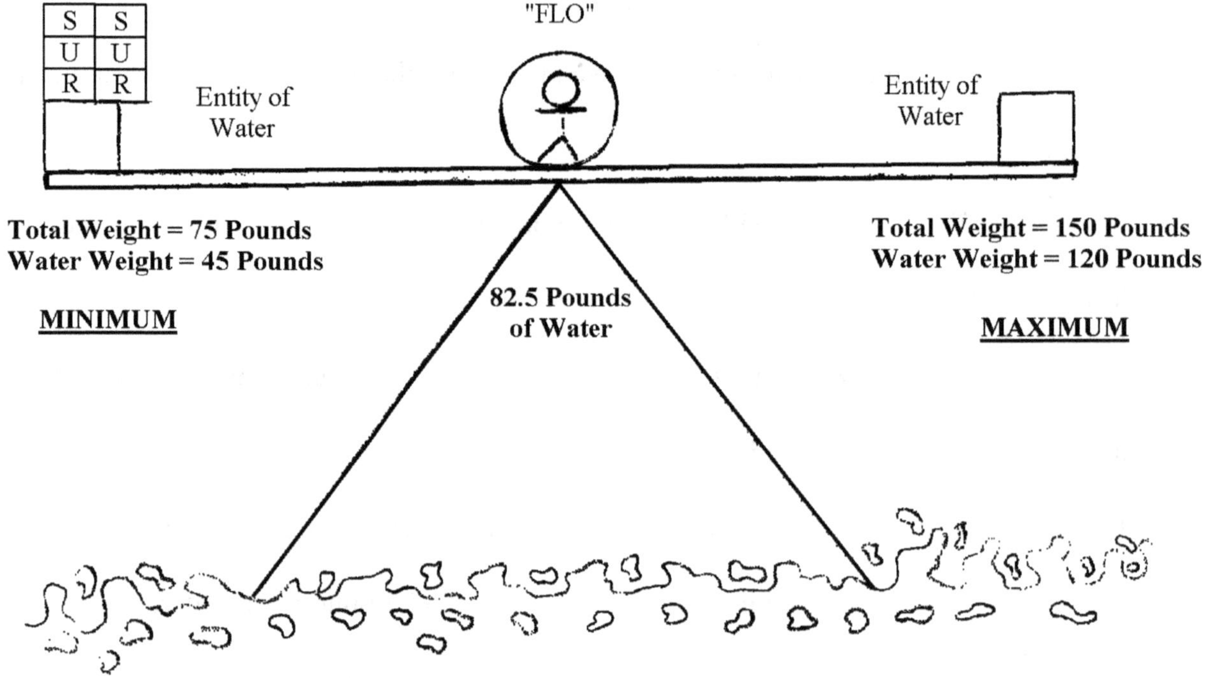

Figure 7.43 If water is being valued for the purpose of raising Flo's water weight or not raising it (and she consumes it at the same rate it is lost), then the fulcrum would be at 82.5 pounds of water weight, 45 pounds of water weight would be the lower bound, and 120 pounds of water weight would be the upper bound. This maintains the 1:1:1:1 ratio established in chapter six, as raising or not raising Flo's water composition are a separate purpose and complementary purpose.

different drives than the first two considered in the example.

There would be a total of 75 pounds of water within the range from 45 to 120, with 37.5 pounds of water lying on each side of the fulcrum. Three units of mass are applied to the drive of raising Flo's water weight while only one unit of mass is applied to the drive of not raising Flo's water weight. Hence, the initial balance between the two drives is as follows.

The drive to raise Flo's water weight equals

3 × 37.5 available pounds of water available = 112.5 units of mass.

The drive to not raise Flo's water weight equals

1 × 37.5 available pounds of water available = 37.5 units of mass.

The ego and roll cage, at a mass of + 4 emotional units, can balance this out however, by moving halfway across the disc and away from the more massive drive to raise Flo's water weight. Hence, after the ego and roll cage move to balance . . .

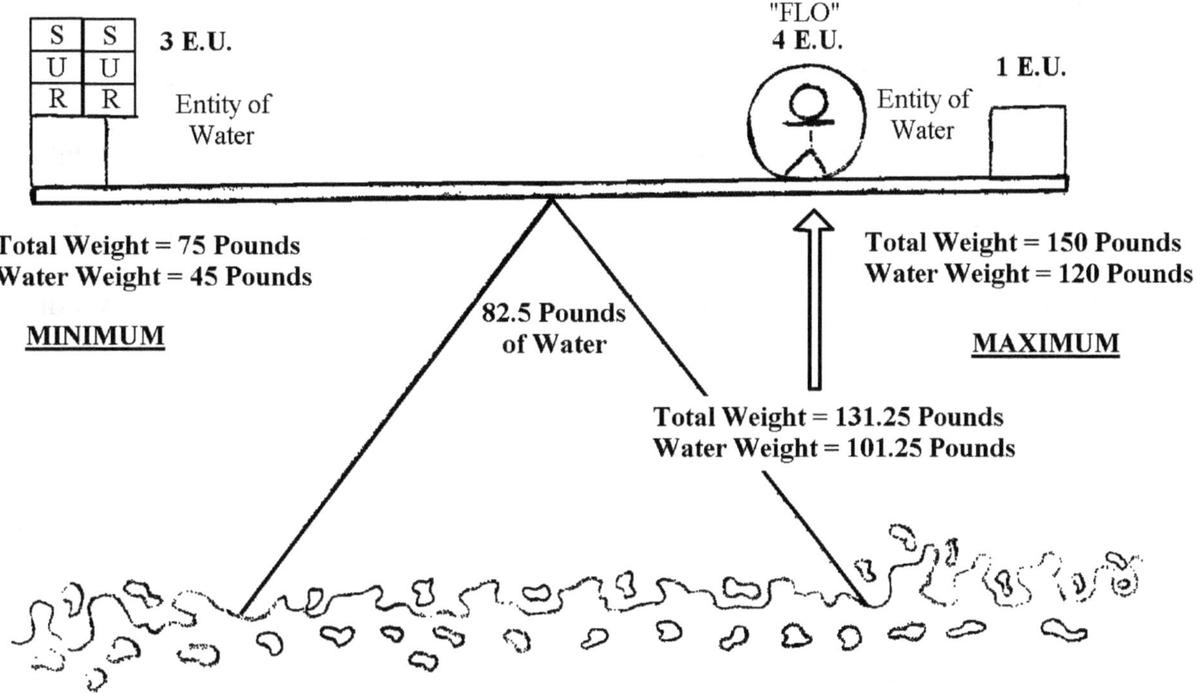

Figure 7.44 Equilibrium is depicted above at a 3:1 idealized success ratio where the emotional unit corresponds to the ability of water to raise Flo's water weight, and not her water composition by percentage point.

The drive to raise Flo's water weight equals

3 × 37.5 available pounds of water available = 112.5 units of mass.

The drive to not raise Flo's water weight:

1 × 37.5 + 4 × [37.5 × (18.75 / 37.5)] = 112.5 units of mass.

The number four value above represents the roll cage's mass, and 18.75 / 37.5, which reduces to one-half, represents the distance traveled by the roll cage as a fraction of the radius of the disc.

37.5 represents the pounds of water that the roll cage's mass in emotional units will be factored by. Hence, 101.25 pounds of water weight, the balance point for equilibrium here, minus 82.5 pounds of water weight at the fulcrum, yields 18.75.

Therefore, if water were being valued in terms of its usefulness to the drive of raising or not raising Flo's water weight, then the two drives, of raising Flo's water weight and of not raising Flo's water weight, would feel balanced if Flo were at a water weight of 101.25 pounds, a total weight of131.25 pounds, and a water composition of 77.1%. This is in accord with maintaining the one to one to one to one ratio, as valuing water for its ability to raise Flo's water weight is a different concept than water's value for raising Flo's water composition.

Measuring Anxiety and Negative Anxiety with the Equation to Start Identifying Emotions

Thus far, only what needs to be measured has been determined, namely, an entity's appraised value judgment with respect to two purposes and not merely one. What remains to be discovered is the relationship between anxiety and specific emotions. Although existing definitions used to specify different emotions are somewhat arbitrary, they will be a helpful starting point nonetheless.

If anxiety is considered to be an indicator for pain and negative anxiety one for pleasure, all that is left to consider is how they can be modulated. Anxiety, for instance, can be altered in one of three ways: it can go up; it can go down; it can stay the same. A change in an entity's appraised value judgment, then, would be indicative of a change in the anxiety resources being used by the individual to mark the entity's value. Likewise, the same would hold true for negative anxiety.

Working backward, it could also be said that a change in anxiety afforded to an entity would be indicative of a change in the appraised valuation of the entity. It would follow that entities perceived as valuable would draw forth a greater amount of anxiety resources than entities valued less, and so evoke a greater amount of pain. The entity itself is not necessarily bad in this case, and may in fact be quite vital to the person's interests or well-being for other purposes. However, the pain associated with the entity, as a result of the high value placed on it for a specific purpose, is often undesirable. Aversion to pain, according to Arnold's logic, would be accompanied by a pattern of "physiological changes organized toward withdrawal" from harmful objects, while an attraction to pleasure would be accompanied by a pattern of "physiological changes organized toward approaching" beneficial objects.[3] She also holds that the "patterns would differ for different emotions."[4] Given, in Arnold's methodology, different appraisals lead to "attraction or aversion," which in turn lead to "specific patterns of physiological changes organized toward approach or withdrawal," which in turn lead to "different patterns for different emotions," it is reasonable to predict that the same would hold true for the equation.[5] For the equation, this means that an entity, with respect to a solitary main purpose, could be appraised as valuable, as having a high amount of anxiety invested into it, and as eliciting pain if its acquisition will lead toward homeostasis, and it will correspond to a specific emotion. Alternatively, that same entity could be appraised as having a negative value, as having a high amount of negative anxiety invested into it, and as eliciting pleasure if its acquisition will lead away from homeostasis, and it will correspond to a specific emotion.

For example, a highly valued entity that induced pain would likely lead to aversive behavior or withdrawal to minimize the pain. If the entity's valuation becomes so high that the resulting anxiety and pain are too much to bear, the individual might be compelled to surrender the possibility of ever acquiring the entity in order to free up anxiety reserves for other tasks required for day to day functioning. Such an occurrence would most likely be indicative of sadness, intense grief, or fear. Sadness might generally be expected to contain some sense of dissatisfaction or misery. Grief is "an intense emotional state associated with the loss of someone (or something) with whom (or which) one has had a deep emotional bond."[6] Fear is "an emotional state in the presence or anticipation of a dangerous or noxious stimulus," that is distinguished from anxiety in that fear involves ". . . specific objects or events while anxiety is regarded as a more general emotional state."[7]

In the same situation, if the entity's value became abnormally high and subsequently evoked a great deal of anxiety, an individual might also strike out at whatever caused the entity's value to elevate. This might be a threat of harm to the entity that puts it in danger of being destroyed. In this case, rather than submit and surrender the entity, an individual may strike against the offender, namely the threat of harm against the entity. This would most likely be indicative of anger or frustration. Anger can be very generally described as ". . . a fairly strong emotional reaction which accompanies a variety of situations such as being physically restrained, being interfered with, having one's possessions removed, observing or hearing of actions or events that one regards as morally repugnant, being attacked or threatened, etc."[8]

Based upon these starting points for definitions, and observing the expected changes in anxiety for the cases of sadness-grief-fear and anger-frustration has yielded that in the case of the former, the appraised value judgment of the entities became elevated but the individual was unable to do anything about it. This may have been because the entity could not be acquired or it could not be protected from some harm. An entity, whose valuation elicits this response, might have a graph resembling figure 7.45. In the Roll Cage Theory of Drives, this would be the equivalent of the roll cage falling backwards.

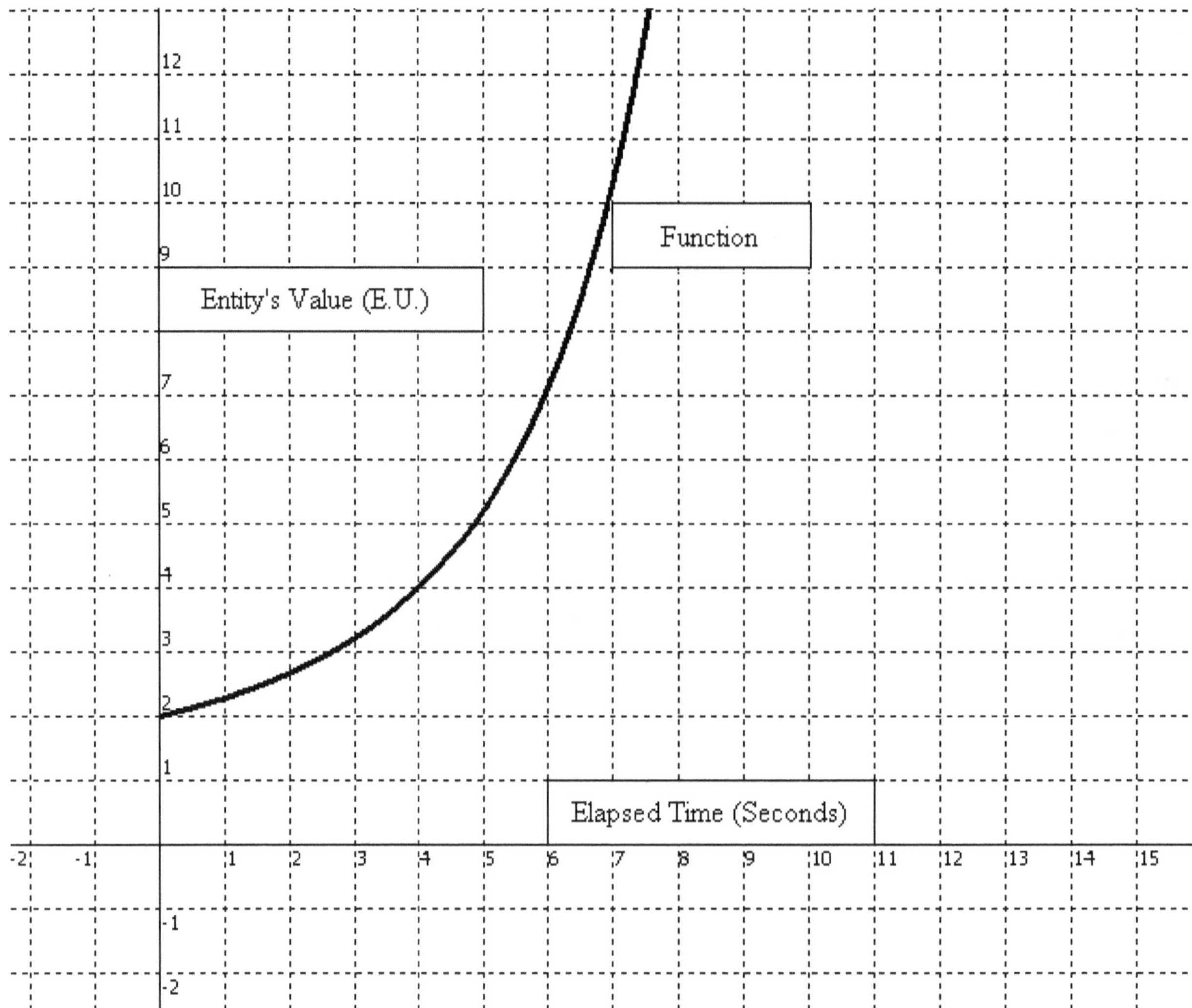

Figure 7.45 The above is a proposal for an individual's evaluation of an entity increasing and staying above the x-axis as it is realized that nothing can be done to save or acquire the entity.

In the case of anger-frustration, the entity's value might become abnormally elevated and then the individual might strike against a harm threatening the entity. If it is assumed that the individual was successful and eliminated the threat of harm to the entity, then the anxiety resources being invested in the entity would be expected to return to a lower level because the threat of harm was prevented. An entity whose valuation indicates this response may have a graph resembling figure 7.46. In the Roll Cage Theory of Drives, this would be the equivalent of the roll cage pushing forward to acquire an entity and restore equilibrium.

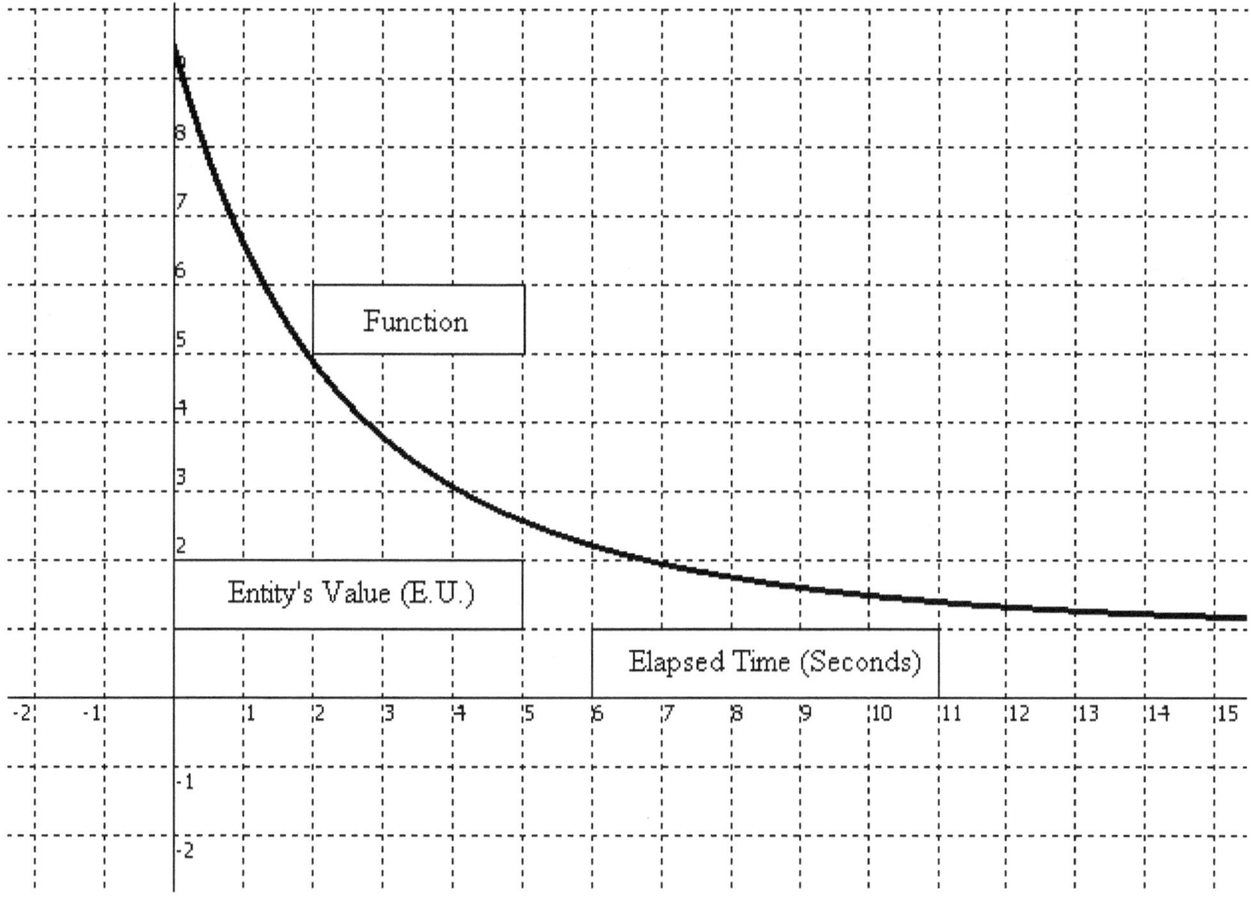

Figure 7.46 The above is a proposal for an individual's evaluation of an entity decreasing and staying above the x-axis, lowering after it is protected from harm or a threat of danger.

Lastly, if an individual's anxiety resources being invested in an entity stayed the same over the course of time, this would most likely be indicative of being content. To be content is to be "satisfied with things the way they are."[9] This is modeled in figure 7.47.

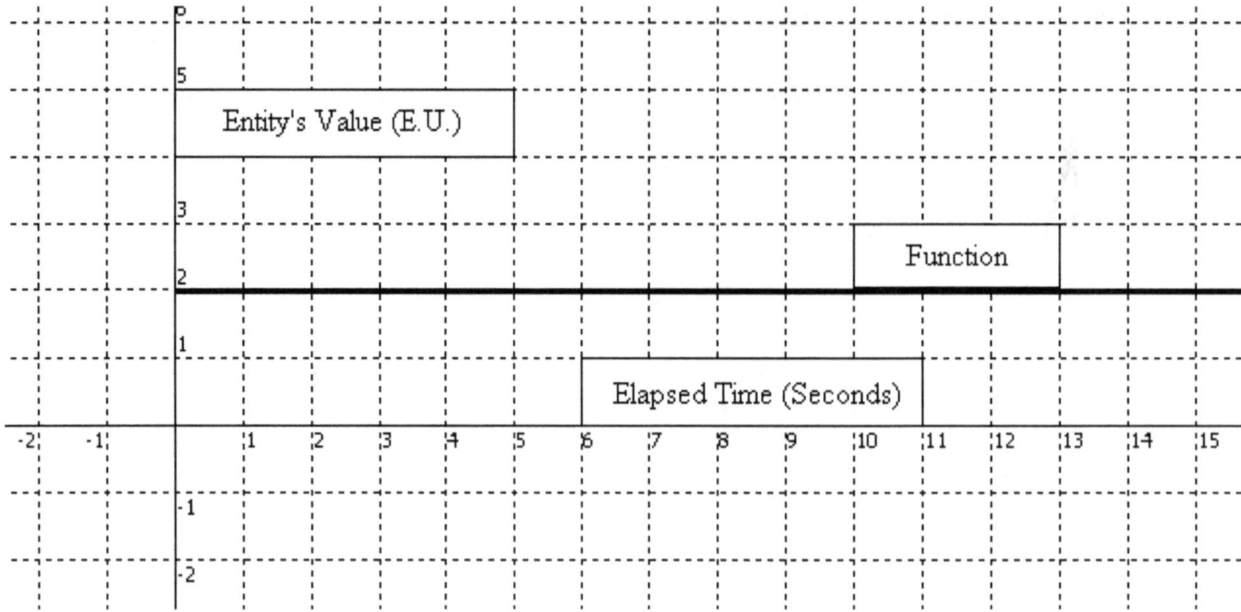

Figure 7.47 The above is a proposal for an individual's evaluation of an entity staying constant above the x-axis

While the different characteristics of the emotions of grief, sadness, fear, frustration, and anger seem to be accessible through the equation, it is still too early to distinguish a specific emotion using the equation (e.g., fear vs. sadness). Other factors will have to be taken into consideration in order to isolate them. Three general patterns can be predicted:

1) Emotions related to an increase in anxiety: fear, sadness, grief

2) Emotions related to a decrease in anxiety: frustration, anger

3) Emotions where anxiety stays the same: content

Emotions Related to Negative Anxiety

On the other side of the coin lies pleasure. Borrowing Arnold's methodology, an entity in the equation whose negative valuation is well below zero should induce pleasure for a single purpose; this should be expected to lead to attractive behavior and approach toward the pleasure. If the entity's value becomes so negatively valenced that it is irresistible, this would most likely be indicative of happiness, euphoria, or courage with respect to a single purpose from a complementary pair. A very general description of happiness would likely include some sense of well being or satisfaction. Euphoria is loosely "a sense of extreme elation generally accompanied by optimism and a deep sense of well-being and heightened activity."[10] Euphoria may also interfere with other purposes and lead to impulse-control disorders, an ailment where an "individual feels a highly increased sense of tension prior to the act and a pleasurable, gratifying feeling afterwards" in which "guilt may or may not be experienced afterwards."[11] Courage is "mental or moral strength to venture, persevere, and withstand danger, fear, or difficulty."[1] With the last emotion of courage, an argument could be made that it should be grouped with frustration and anger because courage sounds as if it concerns anxiety as well. Although the logic behind the decision to categorize it as an emotion concerning negative anxiety is not immediately apparent, the placement of courage here will begin to make sense when harm is introduced into the functions in chapter eight. Courage, happiness, or euphoria might be expected to have a graph resembling figure 7.48. In the Roll Cage Theory of Drives, and for a specific purpose, this would be the equivalent of the roll cage pursuing and successfully acquiring an entity that has already been obtained beyond what is necessary to achieve homeostasis.

---

[1] By permission. From Merriam-Webster's Collegiate® Dictionary, 11th Edition ©2014 by Merriam-Webster, Inc. (www.Merriam-Webster.com).

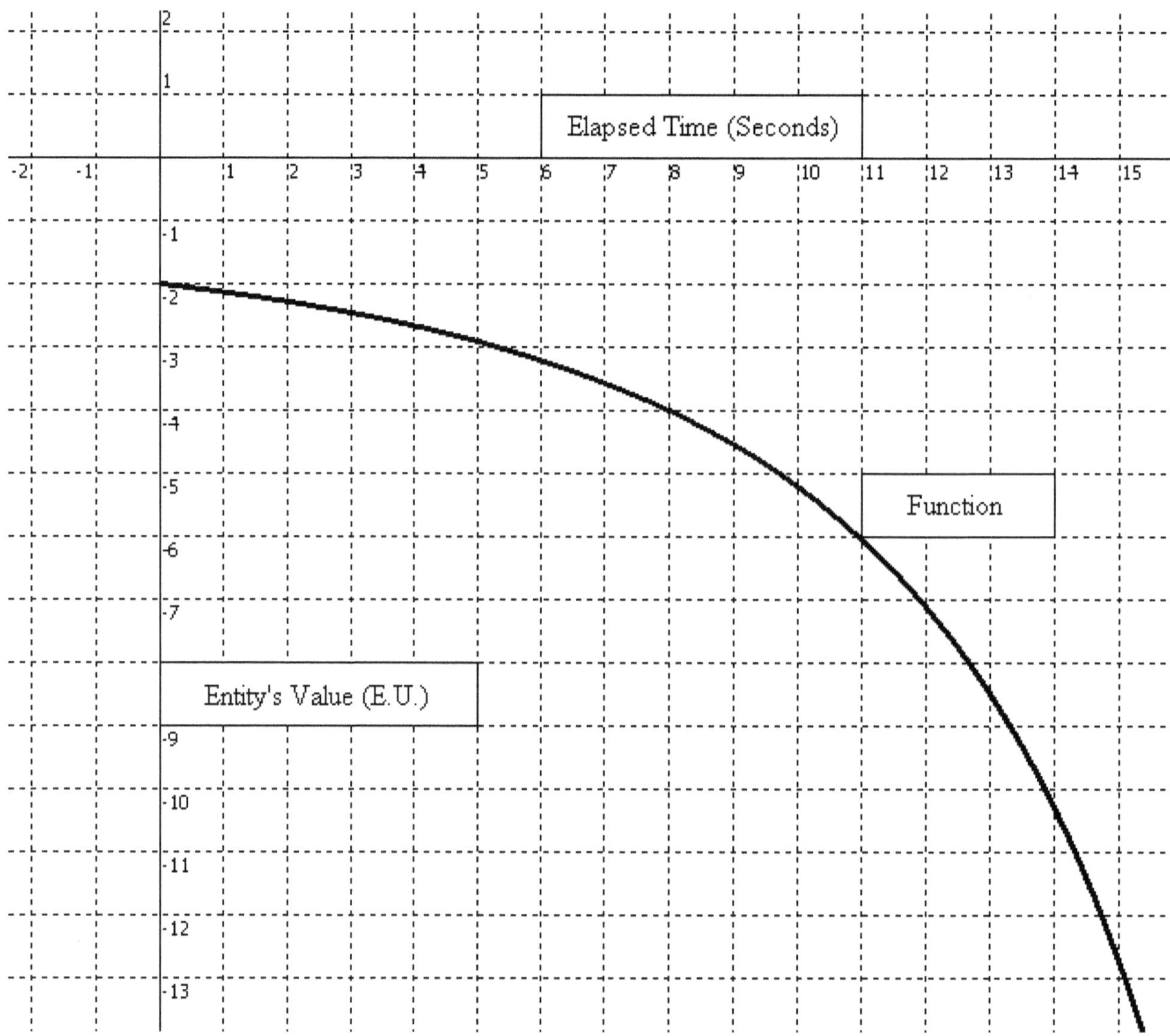

Figure 7.48 The above is a proposal for an individual's evaluation of an entity decreasing and staying below the x-axis as it is realized the entity can be acquired well beyond what is required to maintain homeostasis.

Similar to the consideration for anger and sadness, if an entity is appraised by an individual as having a negative valence well below zero, and through mismanagement of resources or a lack or effort, the entity's valuation begins to increase toward zero thereafter, this might be indicative of guilt or some form of self-effacing disappointment such as despair, self-disgust, or hopelessness. Guilt is defined as "an emotional state produced by the knowledge that one has violated moral standards," and ". . . is, in a sense, a self-administered punishment."[12] To despair is "to lose all hope or confidence."[2] Lastly, disgust is "literally a bad taste," and "by extension, a negative emotional state that follows exposure to a situation that is unpleasant but not an immediate physical threat."[13] Guilt and some other emotions involving self-conscious thoughts might be expected to have a graph resembling figure 7.49 where the individual has negative anxiety invested in an entity, but, through some form of mismanagement or wrongdoing, squanders them. In the Roll Cage Theory of Drives, an emotion like guilt, for a specific purpose, might be represented by the roll cage initially having acquired an entity beyond what was required to balance it against its complement, and then permitting this negative anxiety surplus to slip away. This might result in homeostasis being restored or worse, becoming unbalanced in the other direction, to the detriment of the single purpose being considered. The possibility that it may be viewed as a lost opportunity is a real one.

[2] By permission. From Merriam-Webster's Collegiate® Dictionary, 11th Edition ©2014 by Merriam-Webster, Inc. (www.Merriam-Webster.com).

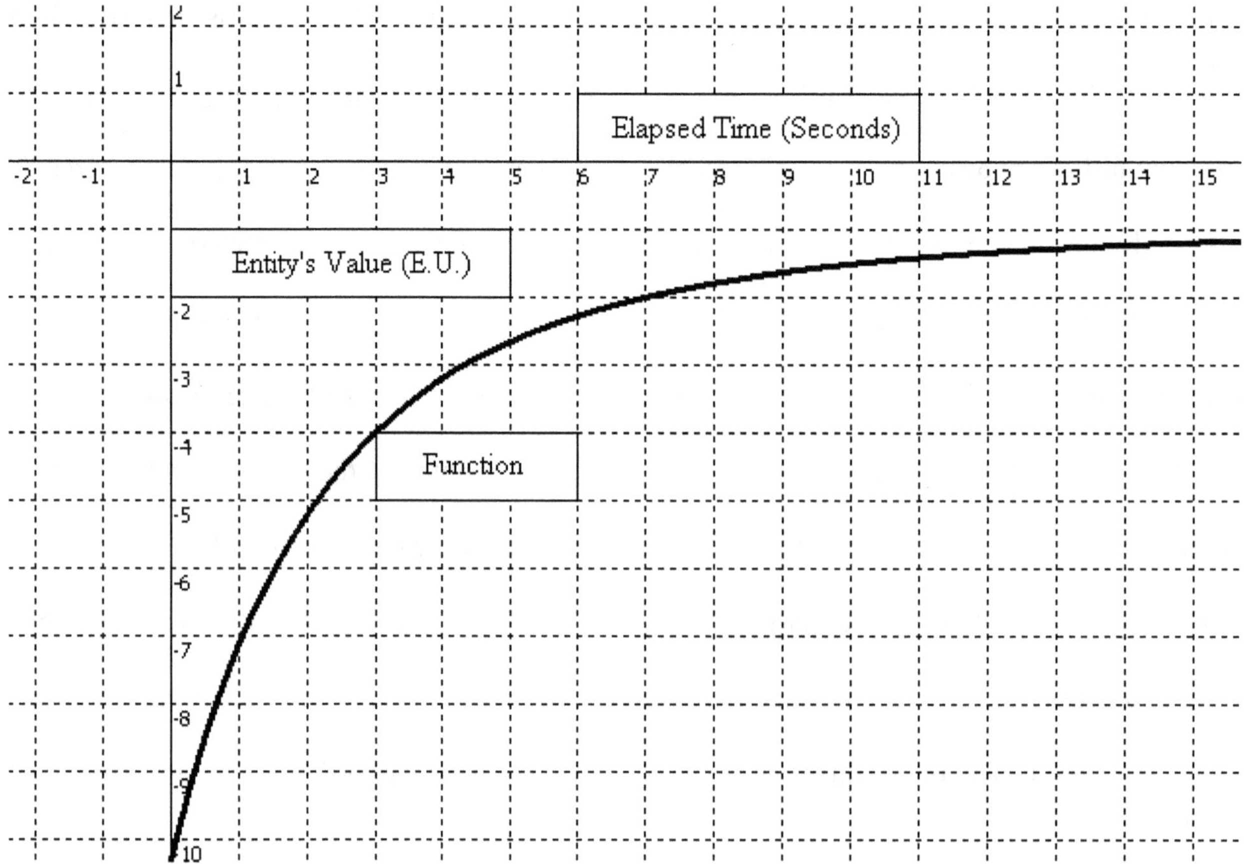

Figure 7.49 The above is a proposal for an individual's evaluation of an entity increasing and staying below the x-axis, ascending once it is realized that the entity is in jeopardy or can not be acquired. The entity, however, was once in ample supply or excess.

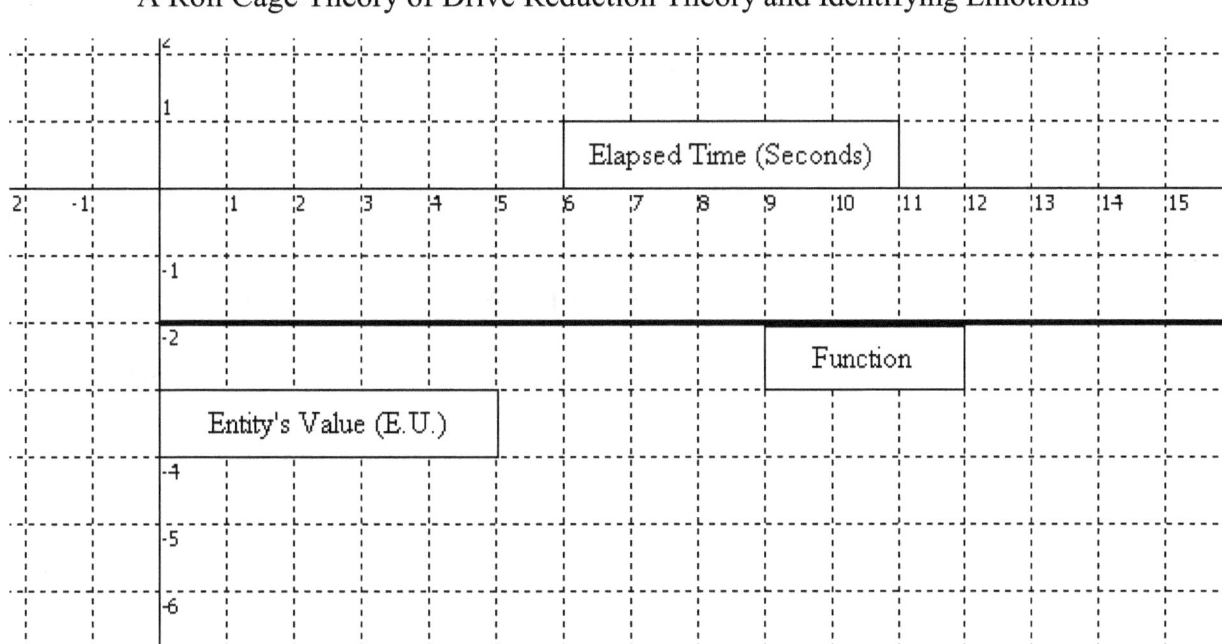

Figure 7.50 The above is a proposal for an individual's evaluation of an entity staying constant below the x-axis

Similarly, content, like in the case where the Appraisal was + 1, might be expected to have a slope of zero.

Theoretical Considerations

A comparison of the different graphs, and the proposals for how an entity's valuation might change when certain emotions are at play, reveals the following predictions. The magnitude of each emotion felt would be assessed by finding the derivative of the entity's valuation at a certain point in time or number of points in time if the slope of the function is changing. Derivatives in calculus will be useful for finding the rate of change in a function, f(x), at a specific x-value.

1) Fear, sadness, or grief are predicted to arise when an entity's appraised value judgment increases and is above the x-axis. The individual would be unable to

acquire or protect the entity from harm. The slope of the equation would be positive.

2) The appraised value judgment of an entity decreasing and above the x-axis:

Caption: Frustration or anger are predicted to arise when an entity's appraised value judgment decreases after an individual protects it by striking at a harm that threatened the entity. The y-value would still be above the x-axis. The slope would be negative. In the case where an individual lashes out at a different, neutral entity, an explanation will have to be found to account for this (i.e., misdirected anger).

3) Happiness, euphoria, or courage is predicted to arise when an entity's appraised y-value decreases and is below the x-axis. The individual would be able to acquire the entity. The slope would be negative.

4) Guilt and other forms of disappointment are predicted to arise when an entity's appraised value increases after an individual fails to acquire or safeguard it but remains negative. The y-value would initially be below the x-axis. The slope would be positive.

Observing the graphs of these predictions also reveals two important parallels. The graphs predicted for fear-sadness-grief and the graph for guilt-self disappointment both show a net increase in the overall anxiety the individual possesses with respect to an entity's value. Because the subtraction of a negative value is similar to adding the same amount, removing the negative anxiety that one harbored would have the same result as adding anxiety. In other words, fear-sadness-grief and guilt-self disappointment, based on the predictions above, would be expected to raise an individual's appraised value of an entity, reduce the ability to allocate anxiety elsewhere, and either increase the overall level of pain felt or reduce the level of pleasure experienced.

The predicted graphs for anger-frustration and happiness-euphoria-courage are modeled to show a net decrease in the overall anxiety the individual possesses. Hence, anger-frustration and happiness-euphoria-courage would be expected to lower an individual's appraised value of an entity, enhance the ability of the individual to allocate anxiety elsewhere, and decrease the amount of pain or increase the level of pleasure possible. This too appears to be in accord with logic as anger-frustration and a subsequent lashing out against one's offender that is successful would be represented by a decrease in the level of anxiety; the reduction in pain would be precipitated by an entity's appraised value being lowered due to the removal of a threat of harm. Meanwhile, happiness-euphoria-courage would be represented by an increase in the amount of pleasure felt; the increase in pleasure would also be precipitated by an entity's appraised value being lowered.

Although these are only early predictions and emotions cannot be individually isolated using the equation thus far, enough information is available to arrange a number of them together into preliminary groups. In the next chapter, harm and benefits will be introduced into the equation along with other concepts relating to how an individual adapts to new circumstances. To aid in this process, another, more recent cognitive appraisal theory, Lazarus's Cognitive-motivational-relational theory will be introduced along with two other concepts: “primary appraisals” and “secondary appraisals.”[14] The equation's form will also take on a new, radical appearance.

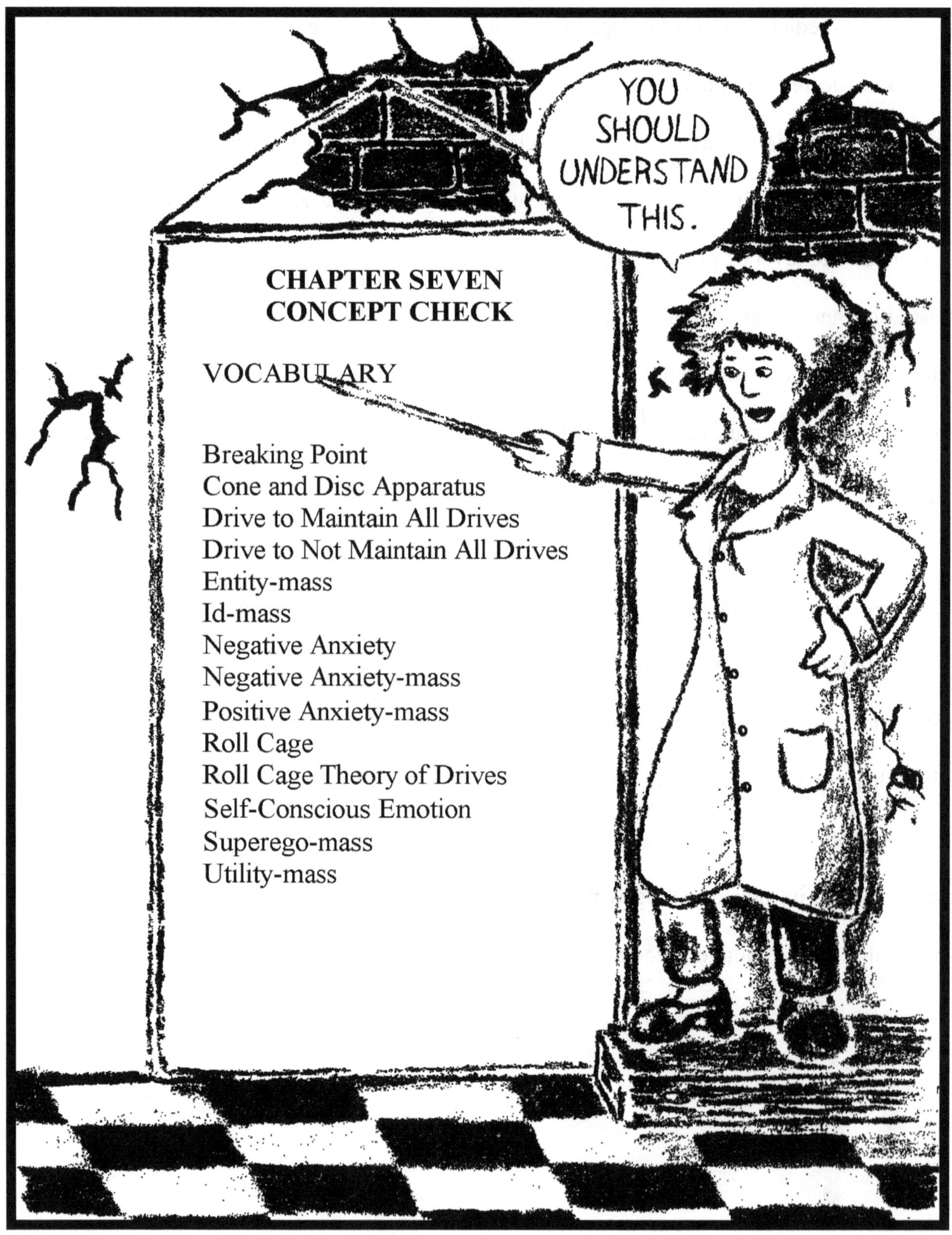
YOU SHOULD UNDERSTAND THIS.
CHAPTER SEVEN
CONCEPT CHECK
VOCABULARY
Breaking Point
Cone and Disc Apparatus
Drive to Maintain All Drives
Drive to Not Maintain All Drives
Entity-mass
Id-mass
Negative Anxiety
Negative Anxiety-mass
Positive Anxiety-mass
Roll Cage
Roll Cage Theory of Drives
Self-Conscious Emotion
Superego-mass
Utility-mass

Notes

1. Wilde, Oscar (2003). *The Picture of Dorian Gray.* New York: Barnes and Nobles Books. Print, p. 201.

2. From *Calculus: An Intuitive and Physical Approach*, second edition, by Kline, Morris. Copyright © 1998. Page 139. Dover Publications (Original work published in 1977 by John Wiley and Sons, Inc., New York). Reprinted with permission of the publisher.

3. From *Emotion and Personality: Vol. 1. Psychological Aspects*, by Arnold, M. B. Copyright © 1960. Page 182. Columbia University Press. Reprinted with permission of the publisher.

4. From *Emotion and Personality: Vol. 1. Psychological Aspects*, by Arnold, M. B. Copyright © 1960. Page 182. Columbia University Press. Reprinted with permission of the publisher.

5. From *Emotion and Personality: Vol. 1. Psychological Aspects*, by Arnold, M. B. Copyright © 1960. Page 182. Columbia University Press. Reprinted with permission of the publisher.

6. Reber, Arthur S., Rhianon Allen, and Emily S. Reber. *Penguin Dictionary of Psychology*. London. Penguin Books, 2009. Print, p. 334.

7. Reber, Arthur S., Rhianon Allen, and Emily S. Reber. *Penguin Dictionary of Psychology*. London. Penguin Books, 2009. Print, p. 292.

8. Reber, Arthur S., Rhianon Allen, and Emily S. Reber. *Penguin Dictionary of Psychology*. London. Penguin Books, 2009. Print, p. 39.

9. Reber, Arthur S., Rhianon Allen, and Emily S. Reber. *Penguin Dictionary of Psychology*. London. Penguin Books, 2009. Print, p. 165.

10. Reber, Arthur S., Rhianon Allen, and Emily S. Reber. *Penguin Dictionary of Psychology*. London. Penguin Books, 2009. Print, p. 273.

11. Reber, Arthur S., Rhianon Allen, and Emily S. Reber. *Penguin Dictionary of Psychology*. London. Penguin Books, 2009. Print, p. 374.

12. Reber, Arthur S., Rhianon Allen, and Emily S. Reber. *Penguin Dictionary of Psychology*. London. Penguin Books, 2009. Print, p. 338.

13. Reber, Arthur S., Rhianon Allen, and Emily S. Reber. *Penguin Dictionary of Psychology*. London. Penguin Books, 2009. Print, p. 223.

14. Cornelius, R. R. *The Science of Emotion: Research and Tradition in the Psychology of Emotion.* Upper Saddle River, NJ. Prentice-Hall, Inc. 1996. Print, p. 125.

"The advantage of the emotions is that they lead us astray, and the advantage of Science is that it is not emotional." - Oscar Wilde[1]

CHAPTER EIGHT

Stimuli, Modifiers of an Entity's Appraisal, and Category I Emotions (Intra-personal)

Front-loading: Avoidance of Pain Equation, Benefits, Bottom-up Processing, Category I Emotions: Intra-Personal, Classical Conditioning, Cognitive-Motivational-Relational Theory, Contingency Theory, Conditioned Stimuli, Cortisol, Distal Stimulus, Excitatory Synapse, Fight-or-Flight Response, Harm, Hormones, Inhibitory Synapse, Negative Reinforcement, Neurotransmitter, Operant Conditioning, Perceived Benefit Intensity, Perceived Benefit Susceptibility, Perceived Response Efficacy, Perceived Self-Efficacy, Perceived Severity of Threat, Perceived Vulnerability to the Threat, Positive Reinforcement, Protection Motivation Theory, Primary Appraisals, Proximal Stimulus, Punishment, Pursuit of Pleasure Equation, Reinforcement, Reward, Secondary Appraisals, Social Constructionism, Top-down Processing, Unconditioned Stimuli

Instigating Stimuli

If it is to be of any practical use, an equation for modeling emotions must also be adept at accounting for instigating stimuli. Generally speaking, a stimulus is "Any thing . . . that has some impact or effect on an organism such that its behavior is modified in some detectable way."[2] The physical conception of a stimulus is already encompassed by the term of entity that is being used for the equation. For practical purposes, an entity that an individual knows exists can also be considered a stimulus in the context of the equation. If an entity is or has been cognized by an individual, meaning the individual is at least aware of the entity's existence, then that entity would be eliciting a response from the individual because it is being given an existential value. A further distinction can be made between a stimulus that an individual is aware of and one that her or she is unaware. For instance, an "adequate stimulus" is described as one that is both "above the threshold for the sensory system under consideration" and an "appropriate energy form for the sense under concern."[3] Contrarily, an inadequate stimulus would be an entity that an individual has no awareness of or it is insufficient to register in the sensory system in its current state to lead to the formation of an existential value. The two classes of stimuli that will be the most important to consider for the equation are unconditioned stimuli and conditioned stimuli.

Two Main Classifications of Stimuli: Unconditioned and Conditioned

An unconditioned stimulus is one that reliably elicits an unconditioned response from an organism, such as pulling back after touching a hot stove.[4] An unconditioned response, or more appropriately, unconditional reflex, is "any response reliably elicited from an organism by a particular unconditioned stimulus."[5]

A conditioned stimulus on the other hand is "any stimulus that, through conditioning, comes to evoke a conditioned response."[6] Additionally, the conditioned stimulus "is an originally neutral stimulus that develops its eliciting power through pairing with an unconditioned stimulus."[7] Within the considerations of conditioning and learning, if a conditioned stimulus becomes paired with an unconditioned stimulus, then the equation will have to offer a means to explain this, regardless of whether or not one entity can actually influence the other's existence, either by harming or by benefitting it, or if the relationship between the two entities is only imagined (e.g., superstition).

As noted earlier, a stimulus could potentially be any object or entity. However, the point of interest in mentioning conditioning for the equation is to highlight the importance of identifying a relationship between two stimuli, namely, the perceived influence of one entity on another. For instance, if the acquisition or existence of one entity, such as a fire truck's siren, influenced the self's valuation of another object, a home, then the self's valuation of the second entity might elevate because a fire is capable of destroying the usefulness of a house by reducing its denominator for Sufficiency to zero. A similar relationship would exist even if one entity did not destroy another but simply permitted the self to gauge the value of another entity by granting access to a new entity, signifying its presence, or by creating it outright. In some cases, the presence of one entity (e.g., an object, act, or event) might create the existence of another entity that an individual actually wants. Before these are considered though, two other important classes of stimuli will be addressed.

Other Classifications of Stimuli: Proximal and Distal

Another distinction can be made between proximal and distal stimuli. A proximal stimulus is ". . . the physical energy that actually impinges upon a receptor."[8] Oppositely, a distal stimulus is "a stimulus away (distant) from the receptor on which it acts."[9] For instance, if a bird flew into a room one was in, the bird itself would be the distal stimulus while light reflected off of the bird onto the retina would be the proximal stimulus.

In example, the proximal stimulus of light enabling an individual to value a distal stimulus or entity, such as print on a book, would be a case of one stimulus granting access to another stimulus. If an individual has set the reading of a word, known to exist in a dark room, as a purpose, he or she at least knows that one printed word is out there and exists despite not being able to see it. Light, gives access to the word, thus lowering the value of the text by making it accessible. Light, therefore, benefits the purpose of reading text in a dark room by facilitating access to it. If there were an infinite number of printed words in the dark room, then the proximal stimulus of light would grant access to a perpetual succession of new printed words or distal stimuli. Darkness, however, inhibits the purpose of reading as it prevents access to the words by the individual.

However, if the entity were something one did not want to know existed in the first place, or at least not be visible such as a stain on a shirt or a monster under a bed, then light would harm the purpose of maintaining the entity's nonexistence, in this case ignorance of the stain or non-acknowledgment of monsters under the bed. Contrarily, darkness would benefit the purpose of maintaining the entity's nonexistence, or ignorance of the stain and monster under the bed.

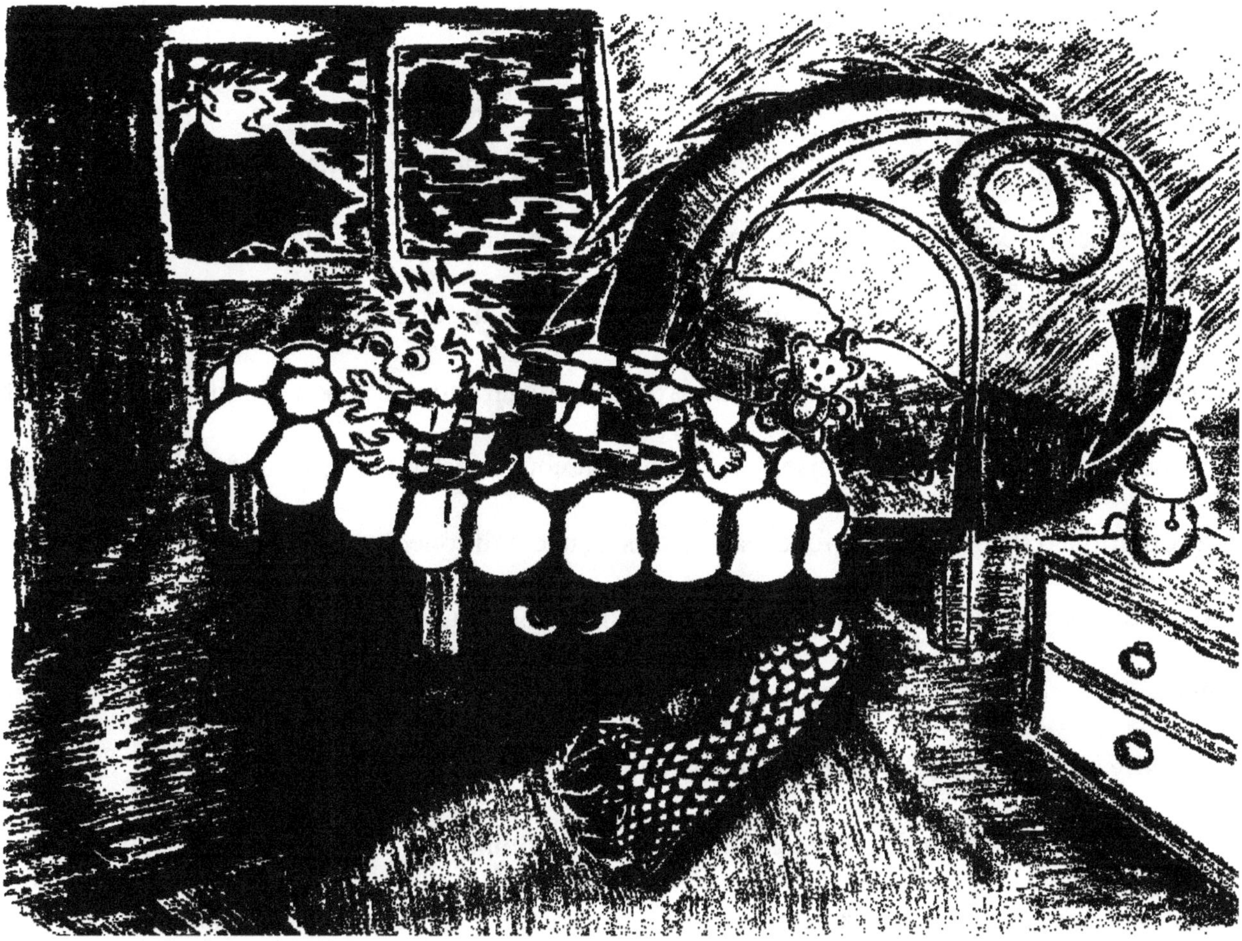

Figure 8.1 Darkness benefits the purpose of remaining ignorant of the very real monsters that may be under the bed or elsewhere, while light is harmful to this aim.

Ultimately, the purpose in question will help an individual determine how two entities are related to one another. If an individual seeks to acquire knowledge of texts in a library, then light would have a beneficial effect on the distal stimulus of printed text by making it an adequate stimulus and revealing an ever growing number of printed texts if an infinite number were available. Alternatively, if an individual sought to remain ignorant of the texts, then light would have a detrimental and harmful effect on the distal stimulus of emptiness by damaging or destroying the void to reveal an ever growing number of printed texts if an infinite number were available.

Stimuli, Modifiers of an Entity's Appraisal, and Category I Emotions (Intra-personal)

Types of Conditioning: Classical or Pavlovian and Operant or Instrumental

Conditioning is a term that generally specifies ". . . the conditions under which associative learning takes place."[10] Generally, there are two types of conditioning, classical and operant.

Pavlovian or classical conditioning occurs when a conditioned stimulus, neutral to an unconditional response at the outset, becomes paired with an unconditioned stimulus and eventually produces the unconditional response.[11] In Pavlov's famous experiment the sound of a tone was used as a conditioned stimulus, while food was used as an unconditioned stimulus and salivation was the unconditioned response.[12] Initially, the presentation of food caused the dog to salivate, whereas the sound of a tone did not. Repeatedly pairing the sound of a tone with the presentation of food eventually lead to the dog salivating if the sound of the tone was presented alone.

Operant conditioning or instrumental conditioning, on the other hand is a type of conditioning where an organism's behavior is reinforced by consequences of an action. Operant conditioning typically involves reward and punishment. These will be discussed below.

Reinforcement

Reinforcement is of two types, positive and negative.[13] A positive reinforcer functions by its presentation, increasing the likelihood of a behavior when present, and if removed functions as negative punishment.[14] On the other side of the coin, a negative reinforcer functions by its removal, increasing the likelihood of a behavior when the aversive stimulus is not present, and if removed functions as positive punishment.[15] Some note that "while it is probably true that pleasurable or satisfying events 'reinforce' behavior, this definitional effort only serves to pass along the problem," as "pleasure and satisfaction are no more definitionally tractable terms than reinforcement, nothing is gained here."[16] However, both pleasure and pain have been mathematically defined within the context of the equation as negative anxiety (negative valuations of an entity toward the fulfillment of a purpose and restoration of equilibrium) and anxiety (positive valuations of an entity toward the fulfillment of a purpose and restoration of equilibrium) respectively. Pain and pleasure can both be quantified in the equation. A stimulus, for all practical purposes in the functions, may be considered the same as an entity. Also, rather than use the concepts of reward and punishment to refer to an entity itself, the closely related terms of benefit and threat will be used to describe an observed relationship between two entities. Benefit will be used to distinguish cases where the self has determined that an original entity's existence experiences a boon due to the existence of another entity or may be promoted by the second entity's presence. Threats of harm will be used to distinguish cases where the self has determined that an original entity's existence is harmed by the existence of another entity and is placed in peril if the second entity is present.

Logical Considerations for the Equation: Reward and Benefit vs. Punishment and Threat

Reward is loosely defined as "any pleasurable or satisfying event or thing that is obtained when some requisite task has been carried out."[17] In the equation, an elevation in pleasure, or similarly a reduction in pain, occurs whenever the valuation of an entity descends toward negative infinity and so this will be the starting point for absorbing the concept of reward.

A reward in the equation will be best thought of as a case where an original entity has its valuation lowered due to another entity's presence. A second entity lowers an individual's valuation of the original entity by making the original entity more plentiful, more available, or more easily accessible. Hence, the second entity promotes, rewards, or benefits an original entity's existence.

Benefit, in the equation, is the extent to which an original entity is enhanced, created, has its growth promoted, or has access to it facilitated by another entity. The maximum value for this will be one, meaning that the original entity will have its existence increased by + 1 if another entity (e.g., an event, an act, an object) is acquired. Benefit is defined in the equation as follows:

> Benefit is where the presence of a second entity (i.e., an event) enhances, promotes, or signals the original entity's existence such as by making it more accessible.

Meanwhile, punishment concerns "the administration of some aversive stimulus contingent upon a particular behavior."[18] Punishment is illustrated in figure 8.2.

Figure 8.2 After being presented food upon pecking a button, a bird's rate of pecking should increase. If a brief electric shock is subsequently presented instead of food and the level of pecking declines, then punishment has occurred.

In the equation, pain, or conversely a reduction in pleasure, occurs whenever the valuation of an entity ascends toward positive infinity, and so this will be the starting point for absorbing the concept of punishment into the equation.

Punishment, in the equation, will concern instances where an original entity has its valuation raised by the presence of a second entity due to a threat of harm. A second entity raises an individual's valuation of another, original entity, by making the original entity less plentiful or less available. Hence, the second entity inhibits, harms the existence, or prevents access to the original entity. A threat of harm is considered to be the extent to which an original entity will be damaged, destroyed, or have access to it denied by another entity. The maximum value for this is one, meaning that the original entity will have its existence totally annihilated and reduced to zero if another entity (e.g., an event, an act, an object, etc.) is acquired. Threat is defined in the equation as follows:

> Threat is where the presence of a second entity (e.g., an event) harms, inhibits, or signals the absence of the original entity's existence, such as by preventing access to it.

List of concerns that will have to be taken into consideration

1) Entities, up to this point, have only been assessed for their value, via usefulness, or lack thereof. Now, the possibility is being considered that some entities may have contingencies with other entities.
2) Threats of harm or promises of a benefit to an entity from another entity must now be considered.
3) Pain and pleasure are being represented by anxiety and negative anxiety and will have to be modified accordingly.
4) Whichever manner the equation is transformed, the concepts of contingencies, harm, and benefit will have to account for the following:
    a) An increase in anxiety felt
    b) A decrease in anxiety felt
    c) An increase in negative anxiety felt
    d) A decrease in negative anxiety felt
    e) Transformations cannot reduce the self's valuation of an entity to zero. For

instance, if a harm destroys an entity completely, some conception of its existence would remain due to memory. For example, in the case where Adam's apple was stolen, the denominator approached zero and sent anxiety toward infinity. Despite having to forsake the apple's acquisition in order to function, the purpose was neither obliterated from Adam's consciousness nor memory, though he chose not to worry about it the purpose of eating an apple anymore by lowering his Sentiment value substantially. Transformations, therefore, must leave the existential value of the entity intact for the self, even if the presence of one entity (e.g., Entity *A*) guarantees the absence or total destruction of the original entity (e.g., Entity *B*).

f) The transformations also must account for the Appraisal of the entity with respect to a purpose and be logically sound. Namely, if the Appraisal's value is modified, then the predictions of the transformation must still hold true to logic for the relationship of harm or benefit that is under consideration.

Lazarus' Cognitive Motivational Relational Theory

Richard Lazarus was a prominent scholar and researcher on cognitive theories of emotion during the second half of the twentieth century who developed a Cognitive-Motivational-Relational theory of emotions.[19] Under Lazarus's Cognitive-Motivational-Relational theory, emotions were viewed as responses to one's perceived environment that enabled an individual to be ready to cope with whatever harm or benefit was appraised as present.[20] A core feature of Lazarus' model of appraisals was that emotions represent “relational meanings” and concern how different aspects of situations influence an individual's well-being.[21] In Lazarus' theory, emotions arose via "personal meanings that people bring to situations that have relevance to their knowledge and aspirations."[22]

Within Lazarus' framework, Cornelius observed that on the "molecular level, individual appraisal components" described an individual's judgments about the harms or benefits present in an environment while at the "molar level individual appraisal components" combined to create "core relational themes" or summaries of the "emotional meaning" of events or situations.[23] In other words, on a microscopic level, an individual would judge a situation for how it might influence him or her while on a larger scale, an individual's judgments merge to create a more general, emotional meaning.

In Lazarus' theory, at the "basic level, a person appraised a situation for the benefit or harm it might hold for him or her.”[24] Appraisals in Cognitive-Motivational-Relational theory were also of two types, primary appraisals and secondary appraisals.[25]

Primary appraisals are concerned with "whether or not an event has any relevance for a person's well-being and if so, how.”[26] Hence, they are merely an assessment of a situation. Secondary appraisals, on the other hand, are concerned with coping.[27] Secondary appraisals, therefore, concern what the individual believes can be done to influence the outcome of a situation.

Figure 8.3 Adam's primary appraisal would concern the threat of the thief stealing his apple. Adam's secondary appraisal would concern coping and might include chasing after the thief to retrieve his apple if the thief does manage to take it. In Adam's case, however, he simply discarded the purpose of eating an apple once it was stolen by modifying the priority of the purpose (Sentiment).

Roger's Protection Motivation Theory

Protection Motivation Theory was a theory concerning fear appeals that was founded by Dr. R.W. Rogers and proposed that individuals respond to threatening information primarily on the appraisal of four variables.[28] These four variables were the severity of the threat, the probability of the threat occurring, the efficacy of a recommended behavior to prevent the negative consequences of the threat, and the individual's belief in performing the recommended behavior.[29]

Stimuli, Modifiers of an Entity's Appraisal, and Category I Emotions (Intra-personal)

Threat: Perceived Threat Severity and Perceived Threat Susceptibility

Generally speaking, a threat is "... any action gesture or response that indicates an intention to attack, harm, or intimidate another."[30] Perceived threat susceptibility or vulnerability refers to how susceptible the individual perceives himself or herself to be to experiencing the negative consequences of the threat.[31] For instance, the threat severity of a meteor shower striking earth and devastating the planet is high, however, the threat susceptibility of this is quite low as it is not likely to happen any time soon.

Efficacy: Perceived Self-Efficacy and Perceived Response-Efficacy

Albert Bandura generally defined perceived self-efficacy as people's sense of their abilities and "capacity to deal with the particular sets of conditions that life puts before them."[32] Similarly, under Roger's Protection Motivation Theory, self-efficacy's definition is restricted to an individual's belief in his or her ability to perform a recommended behavior to prevent a threat.[33] Perceived response-efficacy, or the perceived effectiveness of the recommended behavior at preventing the negative consequences of a threat, comprises the other half of efficacy in Roger's Protection Motivation Theory.[34] In lieu of a meteor strike on earth, the response-efficacy of escaping the planet on a spacecraft to another world hospitable for life might be quite high. However, the likelihood that an individual would be able to personally achieve such a feat might be incredibly small, and the self-efficacy to perform the recommended behavior would be low.

Stimuli, Modifiers of an Entity's Appraisal, and Category I Emotions (Intra-personal)

Common Themes and Elements: Incorporating Cognitive-Motivational-Relational Theory and Protection Motivation Theory into the Equation with a Neurological Network Model

A glance at each of these theories has revealed a few common themes and individual elements that will be vital to incorporate. Harm, benefit, primary appraisal, secondary appraisal, perceived threat severity, perceived threat susceptibility, perceived response-efficacy, and perceived self-efficacy will have to be taken into account by the equation. Similar to the notion of threat severity and threat susceptibility, it would be wise to consider a parallel variable, such as perceived benefit intensity (paralleling threat severity) along with perceived benefit susceptibility. All that would remain then would be the task of determining where each concept would fit into the equation, how these concepts might be gauged, and how they might relate to each other mathematically.

Benefits: Benefit Intensity and Benefit Susceptibility

Two additional concepts, of perceived benefit intensity and perceived benefit susceptibility, are also proposed that would be used for relationships where one entity promotes the growth of, existence of, or access to an original entity. Like its counterparts of harm and threat, the notion of benefit intensity would refer to the extent that a second entity is capable of promoting the existence of an original entity. Similar to threat susceptibility, benefit susceptibility would refer to the likelihood that a benefit would befall the original entity from the second entity.

Proposals for How the Variables Would Interact and Merits for the Use of Exponents as a Means of Incorporating the Two Theories into the Equation.

1) It is proposed that threat severity and threat susceptibility, as perceived by the individual, will have a multiplicative relationship with each other, while benefit intensity and benefit susceptibility will also have a multiplicative relationship with each other. However, only one or the other will be used at once, given that the relationship between two entities can only be such where one entity either threatens another entity's existence with harm, or benefits its existence. It cannot, due to mutual exclusivity, have a negative and positive contingency.
2) Both self-efficacy and response-efficacy to prevent the threat of harm or promise of benefit, as perceived by the individual, are proposed to have a multiplicative relationship with each other. Additionally, the efficacy variables will have a relationship with the threat or benefit variables so that the efficacy components are always in the denominator and the threat of harm or promise of benefit variables are in the numerator to ensure a uniform standard.
3) Of note, because of the flexible nature of the equation, if one desired, the above relationships concerning harm, benefit, and efficacy could also be expressed as additive ones, with each scaled so that the total value was no more than one. A coefficient of one half (e.g., .5) would be logical. Multiplication was deemed the ideal way of combing them here to keep the explanation looking neat and simple.
4) It can be observed that exponents can amplify or shrink a value (distort the self's

appraised value judgment of an entity)

5) An exponent would leave the existential value intact; when taken to powers close to zero, a number moves toward + 1

6) Logic behind the use of exponents:

   a) Relationship of neurons: If it is assumed that it takes one neuron to value an entity, hence produce one emotional unit, and each valuing neuron is potentially linked to every other neuron that can value the entity, then the activation of one neuron could potentially set off a chain reaction with nearby neurons; consequently, an entity's valuation might be expected to grow very quickly. Exponents can increase a number's absolute value very quickly.

   b) For example, if harm were over efficacy as an exponent at a 2:1 ratio, it would suggest that thoughts concerning the destruction or harm of an entity (threat severity factored with threat susceptibility) are taking place at twice the rate of thoughts concerning the likelihood of the entity being protected (response-efficacy factored with self-efficacy). Harm, therefore, would correspond to the elevation of an entity's value while efficacy would correspond to a decrease of an entity's value. A significant threat of harm (harm = threat severity factored against threat susceptibility = + 1) would imply that the threat of harm, more specifically a contingency relationship between the second entity and the original entity, is causing the self to reassess the entity's entire value each time he or she thinks about the two entities. This value is proposed to grow exponentially, as each activation of valuing neurons might trigger nearby valuing neurons that in turn activate all the other valuing neurons they are connected to at a rate much faster than any forces acting against their activation. Conversely, a high amount of efficacy (e.g., Efficacy is response-efficacy factored against self-efficacy) would imply that the self's efficacy, more specifically a contingency relationship between the self and the harm, is causing the self to reassess the entity's entire value each time he or she thinks about the contingency between these two entities. Efficacy in this case would be proposed to shrink the entity's valuation exponentially, as the activation of specific nearby neurons could make nearby ones less likely to fire by raising their activation threshold (e.g., inhibitory synapse and inhibitory post-synaptic potential).

c) The rationale behind the use of harm for the numerator in the exponent and efficacy for the denominator in the exponent is as follows. It is given that four neurons are capable of valuing an entity and they are all connected. If the activation of each one of the neurons is sufficient to trigger the other three, the four neurons, all together, might be activated at least 16 times from the four axons that branch into at least 12 terminal buttons if they are all initially triggered by a fifth, excitatory neuron that can excite each one separately.

Figure 8.4 A Network Model where each neuron is connected to every other neuron in the bundle, both receiving and transmitting information.

d) The above (**Figure 8.4**) would be a network of four neurons linked to one another, though much larger ones may certainly be possible.

□ = Valuing Neuron or Group of Neurons	Each neuron in a network is capable of activating (i.e., triggering an action potential) every other neuron to which it is linked.	
Neurological Model	Number of Neurons and Connections	Number of action potentials if each neuron in the network is activated independently by a neuron outside the network.
 □=□	Two Neurons and Two Connections	**4 Activations**
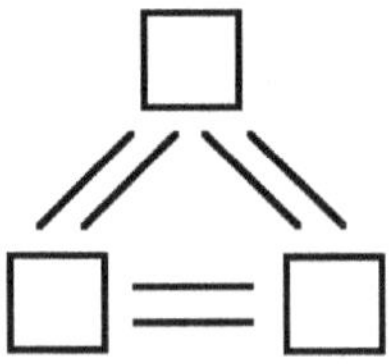	Three Neurons and Six Connections	**9 Activations**
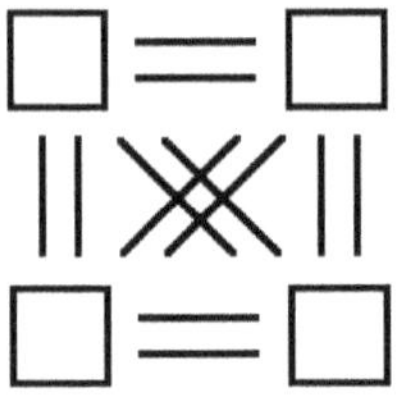	Four Neurons and Twelve Connections	**16 Activations**
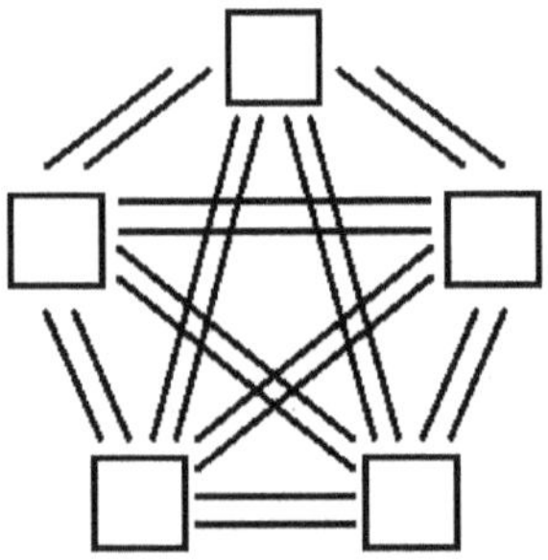	Five Neurons and Twenty Connections	**25 Activations**

Figure 8.5 A Network Model of Valuing Neurons: One neuron correlates to one emotional unit in the above and subsequently, anxiety or negative anxiety.

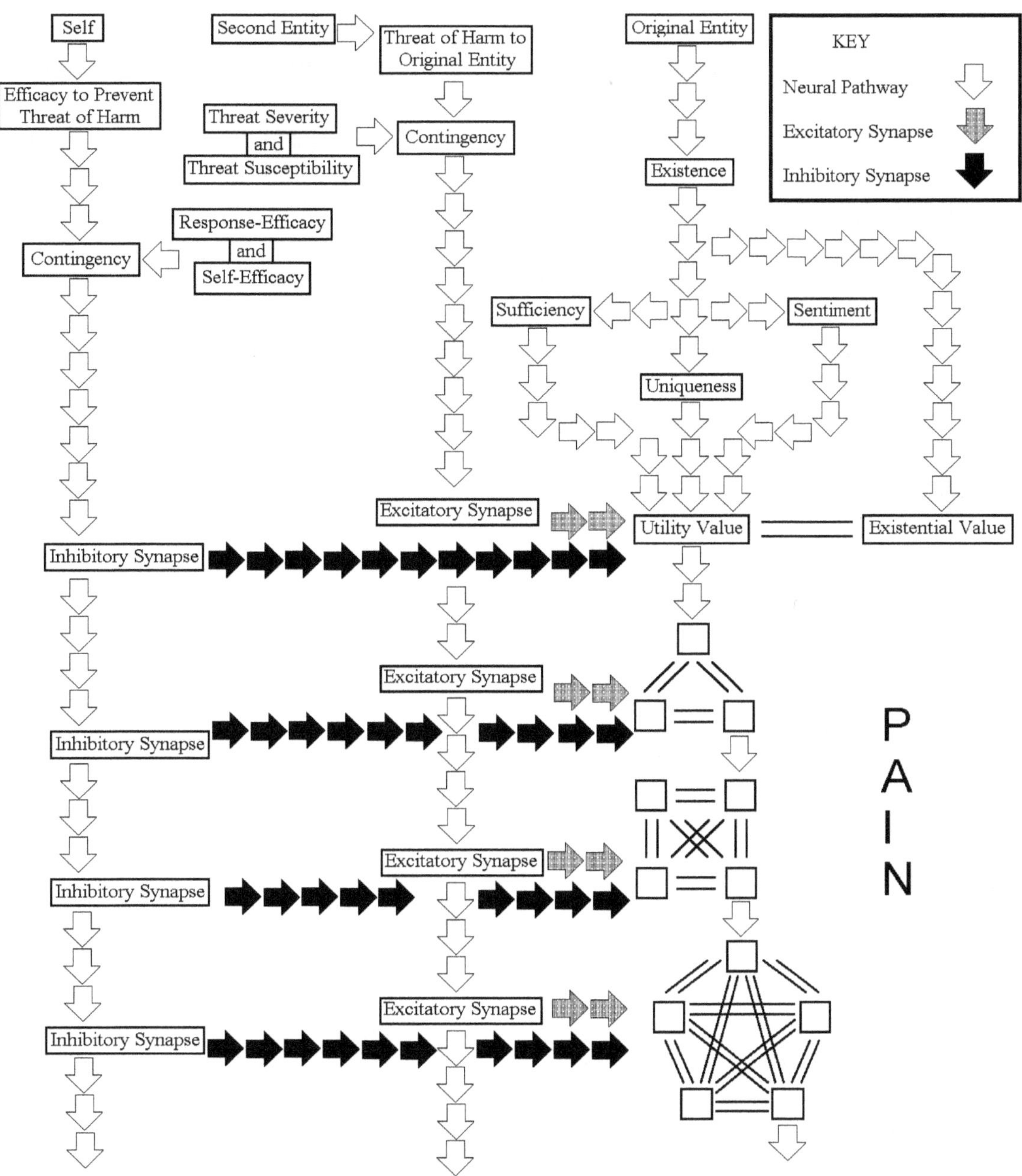

Figure 8.6 Sample Neurological Model for the Avoidance of Pain Equation. A theoretical, neurological network model for positive valuations and the investment of anxiety is above. The contingencies used here are negative ones: the threat of harm posed by the second entity to the original entity; the efficacy of the self to prevent the second entity from harming the original entity. The valuing networked neurons in the bottom right correspond to anxiety in the equation or positive anxiety-mass in the Roll Cage theory (e.g., pain). They would likely be linked to an area of the brain concerned with danger, such as the amygdala.

If efficacy were twice the harm component, then an individual would be activating the excitatory synapses (concerned with valuing the entity highly due to its impending destruction) only half as often as they were activating the inhibitory synapses (concerned with the entity's acquisition or protection) and in descending order from the top. The more active of the two trigger neuron synapses in this case would be an inhibitory synapse connected to the valuing neurons that raises the threshold for activation, thus reducing the likelihood that the valuing neurons become activated. Contrarily, if the harm component were twice the efficacy component, then an individual would be activating excitatory synapses (concerned with valuing the entity due to its impending destruction) twice as often as the inhibitory synapses, meaning that the entity's value would be elevated.

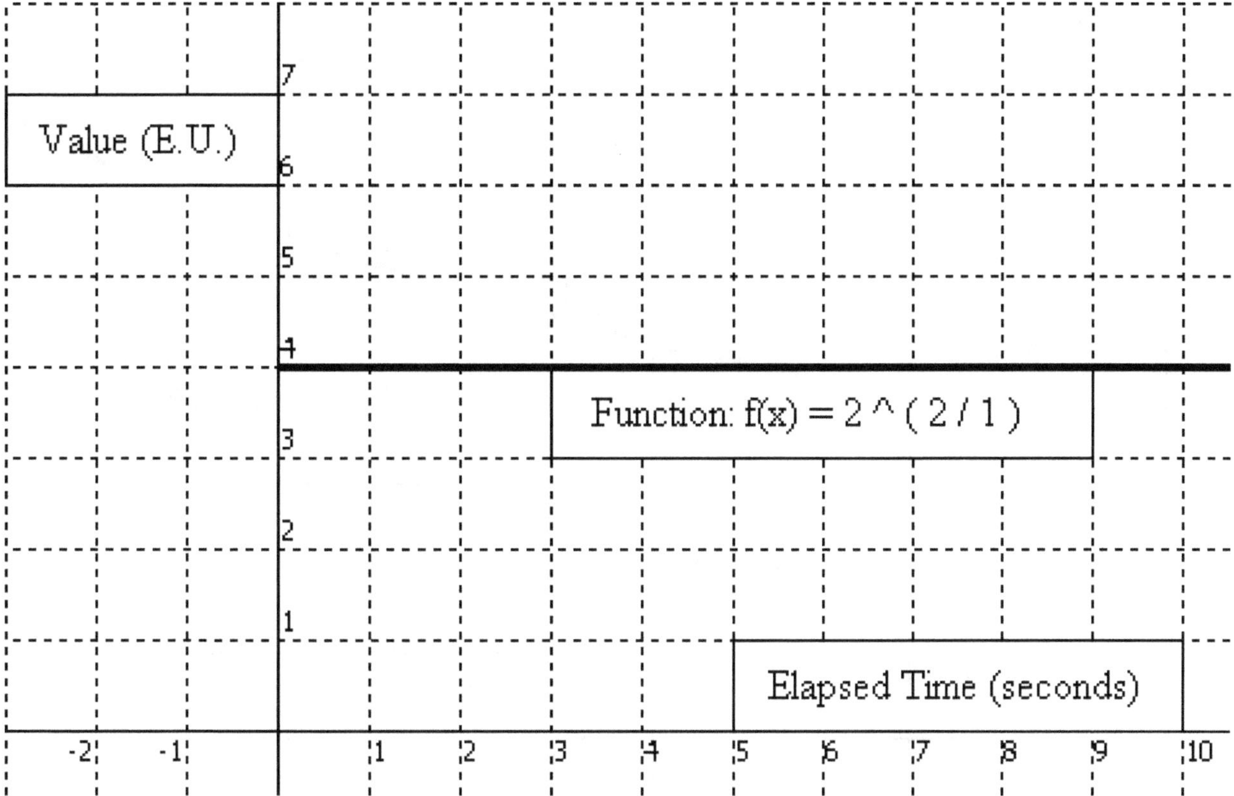

Figure 8.7 Graph of function for anxiety if the entity's original value is + 2 emotional units from the existential and utility components, and the Harm components of Threat Severity and Threat Susceptibility are twice the Efficacy components of Response Efficacy and Self Efficacy. The entity has + 4 emotional units invested in it. The Appraisal coefficient equals + 1.

A check of limits will reveal whether these calculations are in accord with logic. As the exponent approaches positive infinity, meaning that harm goes to the limit of one and efficacy goes to zero, anxiety goes toward positive infinity as well.

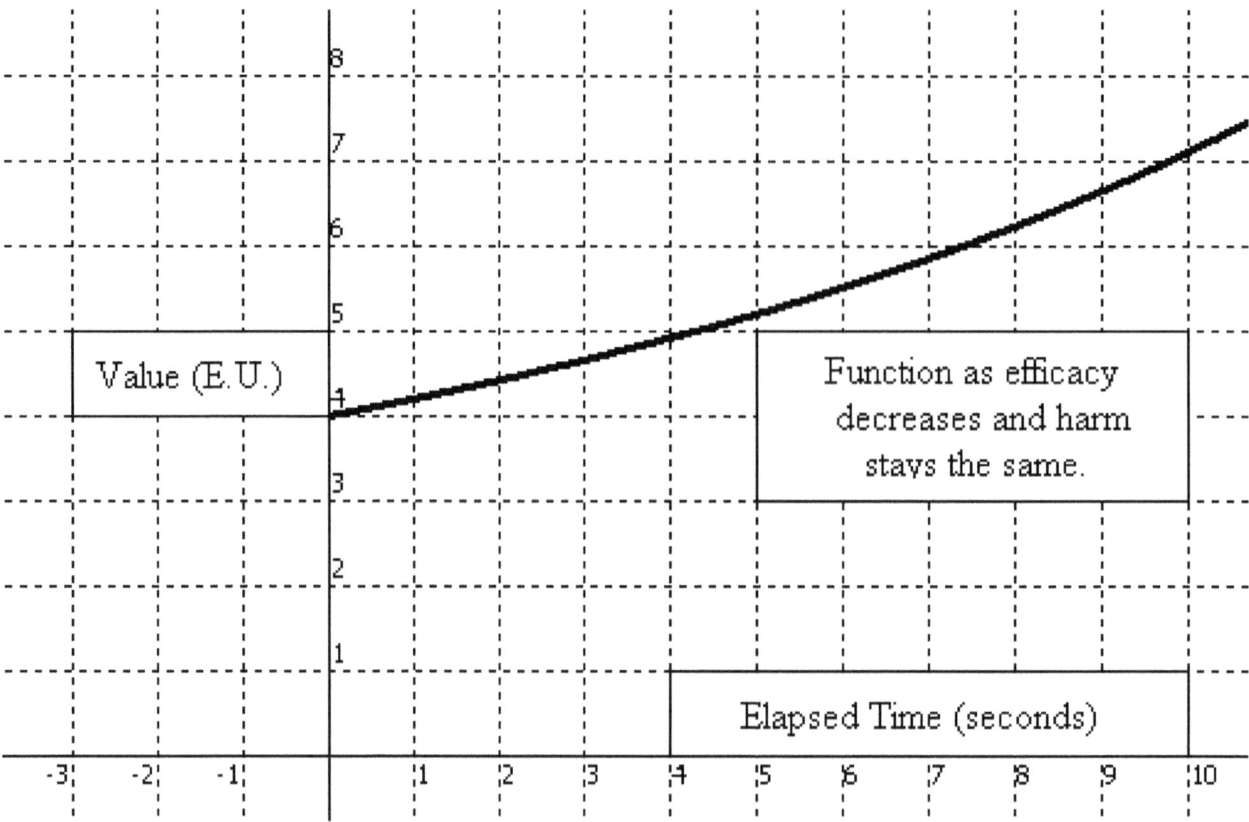

Figure 8.8 Graph of positive anxiety (above the x-axis) with the harm to efficacy variables in the exponent starting at a ratio of 2:1 and then increasing exponentially as it is realized that the self can not safeguard the original entity from a threat of harm.

To hold that the anxiety felt for an entity would be expected to elevate if a significant threat of harm endangered the entity and nothing could be done by the self to protect it is in accord with logic. If an excitatory synapse were causing the valuing neurons to fire more frequently while inhibitory synapses decreased in their potency, then this would be indicative of a flurry of activity amongst the neurons used to value an entity having a positive valence toward the restoration of equilibrium.

On the other side, as the exponent approaches zero, meaning harm goes to zero and efficacy goes toward one, the total anxiety invested would approach one, its existential value. To hold that the anxiety felt for an entity would decrease to its existential value if a small threat or no threat of harm is present, along with the self being able to protect the original entity from the minuscule threat of harm, would be in accord with logic. If inhibitory neurons continue to fire at their maximum rate, raising the threshold for activation between the valuing neurons, eventually none of the valuing neurons will set off the other neurons due to the activation threshold growing too high.

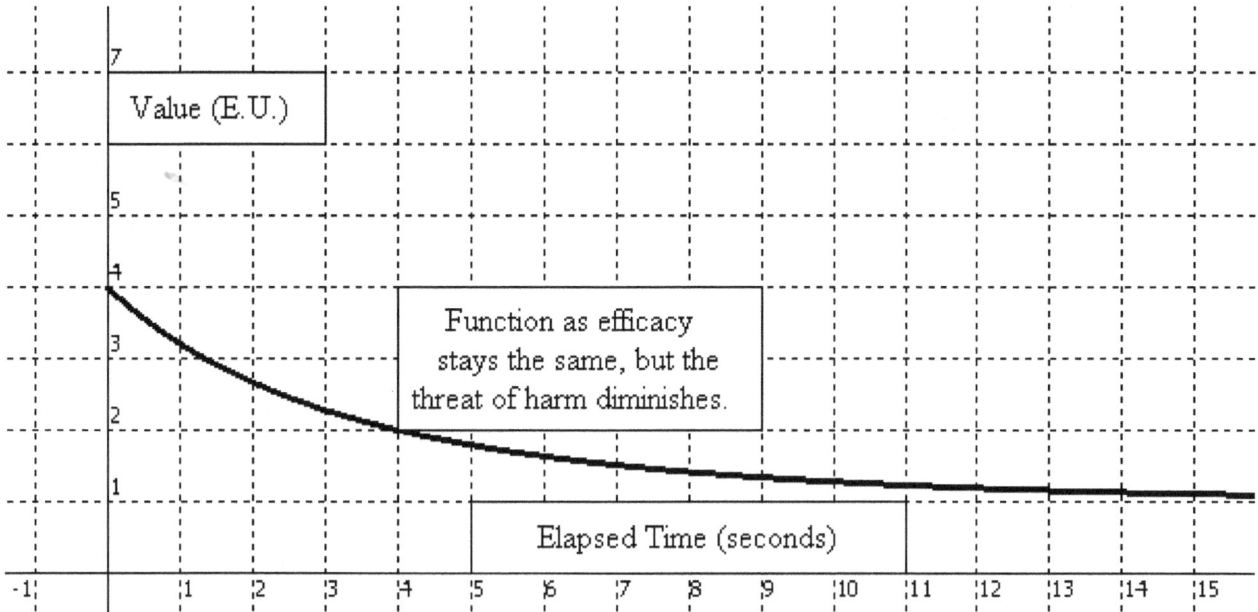

Figure 8.9 Anxiety lowering towards its existential value as a result of the threat of harm becoming less pronounced while efficacy to prevent the threat of harm remains constant. Harm was initially at a 2:1 ratio against efficacy (2 / 1 in the exponent) and then gradually reduced to zero.

Value extinction could be represented by a group of neurons that represent extinction, and they would effectively cut off the circuit between the entity and its existence by raising the activation threshold for existential value. Such a case might occur after the entity's existence had initially been ascertained or postulated. The utility variables of Sufficiency and Uniqueness, if theorized to be primarily driven by bottom-up processing, might feed into the same group of valuing cells that other entities concerning the same purpose feed. Bottom-up processing refers to raw stimuli that work their way up to more ". . . abstract, cognitive representations."[35] Bottom-up processing might be expected to be used when information is new and an individual has no prior conceptions. It would be logical to hold that these two variables, concerning an entity's Sufficiency to do a task, and its Uniqueness for performing that task, might be more dependent on the individual's observations from the outside world than personal expectations. Ultimately, however, this decision is left up to the architect of a function.

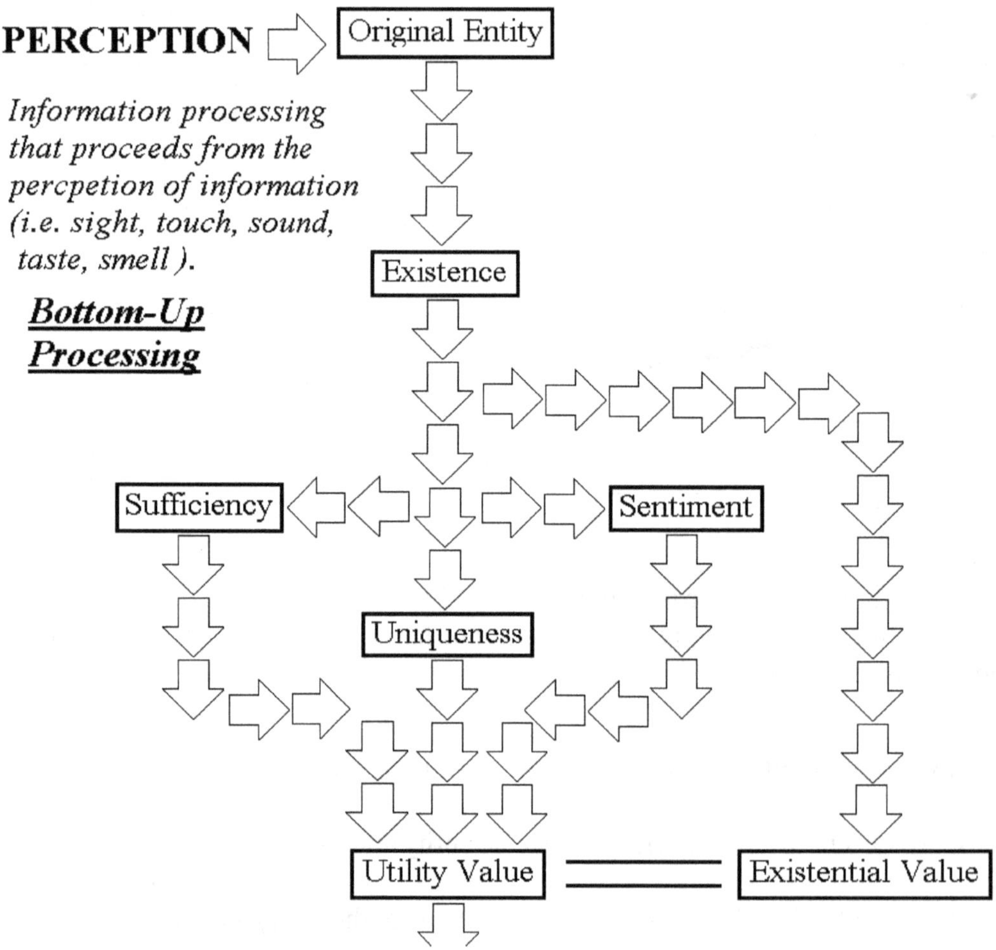

Figure 8.10 Uniqueness and Sufficiency arising from a predominantly bottom-up, neurological route.

Sentiment, on the other hand, is theorized in affect engineering to be primarily driven by top-down processing and would originate from a slightly different route. Top-down processing refers to "... cognitive or perceptual processing in which complex and abstract representations," are used to analyze simple input from stimuli.[36] Top-down processing might be expected to be used when information is familiar. Given that Sentiment is defined in the equation as an individual's absolute ranking of a purpose against the purpose that has the highest ranking, neurologically, Sentiment might be represented by signal strength from a Sentiment core or ethics hub of neurons, as was suggested in chapter five. Those purposes with the highest priority would have the highest signal strength from this core while purposes with lower priorities would have a lower signal strength from the Sentiment core (figure 8.11). The possibility that Sufficiency or Uniqueness may receive influence from a hub of neurons or top-down processes is something that should also be considered by a function's architect.

Stimuli, Modifiers of an Entity's Appraisal, and Category I Emotions (Intra-personal)

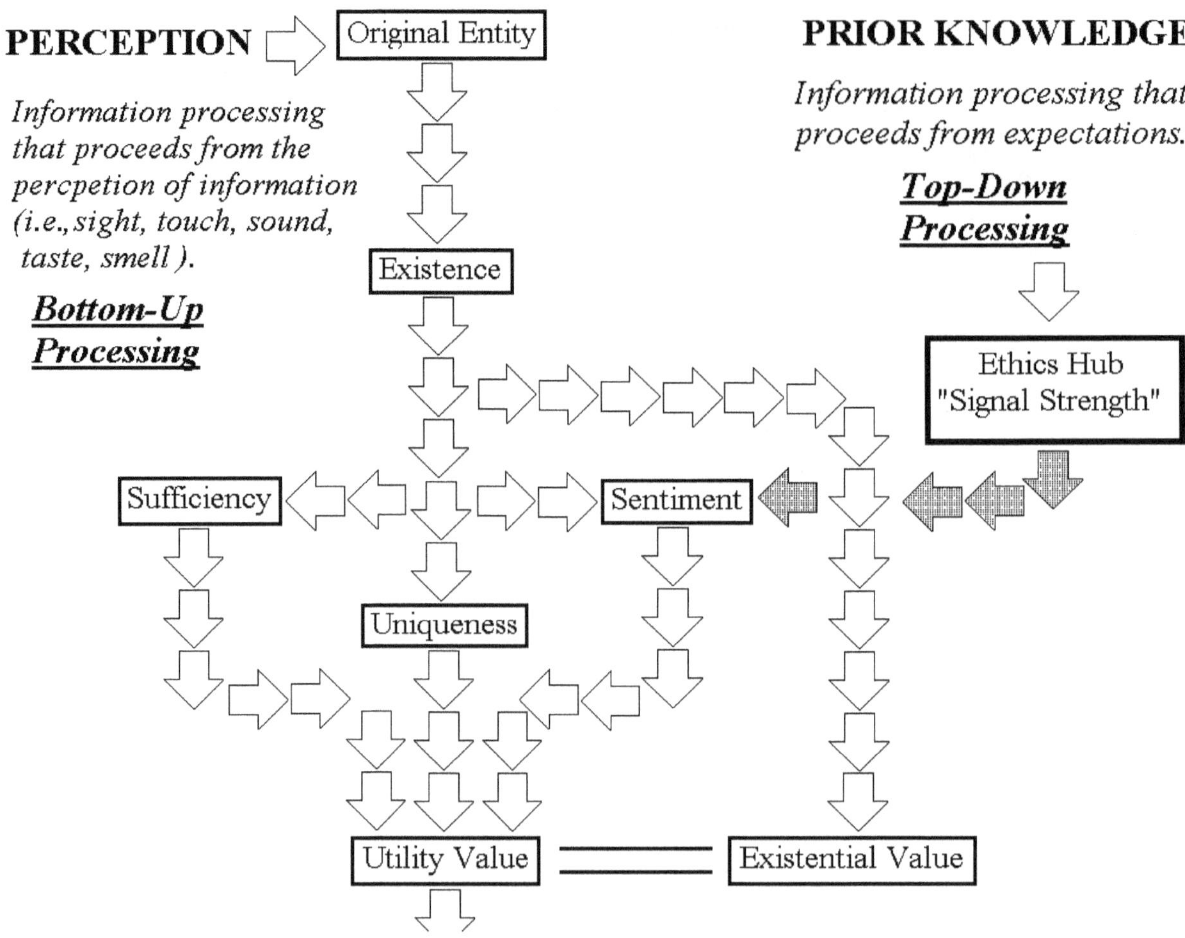

Figure 8.11 Sentiment is also being activated in part by the signal strength felt from top-down processes, such as an ethics hub of neurons with varying firing rates for specific types of entities and their respective purposes.

Neurological Network Model for Negative Anxiety

Alternatively, if the Appraisal value of the entity toward the restoration of equilibrium were negative or became negative, then a different set of neurons would be used for assessing the value of entities with a negative valence, hence negative anxiety. Their structure might look like the image in figure 8.12. The primary difference between the two models would be the role of harm to the original entity, efficacy to prevent the harm, and the functioning of the networked neurons. Harm, instead of exciting the networked neurons, would have an inhibitory effect while efficacy to prevent the harm, instead of inhibiting the networked neurons, would have an excitatory effect. Also, whereas in the avoidance of pain model the networked neurons signaled a positive valuation of an entity, in the pursuit of pleasure model the networked neurons would signal a negative valuation of an entity. These separate networks can respond to two separate valuations of an entity at once, such as between a purpose and a complimentary purpose. The same entity could elicit two valuations for two different purposes in two different pathways simultaneously. If pain is avoided by obtaining an entity for a specific purpose, then for that purpose's complement pleasure acquisition would diminish at the same time.

Stimuli, Modifiers of an Entity's Appraisal, and Category I Emotions (Intra-personal)

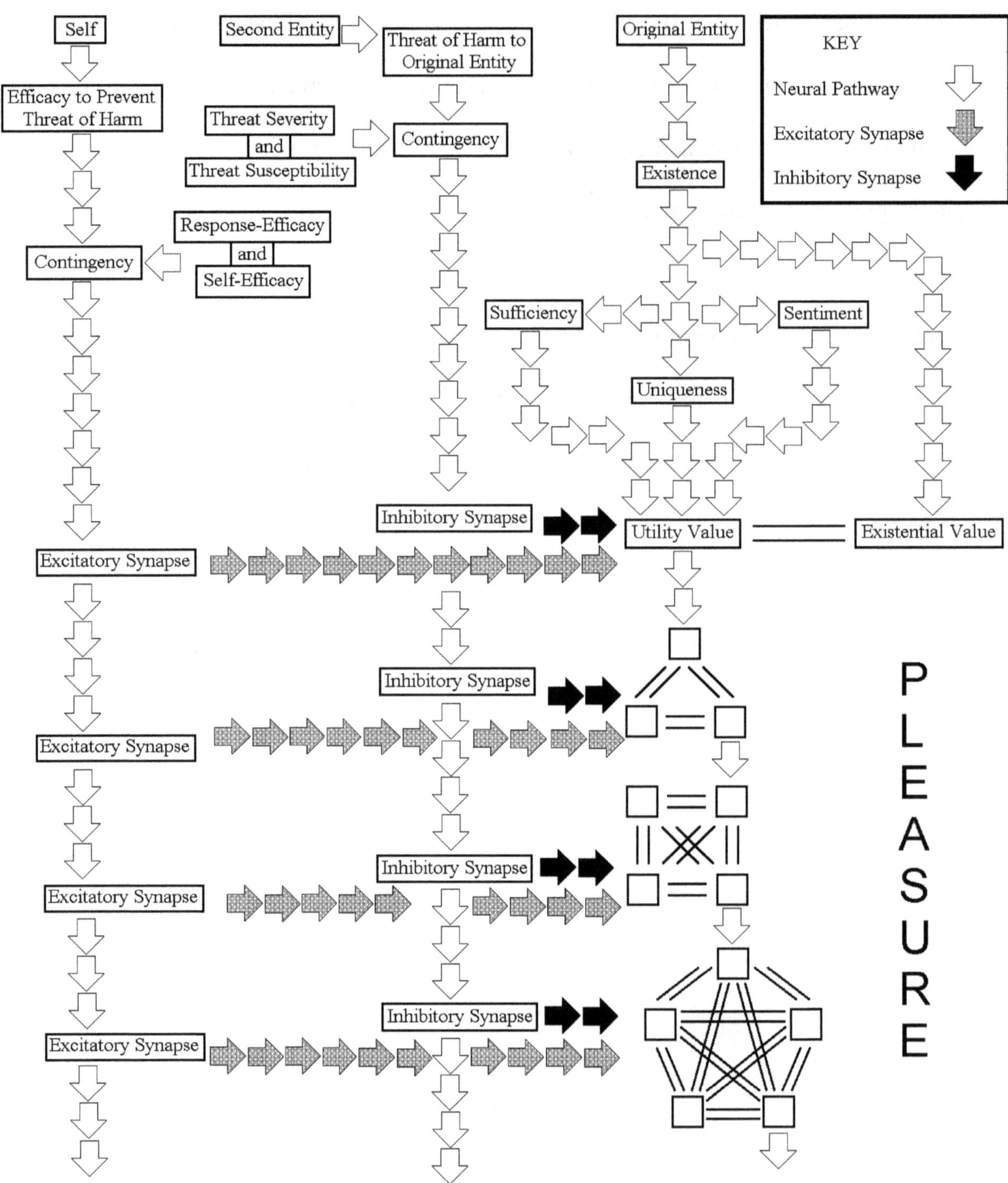

Figure 8.12 Pursuit of Pleasure: A theoretical, neurological network model for negative valuations and the investment of negative anxiety. The contingencies used here are negative ones: the threat of harm posed by the second entity to the original entity; the efficacy of the self to prevent the second entity from harming the original entity. The valuing networked neurons in the bottom right correspond to negative anxiety or negative anxiety-mass in the Roll Cage Theory (e.g., pleasure).

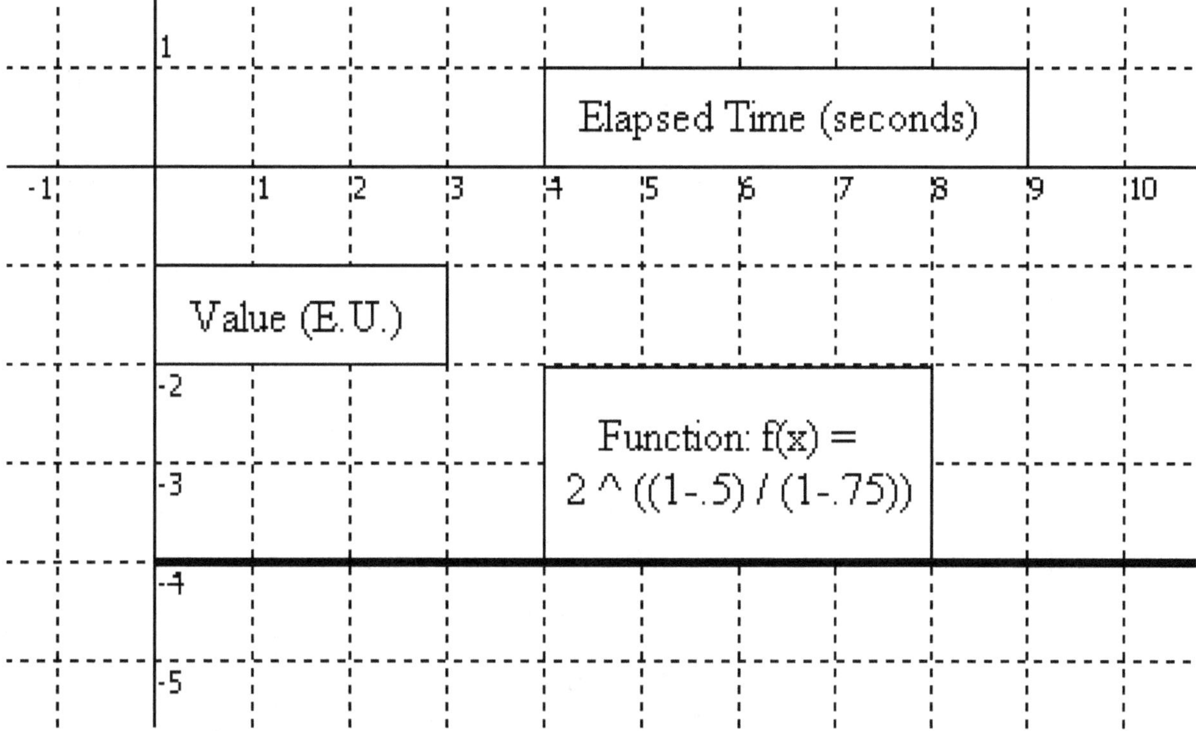

Figure 8.13 Graph of a function for negative anxiety if the entity's original value is + 2 from the existential and utility components. The Harm components of Threat Severity and Threat Susceptibility along with the Efficacy components of Response Efficacy and Self Efficacy result in the exponent being two. The Appraisal value is negative one. To note, the pursuit of pleasure formula has not yet been described, but it is used above.

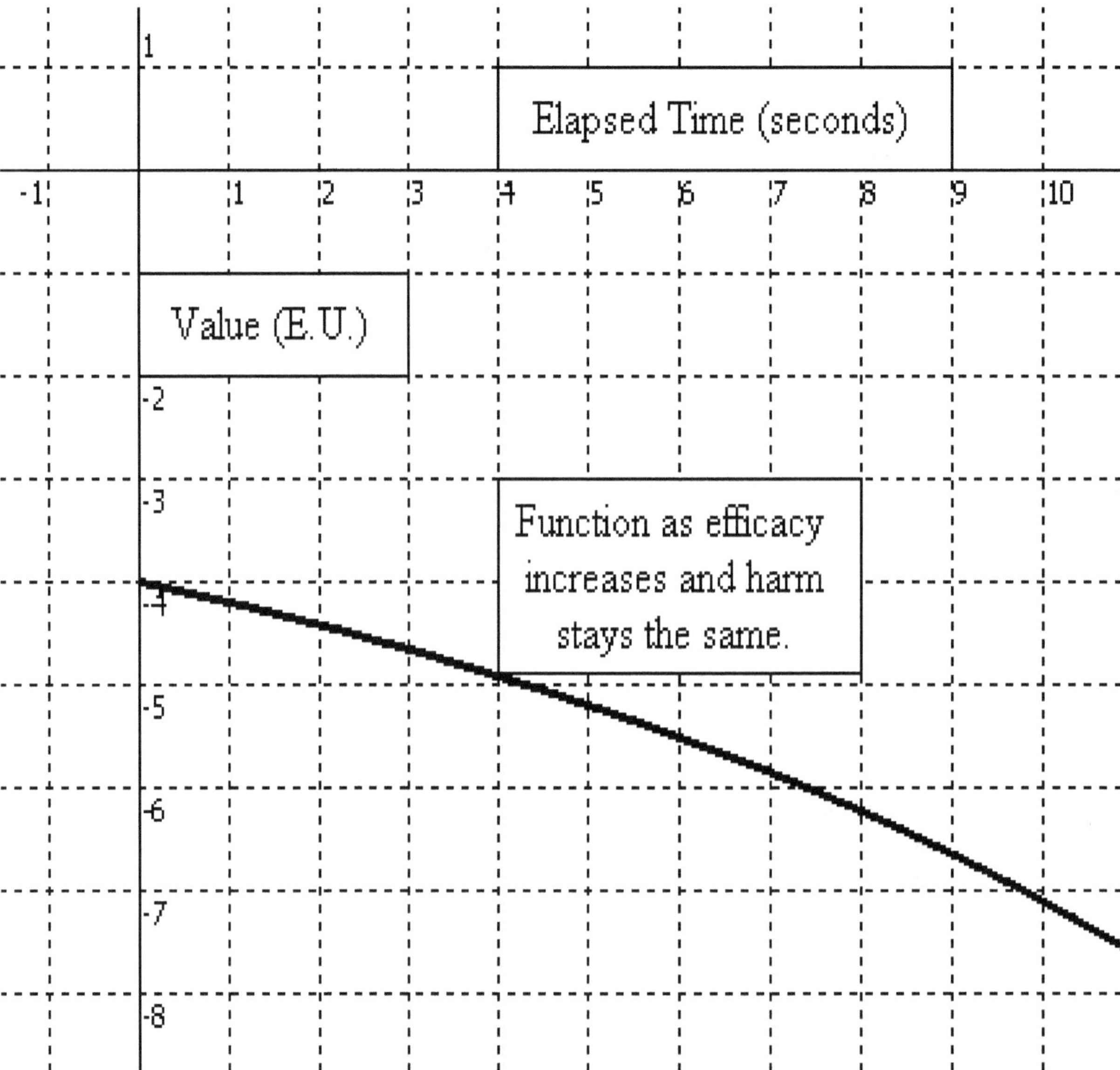

Figure 8.14 Graph of negative anxiety (below the x-axis) with the harm to efficacy variables in the exponent starting at 2:1 if the (i.e. Harm = .5 and 1 - Efficacy = .75) and then increasing exponentially as it is realized that the self can safeguard the original entity from a threat of harm. Negative anxiety will increase towards negative infinity, though a vertical asymptote results if the denominator approaches zero.

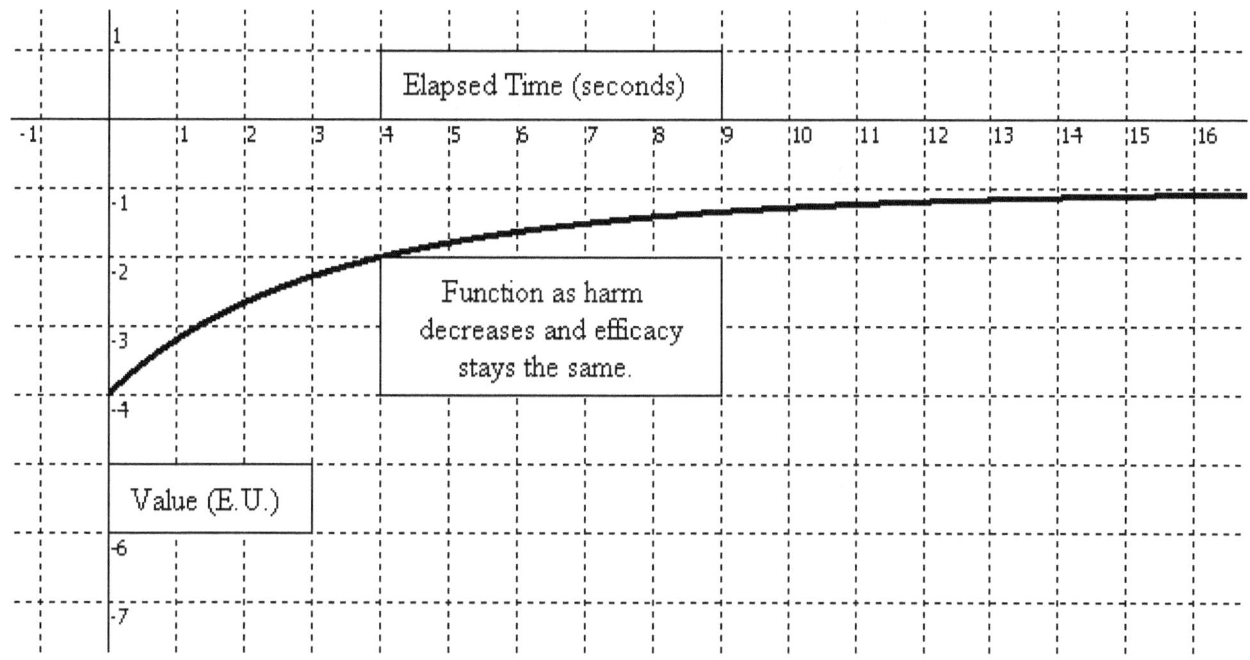

Figure 8.15 Negative anxiety diminishing towards its existential value of - 1 emotional unit as a result of the efficacy to prevent the threat of harm becoming less pronounced while the threat of harm remains constant.

Overview of Neurotransmitters and Hormones

As for how the networked neurons themselves might signal a positive valuation, a negative valuation, an excitatory effect (lowering the activation threshold), and an inhibitory effect (raising the activation threshold), the target receptor cells and the types of neurotransmitters being released by each would offer some clues. The nervous system runs throughout the body; determining whether one area of the brain or body is used for positive valuations and another for negative valuations, for instance, could be assessed using brain scanning equipment, but will not be explored in depth here. Neurotransmitters are "... substances that function as vehicles of communication across the synaptic gap between the terminal buttons of one neuron and the membrane of the receiving cell on the other side."[37] Although there are many types of neurotransmitters, hormones, and chemicals awash in body, a few are worth pointing out here, such as acetylcholine, serotonin, dopamine, and cortisol among others. They will provide the layperson with a basic understanding of neurotransmitters, which will be helpful for comprehending the neurological model.

Acetylcholine is an excitatory neurotransmitter that is emitted at the "neuromuscular junctions of all skeletal muscles," "found diffusely throughout the brain and concentrated specifically in the neurons of the basal ganglia."[38] The basal ganglia themselves are ". . . intimately involved in the initiation and control of movement."[39]

Serotonin is an "inhibitory transmitter" whose actions "have been implicated in various processes, including sleep, pain and the psychobiology of various affective disorders, specifically depression and bipolar disorder."[40] At the time it was written, it was approximated that only about "1-2% of the body's serotonin is found in the nervous system," with most being in the "mucous membranes of the gastrointestinal system and blood platelets."[41]

Dopamine, as a neurotransmitter, "has both excitatory and inhibitory functions depending on the pathway and the properties of the post-synaptic receptors" and is a "precursor of epinephrine and norepinephrine."[42] "Movement, attention, learning, the reinforcing effects of drugs, and various neuropsychiatric disorders" typically involve dopamine in one form or another.[43] If epinephrine is carried in the bloodstream then it is classified as a hormone, but if it is released in a synapse then it is classified as a neurotransmitter.[44] Epinephrine, when in the blood stream, increases oxygen and glucose levels in the brain, boosts the heart rate, and produces vasodilation in skeletal muscles.[45]

Cortisol is a hormone that ". . . suppresses the immune system, has anti-inflammatory properties, increase blood pressure and elevates blood sugar levels."[46] Cortisol levels are highest in the morning.[47] Cortisol plays a major role in the body's response to stress by "suppressing the immune system," "increasing blood pressure," "elevating blood sugar levels," and harboring "anti-inflammatory properties."[48] Cortisol levels that are abnormally high due to "long-term chronic stress" are "associated with muscle wasting, hyperglycemia, and damage to cells in the hippocampus resulting in memory loss."[49]

Lastly, endorphins are "opiate-like peptides are produced by the brain or the pituitary" and are theorized to "play important roles in the control of emotional behaviors such as those associated with pain, anxiety, fear, and other affective states produced by pain."[50] Endorphins are often released during intense exercise.

Implications of Neurotransmitters and Hormones for the Network Model of Affect

The suggested pathways presented in the two presented models differed only slightly in their arrangement, concerning the location of excitatory synapses, inhibitory synapses, and what the networked models represented (i.e., anxiety or negative anxiety). For instance, in the avoidance of pain model, the networked neurons are excited by a threat of harm (or the lack of a benefit if the benefit equations are being used, more on that later). They would ultimately be linked to a receptor site associated with stress or anxiety, for example, one that might trigger the release of large amounts of the hormone cortisol. Alternatively, in the pursuit of pleasure model the networked neurons are excited by the efficacy to prevent harm (or a lack of efficacy to prevent a benefit if the benefit equations are being used). They would ultimately be linked to a receptor site associated with eliminating stress and elevating the concept of negative anxiety, such as one that might trigger the release of large amounts of endorphins or other endogenous opiates associated with pleasure.

Receptor sites for the two types of valuing neurons would be associated as follows:

1) Positive anxiety with the release of stress inducing chemicals, hormones, and neurotransmitters

2) Negative anxiety with the release of stress reducing chemicals, hormones, and neurotransmitters

Dopamine, epinephrine, and norepinephrine, for instance, might be associated with either model for the avoidance of pain or the pursuit of pleasure, depending on whether or not the networked neurons that they are acting on are concerned with anxiety (e.g., cortisol) or negative anxiety (e.g., endorphins). Other neurotransmitters, such as acetylcholine or serotonin that may excite or inhibit the neurons leading up to these different networks, might serve as on-off switches, like breakers in an electrical circuit that protect an appliance's circuit from overloading.

Fight-or-Flight Response: the Avoidance of Pain and the Pursuit of Pleasure

In the context of the equation, the question as to whether a response is of a fight-or-flight nature is somewhat open. It is not the case that all flight-responses are relegated to the pursuit of pleasure; similarly all fight-responses are not relegated to the avoidance of pain equation. Either response may lead to an investment of anxiety (pain) or negative anxiety (pleasure) depending on whether procuring safety (flight response) outweighs its complement of not procuring safety, as judged by the individual. Likewise, striking against the threat (fight response) may lead to an investment of anxiety or negative anxiety depending on whether it is felt to already be balanced with its complementary purpose of not striking against a threat.

Equation for the Avoidance of Pain formula (Where an entity has a positive Appraisal (of + 1) and its acquisition or the safeguarding of the entity restores equilibrium between a purpose and its complementary purposes)

Figure 8.16 The Avoidance of Pain Formula (thus far):
Appraisal equals + 1;
Existence may equal + 1 or 0;
Uniqueness, Sufficiency, and Sentiment may equal any real number between 0 and + 1, though only Sufficiency can equal zero, for instance, for a complementary purpose;
Perceived Threat Severity, Perceived Threat Susceptibility, Perceived Response Efficacy, and Perceived Self-Efficacy may equal any real number between 0 and + 1, so long as the denominator does not equal zero.

1) Where purpose A and purpose B are complementary so that purpose A involves doing something and purpose B involves not doing something.

2) Reversing the Appraisal sign of an entity's value for purpose A indirectly gauges the negative anxiety felt for purpose B.

   a) If an entity is useful to purpose A, then its destruction would be expected to cause pain and an increase in anxiety with respect to that purpose.

b) Reversing the Appraisal value yields a rough estimate of the net negative anxiety that would be invested in the entity cause for the complement, which is purpose B.

c) This would be an indirect way to estimate an entity's value for the complementary purpose B if the entity's value for purpose A is already known. As the entity's value for the complement, purpose B or not doing something, would only approach its existential value of + 1 emotional unit, the difference between the two would yield its net value for purpose A and for purpose B, (close to A).

d) What this also means, however, is that another expression must be found to represent negative anxiety for the current purpose, purpose A, and a different form of the current equation.

Equation for the Pursuit of Pleasure formula (Where an entity has a negative Appraisal (of - 1) and its acquisition or the safeguarding of the entity moves the individual away from equilibrium between a purpose and its complementary purpose)

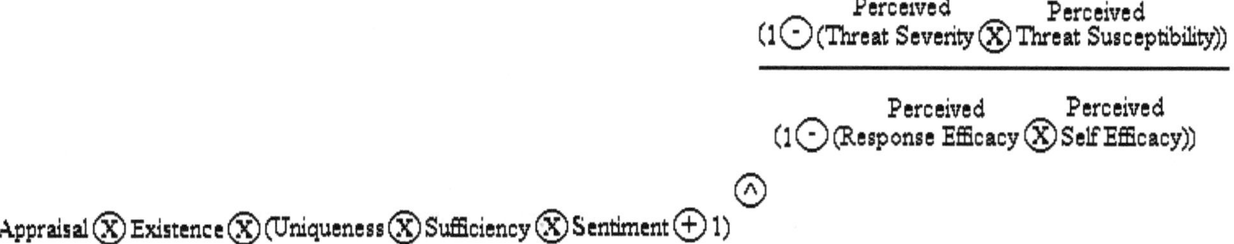

Figure 8.17 The Pursuit of Pleasure Formula (thus far):
Appraisal equals - 1;
Existence may equal + 1 or 0;
Uniqueness, Sufficiency, and Sentiment may equal any real number between 0 and + 1, though only Sufficiency may equal zero;
Perceived Threat Severity, Perceived Threat Susceptibility, Perceived Response Efficacy, and Perceived Self-Efficacy may equal any real number between 0 and + 1, so long as the denominator does not equal zero.

1) If the Appraisal is reversed, then the exponents will need to be modified to reflect the expected nature of pleasure. Both the numerator for the threat of harm components (threat severity and threat susceptibility), and the denominator for the efficacy components (response efficacy and self efficacy) will have to be modified.
   a) If an entity that would provide negative anxiety (pleasure) is harmed and nothing could be done to save it, negative anxiety (pleasure) should diminish.
   b) If an entity that would provide negative anxiety (pleasure) can be acquired with minimal harm done to it, then negative anxiety (pleasure) should increase
2) Representing these concepts
   a) If the exponent is changed to (1 - harm) for the numerator, then for a situation where harm to an entity equals zero and the denominator is held constant, negative anxiety invested would only increase or stay the same,

meaning that the entity's value could potentially approach negative infinity.

b) Likewise, if the denominator is changed to (1 - efficacy), then for a situation where efficacy approaches one, negative anxiety would increase, meaning that the value of the entity approaches negative infinity.

c) Similarly, for the pursuit of pleasure equation, if the Appraisal sign is reversed, then it would reveal a rough estimate of the net anxiety invested in the entity's value for the complementary purpose, purpose B.

Conceptualizing Harm and Efficacy for the Equations Through the Use of Logarithms

In mathematics, logarithms are used to represent the exponents that a specified base must be raised to in order to obtain a particular value. For instance, if the base of a log is 10, then the solutions of the following expressions are as follows:

the logarithm of 100 to base 10 is two, whereas

the logarithm of 1,000 to base 10 is three, whereas

the logarithm of 10,000 to base 10 is four, and so on.

For the avoidance of pain equation, and if the Appraisal coefficient variable of + 1 is affixed afterwards, the equation for the Avoidance of Pain can be rewritten as:

Log of y to base (Sufficiency × Uniqueness × Sentiment + 1) = (Harm) ÷ (Efficacy)

If the utility variables (Sufficiency, Uniqueness, and Sentiment) are all held constant at one, then this expression becomes:

Log of y to base two = (Harm) ÷ (Efficacy).

As the ratio of harm to efficacy grows in favor of harm, meaning that harm > efficacy or, as harm increases and efficacy decreases, an individual's appraised value judgment of an entity, hence the y-value, grows and more anxiety is directed toward it. This is in accord with logic, given that the entity would have a positive Appraisal toward the restoration of equilibrium and its acquisition would be in doubt.

Conversely, as the ratio of harm to efficacy grows in favor of efficacy, meaning that harm < efficacy or, as harm decreases and efficacy increases, an individual's appraised value judgment of an entity, hence the y-value, decreases toward its base or existential value of + 1 and anxiety resources are directed away from it. This is in accord with logic given that the entity would have a positive Appraisal toward the restoration of equilibrium but its acquisition or protection is not in doubt.

For the Pursuit of pleasure equation, and if the Appraisal coefficient variable of − 1 is affixed afterwards, the equation for the pursuit of pleasure equation can be rewritten as

$$\text{Log of y to base (Sufficiency Uniqueness} \times \text{Sentiment} + 1\text{)} =$$

$$(\ 1 - \text{Harm}\ ) \div (\ 1 - \text{Efficacy}\ ).$$

If the utility variables are all held constant at one, then this expression becomes:

$$\text{Log of y to base } 2 = (\ 1 - \text{Harm}\ ) \div (\ 1 - \text{Efficacy}\ ).$$

As the ratio of harm to efficacy grows in favor of efficacy (efficacy > harm), or as harm decreases and efficacy increases, an individual's appraised value judgment of an entity heads first toward infinity, and then toward negative infinity after the appraisal value of negative one is affixed as a coefficient to signal that the entity's acquisition is leading away from equilibrium. It's acquisition leads away from homeostasis between purpose A and its complementary purpose B. This is in accord with logic given that the entity would have a negative Appraisal toward the restoration of equilibrium and its acquisition would not be in doubt.

Conversely, as the ratio of harm to efficacy grows in favor of harm (harm > efficacy), or as harm increases and efficacy decreases, an individual's appraised value judgment of an entity heads first toward positive one and then toward negative one after the Appraisal value (-1) is affixed as a coefficient. This is in accordance with logic given that the entity would have a negative Appraisal toward the restoration of equilibrium (pleasure), but its acquisition is in doubt.

An Explanation of Contingencies and Contingency Theory

One concept that will prove useful in determining the relationship between one entity and another entity is that of contingency theory. An event may be said to be "contingent on another if there exists some demonstrable relationship between the two such that the occurrence of one tends to be accompanied by the occurrence of the other."[51]

The expected harm or benefit that one entity provides another entity can be used by an individual to establish a contingency or change one upon learning new information. Where the contingency between event *A* and event *B* is perfectly negative, it would be interpreted by the individual that the existence of event *A* signals the absence of the existence of event *B*. An individual would hold that the threat severity and susceptibility of event *A* to event *B* is maximized. If the self has only known the contingency between event *A* and event *B* to be perfectly negative, then the self would assess the harm as equaling $+1$ in the equation, meaning that the threat severity and threat susceptibility of event *A* are at such a level that they guarantee the absence of event *B*.

Moreover, the contingency value, which would be understood as harm if negative or as benefit if positive, must account for how event *A* influences the self's valuation of event *B*, in this case raising the value of event *B* toward positive infinity. Conversely, if the contingency between event *A* and event *B* is perfectly positive, benefit may be said to equal one, meaning that the benefit intensity and benefit susceptibility of Event *A* are at such a level that the presence of Event *A* guarantees the existence of Event *B*. The contingency component for benefit would have to properly account for Event *A*'s influence on how the self will value Event *B*, in this case lowering the value of Event *B* toward negative infinity.

The absolute value of the contingency for a relationship between the two entities will be used to assess harm (where $-1 \leq$ Contingency ≤ 0) and for benefit (where $0 \leq$ Contingency $\leq +1$). In the equation, therefore, where event *B* (original entity) has a contingency to event *A* that is negative:

For threats of harm, where ($-1 \leq$ Contingency ≤ 0),

$0 \leq$ the absolute value of event *B*'s contingency to event *A* =

threat severity $\times$ threat susceptibility ≤ 1.

For promises of benefit, where ($0 \leq$ Contingency $\leq +1$):

$0 \leq$ the absolute value of event *B*'s contingency to event *A* =

benefit intensity $\times$ benefit susceptibility ≤ 1.

In order to establish an efficacy variable, contingencies will also be used, with respect to the self and its relationship to the modifying entity, event *A* in this case. For ease of explanation, only negative contingencies for efficacy will be used in demonstrative equations throughout this book (e.g., the first four forms of the exponent). The self is a third entity, and efficacy will be considered to be the combination of an individual's self-efficacy and response-efficacy to prevent the existence of event *A*, regardless of whether or not the relationship between event *A* and event *B* has a negative contingency (between - 1 and zero) or a positive contingency (between zero and + 1). If positive contingencies are used in place of a negative one for efficacy, then these would correspond to the other four forms of the exponent (e.g., the fifth through eighth forms) that will be shown as well.

The absolute value of the contingency for the relationship between the modifying entity or second entity, event *A*, and the entity of the self will be used as a value for the individual's efficacy to prevent event *A* (where $-1 \leq$ Contingency ≤ 0). As an alternative, if a positive contingency between the self and event *A* were used ($0 \leq$ Contingency $\leq +1$), then it would represent the individual's efficacy to guarantee event *A*. Hence:

If efficacy refers to the ability of the self to prevent event *A*, where

($-1 \leq$ Contingency ≤ 0):

$0 \leq$ the absolute value of event *A*'s contingency to the Self =

(response-efficacy $\times$ self-efficacy) ≤ 1.

Alternatively, if, out of personal preference, efficacy were to be used to refer to the ability of the self to guarantee event *A*, where:

($0 \leq$ Contingency $\leq +1$), then there would be no need to distinguish absolute value, and

$0 \leq$ event *A*'s Contingency to Self = Response-Efficacy $\times$ Self-Efficacy ≤ 1).

Contingencies and Learning

Over time, the contingencies for a given relationship between two entities might eventually average out (e.g., from learning) to become an expected contingency. These expected contingencies can ultimately be used by the self to gauge its own predicted response to a novel situation or account for new experiences that are similar to previous scenarios. Like any measurement that utilizes mean, the first few trials would appear to be more indicative of the norm before it averages out. Subsequent trials would not modify the mean nearly as much due to the growing population of the sample. This is the primary manner by which affect engineering would account for learning.

Relationship of Contingencies and Reinforcement to Anxiety (Pain) and Negative Anxiety (Pleasure)

Listed below are scenarios for maximizing pleasure and minimizing pain when an entity is being valued for one purpose and by one person, a second entity has a contingency with the original entity (e.g., a threat of harm or promise of benefit), and the self, a third entity, has a contingency with the second entity (e.g., efficacy to prevent the harm or benefit). Whereas event *B* is the original entity that the self would like to keep, acquire, or maintain the integrity of, event *A* is an entity that has either a positive or negative contingency with event *B*. The self is a third entity that has a contingency with event *A* only. In the case where the self's contingency with event *B* is an issue, the self would be two of the three entities, both event *A* and the self contemplating the prevention of event *A*.

Where a Negative Contingency Exists Between Event A and Event B

For entities (e.g., event *A*) that have a negative contingency toward and inhibit the existence of the original entity by harming it (e.g., event *B*), the avoidance of pain equation is concerned with lowering anxiety, or preventing a harm from occurring, thus reducing anxiety invested in the original entity. The self would take action to ensure that event *A* does not occur so that it cannot harm event *B*. The harm of event *A* to event *B* would be toward + 1 (The absolute value of harm's negative contingency of - 1) while the efficacy of the self to prevent event *A* would ideally be toward positive one to minimize the harm (figure 8.18). Events, it should be remembered, are entities.

$$\frac{\textit{Harm of entity A to entity B}}{\textit{Efficacy to prevent entity A from harming entity B}}$$

Figure 8.18 The exponent for the Avoidance of Pain function and where *B*'s contingency to *A* is negative. The Appraisal coefficient in the base would be + 1.

Meanwhile, the pursuit of pleasure equation is concerned with elevating negative anxiety, or preventing a harm from occurring and thus increasing negative anxiety invested in the original entity. The self would take action to ensure that event *A* does not occur so that it cannot harm event *B*. Ideally, the harm of event *A* to event *B* would be toward zero, the absolute value of harm's negative contingency value, while the efficacy of the self to prevent event *A* would be toward positive one to maximize pleasure (figure 8.19).

$$\frac{1 - \textit{Harm of entity A to entity B}}{1 - \textit{Efficacy to prevent entity A from harming entity B}}$$

Figure 8.19 The exponent for the Pursuit of Pleasure function and where the contingency of *B* to *A* is negative. The Appraisal coefficient in the base would be - 1.

Where a Positive Contingency Exists Between Event A and Event B

For entities (e.g., event *A*) that have a positive contingency toward and promote the existence of the original entity by benefitting it (e.g., event *B*), the avoidance of pain equation is still concerned with lowering anxiety. This would entail causing a benefit to occur and thus reduce anxiety invested in the original entity. The self would not take action to prevent event *A* so that it will benefit event *B*. The benefits of event *A* to event *B* would be between zero and + 1 (Positive contingency value) while the efficacy of the self to prevent event *A* would be toward zero, maximizing benefit.

$$\frac{1 - \textit{Benefit of entity A to Entity B}}{1 - \textit{Efficacy to prevent entity A from benefitting B}}$$

Figure 8.20 The exponent for the Avoidance of Pain function and where the contingency of *B* to *A* is positive. The Appraisal coefficient in the base would be + 1.

Meanwhile, the pursuit of pleasure equation would be concerned with elevating negative anxiety, or causing a benefit to occur to increase negative anxiety invested in the original entity of event *B*. The self would take action to ensure that event *A* occurs so that it will benefit event *B*.

Ideally, the benefit of event *A* to event *B* would be toward + 1 (Positive contingency value) while the efficacy of the self to prevent event A would be toward zero, thus maximizing pleasure.

$$\frac{\textit{Benefit of entity A to B}}{\textit{Efficacy to prevent entity A from benefitting entity B}}$$

Figure 8.21 The exponent for the Pursuit of Pleasure function and where the contingency of *B* to *A* is positive. The Appraisal coefficient would be - 1.

An Adaption of Contingencies with the 1:1:1:1 Ratio to a 1:1:1:1:1 Ratio

The following is a more sophisticated version of the 1:1:1:1 ratio that takes into account contingencies. Hence, a 1:1:1:1:1 ratio results where the self appraises an entity for every contingency available. The sum of all the y-values for the entity against each contingency will yield an overall value for an entity for the specific purpose.

In example, for a situation where two or more entities have a contingency with a single, original entity that is being valued for a single purpose, the contingencies are recognized by the self, and the self assesses what it can do to prevent both, the following results:

1) To uphold a 1:1:1:1:1 ratio, for every entity with which the single and original entity has a contingency that is recognized by the self, there would also have to be a value for the original entity. For example:

a) It is given that *A, B, C,* and *D* are entities and *B* is the original entity. This means that there are three contingencies that exist for *B*: one with *A*; one with *C*; one with *D*. Entity *B*'s original value is + 2 emotional units.

b) *C* and *D* have a positive or negative contingency with *B* while the entity *A* has a zero contingency with *B*. *C*'s contingency with *B* is negative one (- 1), meaning that both Threat Severity and Threat Susceptibility are maximized at one. *D*'s contingency with *B* is positive one (+ 1), meaning that both Benefit Intensity and Benefit Susceptibility are maximized at one. Entity *B* has a zero contingency to entity *A*, meaning there is no relationship. In this example it would be held that entity *B* is being valued with its base value of + 2 emotional units. Efficacy to prevent entities *C* or *D* will be the same at 1 / 2, meaning that the product of Response Efficacy and Self Efficacy will equal .5 in each case.

c) The total appraised value of the original entity B would be represented as the sum of its three values that are separately modified by the contingencies of *A, C*, and *D* at any given point in time.

d) The total value of the original entity *B*, if only its contingency to entity *A* were being considered, would be its original value, as A would have no distorting effect on entity *B*'s valuation (figure 8.22).

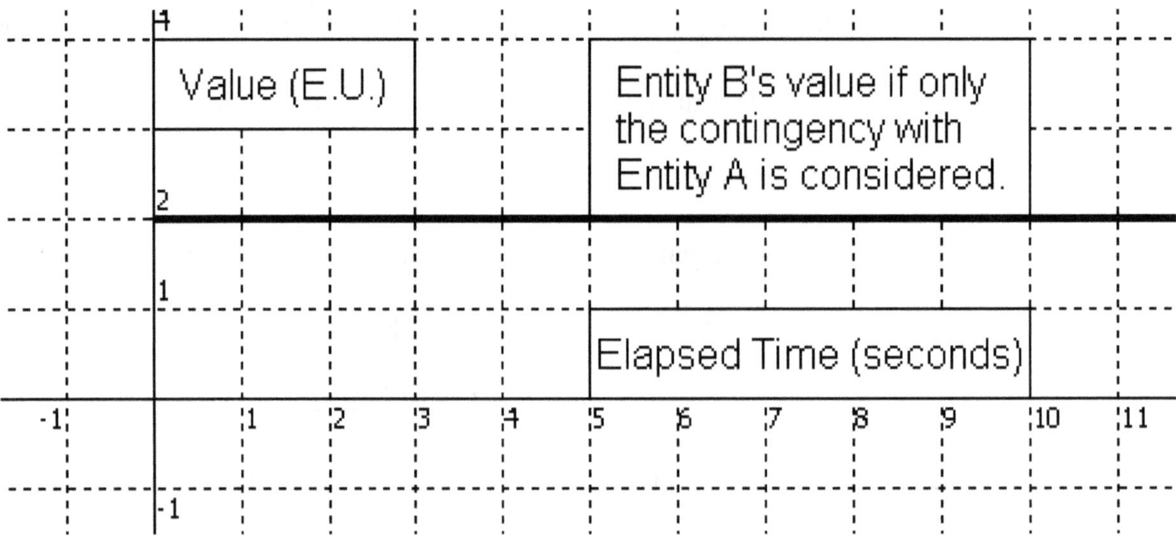

Figure 8.22 If only *B*'s contingency to *A* were considered, then *B*'s valuation would remain at + 2 emotional units, as entity *A* has no influence on entity *B*'s existence.

e) The total value of the original entity *B*, if only *C* were being considered as a contingency, would be higher (closer toward positive infinity) than its original value, because entity *C*'s relationship appears to be harmful to entity *B*'s existence and has a negative contingency (figure 8.23).

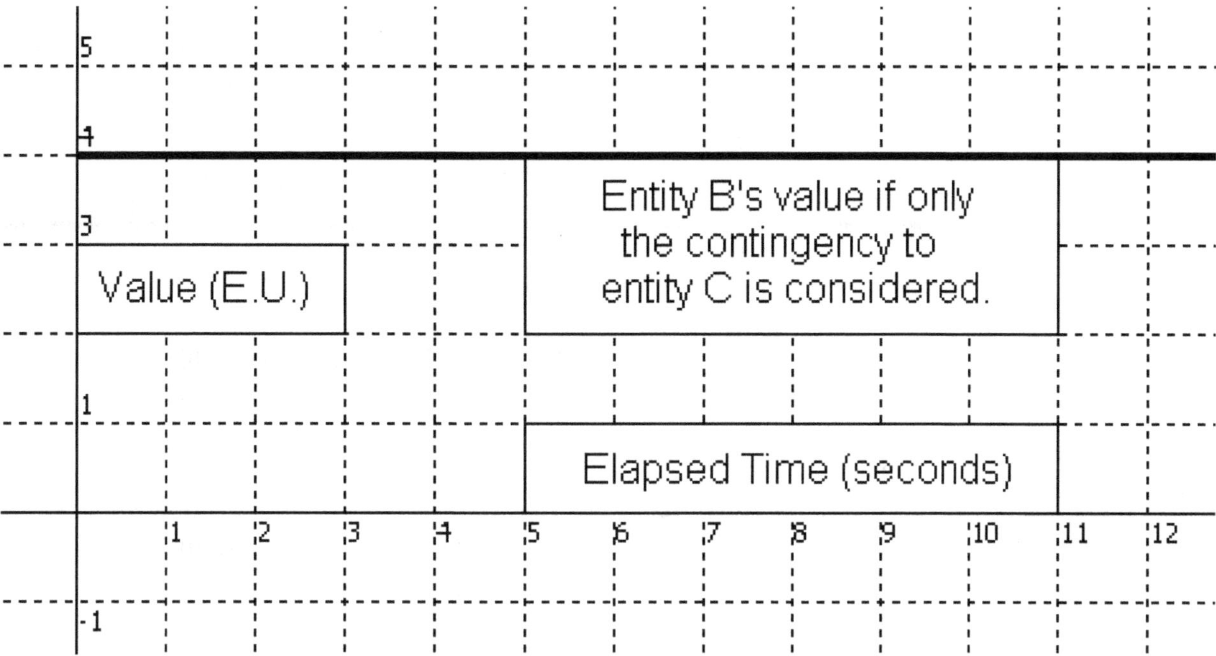

Figure 8.23 If only *B*'s contingency to *C* were considered, then *B*'s valuation would escalate to +4 emotional units, as entity *C* threatens entity *B*'s existence and the efficacy to prevent *C* is half of the harm. $2 ^\wedge (1 / .5) = 4$

f) The total value of the original entity *B*, if only *D* were being considered as a contingency, would be lower (closer toward its existential value) than its original value, because entity *D*'s relationship appears to be beneficial to entity *B*'s existence and has a positive contingency (figure 8.24).

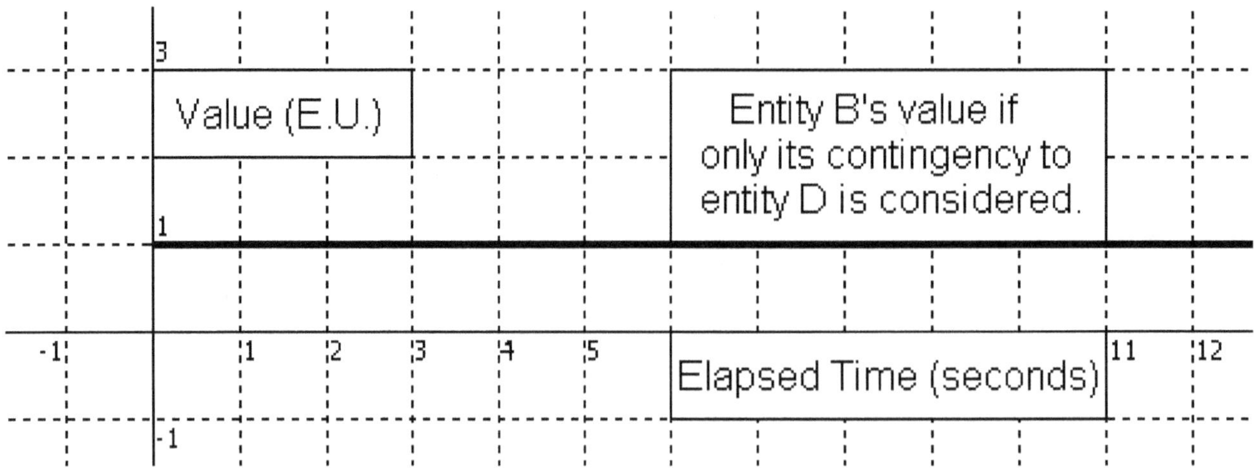

Figure 8.24 If only *B*'s contingency to *D* were considered, then *B*'s valuation would be closer to its existential value of + 1, as entity *D* is an indication of *B*'s existence (e.g., promotes it). 2 ^ ((1 - 1) / (1 - .5)) = 1

g) The combined value of all three contingencies is seven emotional units. If the combined valuations concerning the contingencies of entities *A, C*, and *D* were averaged out (2, 1, and 4 respectively), then it would result in the averaged value of entity *B* balancing out close to its original value at 7 / 3 emotional units, as entity *D* and *C* nearly offset each other. The total value of entity *B*, however, would be higher (closer toward positive infinity) if the acquisition of entity *B* will lead to a restoration of equilibrium due to its Appraisal and the stacking of values. Conversely, it would be lower (closer toward negative infinity) if the acquisition of entity *B* leads away from a restoration of equilibrium due to its Appraisal and the stacking of values.

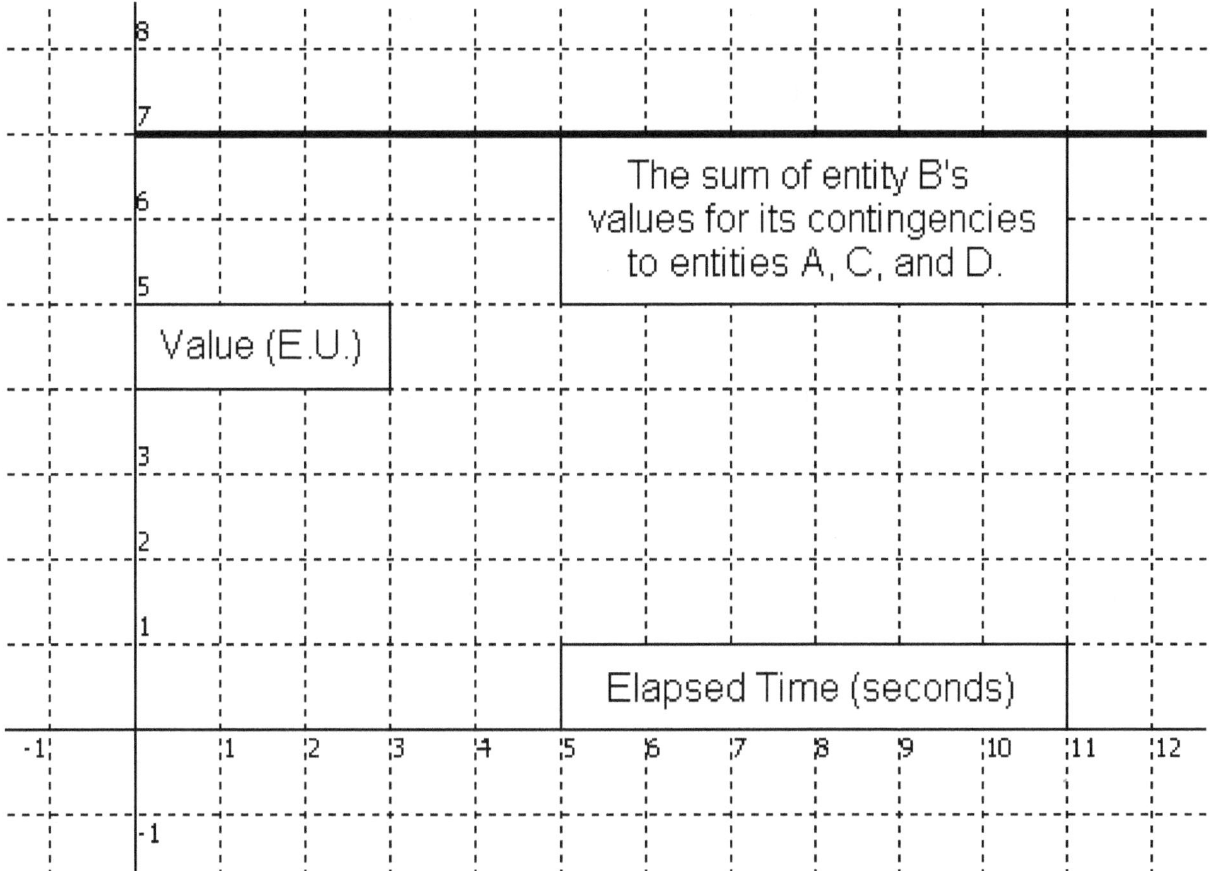

Figure 8.25 The sum of *B*'s contingencies to *A, C,* and *D*. If all three contingencies of *A, C,* and *D* are considered and the Appraisal of *B* towards the restoration of equilibrium is positive, then the total value will be higher than its original value.

2) Attempting to average the values for entity *B* would introduce more problems than it would solve when assessing an entity's value, and is not advised. For instance, if only *B*'s contingency to *A* and *C* are considered, the mean would suggest that the value of *B* is somewhere between its original value when *A* is considered and its value when only the contingency with *C* is considered. Entity *A*, due to having no contingency with *B*, pulls the mean down to its original level and based on its contingency does not indicate either the existence or nonexistence of *B* when *A* is

present. On the other hand, *C* threatens to wipe out entity *B* altogether, and the severity of its influence would be made to appear less so if the contingencies are averaged. It may also be the case that *C* and *D* have a relationship. The alternative to averaging would be to sum of all the contingencies to *B*, namely, *A, C*, and *D* concerning a single purpose, and then treat the sum of *B*'s values, rather than the average of B's values, as the value of entity *B* for whatever purpose or drive it fulfills. In such a case, the overall value of entity *B* would be equal to the sum of the following:

The y-value of entity *B* after accounting for *B*'s contingency to *A* in the equation,

plus

The y-value of entity *B* after accounting for *B*'s contingency to *C* in the equation,

plus

The y-value of entity *B* after accounting for *B*'s contingency to *D* in the equation at a specific point in time for all three (the x-value).

For instance, the following expression will be considered:

$1 + 2 + 3 + 4 + 5 = 15$.

The above can be rewritten using sigma, or $\sum$, a mathematical sign that means summation of or to sum up whatever is after the sign. For example,

$$\sum_{i=1}^{n=5} i = 1 + 2 + 3 + 4 + 5 = 15$$

The addition of the multiple y-values for *B*, with respect to its different contingencies to *A*, *C*, and

D can be expressed by adding the different functions together using sets for the threat and efficacy components concerning *A, C*, and *D*.

For the sets of ***A, C***, and ***D***:

A = No relationship, so no exponent is used and only *B*'s base value is needed;

C = {Threat Severity of *C* to *B*, Threat Susceptibility of *C* to *B*, Response-Efficacy to prevent *C*, Self-Efficacy to prevent *C*} in the exponent;

D = {Benefit Intensity of *D* to *B*, Benefit Susceptibility of *D* to *B*, Response-Efficacy to prevent *D*, Self-Efficacy to Prevent *D*} in the exponent.

For finding the total value of entity *B* if only the contingencies to *A* and *C* are considered, their influence on *B* would have to be added together. It will have to be remembered that the equation:

y = x is equivalent to f(x) = x, and that x is homeless and not the only variable in the equation.

If the fact that the x-variable is still homeless in the equation is ignored, then with all other drive reduction component variables held constant at + 1 (i.e., Uniqueness, Sufficiency, Sentiment):

The sum of the following:

Entity *B*'s base value (as entity *B* has no contingency to entity *A*; this value is + 2)

and

f (x, the threat severity of *C* to *B*, the threat susceptibility of *B* to *C*, response-efficacy to prevent *C*, self-efficacy to prevent *C*) would have a value of + 4 emotional units.

This total would be six emotional units.

This means that the y-value for *B*, concerning the different contingencies to *A* and *C* along with their threat and efficacy components, would be added together. Although *D*'s contingency to *B* is positive, meaning that a different form of the exponent would have to be used, its influence on the valuation of *B* (totaling + 1 emotional unit) would still be added to the total after its calculations had been made to equal the + 7 emotional units shown in figure 8.25.

3) For the sake of simplicity, examples throughout the rest of the book will only consider cases where one contingency is being considered at a time. Though the use of multiple contingencies is perhaps one of the more practical aspects of the equation, the computational effort required to make use of it on any significant scale would unnecessarily complicate the demonstration of other aspects of the equation. Multiple contingencies will be put aside after one final example.

4) An instance where multiple contingencies would be useful is disaster-preparedness:

    a) *A* is an entity that an individual is appraising as valuable and wants to acquire or protect from harm. *D, E,* and *F* are three separate disasters that have a negative contingency with *A*. *X, Y,* and *Z* are actions that the self may take to safeguard entity *A* against the disasters *D, E,* and *F* respectively.

        i) The initial set of [*A, D, E, F, X, Y,* and *Z*] represents all the entities that the self is valuing, aside from the self itself. There are seven functions representing each of these entities.

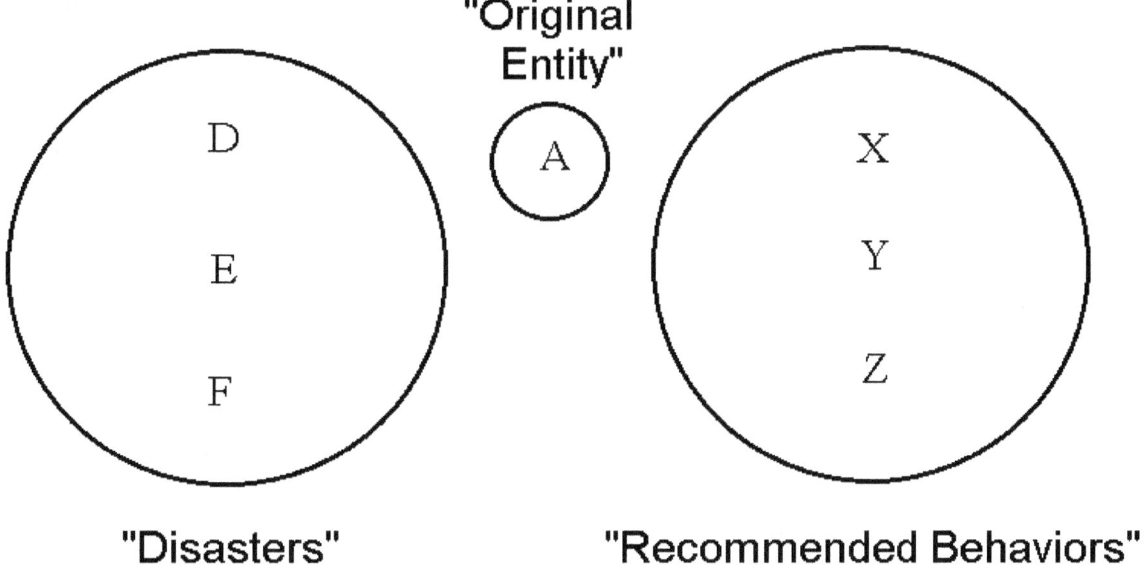

Figure 8.26 All Entities (Excluding the self)
Entity *A*: The self wants to acquire and protect this from *D, E*, and *F*
Entities *D, E,* and *F*: what the self wants to prevent
Entities *X, Y,* and *Z*: what the self wants to acquire to prevent *D, E,* and *F* respectively.

ii) Entities the self wants to protect from destruction = [*A*]

iii) Entities that the self does not want to protect from destruction because they threaten *A* are in the set of [*D, E,* and *F*]. These entities threaten the original entity of *A* and have a negative contingency with entity *A*.

iv) *X, Y,* and *Z* are entities concerning response measures that represent what the self can do to prevent a threat of harm (i.e., *D, E,* or *F*) and have a negative contingency with the entities *D, E,* and *F* respectively. The destruction of the entities *D, E,* and *F* can be accomplished via the self performing actions *X, Y,* and *Z* in order to safeguard entity *A*.

b) If *D, E,* and *F* all have a contingency of - 1 with *A,* and *X, Y,* and *Z* are all behaviors that will prevent *D, E,* and *F* from harming *A*, then the individual only needs to perform (i.e., acquire the entities of action for) the recommended behaviors (*X, Y,* and *Z*) in order to safeguard entity *A*. Ultimately, the self would aim to acquire or safeguard entities *A, X, Y,* and *Z,* while destroying, preventing, or evading entities *D, E,* and *F*.

c) If, however, additional contingencies exist, such as between recommended behaviors, then the model would have to account for changes in the self's value judgment, which it can. These contingencies may exist between any of the entities under consideration [*A, D, E, F, X, Y, Z*] but would most likely occur between recommended behaviors, for instance, where performing one action prohibits the self from engaging in others. If performing action *Y* (i.e., acquiring the entity of a recommended behavior) takes so long to accomplish that it does not leave sufficient time to do either *X* or *Z*, then the acquisitions of *X* and *Z* are themselves in jeopardy, because they are threatened with destruction by entity *Y* and the time or effort needed to acquire it.

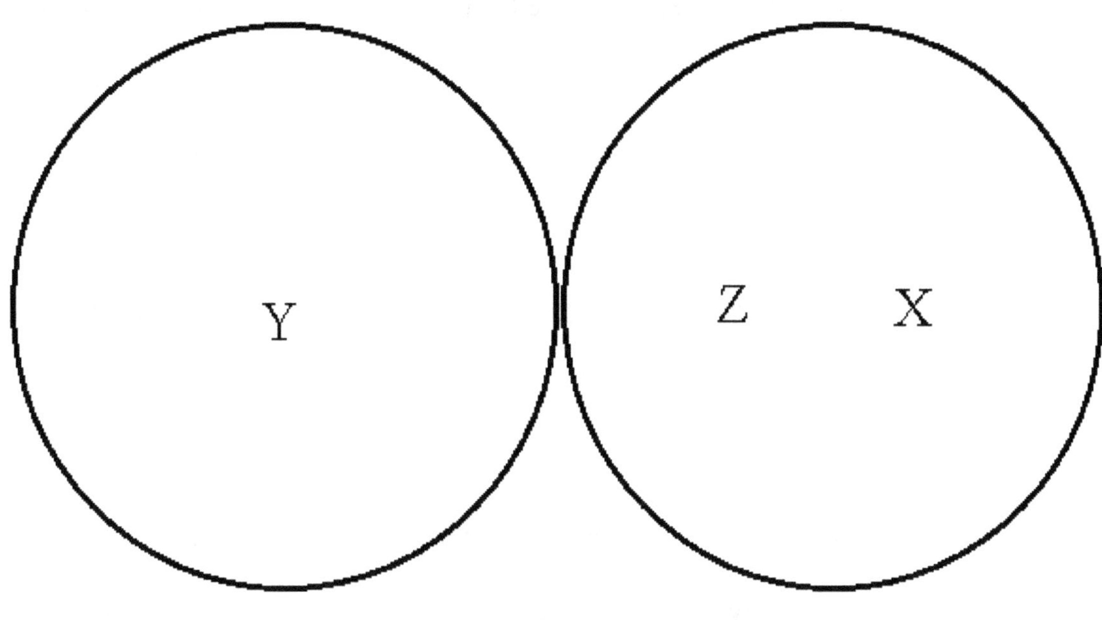

Figure 8.27 Because behavior *Y* takes so long to do, it prohibits performing *X* and *Z*. Performing the recommended behavior *Y* threatens the ability to acquire *X* and *Z*, which in turn jeopardizes entity *A* two fold the amount that performing *Z* and *X* would, which are not mutually exclusive.

If, however, a quicker alternative to entity *X* can be found (i.e., a new, alternative action that prevents harm *D*, such as the entity of behavior *V*) then *X*'s value would lower (e.g., value would be split between *X* and *V*, like a 2-for-1 stock split) if entity *V* can be acquired within the remaining time frame after *Y* is acquired.

Likewise, if harm *F*, which action *Z* is capable of preventing, is reassessed and turns out to be less severe or less probable than originally thought, for instance one quarter or .25 instead of one, then entity *Z*'s importance and value would lower, since it would not be responsible for preventing as much harm to entity *A* as either *Y, X,* or *X*'s alternative *V*. Instead of choosing recommended behaviors *X* and *Z*, the individual may choose recommended behaviors *Y* and *V* to minimize harm to *A*.

Summary of Contingencies and the Eight Forms of the Equation

It has been found that an individual may assess an entity as having one of three types of relationships with another entity:

1) A second entity may have no effect on another valued entity's existence, in which case the entity would have a zero contingency with the stimulus.
2) A second entity may signal the nonexistence of (e.g., harm threatened against) another valued entity, in which case the original entity would be negatively contingent to the second entity. The self, subsequently, would acknowledge that the second entity is harmful to the original entity's existence.
3) A second entity may signal the existence of (e.g., promise of benefit to) another valued entity, in which case the original entity would be positively contingent to the second entity. The self, subsequently, would acknowledge that the second entity is beneficial to the original entity's existence.

In all of the remaining demonstrative scenarios used throughout this book, the second entity will be considered a harm and the original entity will be negatively contingent toward the second entity unless specified otherwise. Also, throughout the book, efficacy (both response-efficacy and self-efficacy) will be thought of as the self's ability to prevent either a harm or benefit (if used) from befalling an entity, meaning the contingency between the self and the second entity will always be a value between negative one and zero, prior to being converted to an absolute value. As mentioned earlier, if one alternatively desired for efficacy to be used instead as a variable to indicate the self's belief in his or her own ability to ensure the existence of a second entity, regardless of whether the second entity has a positive contingency (i.e., benefit) or a negative contingency (i.e., harm) with the original entity, then this could also be arranged. This would additionally mean that the contingency between the second entity and the self would always be positive, between zero and one. However, so long as one keeps the definition of efficacy consistent and modifies the equation accordingly, the decision to employ one or the other lies with the architect's personal preference.

With efficacy being defined as the ability of the self to prevent the existence of a second entity that may harm or benefit an original entity, a total of four equations are possible (to note, if efficacy were being defined as the ability of the self to ensure the existence of a second entity, then these four equations would be slightly modified to yield four more equations for a grand total of eight, albeit measuring the same concepts). However, because the second entity cannot both signal the original entity's existence and nonexistence at the same time, hence, it cannot promote (benefit) and inhibit (harm) its existence simultaneously due to mutual exclusivity, it is only ever necessary to consider two out of the four at any given time. All eight forms of the exponent are listed below, but two are sufficient to gauge the value for any situation concerning a relationship between two entities and the self depending on the contingency between them. For demonstrative purposes, the two for harm will be used in examples throughout the book exclusively, as they are, in the author's opinion, easier to explain and comprehend.

Where efficacy refers to the response-efficacy and self-efficacy to either prevent harm befalling the original entity or to prevent a benefit from befalling the entity, then the exponent would be one of the following four arrangements.

1) Where the contingency between entity *A* and entity *B* is negative, meaning that the existence of entity *B* is harmed by the existence of Entity *A*, the exponents of the function are the following:

   a) **Form I of the exponent: Avoidance of Pain**

      Where the Appraisal value is + 1, the exponent is:

      ([Harm] ÷ [Efficacy to prevent entity *A*])

   b) **Form II of the exponent: Pursuit of Pleasure**

      Where the Appraisal value is - 1, the exponent is:

      ([1 - Harm] ÷ [1 - Efficacy to prevent entity *A*])

   c) The above two forms will be used in the demonstrations throughout the book.

2) Where the contingency between entity *A* and entity *B* is positive, meaning that the existence of entity *B* benefits from the existence of Entity *A*, then the exponents of the function are the following:

   a) **Form III of the exponent: Avoidance of Pain**

      Where the Appraisal value is + 1, the exponent is:

      ([1 - Benefit] ÷ [1 - Efficacy to prevent entity *A*])

   b) **Form IV of the exponent: Pursuit of Pleasure**

      Where the Appraisal value is - 1, the exponent is:

      ([Benefit] ÷ [Efficacy to prevent entity *A*])

If, out of personal preference, efficacy is defined as the self's capacity to ensure that a harm or benefit befalls an entity, then the above would be modified while still reflecting the same principles as before. Where efficacy refers to the response-efficacy and self-efficacy to either ensure that harm befalls the original entity or ensure that a benefit befalls the original entity, the exponent would be one of the following four arrangements described below.

1) Where the contingency between entity *A* and entity *B* is negative, meaning that the existence of entity *B* is harmed by the existence of Entity *A*, the exponents of the function would be:

   a) **Form V of the exponent: Avoidance of Pain**

      Where the Appraisal value is + 1, the exponent is:

      ([Harm] ÷ [1 − Efficacy to ensure entity *A*])

   b) **Form VI of the exponent: Pursuit of Pleasure**

      Where the Appraisal Value is - 1, the exponent is:

      ([1 − Harm] ÷ [Efficacy to ensure entity *A*])

2) Where the contingency between entity A and entity B is positive, meaning that the existence of entity *B* benefits from the existence of Entity *A*, then the exponents of the function would be:

   a) **Form VII of the exponent: Avoidance of Pain**

      Where the Appraisal value is + 1, the exponent is:

      ([1 − Benefit] ÷ [Efficacy to ensure entity *A*])

   b) **Form VIII of the exponent: Pursuit of Pleasure**

      Where the Appraisal Value is - 1, the exponent is:

      ([Benefit] ÷ [1 − Efficacy to ensure entity *A*])

Additional Assessments

In cases where a contingency is positive, meaning that one entity's presence (e.g., entity *A*) ensures the presence of another entity (e.g., Entity *B*), another entity is considered. What this means is that the self must invest more of its anxiety resources into valuing the new, additional entity that is a duplicate of the original. This would be the case if the self already has *B*, and then encounters *A* which signals that another of *B* will soon be available. In this circumstance, the anxiety resources would most easily be derived from the original entity, in much the same manner that shares in a stock would be valued if the stock were split, two for one in this case.

For instance, it is given that a company has 5,000 shares of stock. Rachel has only one share of stock in the company and each share is valued at $100. A two for one split of the stock by the company would result in there now being a total of 10,000 shares of stock and Rachel would now have two shares of stock, each valued at approximately $50 each. Likewise, if the value of Rachel's original, single share ($100), represented her anxiety that she had invested into valuing something else, for instance a set of porcelain chinaware, and by chance she happened to receive a duplicate of the same set of porcelain chinaware as a gift (she only requires one for her purposes), then her $100 valuation of the set of porcelain chinaware would become evenly split between the two entities.

Figure 8.28 Porcelain Chinaware
After obtaining a duplicate of the original, the original $100 worth of anxiety Rachel invested in the chinaware becomes split between the two entities at $50 each.

Accounting for Differences in Semantics and the Social Construction of Emotion

The eight different forms of the emotional equation that are described above comprise the different ways by which the same changes in affect may be construed. This would help to explain why different ways of conveying the same message might enjoy different levels of effectiveness. Social constructionism is an approach to learning, bearing similarities to postmodern thought, that holds "all knowledge, including scientifically obtained knowledge, is a construct of culture, language, and social roles and has no claim to final truth."[52] If this is held true, it would seem that learning bears much in common with postmodernism. However, relationships between two entities can be structured in a multitude of ways with the equation, some an individual may not be familiar with or adept at using.

For instance, it will be given that entity *B*, an ice cube, has a negative contingency with a lit candle, entity *A*, and the self is considering what to do about it in order to balance the temperature of the room. Utilizing the multiple forms of the equation, and because the contingency is a negative one, at least four statements could be used to describe the relationship between the candle, the self, and the ice cube's value for balancing the room's temperature to the individual's preferred equilibrium.

Four constructions are possible to describe the following: prior temperature of the room; what the individual did to change the temperature concerning an interaction between the self and the candle; whether or not temperature equilibrium was achieved. A total of eight possible constructions arise if the complementary purposes are considered in addition to the above. If the original entity were instead the candle, and the ice cube either threatened or promoted the existence of the candle, then another possible eight statements could be made concerning their relationship, depending on whether or not the contingency was negative or positive. Finally, if the contingency of *A* to *B* were actually a positive one, then a separate sixteen statements could be made about the relationship between these three entities, the self, entity *A*, and entity *B*, for a grand total of thirty-two.

For negative contingencies concerning the candle's ability to threaten the ice cube with destruction, the first four statements would be the purposes themselves:

1) The self prevents the candle from melting the ice cube in order to restore equilibrium. (It was hot in the room, and will soon be balanced.)
2) The self prevents the candle from melting the ice cube in order to avert equilibrium.

(It was cold in the room, and will get even colder.)

3) The self ensures that the candle melts the ice cube in order to restore equilibrium. (It was cold in the room, and will soon be balanced.)

4) The self ensures that the candle melts the ice cube in order to avert equilibrium. (It was hot in the room, and will get even hotter.)

The complementary purposes to the above would be the following:

5) The self does not prevent the candle from melting the ice cube in order to restore equilibrium. (It was cold in the room, and will soon be balanced.)

6) The self does not prevent the candle from melting the ice cube in order to avert equilibrium. (It was hot in the room, and will get even hotter.)

7) The self does not ensure that the candle melts the ice cube in order to restore equilibrium. (It was hot in the room, and will soon be balanced.)

8) The self does not ensure that the candle melts the ice cube in order to avert equilibrium. (It was cold in the room, and will get even colder)

An analysis of the above reveals the following:

1) Lines 2, 3, 5, and 8 all convey that it was originally too cold in the room

2) Lines 1, 4, 6, and all convey that it was originally too hot in the room.

3) Lines 1, 3, 5, and 7 all convey that the room's temperature eventually becomes balanced.

4) Lines 2, 4, 6, and 8 all convey that the room's temperature does not become balanced

5) Lines 1 and 7 convey exactly the same idea, but are separate purposes.

6) Lines 2 and 8 convey exactly the same idea, but are separate purposes.

7) Lines 3 and 5 convey exactly the same idea, but are separate purposes.

8) Lines 4 and 6 convey exactly the same idea, but are separate purposes.

The language of math, however, is specific and standardized. Utilizing these attributes then, it is possible to break down differences in word choice or sentence construction in order to explain how an individual might act or respond differently to essentially the same idea presented in an unfamiliar format. An individual, as a result of learning, may be more adept at responding to or interpreting specific styles of sentence construction over another one.

Elements Incorporated into the Equation

Below is a checklist of the variables assessed along with concepts and adaptations that have been incorporated into the equation. Given that it is a math equation, following the standard order of operations for math is primarily all that needs to be considered:

1) Adaptation of Lazarus' notion of "primary appraisal."[53] These are encompassed by the drive reduction component and Appraisal toward the restoration of equilibrium in the equation

   a) Drive Reduction Component

      i) This concerns whether or not an entity is useful to the fulfillment of a purpose.

      ii) If no, then its value for drive reduction or purpose fulfillment approaches zero and it is balanced out by its value for the complement, not reducing a drive or not fulfilling a purpose.

iii) If yes, then its value for drive reduction or purpose fulfillment approaches one and it will outweigh the complementary purpose.

b) Appraisal of the Entity Toward the Restoration of Equilibrium

i) Will acquisition of the entity lead to a restoration of equilibrium between a drive or purpose and its complement?

ii) If yes, then the Appraisal becomes + 1 and the Avoidance of Pain function is used.

iii) If no, then the Appraisal value becomes - 1 and the Pursuit of Pleasure function is used.

iv) If homeostasis is desired, an individual can only avoid pain to the extent that an entity valuable toward restoring equilibrium between complementary purposes or drives is being considered. If the entity's acquisition does not lead toward the restoration of equilibrium, then no felt pain for the specific purpose in question is predicted by the equation.

v) Also, if homeostasis is desired, an individual can only pursue pleasure to the extent that an entity valuable toward not restoring equilibrium between complementary drives or purposes is being considered. If the entity's acquisition does not lead away from the restoration of equilibrium, then no felt pleasure for the specific purpose is predicted by the equation.

2) Adaptation of Lazarus' secondary appraisals is incorporated into the equation in

the exponent. Rogers' Protection Motivation Theory is incorporated as well. These are encompassed in harm, benefit, and efficacy components used in the Avoidance of Pain and Pursuit of Pleasure functions.

a) Order of operations dictates the order in which calculations begin, starting with those items in parentheses and moving outwards, exponents and roots, multiplication and division, addition and subtraction, and working from left to right.

b) Standard mathematical restrictions apply (e.g., cannot divide by zero). This means that for the avoidance of pain equation the efficacy component cannot equal zero. For the pursuit of pleasure equation the efficacy component cannot equal one.

Category I Emotions, the Intra-Personal:
Pairing Emotions with Specific Appraised Value Judgments of Entities Based on the Transformations of an Entity's Value with Harm (or Benefit) and Efficacy Components

The self gives entities an initial appraised value, determined by the drive reduction components, the Appraisal of the entity toward the restoration of equilibrium, threat of harm to an entity (threat severity and threat susceptibility) and the efficacy components (response-efficacy and self-efficacy). Entities then are given a subsequent appraised value at a different time. Using the equation, patterns in these appraised value judgments determine an emotion type.

The following are what will be called Intra-Personal Emotions, or Category I Emotions.

They refer to emotions that only concern an individual and reflect a specific pattern concerning how an entity's value changes or stays the same with respect to a purpose or drive of the individual.

Suggested Appraisal Value Judgments, the Corresponding Emotion, Derivatives, and Integrals

1) The slope of each function indicates both the type of emotion involved and the magnitude. The slope of a function can be determined by finding the derivative, $f'(x)$.

2) Avoidance of Pain Emotions

a) Sadness

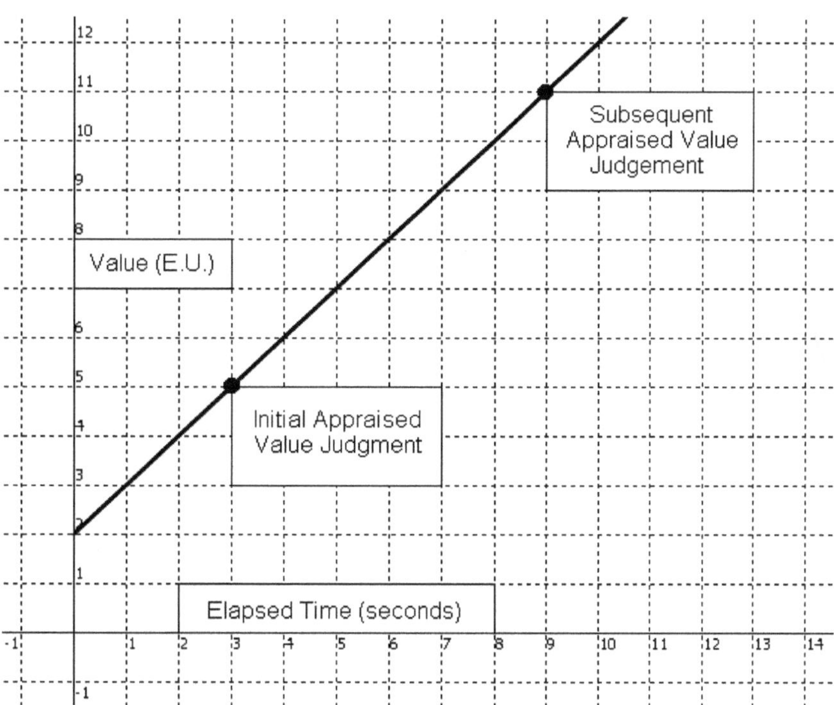

Figure 8.29 The above is a sample of an initial and subsequent appraised value judgement of an entity that is modeling sadness. Positive anxiety (pain) is being measured in emotional units. The magnitude of sadness is constant, as $f'(x)$, the slope of the function, is equal to + 1; anxiety is increasing.

i) Initial appraised value judgment: The value of the function is positive, above the x-axis, and ascending upward at a constant or accelerating rate or positive slope due to Threat Severity and Threat Susceptibility outweighing Response Efficacy and Self Efficacy.

ii) Subsequent Appraised Value Judgment: The value of the function remains positive, above the x-axis, and continues to ascend upward at either a constant or accelerating rate with a positive slope due to Threat Severity and Threat Susceptibility continuing to outweigh Response Efficacy and Self Efficacy.

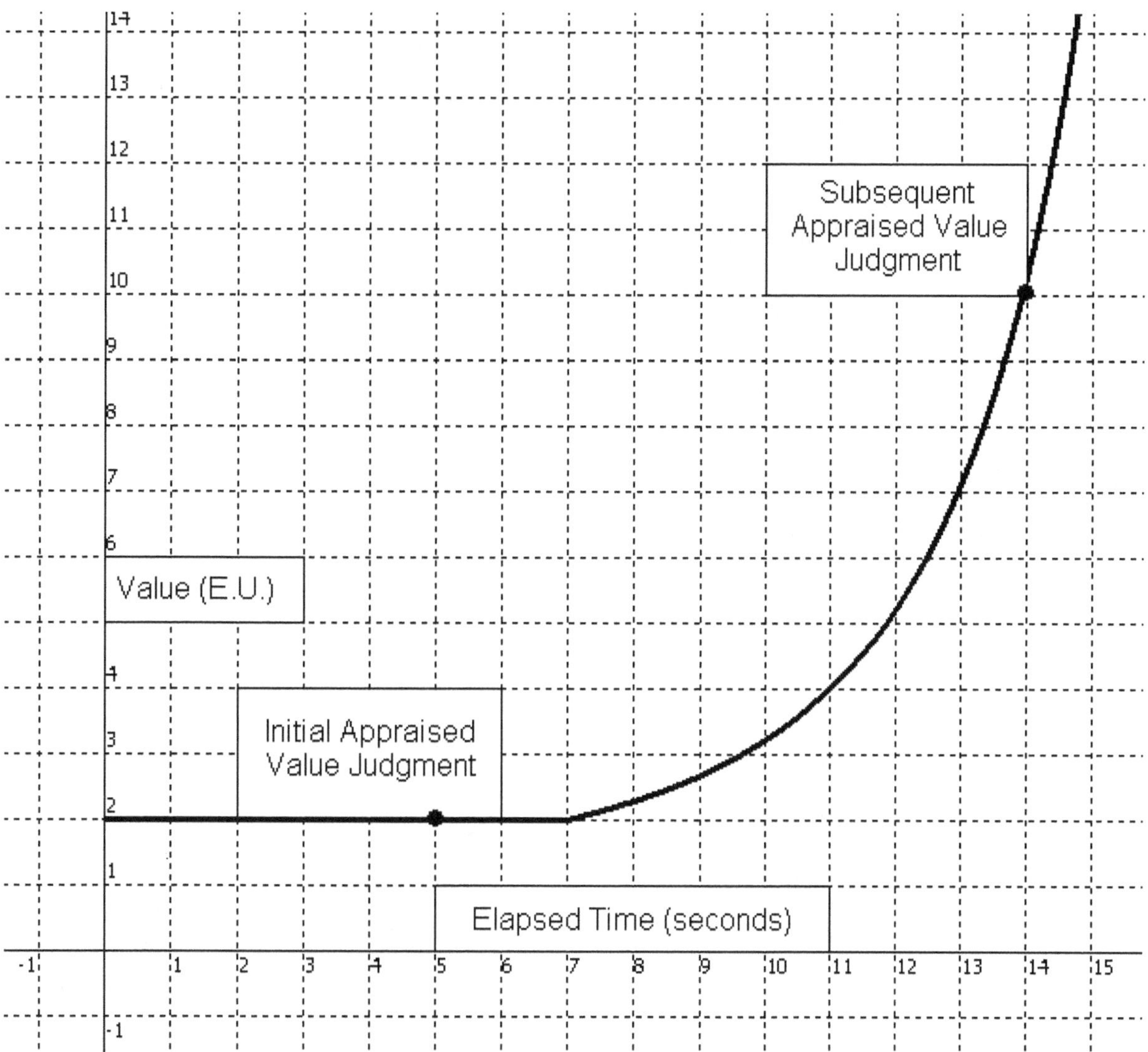

Figure 8.30 The above is a sample of an initial and subsequent appraised value judgement of an entity that models fear or grief. Positive anxiety (pain) is being measured in emotional units. The magnitude of fear is increasing, as f'(x) is elevating.

b) Fear or Grief

i) Initial Appraised Value Judgments: The value of the function is positive, above the x-axis, and neither ascending nor descending with a slope of zero. Threat Severity and Threat Susceptibility are in an equal proportion to Response Efficacy and Self Efficacy, so no distortion occurs.

ii) Subsequent Appraised Value Judgment: The value of the function is positive, above the x-axis, and begins to ascend upward at either an accelerating rate with a positive slope, due to Response Efficacy and Self Efficacy beginning to diminish, while Threat Severity and Threat Susceptibility remain elevated.

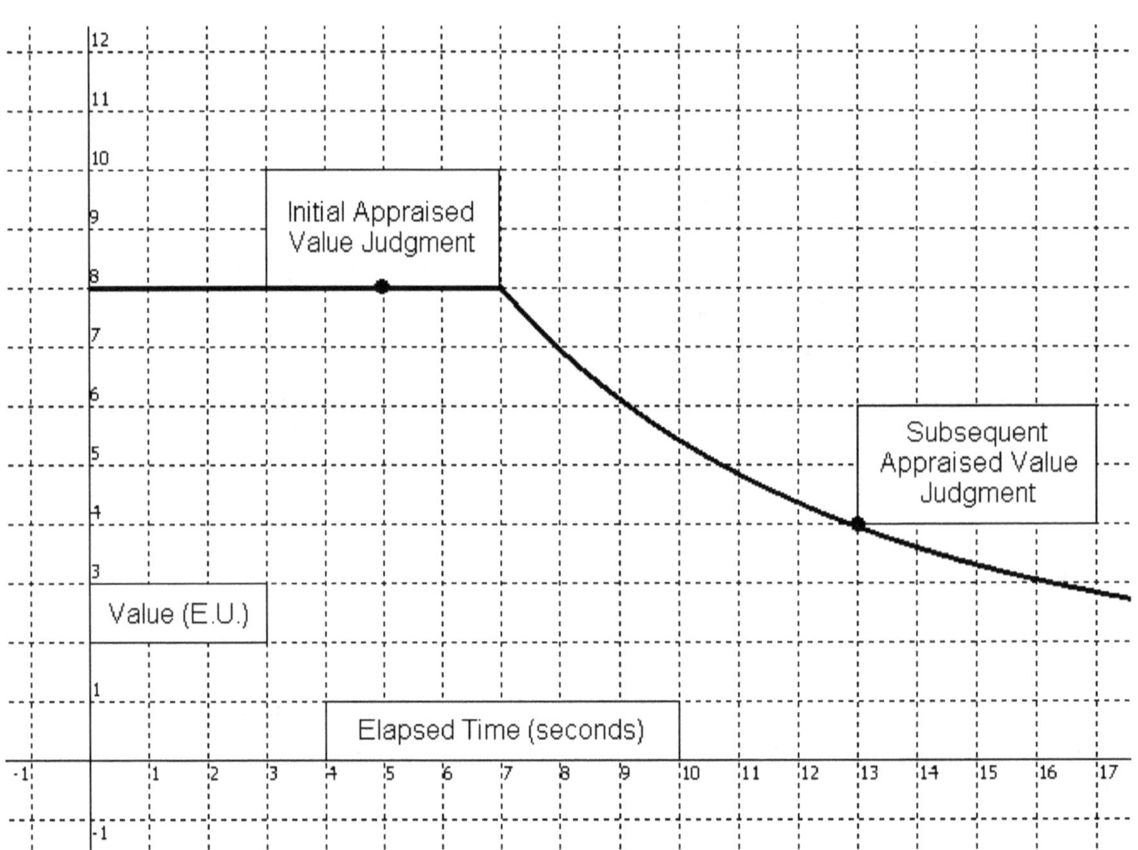

Figure 8.31 The above is a sample of an initial and subsequent appraised value judgement of an entity that models anger and aggression. Positive anxiety (pain) is being measured in emotional units. The magnitude of anger is decreasing, as f'(x) is lowering.

c) Anger, Aggression, Righteous Indignation

i) Initial Appraised Value Judgments: The value of the function is positive, elevated well above the x-axis, and initially constant or increasing due to Threat Severity and Threat Susceptibility initially outweighing Response Efficacy and Self Efficacy.

ii) Subsequent Appraised Value Judgment: The value of the function is positive, above the x-axis, and begins to descend at a constant or decelerating rate with a negative slope due to the Response Efficacy of a solution and the Self Efficacy of an individual to perform it becoming more pronounced. The individual begins to feel capable of thwarting whatever threat was directed at the original entity, such as by successfully striking against it.

3) Pursuit of Pleasure Emotions

a) Happiness

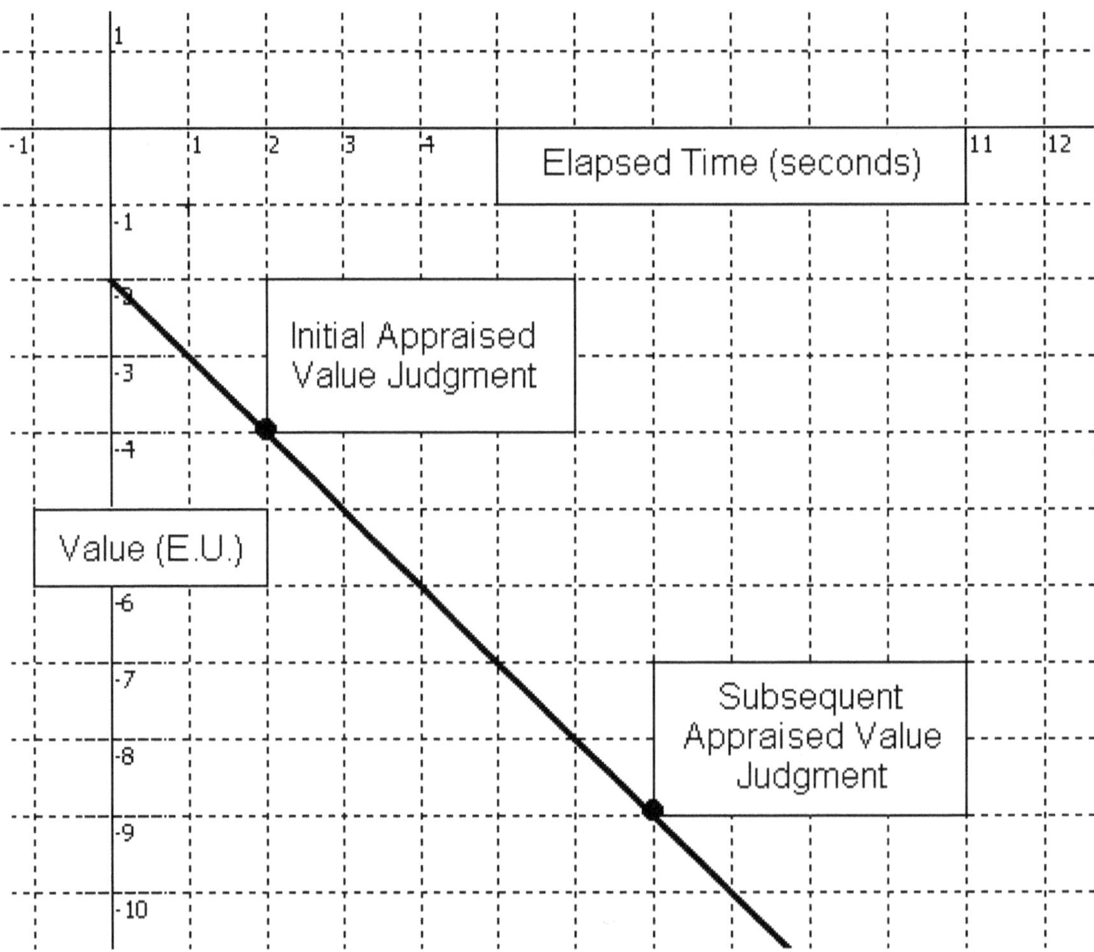

Figure 8.32 The above is a sample of initial and subsequent appraised value judgement of an entity that models happiness. Negative anxiety (pleasure) is being measured in emotional units. The magnitude of happiness is constant, as f'(x) is constant at -1; negative anxiety is increasing.

i) Initial Appraised Value Judgments: The value of the function is negative, below the x-axis, and descending downward at a constant rate with a negative slope due to Response Efficacy and Self Efficacy outweighing the Threat Severity and Threat Susceptibility imposed on the original entity by a second entity.

ii) Subsequent Appraised Value Judgment: The value of the function is negative, below the x-axis, and continues to descend downward at a constant or accelerating rate with a negative slope.

b) Courage or Euphoria

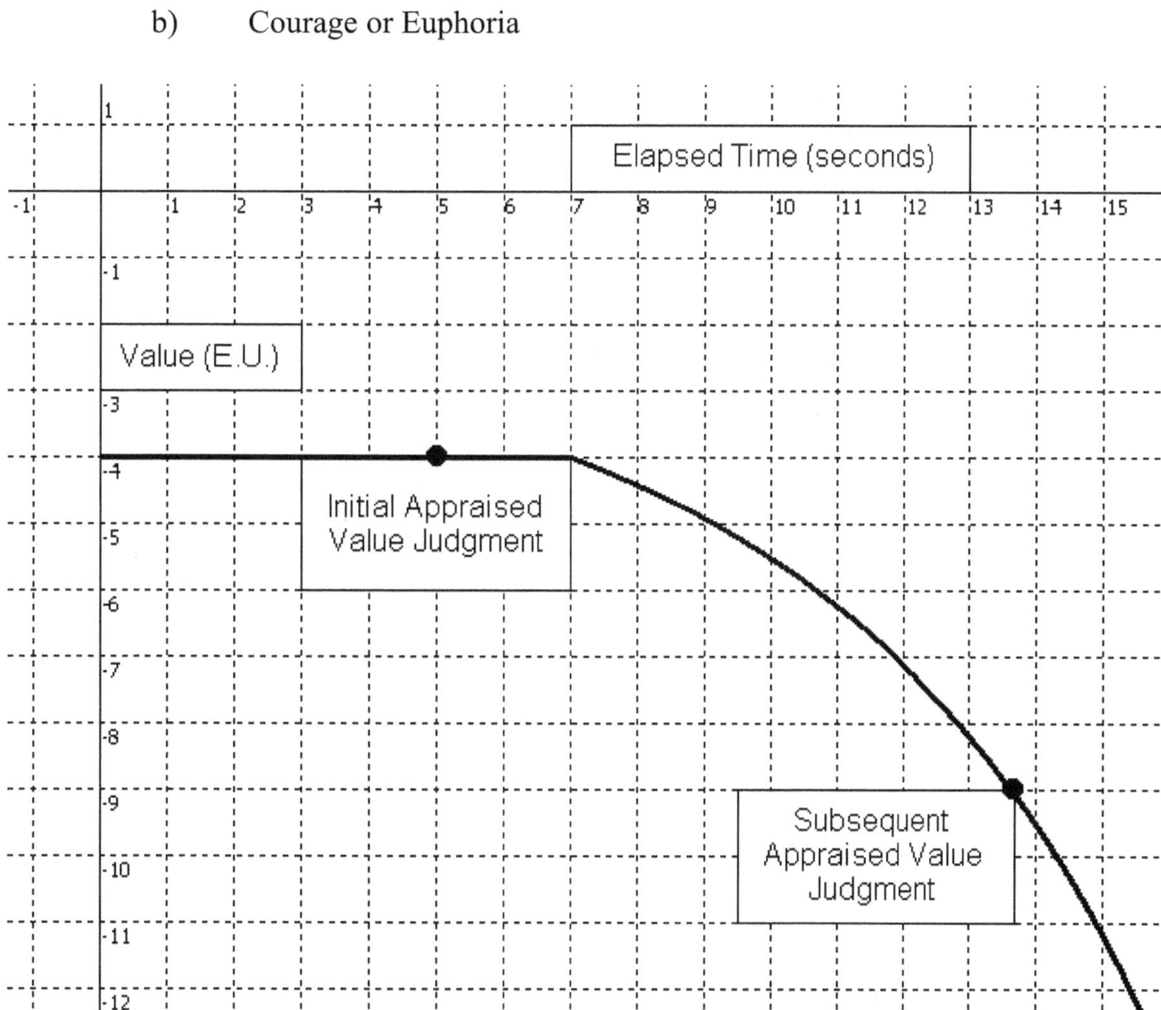

Figure 8.33 The above is a sample of initial and subsequent appraised value judgement of an entity that models courage or euphoria. Negative anxiety (pleasure) is being measured in emotional units. The magnitude of courage is increasing along with f'(x).

i) Initial Appraised Value Judgments: The value of the function is negative, below the x-axis, and constant with a slope of zero. Response Efficacy and Self Efficacy are in equal proportion to Threat Severity and Threat Susceptibility so no distortion is initially evident.

ii) Subsequent Appraised Value Judgment: The value of the function is negative, below the x-axis, and begins to descend downward at a constant or an accelerating rate with a negative slope. Response Efficacy and Self Efficacy begin to elevate against Threat Severity and Threat Susceptibility and remain high as the individual feels more capable of acquiring an entity.

c) Guilt, other forms of self-effacing disappointment

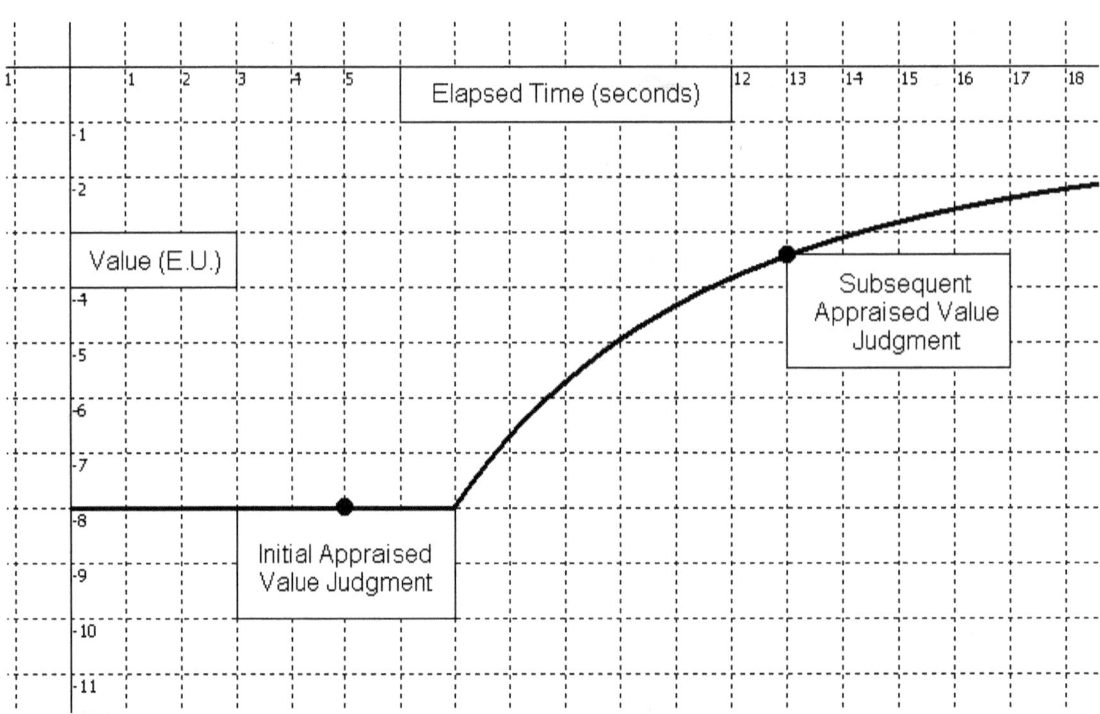

Figure 8.34 The above is a sample of an initial and subsequent appraised value judgement of an entity that models guilt. Negative anxiety (pleasure) is being measured in emotional units. The magnitude of guilt is decreasing, as f'(x) tapers to zero over time.

i) Initial Appraised Value Judgments: The value of the function is negative, well below the x-axis, and constant or descending with a negative slope due to Response Efficacy and Self Efficacy initially outweighing Threat Severity and Threat Susceptibility.

ii) Subsequent Appraised Value Judgment: The value of the function is negative, below the x-axis, and begins ascending at a decelerating rate with a positive slope toward its existential value of - 1. The Response Efficacy of a solution and the Self Efficacy of an individual to perform it become less pronounced while the Threat Severity and Threat Susceptibility may become more pronounced. The individual begins to feel incapable of thwarting whatever threat was directed at the original entity when initially this was not the case.

4) Finally, the area between the x-axis and each function gives an indication of the amount of anxiety or negative anxiety invested in an entity over a given point of time. The area between a function and the x-axis can be determined by finding the integral, $\int f(x)$.

Accounting for Threat and Efficacy in the Roll Cage Model

Threat and efficacy can also be depicted in the roll cage model, but with magnetic charge from the concerned entity-mass. High threat and low efficacy, if the entity-mass is below the fulcrum (positive anxiety) will draw forth more anxiety-mass from the stockpile. If the entity-mass is above the fulcrum (negative anxiety) then high efficacy and low threat will draw forth more anxiety mass from the stockpile. An entity-mass, threatened with destruction, is depicted in figure 8.35 as having a heightened magnetic charge and drawing forth more anxiety-mass from the stockpile. With more mass on the disc, the ego's valuation of the drive to maintain all drives

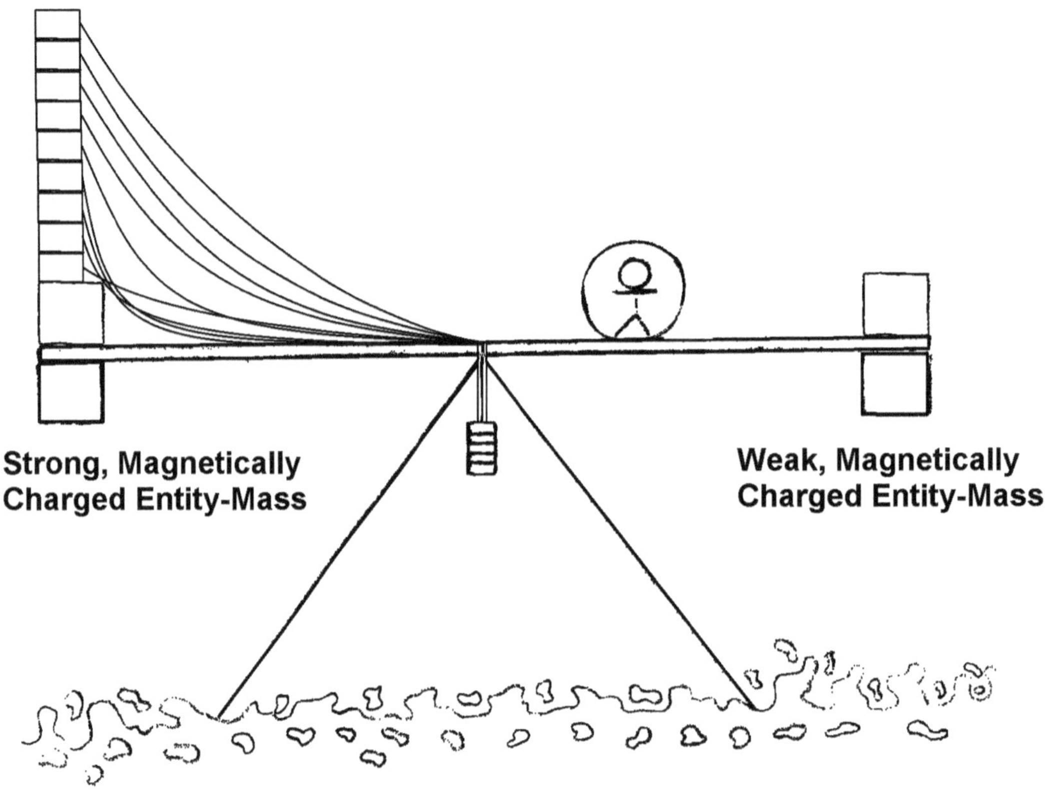

Figure 8.35 Threat and Efficacy in the Roll Cage Theory as Magnetic Charge. An increase in threat over efficacy when an entity-mass is *below* the fulcrum (pain and anxiety) will increase its magnetic charge and draw forth more mass from the stockpile. Conversely, an increase in efficacy over threat when an entity-mass is *above* the fulcrum (pleasure and negative anxiety) will also increase its magnetic charge and draw forth more anxiety-mass from the stockpile.

elevates. Having more mass on the disc threatens the ego's ability to cope and heightens anxiety invested in the purpose of maintaining equilibrium.

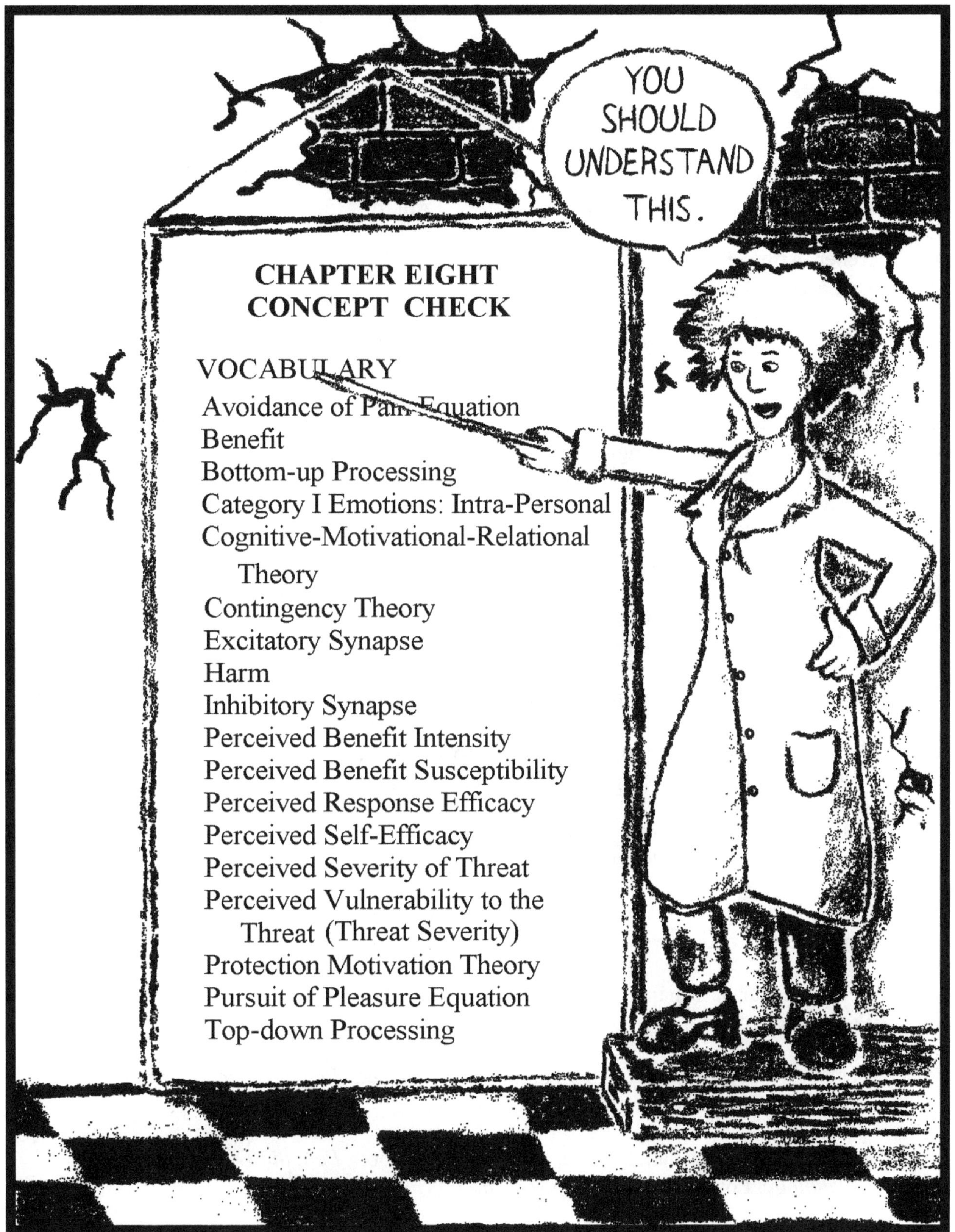

YOU SHOULD UNDERSTAND THIS.
CHAPTER EIGHT
CONCEPT CHECK
VOCABULARY
Avoidance of Pain Equation
Benefit
Bottom-up Processing
Category I Emotions: Intra-Personal
Cognitive-Motivational-Relational Theory
Contingency Theory
Excitatory Synapse
Harm
Inhibitory Synapse
Perceived Benefit Intensity
Perceived Benefit Susceptibility
Perceived Response Efficacy
Perceived Self-Efficacy
Perceived Severity of Threat
Perceived Vulnerability to the Threat (Threat Severity)
Protection Motivation Theory
Pursuit of Pleasure Equation
Top-down Processing

Notes

1. Wilde, Oscar (2003). *The Picture of Dorian Gray.* New York: Barnes and Nobles Books. Print, p. 43. (Original work published 1890).

2. Reber, Arthur S., Rhianon Allen, and Emily S. Reber. *Penguin Dictionary of Psychology*. London. Penguin Books, 2009. Print, p. 774-775.

3. Reber, Arthur S., Rhianon Allen, and Emily S. Reber. *Penguin Dictionary of Psychology*. London. Penguin Books, 2009. Print, p. 775.

4. Reber, Arthur S., Rhianon Allen, and Emily S. Reber. *Penguin Dictionary of Psychology*. London. Penguin Books, 2009. Print, p. 842.

5. Reber, Arthur S., Rhianon Allen, and Emily S. Reber. *Penguin Dictionary of Psychology*. London. Penguin Books, 2009. Print, p. 842.

6. Reber, Arthur S., Rhianon Allen, and Emily S. Reber. *Penguin Dictionary of Psychology*. London. Penguin Books, 2009. Print, p. 155.

7. Reber, Arthur S., Rhianon Allen, and Emily S. Reber. *Penguin Dictionary of Psychology*. London. Penguin Books, 2009. Print, p. 155.

8. Reber, Arthur S., Rhianon Allen, and Emily S. Reber. *Penguin Dictionary of Psychology*. London. Penguin Books, 2009. Print, p. 627.

9. Reber, Arthur S., Rhianon Allen, and Emily S. Reber. *Penguin Dictionary of Psychology*. London. Penguin Books, 2009. Print, p. 226.

10. Reber, Arthur S., Rhianon Allen, and Emily S. Reber. *Penguin Dictionary of Psychology*. London. Penguin Books, 2009. Print, p. 155.

11. Reber, Arthur S., Rhianon Allen, and Emily S. Reber. *Penguin Dictionary of Psychology*. London. Penguin Books, 2009. Print, p. 133.

12. Reber, Arthur S., Rhianon Allen, and Emily S. Reber. *Penguin Dictionary of Psychology*. London. Penguin Books, 2009. Print, p. 133.

13. Reber, Arthur S., Rhianon Allen, and Emily S. Reber. *Penguin Dictionary of Psychology*. London. Penguin Books, 2009. Print, p. 670 and p. 671

14. Reber, Arthur S., Rhianon Allen, and Emily S. Reber. *Penguin Dictionary of Psychology*. London. Penguin Books, 2009. Print, p. 673.

15. Reber, Arthur S., Rhianon Allen, and Emily S. Reber. *Penguin Dictionary of Psychology*. London. Penguin Books, 2009. Print, p. 673.

16. Reber, Arthur S., Rhianon Allen, and Emily S. Reber. *Penguin Dictionary of Psychology*. London. Penguin Books, 2009. Print, p. 669

17. Reber, Arthur S., Rhianon Allen, and Emily S. Reber. *Penguin Dictionary of Psychology*. London. Penguin Books, 2009. Print, p. 687

18. Reber, Arthur S., Rhianon Allen, and Emily S. Reber. *Penguin Dictionary of Psychology*. London. Penguin Books, 2009. Print, p. 641.

19. Cornelius, R. R. *The Science of Emotion: Research and Tradition in the Psychology of Emotion.* Upper Saddle River, NJ. Prentice-Hall, Inc. 1996. Print, p. 124

20. Cornelius, R. R. *The Science of Emotion: Research and Tradition in the Psychology of Emotion.* Upper Saddle River, NJ. Prentice-Hall, Inc. 1996. Print, p. 124

21. Cornelius, R. R. *The Science of Emotion: Research and Tradition in the Psychology of Emotion.* Upper Saddle River, NJ. Prentice-Hall, Inc. 1996. Print, p. 124

22. Cornelius, R. R. *The Science of Emotion: Research and Tradition in the Psychology of Emotion.* Upper Saddle River, NJ. Prentice-Hall, Inc. 1996. Print, p. 125

23. Cornelius, R. R. *The Science of Emotion: Research and Tradition in the Psychology of Emotion.* Upper Saddle River, NJ. Prentice-Hall, Inc. 1996. Print, p. 125

24. Cornelius, R. R. *The Science of Emotion: Research and Tradition in the Psychology of Emotion.* Upper Saddle River, NJ. Prentice-Hall, Inc. 1996. Print, p. 125

25. Cornelius, R. R. *The Science of Emotion: Research and Tradition in the Psychology of Emotion.* Upper Saddle River, NJ. Prentice-Hall, Inc. 1996. Print, p. 125

26. Cornelius, R. R. *The Science of Emotion: Research and Tradition in the Psychology of Emotion.* Upper Saddle River, NJ. Prentice-Hall, Inc. 1996. Print, p. 125

27. Cornelius, R. R. *The Science of Emotion: Research and Tradition in the Psychology of Emotion.* Upper Saddle River, NJ. Prentice-Hall, Inc. 1996. Print, p. 125

28. Rogers. R.W. *Cognitive and Physiological Processes in Fear Appeals and Attitude Change: A Revised Theory of Protection Motivation.* In J. Cacioppo & R. Petty (Eds.), Social Psychophysiology (1983), p. 153-176. New York: Guilford Press. Reprinted with permission of The Guilford Press.

29. Rogers. R.W. *Cognitive and Physiological Processes in Fear Appeals and Attitude Change: A Revised Theory of Protection Motivation.* In J. Cacioppo & R. Petty (Eds.), Social Psychophysiology (1983), p. 153-176. New York: Guilford Press. Reprinted with permission of The Guilford Press.

30. Reber, Arthur S., Rhianon Allen, and Emily S. Reber. *Penguin Dictionary of Psychology*. London. Penguin Books, 2009. Print, p. 816.

31. Rogers. R.W. *Cognitive and Physiological Processes in Fear Appeals and Attitude Change: A Revised Theory of Protection Motivation.* In J. Cacioppo & R. Petty (Eds.), Social Psychophysiology (1983), p. 153-176. New York: Guilford Press. Reprinted with permission of The Guilford Press.

32. Reber, Arthur S., Rhianon Allen, and Emily S. Reber. *Penguin Dictionary of Psychology*. London. Penguin Books, 2009. Print, p. 719

33. Rogers. R.W. *Cognitive and Physiological Processes in Fear Appeals and Attitude Change: A Revised Theory of Protection Motivation.* In J. Cacioppo & R. Petty (Eds.), Social Psychophysiology (1983), p. 153-176. New York: Guilford Press. Reprinted with permission of The Guilford Press.

34. Rogers. R.W. *Cognitive and Physiological Processes in Fear Appeals and Attitude Change: A Revised Theory of Protection Motivation.* In J. Cacioppo & R. Petty (Eds.), Social Psychophysiology (1983), p. 153-176. New York: Guilford Press. Reprinted with permission of The Guilford Press.

35. Reber, Arthur S., Rhianon Allen, and Emily S. Reber. *Penguin Dictionary of Psychology*. London. Penguin Books, 2009. Print, p. 107.

36. Reber, Arthur S., Rhianon Allen, and Emily S. Reber. *Penguin Dictionary of Psychology*. London. Penguin Books, 2009. Print, p. 823.

37. Reber, Arthur S., Rhianon Allen, and Emily S. Reber. *Penguin Dictionary of Psychology*. London. Penguin Books, 2009. Print, p. 509.

38. Reber, Arthur S., Rhianon Allen, and Emily S. Reber. *Penguin Dictionary of Psychology*. London. Penguin Books, 2009. Print, p. 7

39. Reber, Arthur S., Rhianon Allen, and Emily S. Reber. *Penguin Dictionary of Psychology*. London. Penguin Books, 2009. Print, p. 88.

40. Reber, Arthur S., Rhianon Allen, and Emily S. Reber. *Penguin Dictionary of Psychology*. London. Penguin Books, 2009. Print, p. 729

41. Reber, Arthur S., Rhianon Allen, and Emily S. Reber. *Penguin Dictionary of Psychology*. London. Penguin Books, 2009. Print, p. 729

42. Reber, Arthur S., Rhianon Allen, and Emily S. Reber. *Penguin Dictionary of Psychology*. London. Penguin Books, 2009. Print, p. 232

43. Reber, Arthur S., Rhianon Allen, and Emily S. Reber. *Penguin Dictionary of Psychology*. London. Penguin Books, 2009. Print, p. 232

44. Reber, Arthur S., Rhianon Allen, and Emily S. Reber. *Penguin Dictionary of Psychology*. London. Penguin Books, 2009. Print, p. 266

45. Reber, Arthur S., Rhianon Allen, and Emily S. Reber. *Penguin Dictionary of Psychology*. London. Penguin Books, 2009. Print, p. 266.

46. Reber, Arthur S., Rhianon Allen, and Emily S. Reber. *Penguin Dictionary of Psychology*. London. Penguin Books, 2009. Print, p. 175.

47. Reber, Arthur S., Rhianon Allen, and Emily S. Reber. *Penguin Dictionary of Psychology*. London. Penguin Books, 2009. Print, p. 175.

48. Reber, Arthur S., Rhianon Allen, and Emily S. Reber. *Penguin Dictionary of Psychology*. London. Penguin Books, 2009. Print, p. 175

49. Reber, Arthur S., Rhianon Allen, and Emily S. Reber. *Penguin Dictionary of Psychology*. London. Penguin Books, 2009. Print, p. 175

50. Reber, Arthur S., Rhianon Allen, and Emily S. Reber. *Penguin Dictionary of Psychology*. London. Penguin Books, 2009. Print, p. 262

51. Reber, Arthur S., Rhianon Allen, and Emily S. Reber. *Penguin Dictionary of Psychology*. London. Penguin Books, 2009. Print, p. 166

52. Reber, Arthur S., Rhianon Allen, and Emily S. Reber. *Penguin Dictionary of Psychology*. London. Penguin Books, 2009. Print, p. 747

53. Cornelius, R. R. *The Science of Emotion: Research and Tradition in the Psychology of Emotion.* Upper Saddle River, NJ. Prentice-Hall, Inc. 1996. Print, p. 125

"When one is in love, one always begins by deceiving one's self, and one always ends by deceiving others. That is what the world calls a romance." - Oscar Wilde[1]

CHAPTER NINE

Neurons, Empathy, and Category II Emotions (Inter-personal)

Front-loading: Antipathetic Mercy, Antipathy, Category II Emotions: Inter-Personal, Companionate Love, Empathy, Four Degrees of Empathy, Gamma-aminobutyric acid, Glutamate, Hateful Humiliation, Hatred, Imaginary Numbers, Indifference, Limerence Loneliness, Love, Loving Pride, Mercy, Mirror Neurons, Neutrality, Other, Pride Romantic Love, Self, Self-Distinction, Shame, Sympathetic Shame, Sympathy, Vicarious Humiliation, Vicarious Mercy, Vicarious Pride, Vicarious Shame, Vicarious Loneliness

The Role of Neurons and the Nervous System in the Valuation of an Entity

If the body is thought of as a symphony, then the nervous system would be the conductor leading the orchestra. Neurons, the basic cellular units of the nervous system, are generally of three types: motor, sensory, and interneurons. Motor neurons are nerve cells that activate an effector, such as muscle fibers.[2] Sensory nerves convey ". . . sensory information to the central nervous system."[3] Interneurons are connecting neurons and lie ". . . between sensory (afferent) and motor (efferent) neurons."[4] The excitatory or inhibitory effect that some neurons may have on other neurons or target cells was only mentioned briefly in chapter eight, but will be given more consideration here.

Neurons, Empathy, and Category II Emotions (Inter-personal

Excitatory Synapses and Inhibitory Synapses

A postsynaptic potential is a "... change in the membrane potential of a postsynaptic neuron," and may be either an excitatory one or an inhibitory one.[5] Depolarizations occur at excitatory post-synaptic potentials, which "... lower the threshold of the neuron and increase its likelihood of firing."[6] One such excitatory neurotransmitter is the amino acid glutamate, which binds with glutamate receptors.[7] Hyperpolarizations occur at inhibitory post-synaptic potentials, which raise the threshold of the neuron and decrease the likelihood of firing.[8] One such inhibitory neurotransmitter is Gamma-aminobutyric acid (i.e., $GABA_A$, $GABA_B$, or $GABA_C$) and is "... found throughout the grey matter."[9]

What About Shared Feelings and Empathy?

If the equation is to prove itself useful as a tool by modeling all that it claims it can, then ultimately it will have to account for empathy and shared emotions. Empathy entails "... cognitive awareness of the emotions and feelings of another person."[10] Empathy is also described as a "vicarious affective response to the emotional experiences of another person that mirrors or mimics that emotion" and is also described as "assuming, in one's mind, the role of another person."[11] A means of representing empathy in the equation will have to be found, but first a physiological source for empathy will be identified.

A strong candidate for a physiological source for empathy is a specific type of neuron in primates called the mirror neuron. Mirror neurons are "neurons that respond similarly whether one merely observes an action or event, or experiences it oneself."[12] In essence, an individual primate imagines himself or herself to be in the place of another and responds in the same manner as if he or she were performing the action. The extension of vicarious experience from other primates to other animals, life forms, or even inanimate objects that one has personified would not be improbable (e.g., vicariously experiencing the emotions of a family pet or a beloved car that is treated as human or alive after being exposed to hail damage). Hereafter, the task for the equation will be to find a means by which to mathematically express the concept of empathy or vicarious experience.

Empathy in the Equation

Given that the equation is concerned with how an individual values an entity, empathy in the equation will entail considering how another person values an entity. Hence, empathy will involve assessing another person, life form, or personified object's appraised judgment value as it relates to a purpose harbored by another (person, thing, or personified object) and as if the *self* were the *other*.

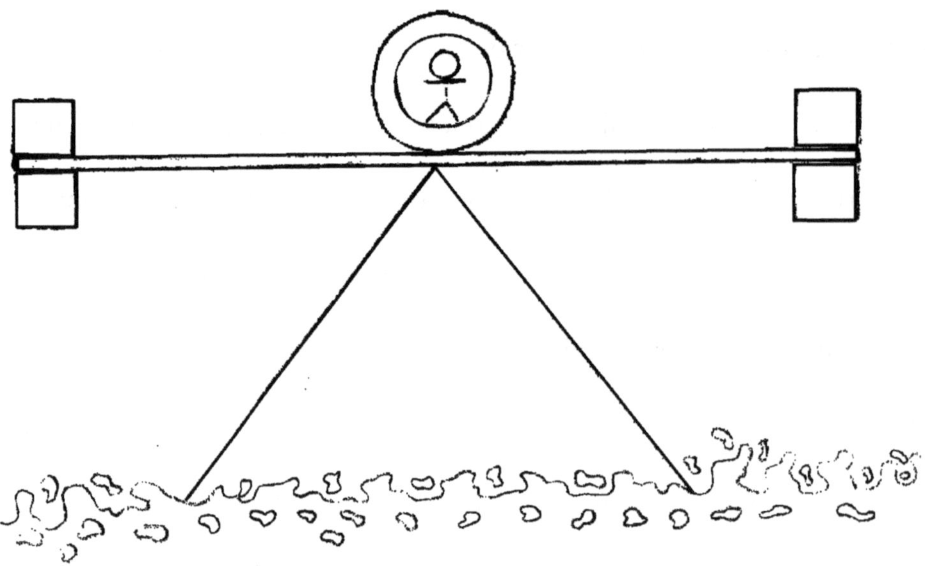

Figure 9.1 Vicarious experience in the Roll Cage Theory of Drives will be represented as a double sphere, hence, the self imagining itself as an other (person, place, or thing).

Fortunately, representing this concept mathematically will be less perplexing. The use of some complex math and imaginary numbers will make the task feasible. Before proceeding, some of the basics of imaginary numbers will be covered. The first order at hand would be defining what a square root is. A square root is a number that, when multiplied by itself, yields another value; the process of finding a number's root is known as root extraction. For instance, the square root of four is two. The square root of nine is three. Negative two and negative three are also square roots of four and nine respectively. The roots of negative numbers, however, can only be real numbers if the root is an odd number. For instance, the cubed root of negative eight is negative two,

$$(- 8) \wedge (1 / 3) = - 2, \text{ because } (- 2) \times (- 2) \times (- 2) = -8$$

However, the root of a negative number where the root itself is an even number always results in an imaginary number. For instance, the square root of negative one is represented by *i*:

$\sqrt{-1} = i$

Roots of larger negative numbers can be written with the coefficient *i* next to them. For instance, the square root of negative four is 2i, hence, $\sqrt{-4}$ can be rewritten as:

$\sqrt{4} \times \sqrt{-1} = 2 \times \sqrt{-1} = 2 \times i = 2i$

The square root of negative nine $\sqrt{-9}$ can be rewritten as:

$\sqrt{9} \times \sqrt{-1} = 3 \times \sqrt{-1} = 3 \times i = 3i$

What this means for the equation is that imaginary numbers can be used to signify when an individual is imagining himself or herself in the place of another person, life form, or personified object. Alternatively, real numbers would signify when an individual is not imagining himself or herself in the place of another person, life form, or personified object. For example, the coefficient i, or $\sqrt{-1}$, could be affixed to the front of the equation to signify the individual is imagining himself or herself as another person, life form, or personified object valuing an entity with respect to a purpose harbored by the other (person, life form, or personified object). Alternatively, the coefficient 1 or $\sqrt{1}$ would be affixed to the front of the equation to signify the individual is assessing the entity's value for a purpose harbored by the self. The terms self and other will be used to distinguish whether a valuation concerns a purpose harbored by an individual or is an instance of empathy.

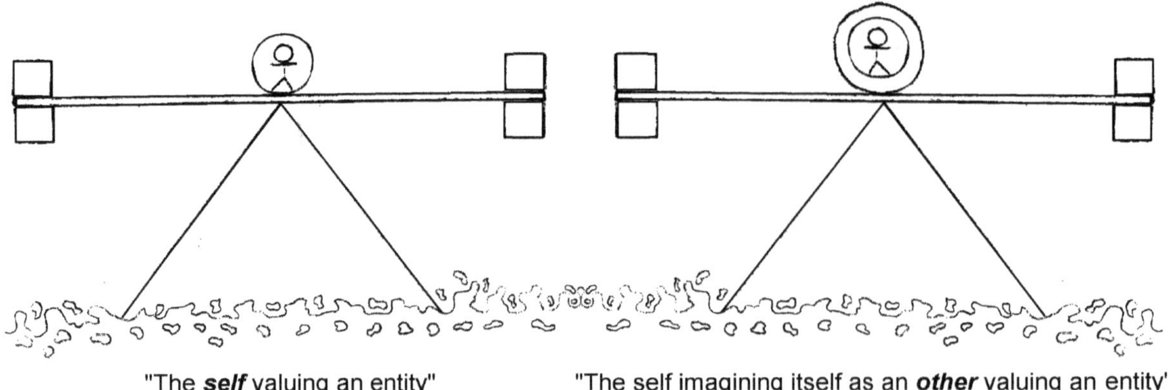

Figure 9.2 The self valuing an entity vs. the self imagining itself as an other valuing an entity.

Self-Distinction, therefore, will be the variable used to indicate whether the individual has distinguished the appraised value-judgment of the entity to be corresponding with a purpose harbored by the self or harbored by the other. This will be represented by the following coefficient:

$\sqrt{}$(Self-Distinction),

Self-Distinction is an integer that only equals either + 1 or − 1. A value of $\sqrt{1}$ or one signifies the purpose belongs to the self, or that the above for the Self-Distinction variable is real and the valued entity is distinguished as belonging to the self. A value of $\sqrt{-1}$ or i signifies that the purpose belongs to something or someone else. In other words, the link is imaginary and the valuation of an entity is perceived to belong to an other. Lastly, it is proposed that Self-Distinction can modify the equation to account for when the self is considering the valuations of an other in the following manner:

1) The other's appraised value judgments have drive reduction components, appraisal (appraisal towards the restoration of equilibrium) and coping (harm or

Neurons, Empathy, and Category II Emotions (Inter-personal benefit and efficacy to prevent) components. The self imagines itself in the other's place, harm, efficacy and all.

2) However, the self's appraised value judgment of the entity of the other's appraised value judgment can only have drive reduction and appraisal components (utility, existence, and appraisal towards the restoration of equilibrium) *if the self can do nothing to influence the outcome of the other's appraised value judgment.* The self in this circumstance is a spectator and only passive empathy is felt; it is fatalistic in the sense that the self has no influence on the outcome of the other's appraised value judgment and is influenced by the other's actions.

3) In essence, the individual imagines himself or herself as being another person. This permits the necessary flexibility under the 1:1:1:1 ratio. Because the self considers itself to be two people, itself and the other, an individual may give a single entity two different values for the same purpose if one of the values is for a purpose harbored by the self and the other value is for a purpose harbored by the other (vicariously experienced).

In sum, based on the two equations, the Pursuit of Pleasure and the Avoidance of Pain functions can be observed to vary in one of four ways, five if the possibility of a constant function is considered. Empathy, in the equation, will have to adequately address situations where the other's appraised valuation of an entity, based on predicted changes in affect, undergoes the following.

1) An increase in anxiety, as in the case of sadness, grief, or fear.

2) A decrease in anxiety, as in the case of anger or hostility.

3) An increase in negative anxiety, as in the case of happiness, euphoria, or courage.

4) A decrease in negative anxiety, as in the case of guilt or self-effacing disappointment.

Additionally, empathy, in the equation, will have to address situations where the self's valuation of the other's valuation of an entity undergoes one of the following.

5) A decrease in anxiety (or increase in negative anxiety), if the self wanted the other to successfully acquire the entity and the other's valuation of the entity lowered due to success --- loving pride (love and vicarious pride).

6) An increase in anxiety (or decrease in negative anxiety), if the self wanted the other to successfully acquire the entity but the other's valuation of the entity rose due to failure --- sympathetic shame (sympathy and vicarious shame).

7) A decrease in anxiety (or increase in negative anxiety), if the self wanted the other to fail to acquire the entity and the other's valuation of the entity elevated due to failure --- hateful humility (hatred and vicarious humility).

8) An increase in anxiety (or decrease in negative anxiety), if the self wanted the other to fail to acquire the entity but the other's valuation of the entity lowered due to success --- Antipathetic Mercy (antipathy and vicarious mercy).

9) No change in either anxiety or negative anxiety, regardless of whether the other's valuation of an entity lowers or elevates due to success or failure, respectively — Indifference (neutrality and vicarious loneliness).

Category II Emotions: The Inter-Personal Emotions (The Four Degrees of Empathy)

Empathy, then, will primarily be distinguished in one of five ways along two factors. One factor will concern whether the other's anxiety invested is elevating (alternately, negative anxiety invested may be decreasing) or if the other's anxiety invested is lowering (alternately, negative anxiety invested may be increasing). The other factor will concern the self's purpose of acquiring a vicarious experience concerning the other's valuation of an entity. This will consist in the self desiring that the other's valuation of an entity decrease or elevate and a corresponding valuation of the self's purpose will decrease or elevate. Together, these two factors will create the four degrees of empathy in the Category II or Inter-Personal Emotions. A fifth degree might be a lack of empathy, or indifference. Indifference would be distinguished by the self's valuation not becoming displaced from the x-axis over a given period of time while the other's valuation of an entity, as conceived by the self, oscillates up or down.

A working definition of self and other must first be established. In its broadest terms, other refers to "everyone and everything but oneself," which "encompasses the entire matrix of events, stimuli, persons, etc. that make up the psychological environment."[13] Anything that is not oneself can potentially become distinguished as an *other*. The self, on the other hand, will refer only those items that concern oneself. *Other*, in italics, will be used to indicate vicarious valuations.

Neurons, Empathy, and Category II Emotions (Inter-personal

Love, Pride, Vicarious Love, Vicarious Pride, and Loving Pride

Of all the emotions, love is perhaps one of the most talked and written about, the most complex, the most ubiquitous, and the least understood. Generally, love is described as "an intense feeling of strong liking or affection for some specific thing or person" and as "an enduring sentiment toward a person producing a desire to be with that person and a concern for the happiness and satisfaction of that person."[14] It is also understood to be "an enduring sentiment toward a person producing a desire to be with that person and a concern for the happiness and satisfactions of that person."[15] A few types of love are worth mentioning here and they will help establish a mathematical conception of love: agape, companionate, and romantic.

Agape love is "...selfless, unconditional love," companionate love is "...based on a secure and trusting relationship," and romantic love is the "... type of love hypothesized to exist between opposite-sexed peers."[16] Love that is of the passionate type and tending toward the more carnal inclinations must not also be forgotten here. With so many different interpretations of love to consider, whatever is decided upon will have to be simple enough to encompass them all.

Core of Love

Although the definition of love that will be described in this chapter falls predominantly along the lines of companionate love and is more equivalent to familial love or of that between close friends, some general comparisons and inferences can be made. At their cores, most of the above types of love have some common elements that can be combined to create a mathematical theme. A common element of agape, companionate, and romantic love is that they include a concern for the *other's* desire, in that the self wants the *other's* well-being to be maintained or enhanced. Although desiring that someone else feel the same toward oneself is sometimes a feature of passionate and romantic love, it will not be included here in the definition of love here because the 1:1:1:1 ratio would become much more complicated. Wanting another to be well-off, for instance, and also wanting the other to want the self to be well-off, entails the appraisal of three or four main purposes by the self and is more complex. However, if the self only wants someone else to be well-off, and is also imagining himself or herself as the *other*, then the self only imagines itself to be both itself and the object of affection. In such a case, the 1:1:1:1 ratio would become 2:2:2:2, because the self is itself and imagines itself as the *other*; each of these conceptions, the self and the *other*, has its own values for entities for its own purposes. The reciprocation of feeling that is frequently desired in passionate love is indicative of a multitude of purposes, such as in Dorothy Tennov's conception of limerence that she described in *Love and Limerence* (1979). Limerence is an involuntary state in which an individual desires for a return of feeling from another person. Limerence entails ratios with numbers larger than the 2:2:2:2 one mentioned, and will be given more consideration later on in the book. Likewise, the notion of romantic love, having more in common with Tennov's description of limerence and consisting of elements from both companionate and passionate love, will be put aside until Category IV Emotions are addressed in chapter twelve.

Fitting Love and Vicarious Pride into the Equation: Loving Pride

Love, henceforth, will be mathematically defined as an instance where the self desires for an *other's* valuation of an entity to lower, and the *other's* valuation of the entity (as conceived by the self) lowers between an initial and subsequent appraisal because the *other* successfully achieves a purpose. The *other's* valuation, as conceived by the self, has variables for the Appraisal of an entity toward the restoration of equilibrium, variables for the drive reduction component, and variables for both the harm and efficacy components. Meanwhile, the self's valuation of the *other's* purpose only has variables for the Appraisal of an entity (the *other's* purpose) toward the restoration of equilibrium and values for the drive reduction components; in this case there are no harm or efficacy components for the self because it takes no action to influence the outcome of the other's purpose. Rather, the self's valuation is influenced by the threat or benefit and efficacy components of the *other*.

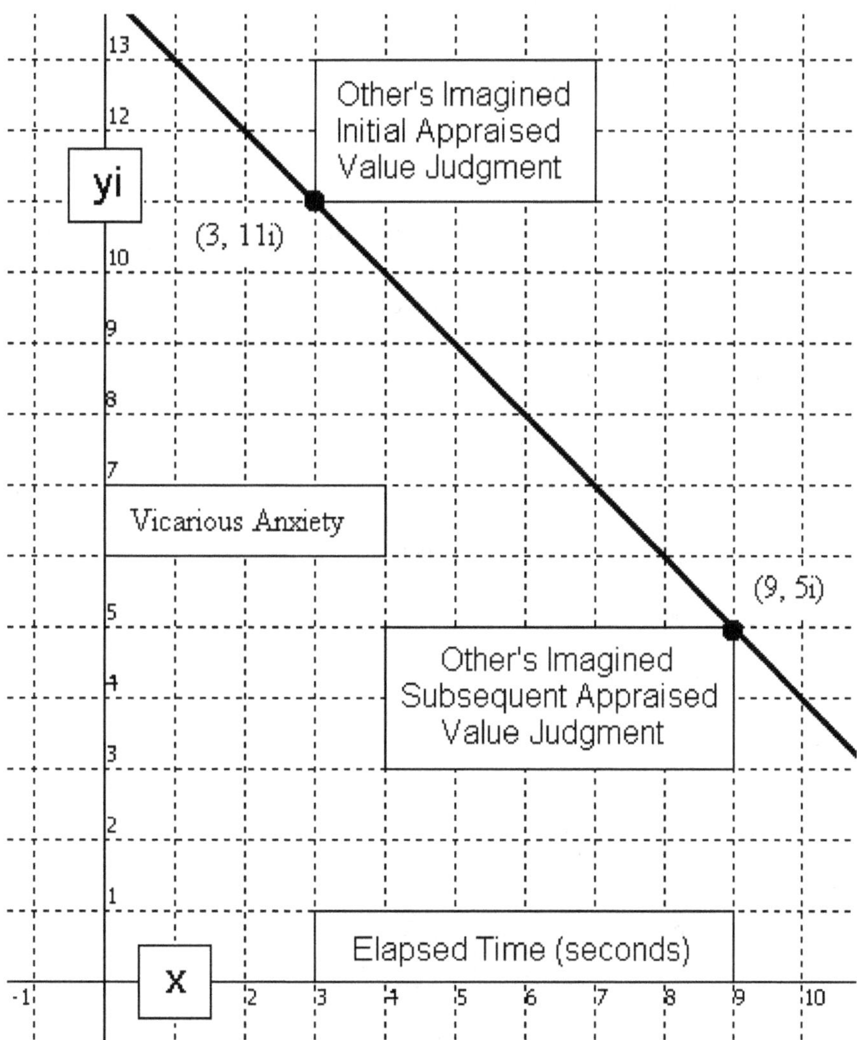

Figure 9.3 *Other's* Initial Appraised Value Judgment, as conceived by the self, and the *Other's* Subsequent Appraised Value Judgment as conceived by the self. The Self-Distinction variable in both cases equals √-1 or i.

In order to maximize this conception of love, an *other's* initial appraised value judgment of an entity, as conceived by the self, would have to have a harm variable that approaches zero and an efficacy variable approaching positive one. Consequently, for the *other's* subsequent appraised value judgment of an entity (as conceived by the self), the harm variables would have to continue to approach or remain at zero while the efficacy variables continue to move toward positive one. This means that the *other's* anxiety would head toward positive one. Alternatively, the *other's* appraised value judgment would head toward negative infinity if negative anxiety is at stake, if the further acquisition of the entity is leading away from a restoration of equilibrium, and if only one purpose in a complementary pair is considered. This is a direct relationship between the *other's* valuation of an entity, as conceived by the self, and the self's valuation of the *other's* purpose.

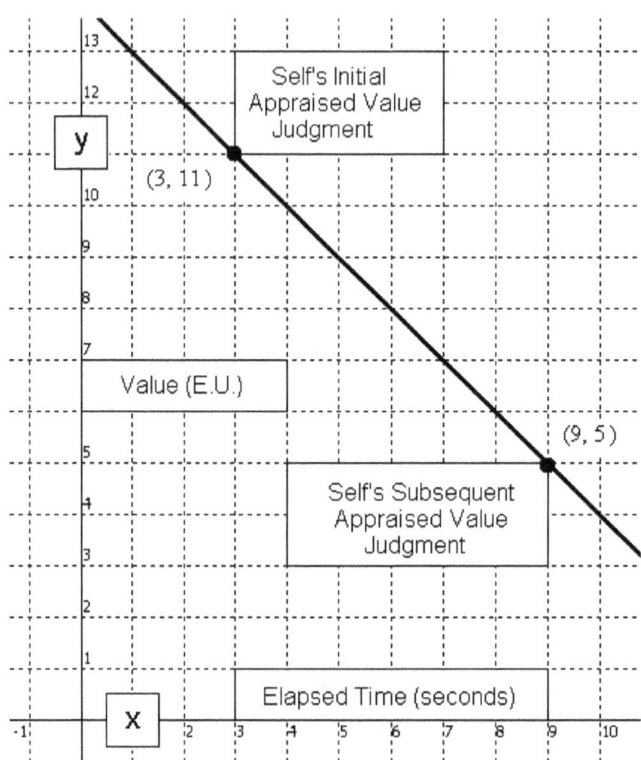

Figure 9.4 Self's initial Appraised Value Judgment of the other's valuation and Self's subsequent Appraised Value Judgment of the other's valuation. The Self-Distinction variable equals $\sqrt{1}$ in each case.

Additionally, in order to ensure that this conception of love is both being gauged and maximized, the self's initial appraised value judgment of the following entity, the *other's* valuation of the *other's* own success as conceived by the self, would have to already be approaching negative infinity if the *other* is successfully acquiring his or her entity and continues to do so. The self's subsequent appraised value judgment of its desired entity, namely, the *other's* valuation as conceived by the self, would either continue to approach one or remain at positive one because the other is successfully acquiring his or her entity. Alternatively, the self's appraised value judgment of its desired entity may head toward negative infinity if negative anxiety is at stake and if the further acquisition of the entity begins to lead away from a restoration of equilibrium.

A Closer Look at Love and Other Empathetic Emotions: Distinguishing Between Category II Emotions (Inter-personal), and Category III Emotions (Compound Interactive).

The self's desired entity is vicarious experience of the *other's* success. This entity, the vicarious experience of the *other's* success, either exists internally in the self or does not exist and it is vital either to a purpose harbored by the self (in the cases of love and shame), antagonistic to a purpose harbored by the self (in the cases of hate and mercy), or of no relevance to a purpose harbored by the self (no empathy or indifference). In each case, because the self's appraised value judgment of the other's valuation is passive, the self's desired purpose lacks harm and efficacy components and merely mirrors the *other's* valuation (e.g., in the instances of love and sympathetic shame), or mirrors and inflects the *other's* valuation over the x-axis (e.g., in the instances of hate and antipathetic mercy). The self's desired entity, vicarious experience of the *other's* success or alternatively the *other's* failure, arises internally within the self and is ultimately dependent on the self's conception of the *other's* situation (e.g., threat or benefit and efficacy components).

In essence, for Category II Inter-Personal Emotions or the four degrees of empathy, the self can make no effort to influence the outcome of the *other's* valuation, either externally by changing the environment to help the *other* acquire the entity, or internally by changing the self's conception of the *other's* situation. If the self changes the external or internal environment to influence its conception of the *other's* situation, then the dynamics of the interaction change to become what will be eventually be classified as a Compound-Interactive or Category III Emotion. They will be addressed with greater detail in chapter eleven.

No External or Internal Change by the Self

The Category II Emotion of Love would occur where the self makes no effort to influence the *other's* valuation of an entity either internally or externally and the *other* is successful. This is depicted in the neurological model in Figure 9.6. Vicarious affect (anxiety) is depicted in Figure 9.5.

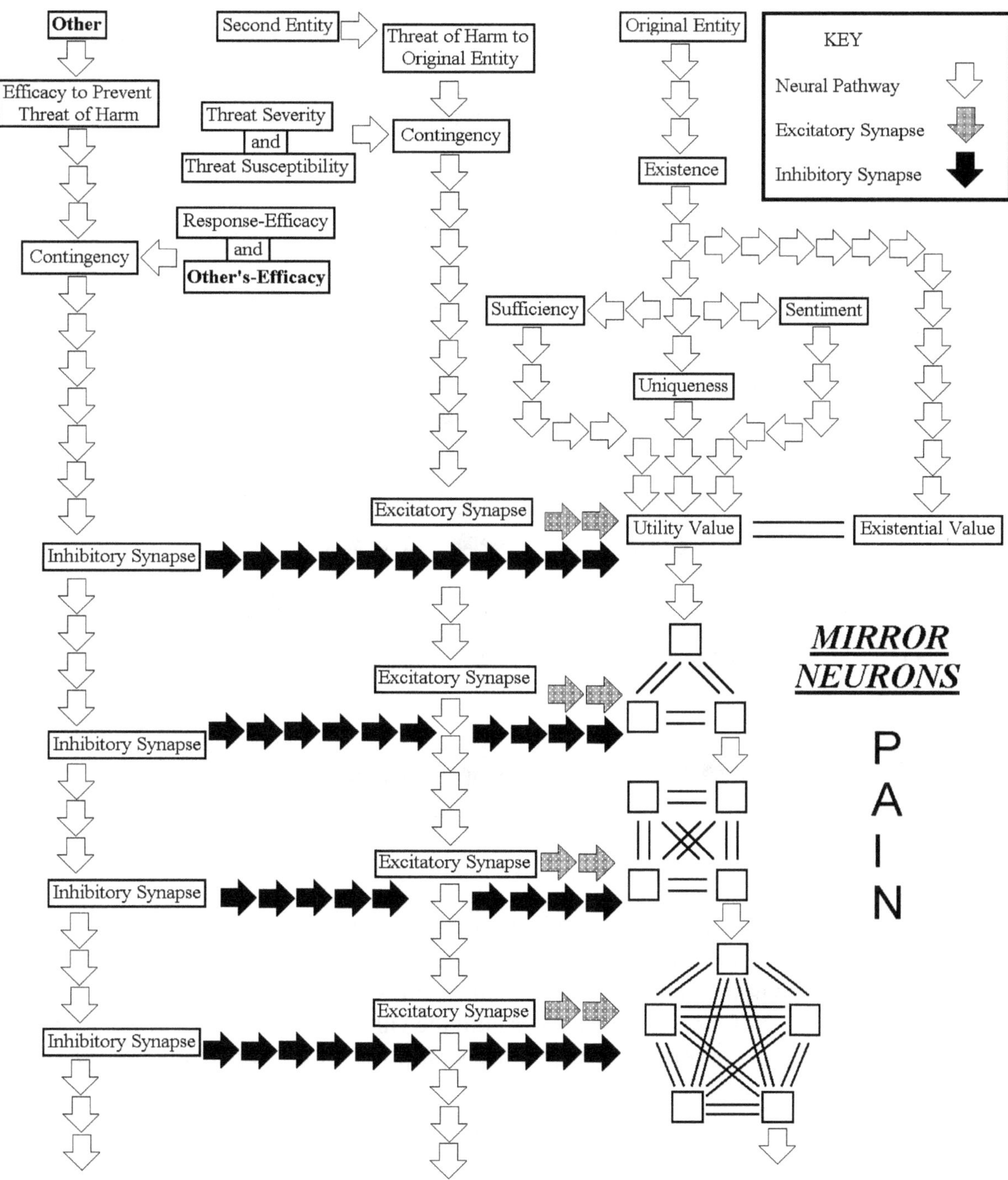

Figure 9.5 Above is a neurological model depicting vicarious valuation of an entity. The self imagines itself as an *other* and values an entity through the use of mirror neurons. The Self-Distinction variable in the above model is √-1 and represented by **Other** in the top left corner. If the Self-Distinction variable were √1, then it would be **Self**. Anxiety is being invested.

Neurons, Empathy, and Category II Emotions (Inter-personal

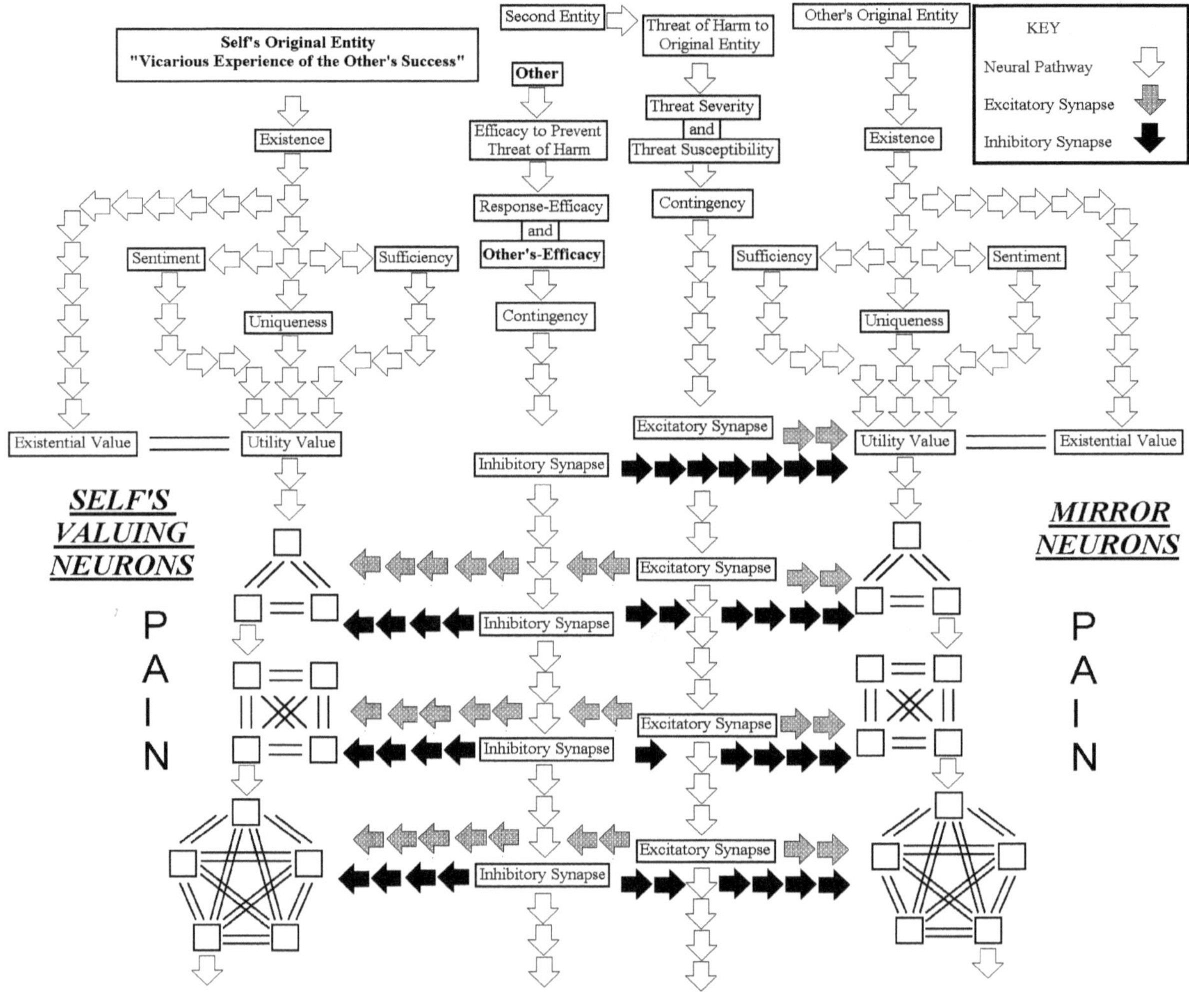

Figure 9.6 Above is a neurological model depicting vicarious valuation of an entity where the self also wants to vicariously experience the *other's* success (Category II Emotion of love or sympathy). The self imagines itself as an *other* and values an entity. The imagined Threat versus Efficacy firing rate of the *other* influences both the valuations imagined for the *other* by the self and the valuations the self has for the other's purpose. Anxiety is being invested.

External Change by the Self

A Category III Emotion of Benevolence, where the self makes an effort to influence the *other's* valuation of an entity by changing the external environment is depicted in figure 9.7. For instance, if an *other* desired a chocolate bar and the self helped the *other* to acquire the chocolate bar directly or indirectly through an assistant, then the non-passive, compound interactive emotion of benevolence is said to have occurred. In a Category III, Compound Interactive Emotion, the self performs a recommended behavior that it believes will enable the *other* to successfully achieve a purpose.

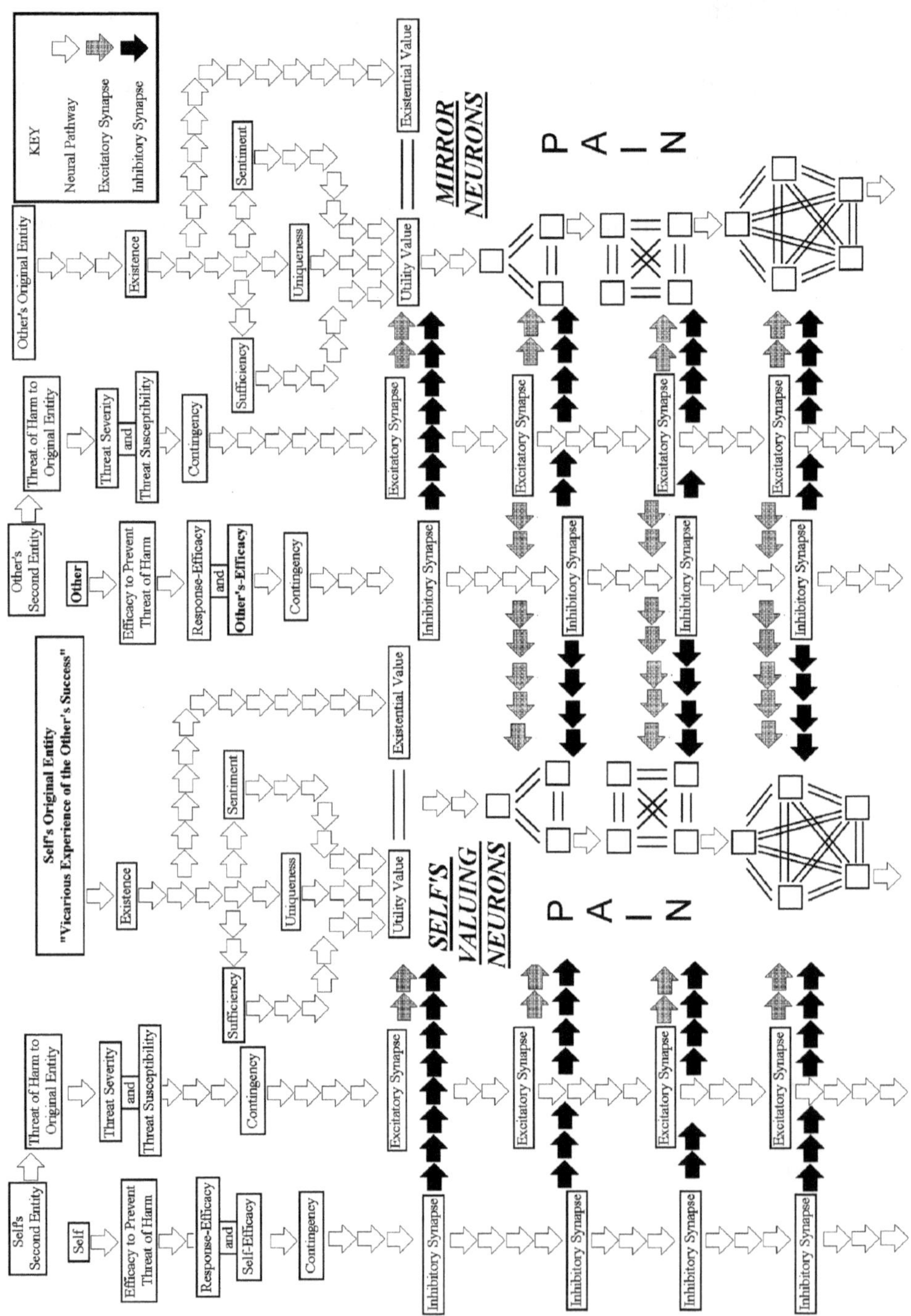

Figure 9.7 Sample neurological model depicting the relationship between the mirror neurons and the self's valuing neurons for a Category III Emotion where the self acts to influence the *other's* success.

Internal Change by the Self

A Category III Emotion of Benevolence, where the self makes an effort to influence the *other's* valuation of an entity by changing the internal environment, is depicted in figure 9.8. More specifically stated, the self modifies how it conceives the *other's* situation. In this scenario, although the self does not influence the external world so that the *other* actually acquires a desired entity, the self modifies internally how he or she conceives the *other's* valuation of an entity. If, for instance, an *other* desired a chocolate bar, and the self, being unable to externally influence the *other's* acquisition of the chocolate bar by itself or through assistance, modified its conception of the *other's* valuation of the chocolate bar, then a Category III Emotion has occurred. One such modification by the self to change the *other's* valuation of the chocolate bar could be to alter the self's conception of the *other's* variables that influence the *other's* valuation of the chocolate bar:

1) One internal modification by the self would be to alter the self's conception of how the *other* interpreted the harm components. For instance, if the self internally modified its conception of the threat of harm to the *other's* original entity (e.g., lowering it), then this would distort the self's estimation of the *other's* anxiety, for instance, making it appear as if the *other's* anxiety lowered when in reality it may have elevated. This would be a distortion of the harm variable and would give the semblance of benevolence (figure 9.8).

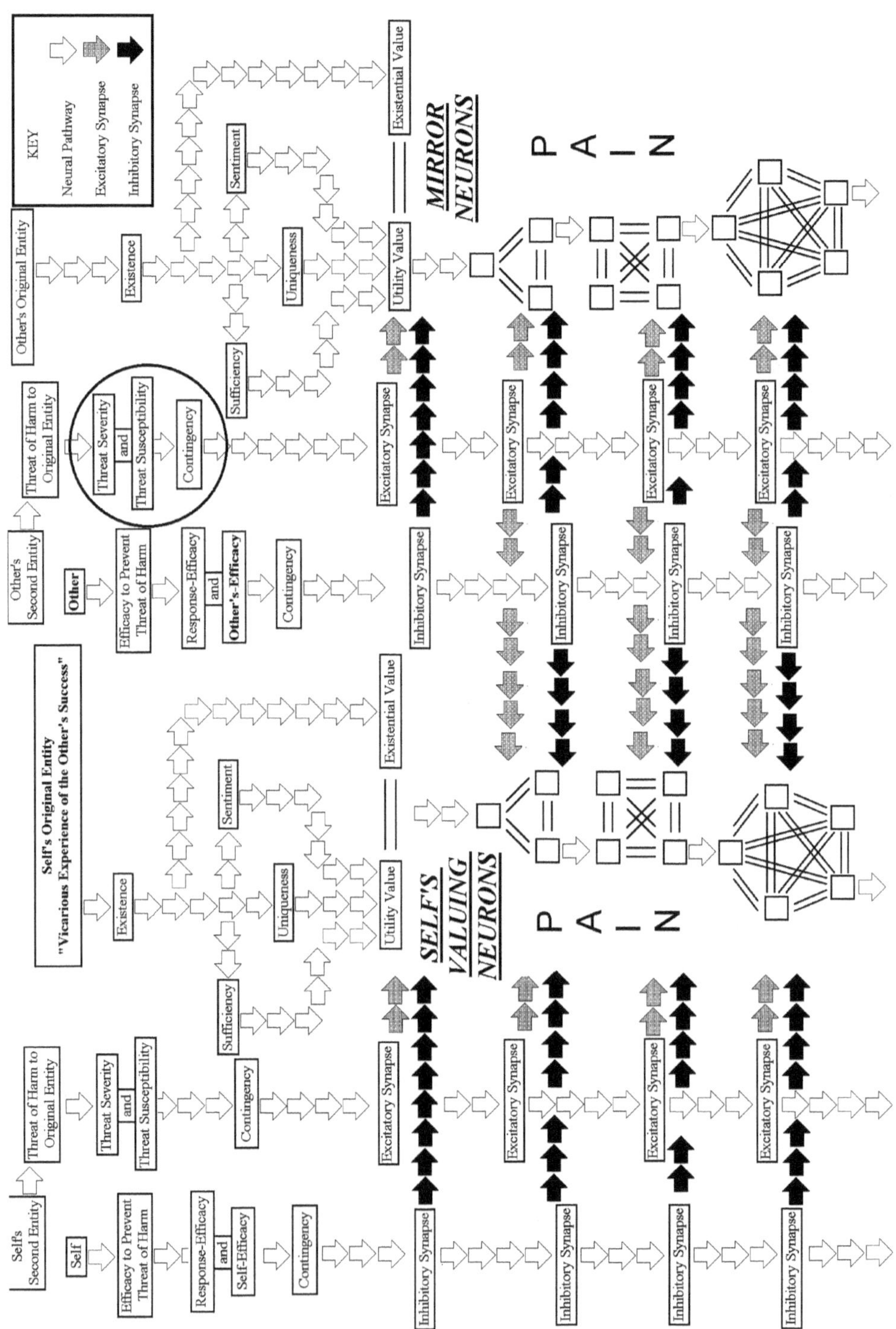

Figure 9.8 Sample neurological model depicting internal change in a Category III emotion. The self is inhibiting the circled threat components.

From a neurological standpoint, this might look like neurons from the self's efficacy inhibiting the contingency between the other's desired entity and a threat to the other's desired entity. If this connection served to inhibit the self's interpretation of the threat to the other's original entity (e.g., through an inhibitory synapse) then those neurons' firing rate would be weakened against the firing rate for the self's interpretation of the other's efficacy to protect the original entity. The self may then incorrectly deem the threat of harm is not as severe or likely to occur against the other's original entity and mistakenly believe the other is achieving success when they are not.

2) Another modification may be to alter the self's conception of the *other's* Sentiment ranking of the chocolate bar. For instance, if the self lowered its conception of the *other's* ranking of the purpose of acquiring the chocolate bar, this would distort the self's estimation of the *other's* anxiety to make it appear as if the *other's* anxiety lowered when in reality it may have elevated. This would be a distortion of the Sentiment variable and would give the semblance of benevolence by lowering the significance of the chocolate bar (figure 9.9.).

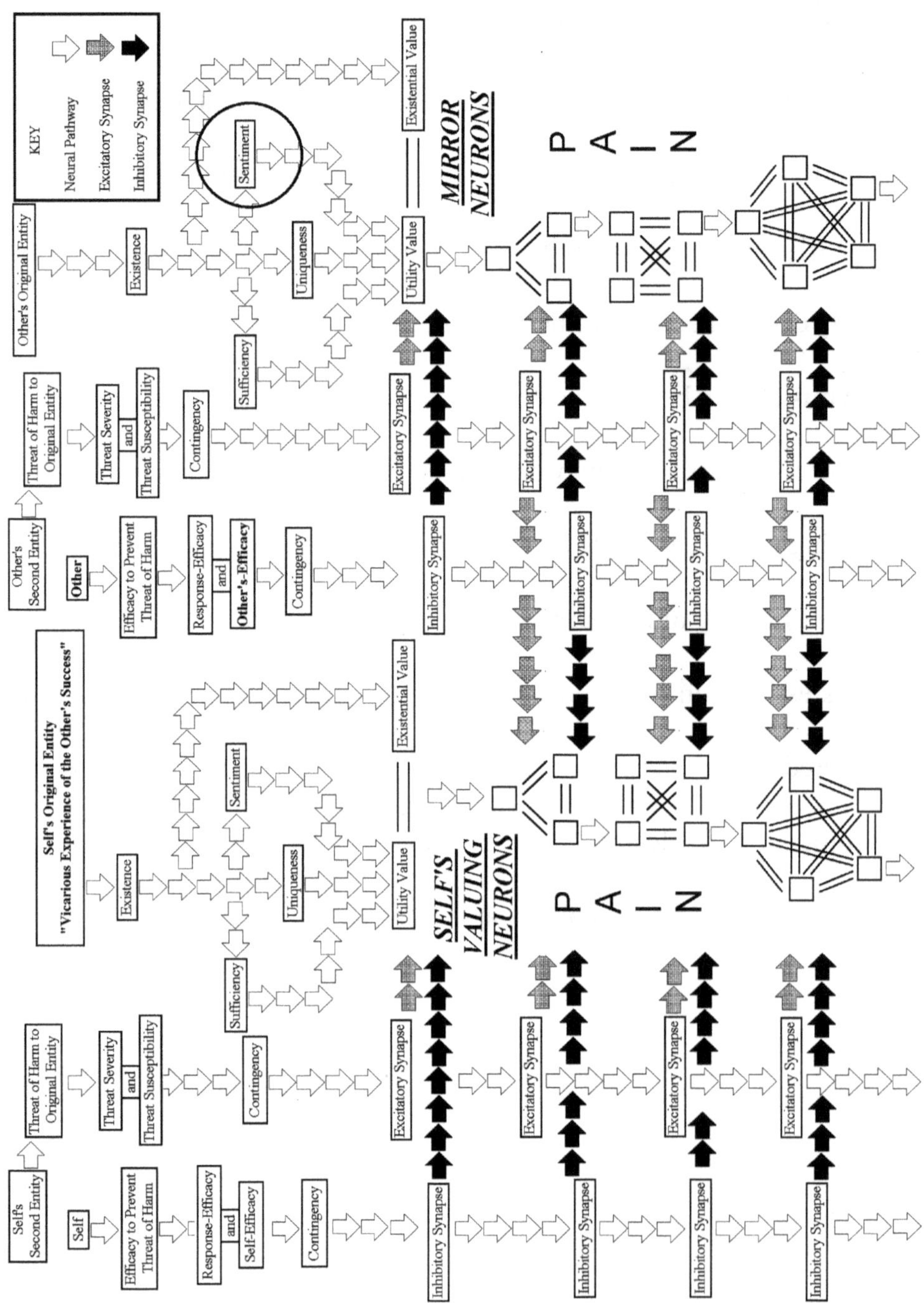

Figure 9.9 Sample neurological model depicting internal change in a Category III emotion. The self has modified the *other's* Sentiment and changed how important it believes the entity is to the *other*.

This connection might act to inhibit through an inhibitory synapse, or excite through an excitatory synapse, the group of neurons that represent the *other's* Sentiment. An inhibitory synapse would lower the self's estimate of how important the original entity is to the other. Conversely, an excitatory synapse would increase the self's estimation of how important the original entity is to the *other*.

The relatively far reach of Sentiment would suggest that neurons concerning Sentiment, in addition to being linked to the entity's existence, may also be linked to a sort of Sentiment hub (e.g., an ethics or moral compass). For instance, a purpose's Sentiment ranking, described in chapter five as its absolute rank, might be determined by the strength of its signal relative to the sentiment-related neurons of other purposes. In example, if an individual has 20 total purposes that he or she is considering, and each purpose has a different Sentiment value or absolute rank, then there will be 20 different signal strengths (i.e., firing rates of neurons) that correspond to 20 separate purposes.

3) An additional modification may be to alter the self's conception of the *other's* efficacy components. For instance, raising the *other's* efficacy variables would distort the self's estimation of the *other's* anxiety to make it appear as if the *other's* anxiety lowered when in reality it may have elevated or stayed the same. This distortion of the efficacy variable would also give the semblance of benevolence, making it appear as if the *other* were more capable of preventing a threat than is the case. It follows then, that the rest of the variables in the *other's* valuation could be distorted by the self in a similar manner.

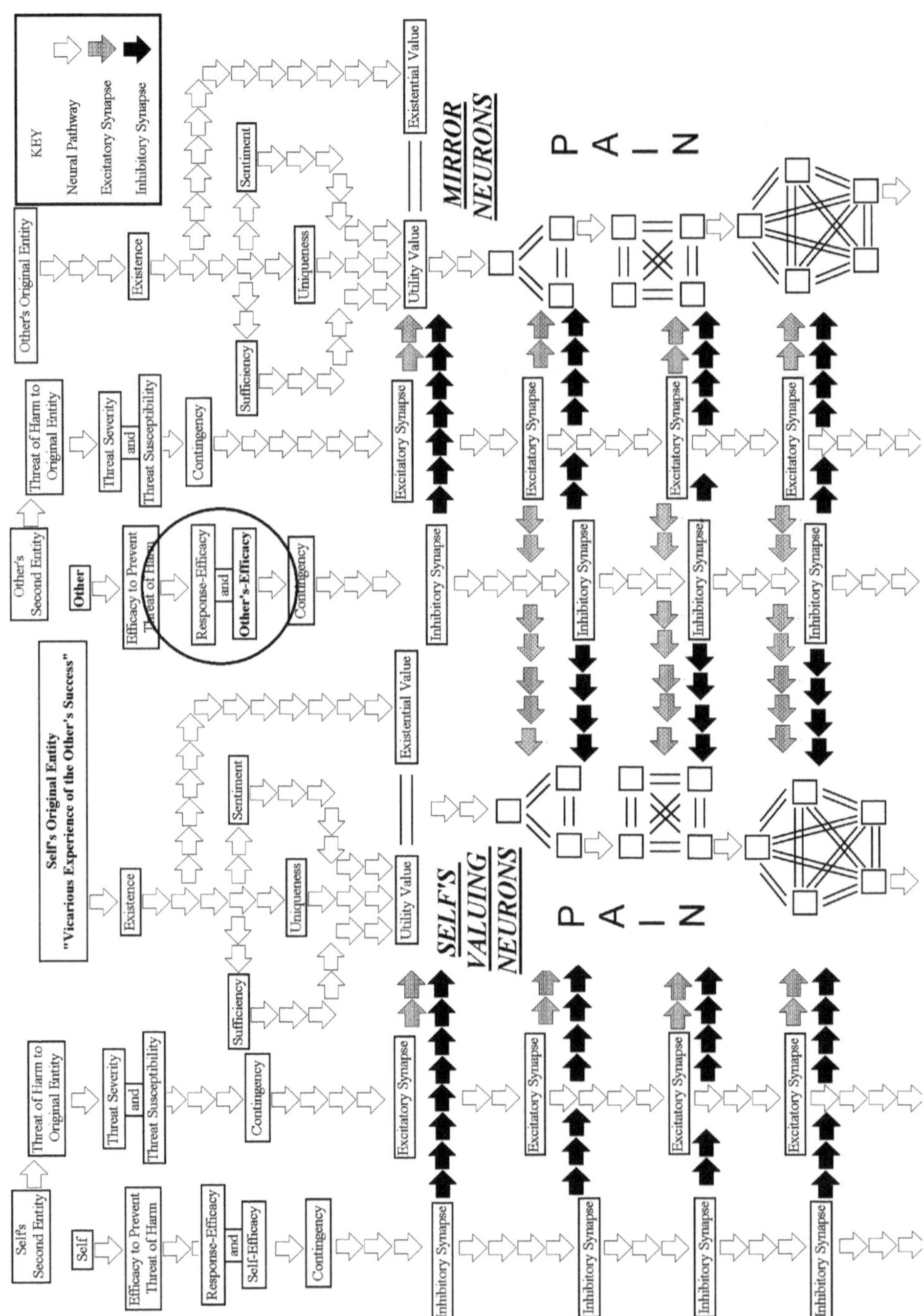

Figure 9.10 Sample neurological model depicting internal change in a Category III emotion. The self has modified the *other's* Efficacy (circled).

Valuation End Result for Loving Pride (Category II Emotion)

In the case of the Category II emotion of love, the *other* is successful. The anxiety for both the self and the *other* lowers. Alternatively, the negative anxiety for both the self and the *other* would increase if further acquisition of the concerned entity leads away from a restoration of equilibrium. Ultimately, this definition of love consists in the self's ability to imagine someone else's pleasure felt, or reduction in pain felt for a single purpose and to have a personal response to it that is the same as the *other*.

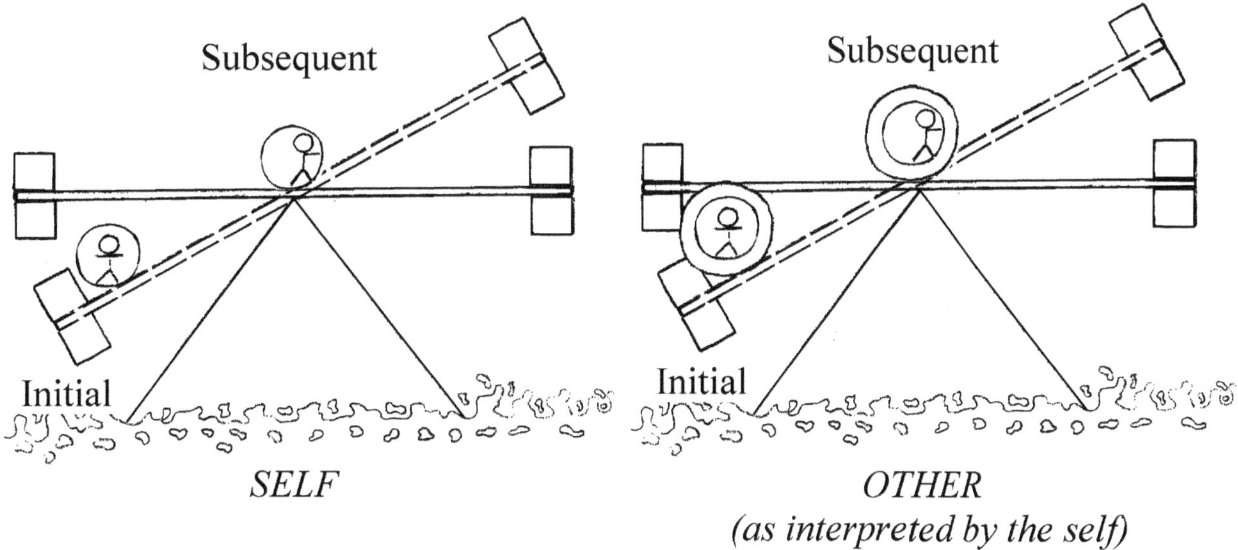

Figure 9.11 Sample End Result for Loving Pride (Roll Cage Theory of Drives Model)
The self's desired entity is vicarious experience of the *other's* success. First, the self interprets that the *other* is succeeding at his or her purpose. Thereafter, the self is able to acquire the desired entity of vicarious experience of the *other's* success. The *other* succeeds and as a result the self acquires its desired entity, namely, vicarious experience of the *other's* success. Success for the *other* in this case, is the aim of achieving homeostasis.

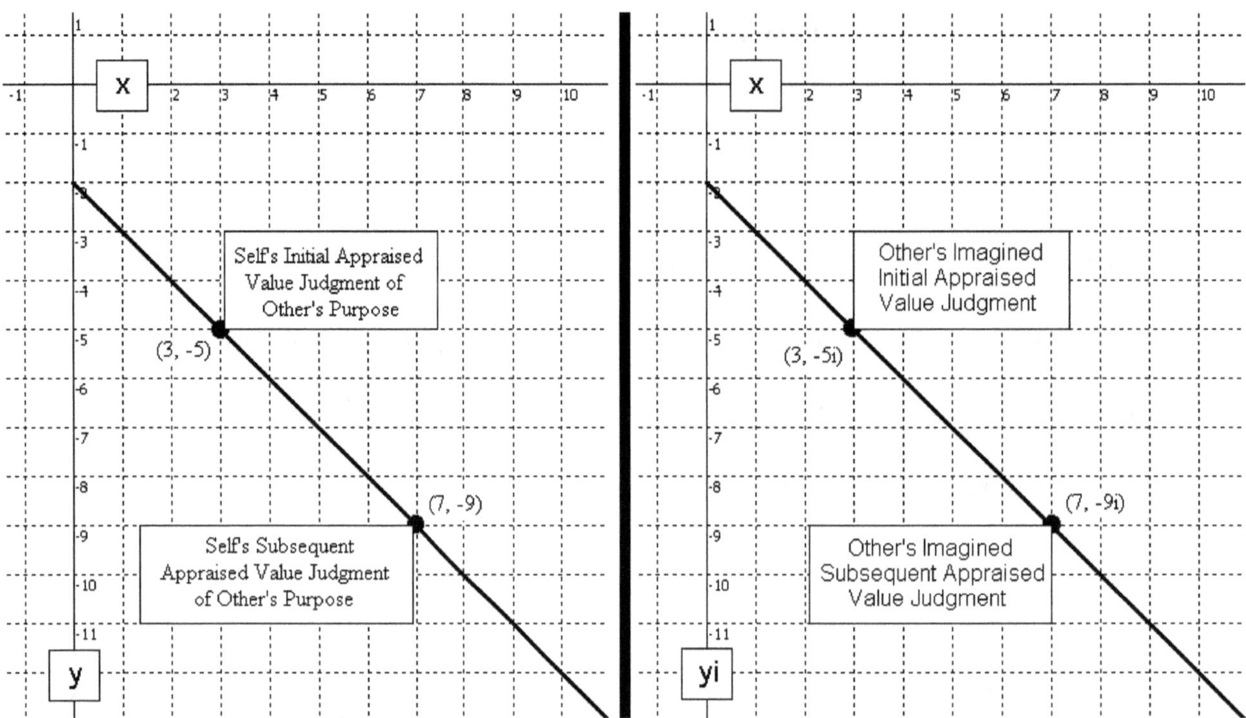

Figure 9.12 Sample End Result for Loving Pride (Cartesian Model)
The self's desired entity is vicarious experience of the *other's* success. In this case, the *other* has acquired an entity in an amount that is greater than is required for achieving homeostasis, so negative anxiety is used with respect to a single purpose. Because the *other* is succeeding (e.g., pleasure felt with respect to a single purpose from a complementary pair is increasing), the self also feels pleasure, as the self's desired entity was vicarious experience of the *other's* success for the specific purpose. Regardless of whether pain or pleasure are concerned (anxiety or positive anxiety), the slope of both the graph for the self and the *other* is negative in the case of loving pride where a single purpose is considered.

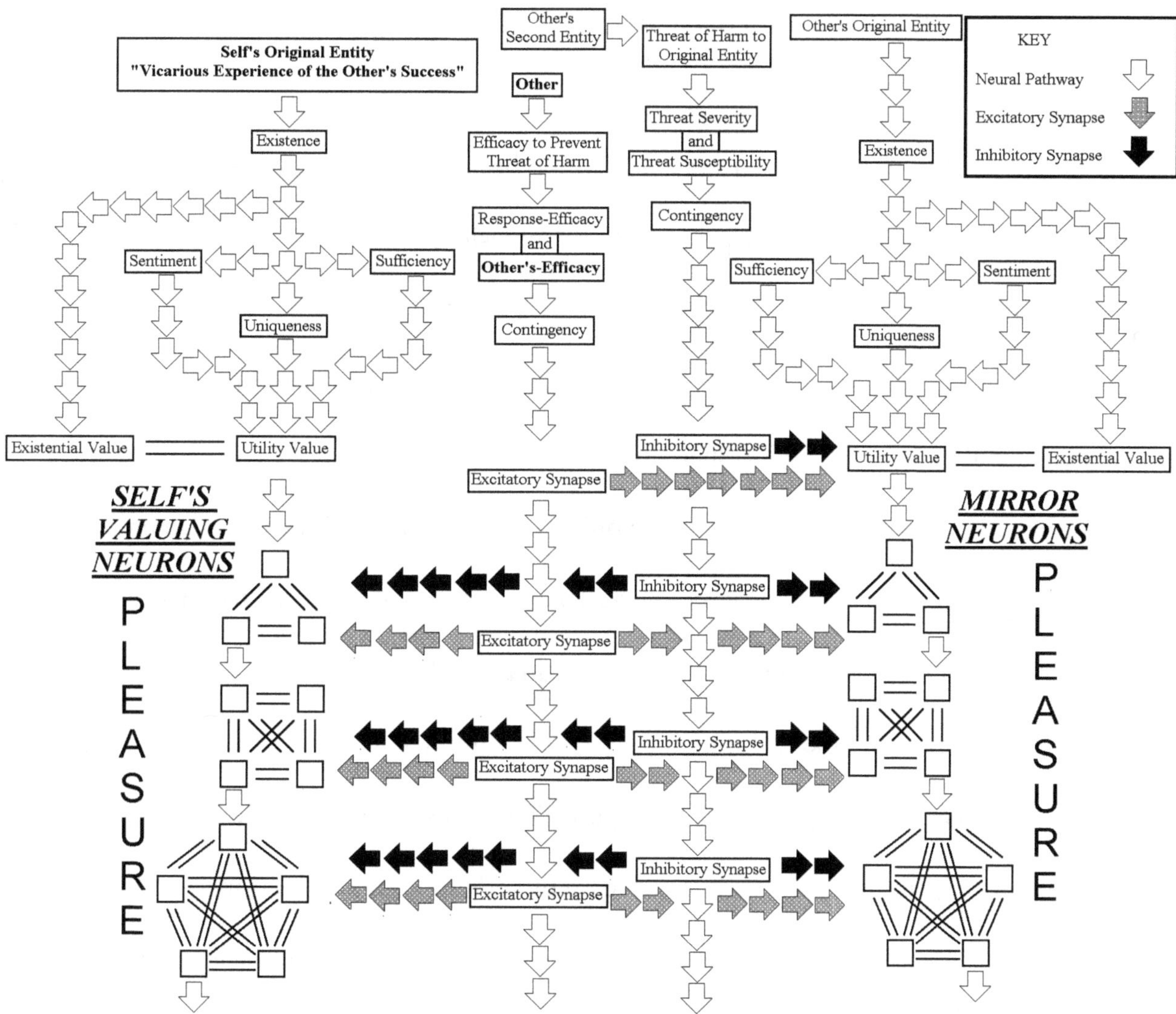

Figure 9.13 Sample End Result for Loving Pride (Neurological model): Threat < Efficacy. The self's desired entity is vicarious experience of the *other's* success. Negative anxiety (pleasure) is depicted in the model. The self has no efficacy or threat components because it has no means for influencing the outcome of the *other's* situation. The self approves of the *other's* purpose and wants to witness and vicariously experience the *other's* success. Necessarily, because the self approves of the *other's* purpose, the self would also feel vicarious pride.

Breakdown of Loving Pride: Two Separate Valuations

The 1:1:1:1 ratio is maintained as a 2:2:2:2 ratio because the individual imagines himself or herself being in the place of the *other* person. The expression:

√Self-Distinction = 1, represents the self;

√Self-Distinction = root of − 1, represents the individual's ability to imagine that he or she is valuing an entity with respect to a purpose from another person's perspective (or animal, plant, or personified inanimate object). Hence, the self is valuing an entity vicariously. In the neurological model, Self-Distinction is depicted with either the term Self if it equals √(1), or with the term Other if it equals √(-1) .

1) Experiencing the fulfillment of the *other's* purpose is the entity that the self is concerned with and wants to acquire. The self's desired entity can be thought of as an experience, namely, experiencing the *other's* success vicariously. More specifically, this would be the lowering of anxiety or an increase in negative anxiety that is vicariously felt for the specific purpose. If the self were feeling pride for itself, then the self would simply imagine the *other* approving of the self's success. The neurological model for pride, when felt by the self if it is active, would look like the following:

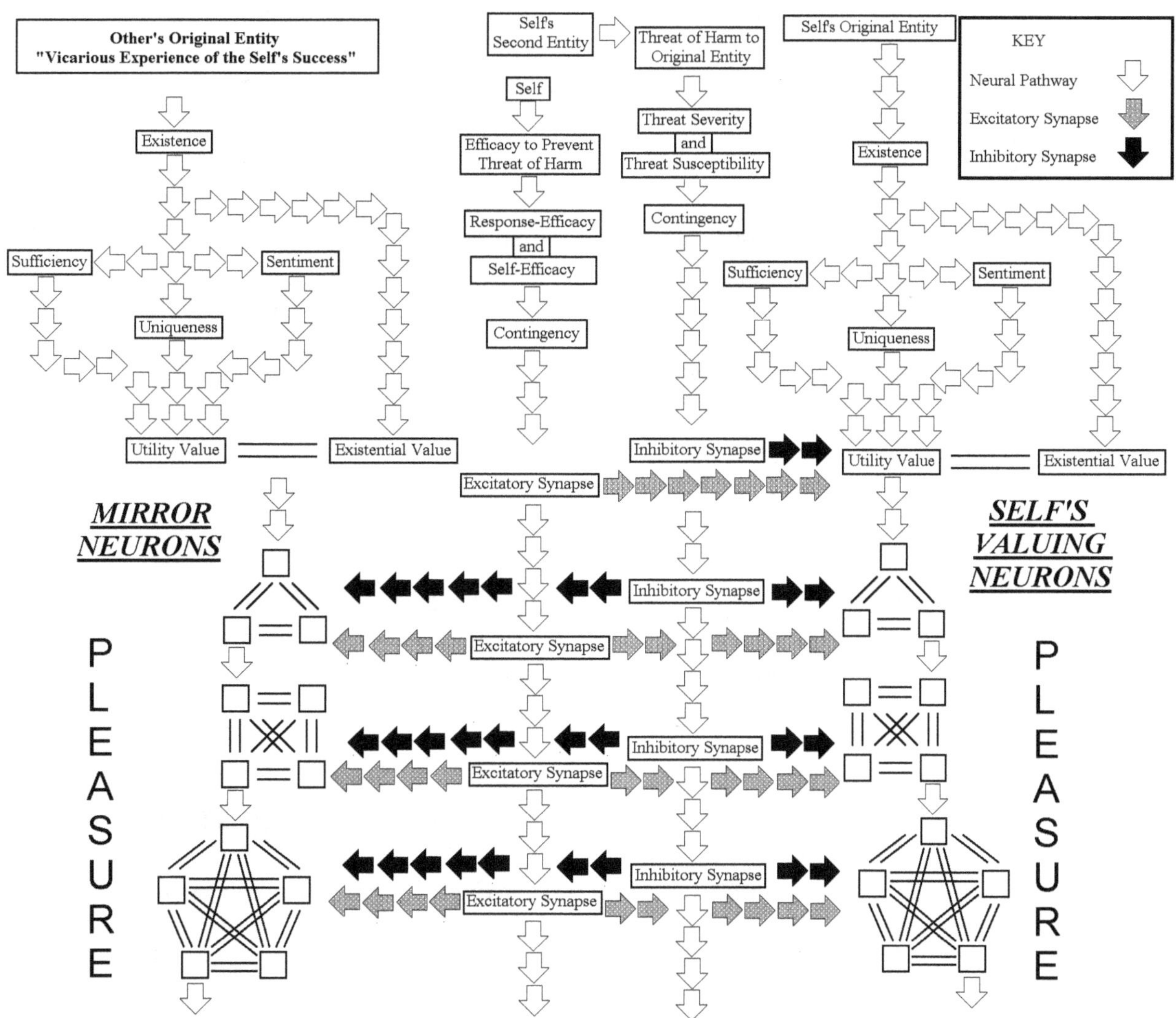

Figure 9.14 Sample End Result for Pride (Neurological model): Threat < Efficacy. Neurological model of pride felt by the self (right) for having its accomplishments acknowledged by an *other* (left). Necessarily, vicarious love is also felt by the self. Pleasure increases, as Efficacy is greater than Threat.

2) In the case of love felt by the self, the self's appraised judgment value only has a mimicked value of the *other's* appraised judgment valuation (as conceived by the individual). In the case of pride felt by the self, it is the *other's* valuation (as

conceived by the individual) that mimics the self's appraised judgment valuation.

In short, for Category II Emotions, Inter-Personal Emotions or the Four Degrees of Empathy, threat and efficacy components do not exist for either the self or the other. For Category III Emotions: Compound Interactive Emotions, the threat and efficacy components exist for both the self and the *other,* and an active role is taken by the self to modulate the *other's* valuation, either externally in the world or internally through introspection.

3) The *other*, if active, has components for the Appraisal of the entity toward the restoration of equilibrium, drive reduction components, harm, and efficacy. Both the *other's* appraised value judgments as conceived by the self, and the self's mirrored values have an initial and subsequent appraisal.

4) Pattern

   a) The self's purpose, to vicariously experience the *other's* success, outweighs its complement, to not vicariously experience the *other's* success. The self desires to acquire the following entity: vicarious experience of the *other's* success. The *other* successfully fulfills the purpose in question. Success, for the *other*, is measured as the lowering of anxiety or an increase in negative anxiety.

   b) Examples

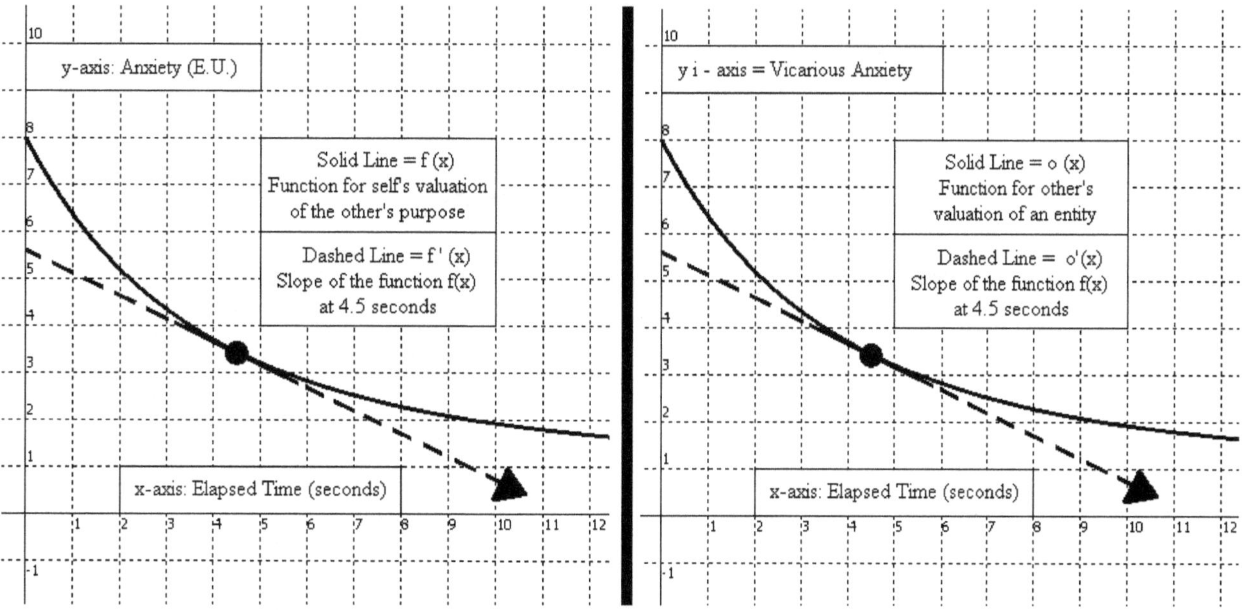

Figure 9.15 Above is a graph of love with anxiety lowering. The function f(x) represents the self's valuation of the *other's* purpose, on the left. The *other's* valuation of an entity is represented by the function o(x) on the right. The dashed lines of f '(x) and o'(x) correspond to the slopes of each function at time 4.5 seconds and the magnitude of a specific emotion felt at time 4.5 seconds.

i) Anxiety Lowering:

Graph of anxiety lowering for the *other* and anxiety lowering for the self's estimation of the *other's* purpose. The predicted slope for the *other* is negative and for the self also negative. Both functions are above the x-axis.

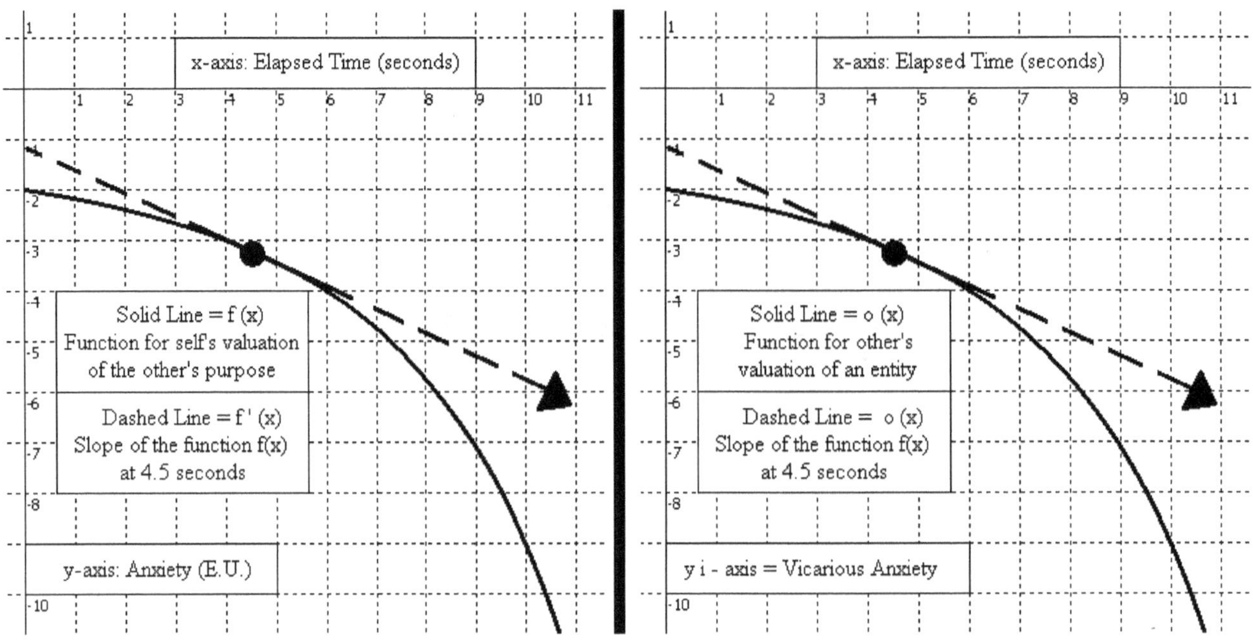

Figure 9.16 The above is a graph of love with negative anxiety increasing. The function f(x) represents the self's valuation of the *other's* purpose, on the left. The *other's* valuation of an entity is represented by the function o(x) on the right. The dashed lines of f '(x) and o'(x) correspond to the slopes of each function at time 4.5 seconds and the magnitude of a specific emotion felt at time 4.5 seconds.

ii) Negative Anxiety increasing:

Graph of negative anxiety increasing for the *other* and negative anxiety increasing for the self's estimation of the *other's* purpose. The predicted slopes for the *other* and for the self are negative. Both functions are below the x-axis.

c) Below is a description for a situation where the self wants the *other's* anxiety to lower or negative anxiety to increase and the *other's* appraised judgment value of an entity for a purpose is the heavier of a complementary pair. Additionally, the *other* is successful and the other's anxiety lowers or alternatively negative anxiety may increase.

i) A spectator at a baseball game is a fan of the home team and they are playing their rivals; he wants the home team to overcome a deficit (harm) to tie the game (restore equilibrium). The home team overcomes the deficit to tie the game and the anxiety for both the *other* and for the self is lowered.

ii) Similarly, if the game were already tied (at equilibrium) and the same fan wanted the home team to demolish the away team, negative anxiety for both the *other* and for the self would increase if the home team took the lead and went ahead by a wide margin.

Sympathy, Shame, Vicarious Sympathy, Vicarious Shame, and Sympathetic Shame

Whenever someone experiences tragedy or misfortune, typically those closest to the victim can imagine themselves in the victim's position and they may offer their condolences or their support. Sympathy is "a sense of compassion or understanding which allows one to interpret or justify the actions and/or feelings of another."[17] Sympathy, from its Greek roots, literally means "sharing in the feelings of another," however, it is generally only used in reference to "painful or unpleasant emotions."[18] One such unpleasant emotion is shame.

Shame is "an emotional state produced by the awareness that one has acted dishonorably or ridiculously."[19] Moreover, shame is generally "reserved for situations in which one's actions are publicly known or exposed to real or potential ridicule."[20] Hence, the self is aware that someone else is judging his or her actions.

A shrewd analysis of shame reveals that at its base level, shame entails a sense of awareness by the self that an *other* has failed to acquire the entity of approval because of the self's actions. Accordingly, the entity in each of these cases that the *other* desired to obtain was the vicarious experiencing of the self's success at a particular endeavor, and ultimately the self was unsuccessful. Shame can be thought of as taking place from the perspective of the *self* or from the perspective of the *other*.

Generally speaking, pride is described as a "state of satisfaction with oneself for efforts made and gains accomplished."[21] Moreover, some type of recognition from others is usually considered a criterion for pride to exist. It follows that pride, being associated with feelings of success, should be expected to have more in common with the definition of love being used for the equation where the onlooker responds to the active party's success rather than sympathy, where the onlooker responds to the active party's misery or woe.

Core of Vicarious Shame and Vicarious Pride (Concerning Love in the Latter Case)

To do something vicariously is equivalent to sharing in another's experience. At first glance it may seem perplexing to include within the definition of vicarious shame a sense of wrongness on the part of the *other's* actions while the self is vicariously imagining itself to be in the other's situation. However, in its broader context, shame may be used to refer to situations where an individual, acting on behalf of a whole whether knowingly or otherwise, might bring disgrace, pain, or suffering to those from the whole that have designated the individual as their representative. Those from the whole who are vicariously experiencing the situation of the individual may feel ashamed by proximity if they desired for the individual to achieve a purpose or acquire an entity and the individual did not succeed. The members of the whole may also feel that the individual is acutely aware both of the shame brought to the group, as if the individual was the group, regardless of whether or not the individual feels so. In such circumstances, there may have been nothing that the observers could have done to influence the outcome of the situation, yet they may feel as if they have personally failed, done something wrong, or brought disgrace to themselves by virtue of believing in the efficacy of someone else with whom they imagined themselves to be and with whom they have identified as an extension of themselves.

For instance, it will be held that a son or daughter who was hoping to get accepted into a specific medical school fails to do so. Family, close connections, and supporting relatives might vicariously experience his or her sense of agony and disgrace at having unsuccessfully gained admittance. However, distant relatives, casual acquaintances, or total strangers who, despite only hearing of the news secondhand, despite having never met the candidate, and despite possessing no influence over the outcome whatsoever, may vicariously gather a sense of shame and disgrace via the interaction between their personal desire for the candidate to succeed and their vicarious interpretation of the *other's* situation. This may hold true even if the outsiders are not aware of any person or group of persons who have ridiculed the candidate's failure save the outsiders themselves.

To demonstrate this idea in another scenario, an optimistic, celebrity athlete who expects to win an Olympic race and loses in the medal round may feel proud of all the hard work put in just to get a chance to represent his or her country, and may modify his or her Sentiment ranking of winning a gold medal against just participating in the Olympics accordingly. However, the millions of fans and supporters he or she had may feel vicarious shame for believing in their idol; even simpler, they may feel ashamed for their idol in the sense that the self is shaming the athlete (*other*) while simultaneously imagining itself in the *other's* situation, and not due to any shame being affixed to the idol by the rest of the world.

In another case, an ambassador of a country who becomes rowdy after getting inebriated may have no recollection of the event and feel no remorse for his actions. Some members of his country may feel ashamed both for him, his actions, and because of their own judgment of his behavior. However, different members of his country may only sense the ambassador should feel ashamed for his actions because they can sense the disapproval from onlookers, but they might not personally pass any judgment on him whether good or bad. Vicarious experience of an *other's* situation that lacks a personal judgment by the self is in essence isolated from the self and cannot move the self. The self may be able to sense what the *other* should feel, but that is all. No judgment would be made by the self and no personal purposes would have relevance to the *other's* situation. It would be akin to an eye witness watching a theft, imagining himself in either the thief's or the victim's place, and not attaching any personal significance to the event. Though this kind of neutrality may be hard to imagine, if such spectacles happened on a daily occurrence then one might eventually grow numb to them. Neutrality, or indifference, refers to a state where "one has no preferences between alternative choices or courses of action" and embodies this idea.[22] Indifference and neutrality are a special case of Category II Emotions, as the signal the absence of empathy.

In a similar manner to vicarious shame, pride may also be felt vicariously as well. If the *other* were to succeed when the self wanted to vicariously experience the *other's* success, then the self could imagine itself in the *other's* situation receiving approval from none else but the self. Likewise, vicarious pride would correspond with love, whereas vicarious shame would correspond with sympathy. In short, unless the self is indifferent to an *other's* valuation of an entity related to a purpose or feels antipathetic toward it, then sympathy felt for an *other's* failure will always be accompanied by a judgment from the self where it was originally desired for the *other* to have succeeded but the *other* fails. The following three concepts encompass the entire range of empathy:

1) The self's desire for the *other* to succeed, from which may arise:
   a) love and vicarious pride
   b) sympathy and vicarious shame
   c) Likewise, the reciprocals of the above, if the *other* desired for the self to succeed instead, would be pride and vicarious love (a), along with shame and vicarious sympathy (b).

The following two have not yet been considered:

2) The self's desire for the *other* to not succeed, from which may arise:
   a) hate and vicarious humility
   b) antipathy and vicarious mercy
   c) Likewise, the reciprocals of the above, if the *other* desired for the self to not succeed instead, would be humility and vicarious hate (a), along with mercy

Neurons, Empathy, and Category II Emotions (Inter-personal

and vicarious antipathy (b).

3) Indifference or having no purpose relevant to the *other's* success or lack of success (e.g., acquiring vicarious experience of the other's success), regardless of whether or not the other's success or failure is vicariously felt.

Fitting Sympathy and Vicarious Shame into the Equation: Sympathetic Shame

The notion of sympathetic shame used in the equation encompasses both the definition of sympathy and the notion of vicarious shame, or alternatively vicarious sympathy and shame. In the case where the self is passive and the *other* is active, then the *other* would feel shame in its established meaning. In order to maximize this conception of sympathetic shame, an *other's* initial appraised value judgment of an entity, as conceived by the self, would have to have a harm variable that begins to approach positive one and an efficacy variable beginning to approach zero. Consequently, for the other's subsequent appraised value judgment of an entity (as conceived by the self), the harm variables would have to continue to approach or remain at positive one while the efficacy variables continue to move toward zero. This means that the *other's* anxiety would head toward infinity. Alternatively, the *other's* appraised value judgment may head toward negative one if negative anxiety is at stake and further acquisition of the entity would have lead away from equilibrium.

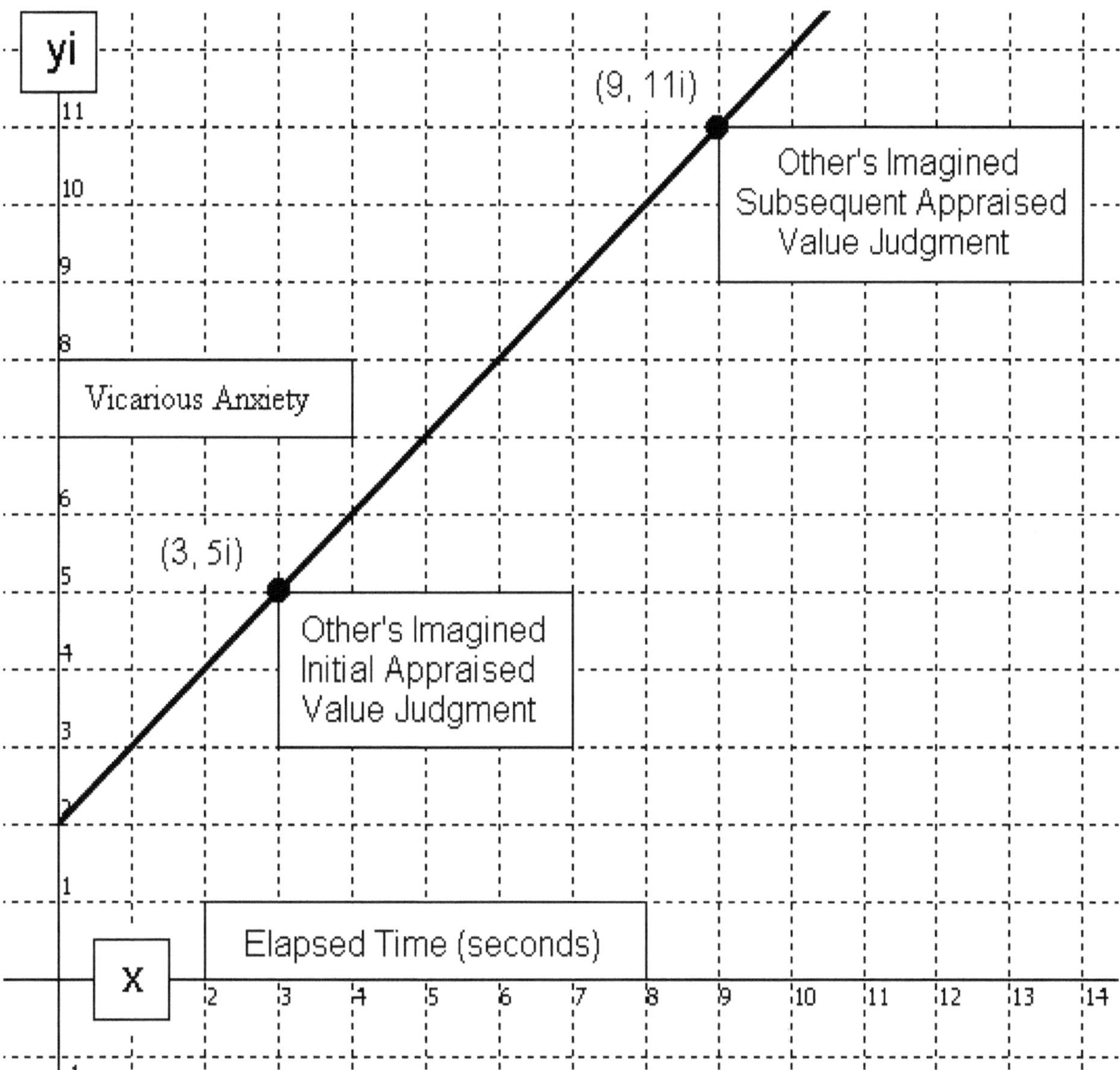

Figure 9.17 *Other's* Initial Appraised Value Judgment conceived by the self and the *Other's* Subsequent Appraised Value Judgment as conceived by the self. The Self-Distinction variable in both cases equals √-1 or I.

Additionally, in order to ensure that this conception of sympathetic shame is both being gauged and maximized, the self's initial appraised value judgment of the *other's* success, based on the other's valuation of an entity as conceived by the self, would be beginning to approach positive infinity if the *other* is not successfully acquiring his or her entity. The self's subsequent appraised value judgment of its desired entity, namely, the *other's* valuation as conceived by the self, would continue to approach positive infinity because the *other* would not be successfully acquiring his or her entity. Alternatively, the self's appraised value judgment of its desired entity may head toward negative one if negative anxiety is at stake and further acquisition of the entity would lead away from a restoration of equilibrium. This is a direct relationship between the *other's* valuation of an entity, as conceived by the self, and the self's valuation of the *other's* purpose.

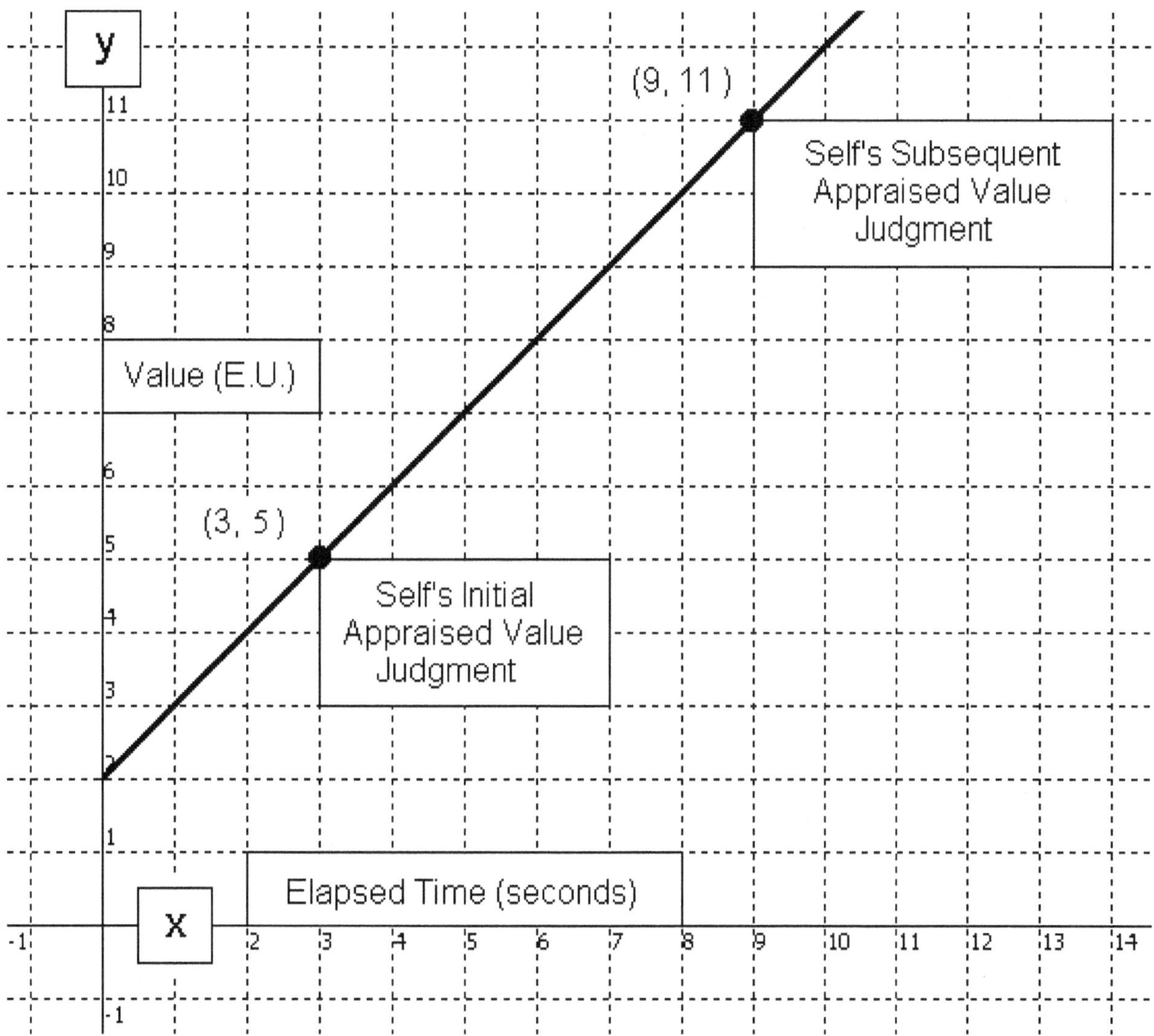

Figure 9.18 Self's initial Appraised Value Judgment of the *other's* valuation and Self's subsequent Appraised Value Judgment of the *other's* valuation. The Self-Distinction variable equals √1 in each case.

Valuation End Result for Sympathetic Shame

The *other* is not successful. The anxiety for both the self and the *other* elevates. Similarly, the negative anxiety for both the self and the *other* would decrease if further acquisition of the concerned entity would have lead away from a restoration of equilibrium. Ultimately, this definition of sympathetic shame consists in the self's ability to imagine the *other's* pain increasing or pleasure decreasing and having a personal response to it that is the same.

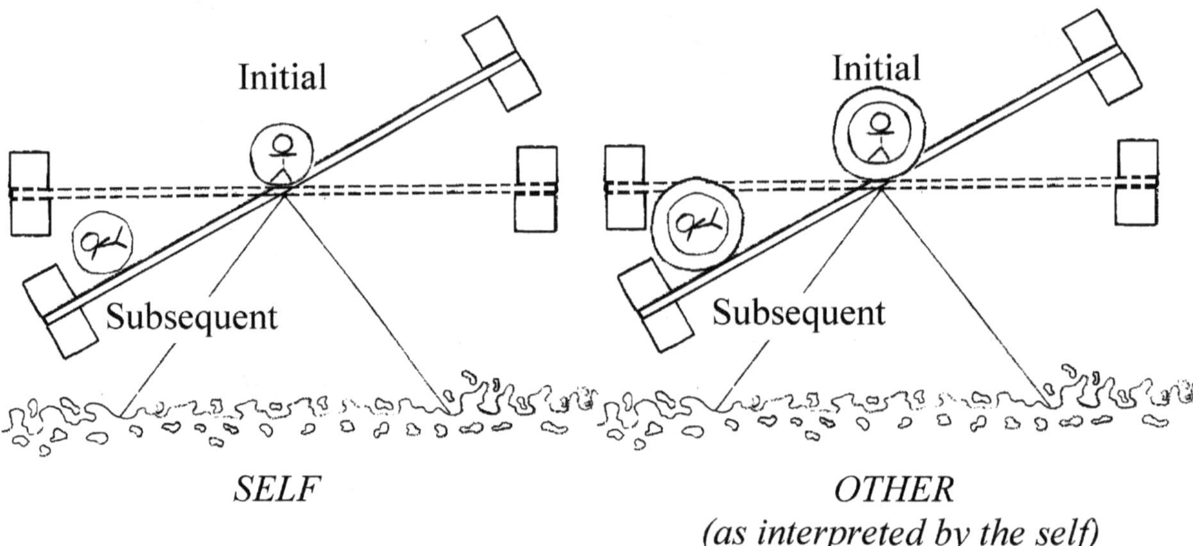

Figure 9.19 Sample End Result for Sympathetic Shame (Roll Cage Theory of Drives) The self's desired entity is vicarious experience of the *other's* success. First, the self interprets that the *other* is either not succeeding at his or her purpose or is initially well off. However, the self is ultimately not able to acquire the desired entity of vicarious experience of the *other's* success. The *other* eventually fails and as a result the self does not acquire its desired entity, namely, vicarious experience of the *other's* success. Success for the *other* in this case, would be the aim of achieving homeostasis.

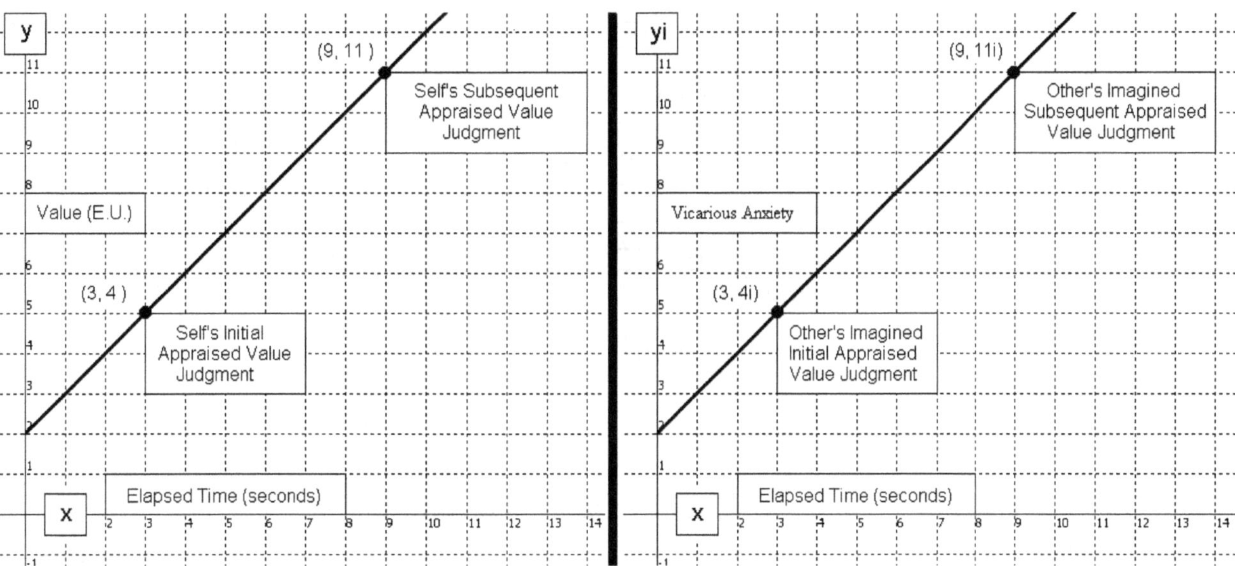

Figure 9.20 Sample End Result for Sympathetic Shame (Cartesian Model)
The self's desired entity is vicarious experience of the *other's* success. In this case, the *other* must acquire an entity to achieve homeostasis, so positive anxiety is used with respect to a single purpose. Because the *other* is not succeeding (e.g., pain felt with respect to a single purpose from a complementary pair is increasing), the self also feels pain, as the self's desired entity was vicarious experience of the *other's* success. Regardless of whether pain or pleasure are concerned (anxiety or positive anxiety), the slope of both the graph for the self and the graph for the *other* is positive in the case of sympathetic shame.

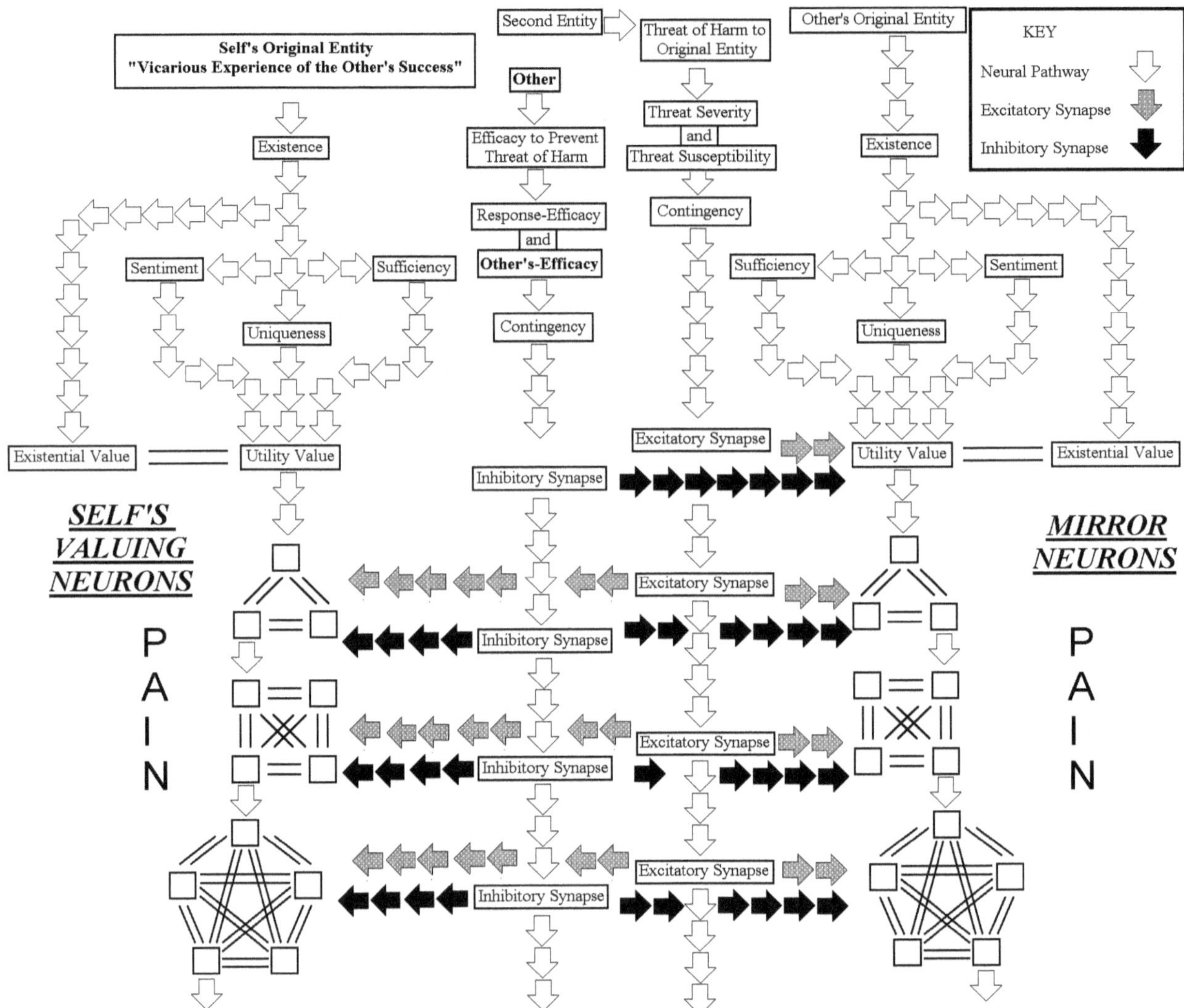

Figure 9.21 Sample End Result, Sympathetic Pride (Neurological model): Threat > Efficacy. The self's desired entity is vicarious experience of the *other's* success. Positive anxiety (pleasure) is depicted in the model. The self has no efficacy or threat component because it has no means for influencing the outcome of the *other's* situation. The self approves of the *other's* purpose and wants to witness and vicariously experience the *other's* success. Unfortunately, the *other* is not able to succeed (e.g., the *Other's* Threat neurons fire more frequently than Efficacy neurons) and the self does not vicariously experience the *other's* success. Necessarily, because the self approves of the *other's* purpose but is disappointed in the *other's* execution of it, the self would also feel vicarious shame.

Breakdown of Sympathetic Shame

The 1:1:1:1 ratio is maintained as a 2:2:2:2 ratio because the individual imagines himself or herself as two people. The expression:

√Self-Distinction = + 1 indicates the purpose belongs to the self;

√Self-Distinction = root of − 1 represents the individual's ability to imagine that he or she is valuing an entity with respect to a purpose while in another person's perspective (or animal or personified inanimate object), hence, experience it vicariously.

1) Experiencing the fulfillment of the *other's* purpose is the entity that the self is concerned with and wants to acquire. The self's desired entity can be thought of as an experience, namely, experiencing the *other's* success vicariously. More specifically, this would be the lowering of anxiety or a decrease in negative anxiety that is vicariously felt for a specific purpose. If the self were feeling shame for itself, then the neurological model would look like the following (figure 9.22):

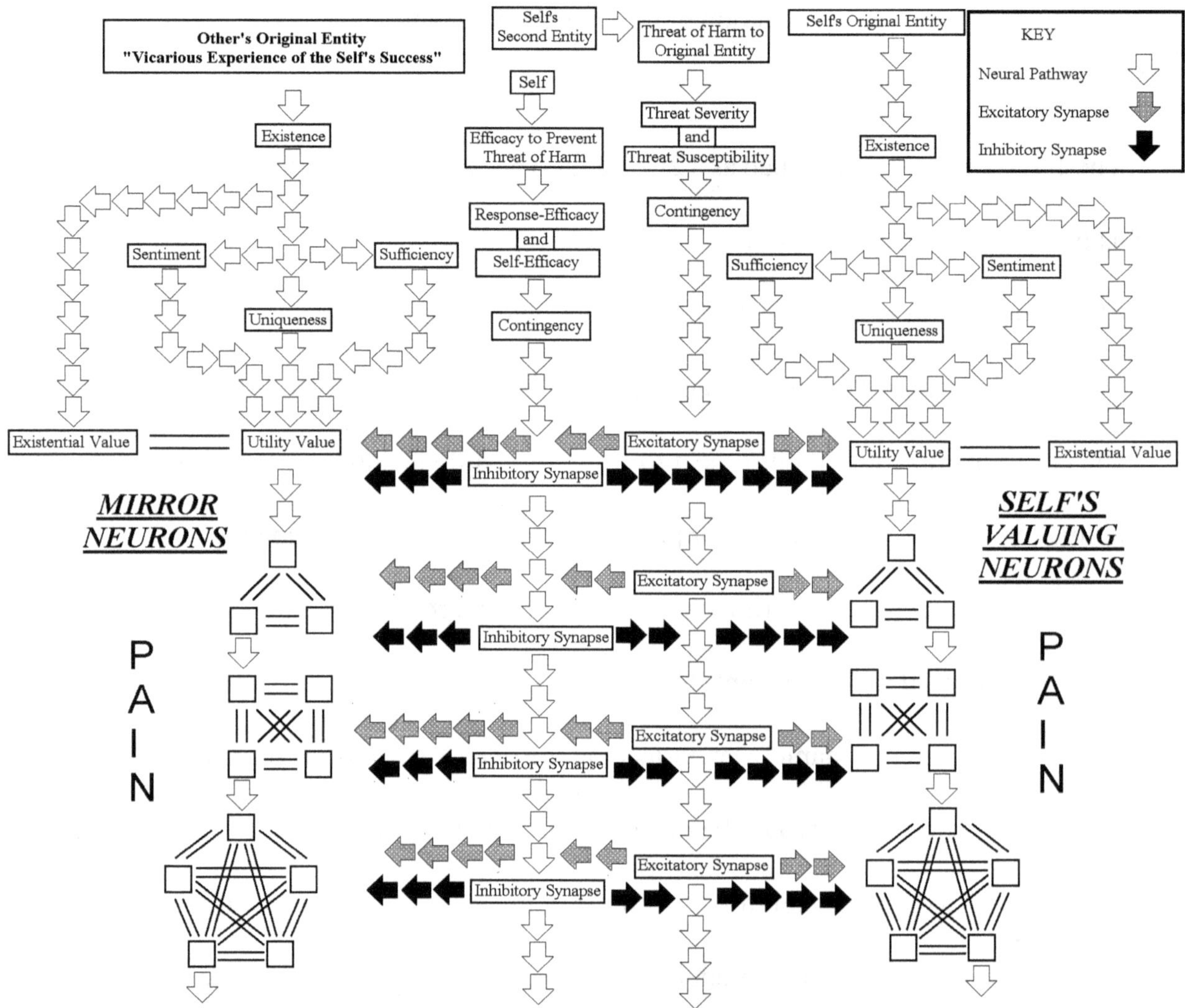

Figure 9.22 Sample End Result for Shame (Neurological model): Threat > Efficacy. Neurological model of shame felt by the self (right) for acknowledging sympathy felt by an *other* (left). Although the self did not successfully achieve its purpose, it is aware that an *other* approved of the self's purpose in the first place and wanted the self to succeed. Necessarily, vicarious sympathy is also felt by the self, as the self imagines itself as an *other* sympathizing for the self. Pain increases, as Efficacy is lesser than Threat.

2) In the case of sympathetic shame, The self's appraised judgment value only has a mimicked value of the *other's* valuation, as conceived by the self. Conversely, in the case of shame, the *other's* appraised judgment value only has a mimicked value

of the self's valuation.

3) The other, if active , has components for the Appraisal of the entity toward the restoration of equilibrium, drive reduction components, harm, and efficacy. Both the *other's* appraised value judgments as conceived by the self, and the self's mirrored values have an initial and subsequent appraisal.

4) Pattern

   a) The self's purpose, to vicariously experience the *other's* success, outweighs its complement, to not vicariously experience the *other's* success. The self desires to acquire the following entity: vicarious experience of the *other's* success. However, the *other* fails to fulfill the purpose in question. Failure, for the *other*, is measured as the elevation of anxiety or a decrease in negative anxiety.

   b) Sample graphs (figure 9.23 and 9.24)

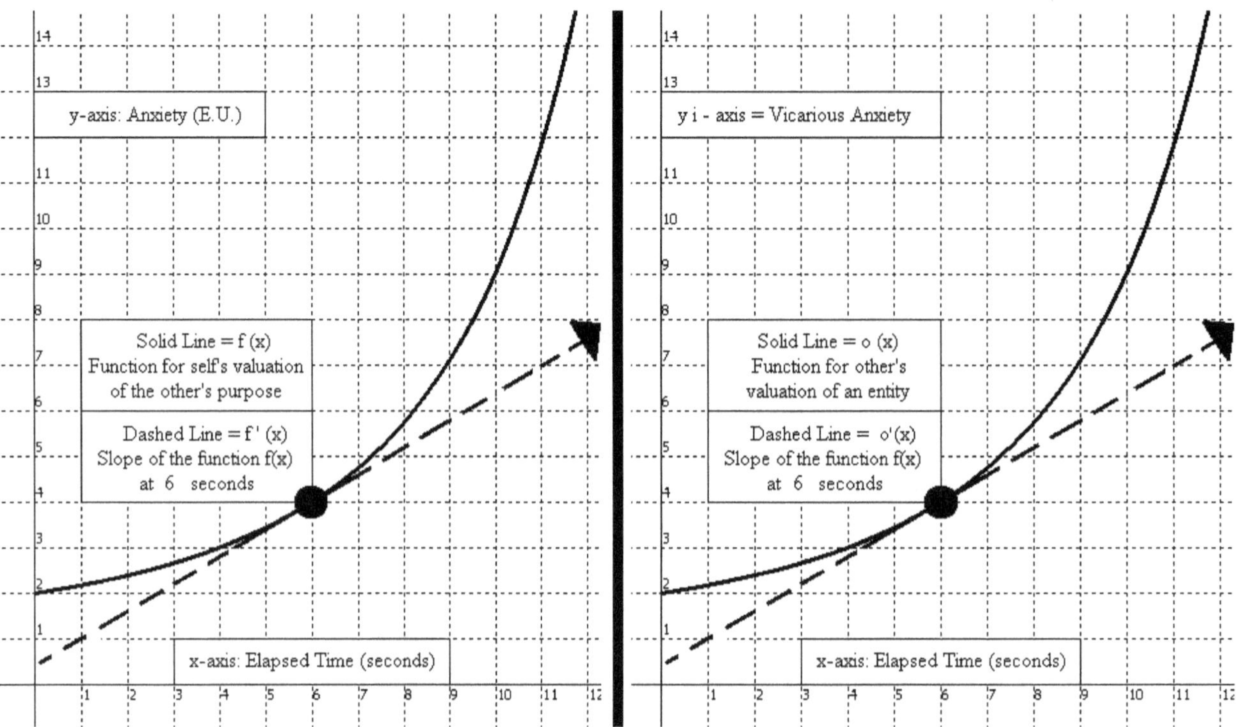

Figure 9.23 The above is a graph of sympathetic shame with anxiety elevating for the self and *other.* The function f(x) represents the self's valuation of the *other's* purpose, on the left. The *other's* valuation of an entity is represented by the function o(x) on the right. The dashed lines f '(x) and o'(x) correspond to the slopes of each function at time 6 seconds and the magnitude of a specific emotion felt at time 6 seconds.

i) Anxiety elevating:

Graph of anxiety elevating for the *other* and anxiety elevating for the self's estimation of the *other's* purpose. The predicted slope for the other is positive and for the self also positive. Both functions are above the x-axis.

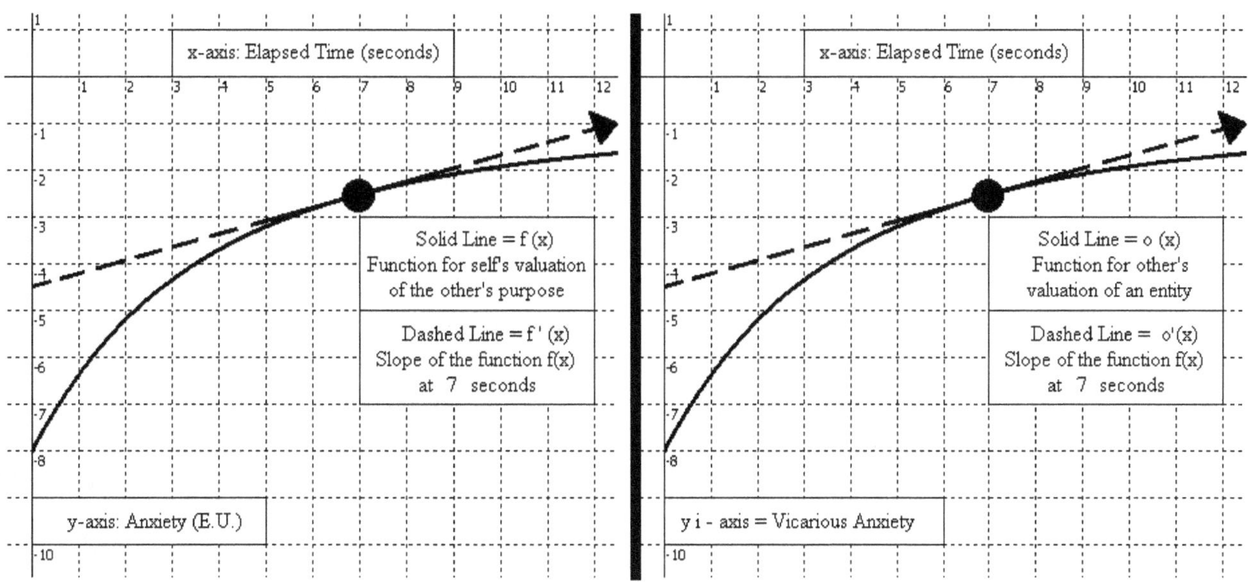

Figure 9.24 The above is a graph of sympathetic shame with negative anxiety decreasing. The function f(x) represents the self's valuation of the *other's* purpose, on the left. The *other's* valuation of an entity is represented by the function o(x) on the right. The dashed lines of f '(x) and o'(x) correspond to the slopes of each function at time 7 seconds and the magnitude of a specific emotion felt at time 7 seconds.

ii) Negative Anxiety decreasing:

Graph of negative anxiety decreasing for the *other* and negative anxiety decreasing for the self's estimation of the *other's* purpose. The predicted slopes for the *other* and for the self are both positive. Both functions are below the x-axis.

c) The following is an example for a situation where the self wants the *other's* anxiety to lower or negative anxiety to increase and the *other's* appraised judgment value of an entity for a purpose is the heavier of a complementary pair. Additionally, the *other* fails and the *other's* anxiety elevates, or alternatively negative anxiety decreases.

Neurons, Empathy, and Category II Emotions (Inter-personal

i) For example, at the same baseball game, if the spectator, who is a fan of the home team, was cheering for them to overcome a deficit to tie the game (restore equilibrium) and they are unsuccessful, then the *other's* anxiety elevates and the self's anxiety elevates as well.

ii) Similarly, if the home team were already winning and the same fan wanted the home team to demolish the away team, but the away team ends up scoring enough to tie the game and restore equilibrium, then the negative anxiety for both the *other* and for the self would decrease.

Hatred, Humiliation, Vicarious Hatred, Vicarious Humiliation, and Hateful Humiliation

With regard to conflict management, an understanding of the third empathetic emotion, hatred, would be the most useful. Hatred is considered to be a "deep, enduring, intense emotion expressing animosity, anger and hostility toward a person, group, or object."[23] Additionally, it is "characterized by both a desire to (a) harm or cause pain to the object of the emotion and (b) feelings of pleasure at the object's misfortunes."[24] Given that hatred is often considered to be the opposite of love, whatever model chosen for it should be diametrically opposed to the model established for love. For instance, if love was considered to be a situation where the self wanted the *other* to succeed and the *other* succeeds, then hatred must be the opposite of this. Hatred will be described as an instance where the self wants the *other* to not succeed and the *other* fails, which gives the self pleasure, a feeling opposite to the *other's* pain.

Additionally, as vicarious pride was postulated to coincide with empathetic love (e.g., companionate), so too must the opposite of vicarious pride. One antonym of pride and a good candidate is humiliation. Henceforth, it should be expected that the model chosen for hatred should also include the possibility for the experiencing of vicarious humiliation. Closely related to humiliation is humility, which evokes the notion of humbleness, though in a different connotation.

Although the term shame is sometimes used synonymously with humiliation, a strong and clear effort will be made to distinguish the two. Humiliation carries with it much heavier undertones than shame. Whereas shame merely implies defeat in a battle, humiliation implies a total rout and full retreat. Vicarious shame will coincide with sympathy throughout affect engineering, while vicarious humility will coincide with hatred. The reasoning for this is as follows.

In the context of the equation, vicarious shame, being associated with sympathy, referred to instances where the self wanted the *other* to succeed, but the *other* failed. Notwithstanding, in such instances the self reasons that the *other*, as conceived by the self, is at least aware that the self wanted the *other* to succeed. The disapproval, in this case is merely of the *other's* performance, in essence, disapproval of the *other's* execution or efficacy to prevent a harm.

Conversely, if vicarious humiliation, being associated with hatred, referred to instances where the self wanted the *other* to fail and the *other* fails, then a starkly different scenario arises. The self reasons that the *other*, as conceived by the self, is at least aware that the self wanted the *other* to fail. The disapproval, in this case, is a disapproval of the *other's* purpose and, in essence, an insult or attack on the *other's* character, henceforth, the *other's* Sentiment, ethical code, and value system.

What results, then, are two types of disapproval. In the case of sympathy and vicarious shame, the self's disapproval was of the *other's* efficacy to prevent a threat of harm and to achieve the purpose. In the case of hate and vicarious humiliation, the self's disapproval is of the *other's* purpose; to state it differently, the self approves of the *other's* failure.

Core of Hatred and Vicarious Humiliation

At its core, hatred consists in the ability of an individual to harbor animosity and a desire to witness an *other* suffer. Moreover, the self would delight in the misfortune (e.g., pain) of the *other*. Hence, the self, in the instance of hate, imagines itself feeling the *other's* pain for not succeeding and rather than mimic this feeling, the self feels an opposite emotion proportional to the *other's* plight. If the *other's* anxiety is increasing because of the inability to acquire an entity, then the self's negative anxiety, with respect to its valuation of the *other's* purpose, would increase. Likewise, if the *other's* negative anxiety were decreasing due to the inability to acquire an entity, then the self's anxiety, with respect to its valuation of the *other's* purpose, would decrease. Along the same lines, the self would also feel vicarious humiliation as a result of imagining itself in the *other's* situation, because the self is taking pleasure in the *other's* misfortune.

Fitting Hatred and Vicarious Humiliation into the Equation: Hateful Humiliation

The notion of hatred used in the equation encompasses both the definitions of hate and the notion of vicarious humiliation. In order to maximize this conception of hatred, an *other's* initial appraised value judgment of an entity, as conceived by the self, would have to have a harm variable that begins to approach positive one and an efficacy variable beginning to approach zero. Consequently, for the *other's* subsequent appraised value judgment of an entity, as conceived by the self, the harm variables would have to continue to approach or remain at positive one while the efficacy variables continue to move toward zero. This means that the *other's* anxiety would head toward infinity. Alternatively, the *other's* appraised value judgment may head toward negative one if negative anxiety is at stake and further acquisition of the entity would lead away from a restoration of equilibrium.

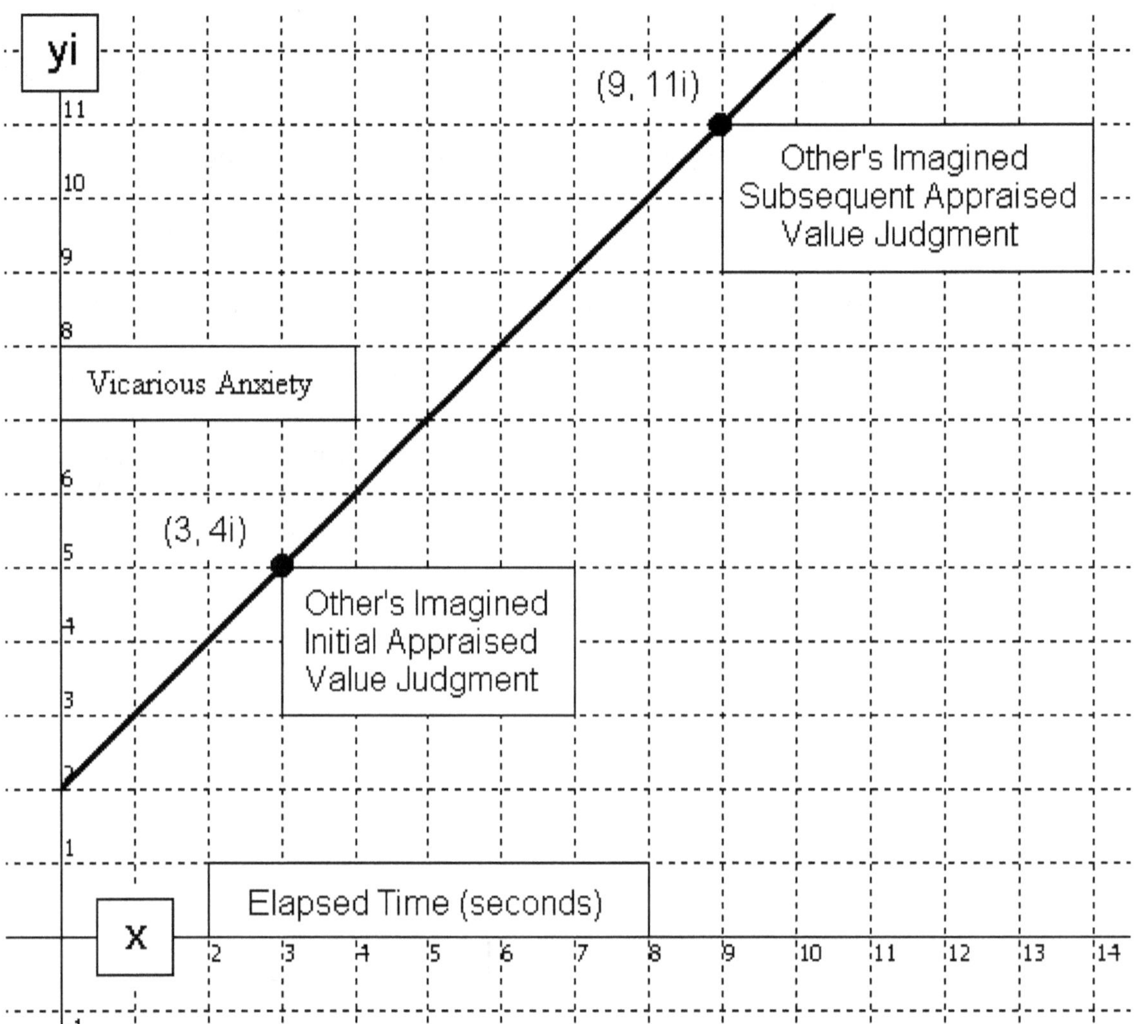

Figure 9.25 An *other's* Initial Appraised Value Judgment as conceived by the self and an *other's* Subsequent Appraised Value Judgment √-1, as conceived by the self.

Additionally, in order to ensure that this conception of hatred is both being gauged and maximized, the self's initial appraised value judgment of the *other's* failure, based upon the *other's* valuation of an entity as conceived by the self, should be approaching negative infinity if the other is not successfully acquiring his or her entity and continues to not acquire it. Therefore, the self's subsequent appraised value judgment of its desired entity, namely, the *other's* failure as conceived by the self, would continue to approach negative infinity because the *other* is not successfully acquiring his or her entity and this gives the self pleasure. Alternatively, the self's appraised value judgment of its desired entity may head toward positive one if positive anxiety is at stake and the self's further acquisition of its desired entity would lead toward a restoration of equilibrium. This is an indirect relationship between the *other's* valuation of an entity, as conceived by the self, and the self's valuation or vicarious experiencing of the *other's* purpose.

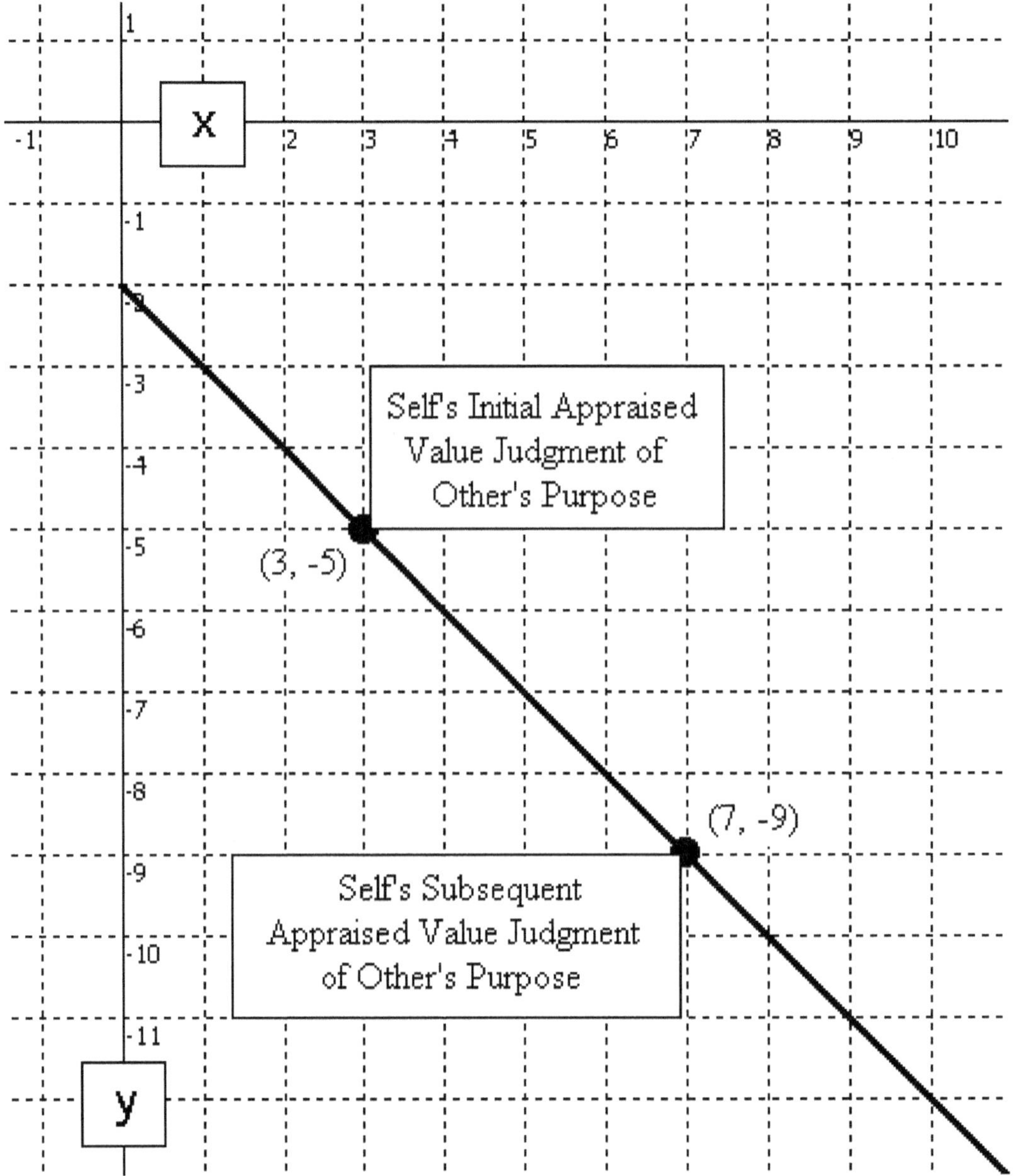

Figure 9.26 Self's initial Appraised Value Judgment of the *other's* valuation and self's subsequent Appraised Value Judgment of the *other's* valuation.

Valuation End Result for Hateful Humiliation

The *other* is not successful. The anxiety for the *other* elevates while the negative anxiety for the self increases. Alternatively, the negative anxiety for the *other* may decrease while the anxiety for the self lowers. Ultimately, this definition of hate consists in the self's ability to imagine another's pain and have a personal response to it that is the opposite, in this case pleasure.

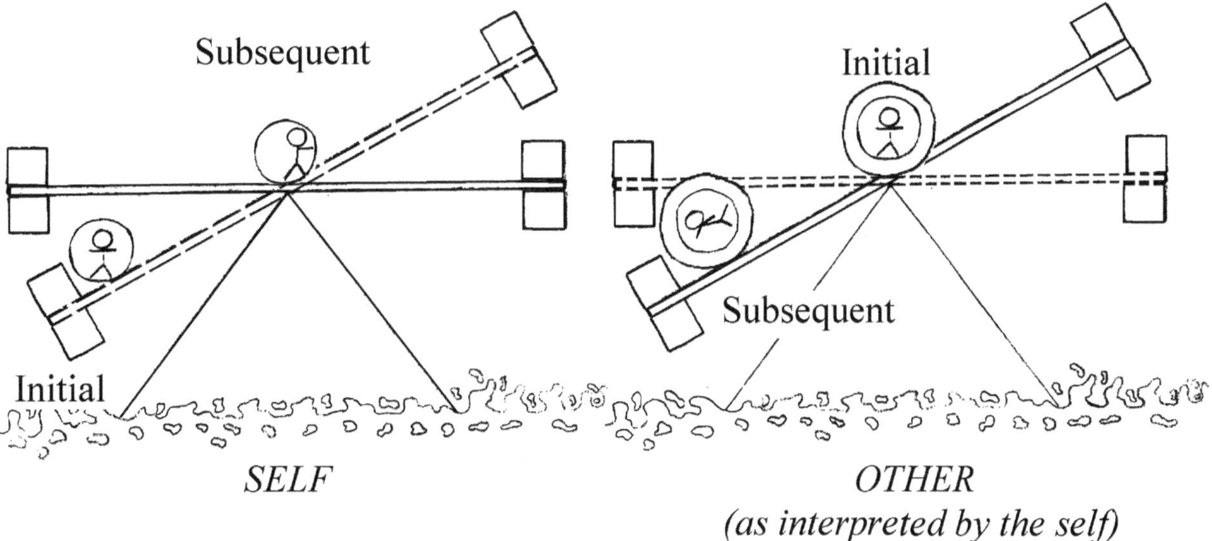

Figure 9.27 Sample End Result for Hateful Humiliation (Roll Cage Theory of Drives Model). The self's desired entity is vicarious experience of the *other's* failure. First, the self interprets that the *other* is not succeeding at his or her purpose. Thereafter, the self is able to acquire the desired entity of vicarious experience of the *other's* failure. The *other* fails and as a result the self acquires its desired entity, namely, vicarious experience of the *other's* failure. Failure for the *other* in this case, is the act of failing to achieving homeostasis.

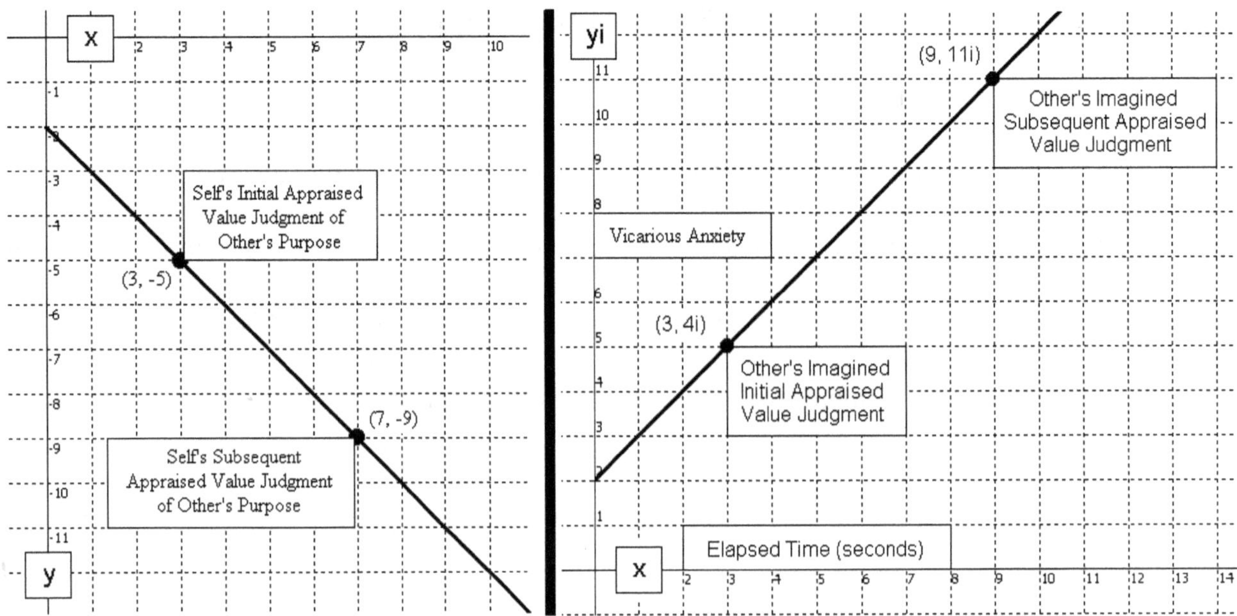

Figure 9.28 Sample End result for Hateful Humiliation (Cartesian Model)
The self's desired entity is vicarious experience of the *other's* failure. In this case, the *other* must acquire an entity in order to restore homeostasis, so positive anxiety is used with respect to a single purpose. Because the *other* is failing, the self feels pleasure, as the self's desired entity is vicarious experience of the *other's* failure. The slope of the graph for the self is negative, but the slope of graph for the *other* is positive in the case of Hateful Humiliation.

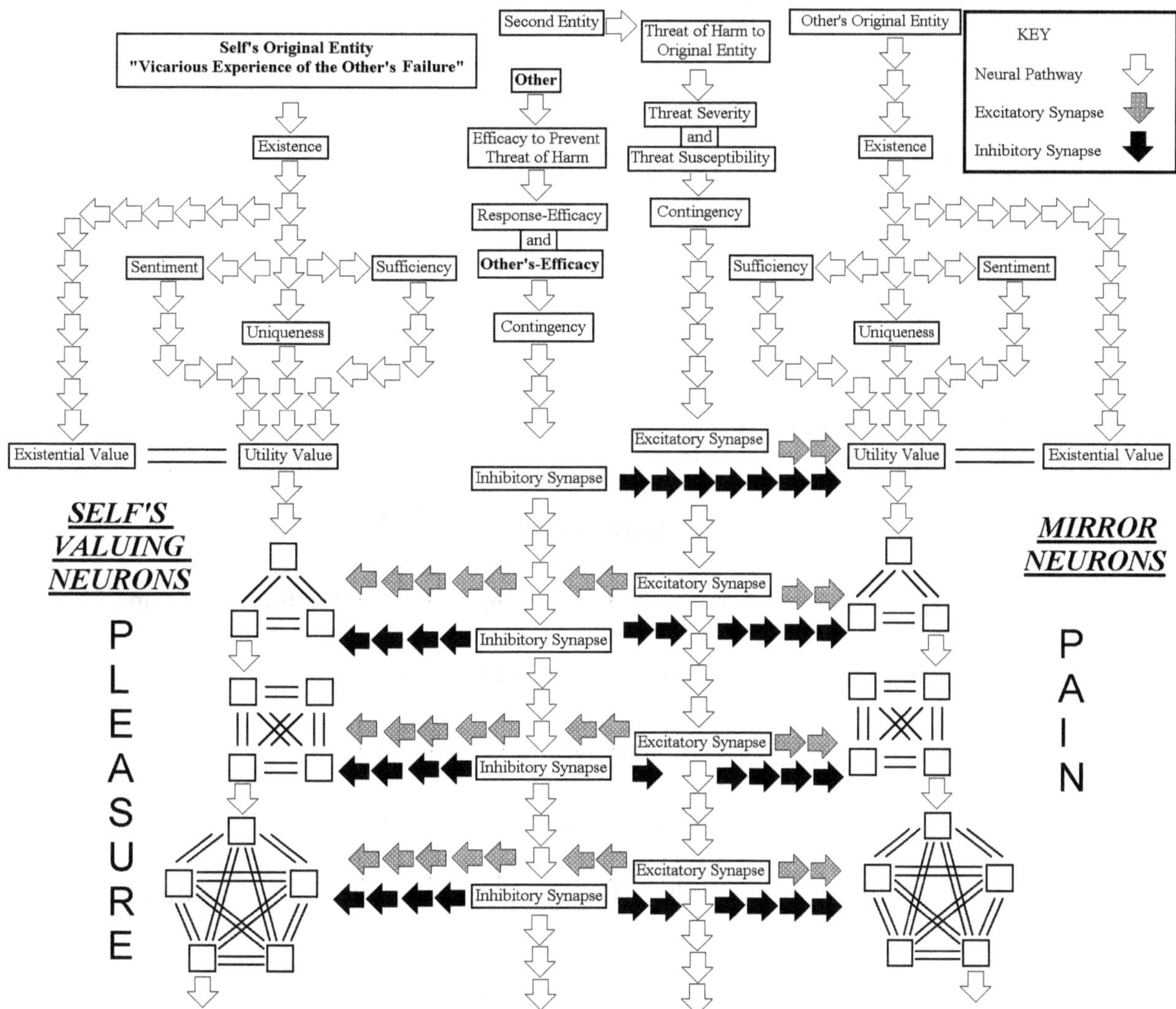

Figure 9.29 Sample End Result for Hateful Humiliation (Neurological model): Threat > Efficacy. The self's desired entity is vicarious experience of the *other's* failure. Anxiety is depicted in the model. The self has no efficacy or threat components because it has no means for influencing the outcome of the *other's* situation. The self disapproves of the *other's* purpose and wants to witness and vicariously experience the *other's* failure. Because the self disapproves of the *other's* purpose, the self would also feel vicarious humiliation.

Breakdown of Hateful Humiliation

The 1:1:1:1 ratio is maintained as a 2:2:2:2 ratio because the individual imagines himself or herself as two people. The expression:

√Self-Distinction = 1 represents the self, and

√Self-Distinction = root of − 1 represents the individual's ability to imagine that he or she is valuing an entity with respect to a purpose while in another person's perspective (or animal or personified inanimate object), hence, experiencing it vicariously.

1) The nonfulfillment of the *other's* purpose is the entity that the self is concerned with and wants to acquire. The self's desired entity can be thought of as an experience, namely, experiencing the *other's* failure vicariously. More specifically, this would be the elevation of anxiety or the decrease in negative anxiety that is vicariously felt for a specific purpose. If the self were feeling humiliation for itself, then the graph and neurological model would look like the following (figure 9.30):

Neurons, Empathy, and Category II Emotions (Inter-personal

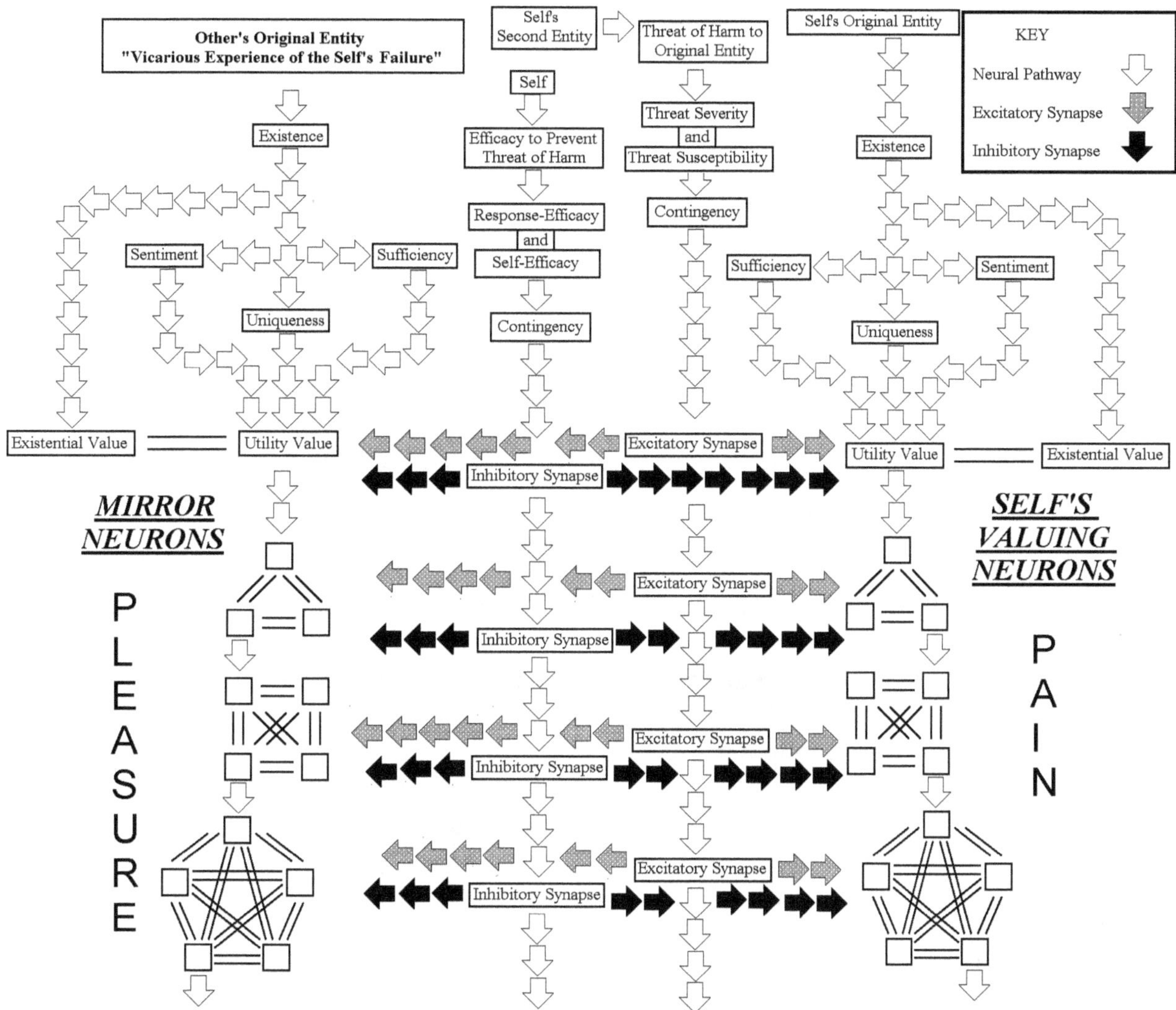

Figure 9.30 Sample End Result for Humiliation (Neurological model): Threat > Efficacy.
The above is a neurological model of humiliation felt by the self (right) for acknowledging vicarious hatred felt by an *other* (left). The self did not successfully achieve its purpose. It is aware that an *other* disapproved of the self's purpose in the first place and wanted the self to fail or not succeed. Necessarily, vicarious hatred is also felt by the self, as the self imagines itself as an *other* deriving pleasure from the inability of the self to achieve its aims. Pain increases for the self, but pleasure increases for the *other*. Efficacy is lesser than Threat.

2) The self's appraised judgment value only has a mirrored and inflected value of the *other's* valuation, as conceived by the self.

3) The *other*, if active, has components for the Appraisal of the entity towards the restoration of equilibrium, drive reduction components, harm, and efficacy. Both the *other's* appraised value judgments as conceived by the self's, and the self's mirrored values have an initial and subsequent appraisal.

4) Pattern

   a) The self's purpose, to vicariously experience the *other's* failure, outweighs its complement, to not vicariously experience the *other's* failure. The self desires to acquire the following entity: vicarious experience of the *other's* failure or not succeeding. The *other* does not successfully fulfill the purpose in question. Failure, for the other, is measured as the elevation of anxiety or a decrease in negative anxiety.

   b) Example graphs (figure 9.32 and figure 9.33)

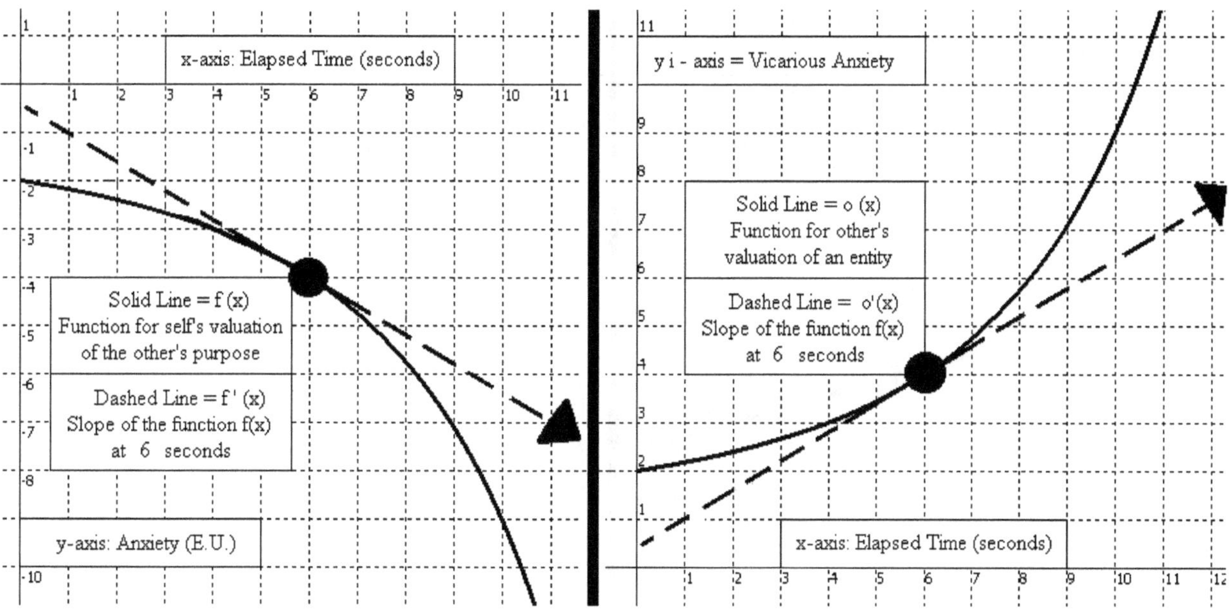

Figure 9.31 The above is a graph of hateful humiliation with anxiety increasing for the *other* and negative anxiety increasing for the self. The function f(x) represents the self's valuation of the *other's* purpose, on the left. The *other's* valuation of an entity is represented by the function o(x) on the right. The dashed lines of f '(x) and o'(x) correspond to the slopes of each function at time 6 seconds and the magnitude of a specific emotion felt at time 6 seconds.

i) Anxiety elevating:

Graph of anxiety elevating for the *other* and negative anxiety increasing for the self's estimation of the *other's* purpose. Predicted slope for the *other* is positive and for the self negative. The *other's* function is above the x-axis while the self's function is below the x-axis.

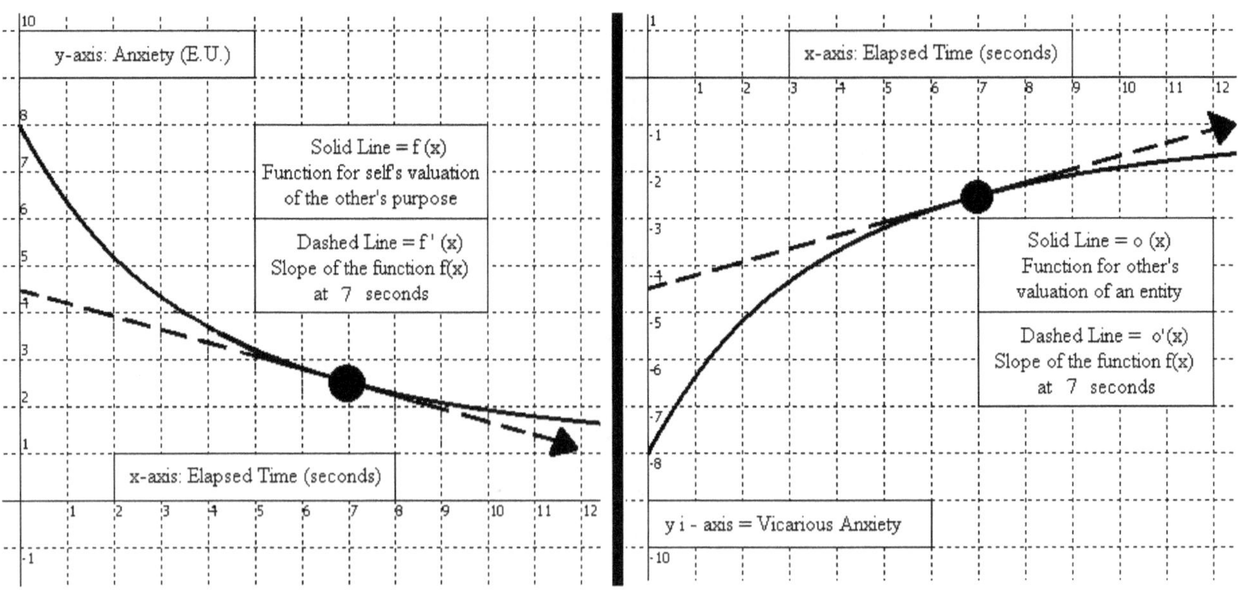

Figure 9.32 The above is a graph of hateful humiliation with negative anxiety decreasing for the *other* and anxiety decreasing for the self. The function f(x) represents the self's valuation of the *other's* purpose, on the left. The *other's* valuation of an entity is represented by the function o(x) on the right. The dashed lines of f '(x) and o'(x) correspond to the slopes of each function at time 7 seconds and the magnitude of a specific emotion felt at time 7 seconds.

ii) Negative anxiety decreasing:

Graph of negative anxiety decreasing for the *other* and anxiety lowering for the self's estimation of the *other's* purpose. The predicted slope for the *other* is positive while the predicted slope for the self is negative. The *other's* function is below the x-axis while the self's function is above the x-axis.

c) Below is a description of a situation where the self wants the *other's* anxiety to elevate or negative anxiety to decrease and the *other's* appraised judgment value of an entity for a purpose is the heavier of a complementary pair. Moreover, the *other* is not successful.

i) A movie-goer is at a theater watching a movie and cheering against the villain. The villain is deadlocked against an opponent (at equilibrium) and the moviegoer wants the villain to fail. The villain ultimately fails. The *other's* anxiety elevates while the self's negative anxiety increases.

ii) Alternatively, if the moviegoer is at the theater watching a movie, cheering against the villain, and the villain has a plot that looks as if it will succeed but ends up being thwarted and everything returns to equilibrium, then the *other's* negative anxiety will decrease and the self's anxiety will lower.

Antipathy, Mercy, Vicarious Antipathy, Vicarious Mercy, and Antipathetic Mercy

The last of the four degrees of empathy that will be considered concerns antipathy and vicarious mercy. Antipathy is a general dislike of something, and, generally speaking, it can be considered the opposite of sympathy in the sense that it refers to a distaste for something else. Disgust, which literally means "a bad taste," is often used synonymously with distaste and bears a number of similarities to the concept being described.[25] It should be kept in mind.

Mercy, on the other hand, has a number of slightly different meanings, but the one of concern for the equation is "compassion or forbearance shown especially to an offender or to one subject to one's power . . ."[1] One such example would be "imprisonment rather than death imposed as penalty for first-degree murder."[2] The connotation of mercy in the equation is less than favorable, however, in the sense that the self perceives the one receiving mercy to have escaped justice, or been unfairly blessed. Vicarious mercy, therefore, would be an instance where the self imagines itself to be in an *other's* situation escaping justice, a deserved threat of harm, or evading retribution.

Core of Antipathy and Vicarious Mercy

At its core, antipathy, like hatred, includes the capacity of an individual to harbor animosity and a desire to see an *other* suffer. Although the self would like to delight in the misfortune (e.g., pain) of the *other*, this is prevented by the *other* being successful in his or her endeavor. Hence, the self, in the instance of antipathy, imagines itself feeling the *other's* pleasure for succeeding and rather than mimic this feeling, the self feels an opposite emotion proportional to the *other's* pleasure. If the *other's* negative anxiety is increasing because the other is successfully acquiring an entity, then the self's anxiety, with respect to its valuation of the *other's* purpose, would increase. Likewise, if the *other's* anxiety were decreasing due to the ability to acquire an entity, then the self's negative anxiety, with respect to its valuation of the *other's* purpose, would decrease toward negative one. Along the same lines, the self would also feel vicarious mercy as a result of imagining itself in the

---

[1] By permission. From Merriam-Webster's Collegiate® Dictionary, 11th Edition ©2014 by Merriam-Webster, Inc. (www.Merriam-Webster.com).

[2] By permission. From Merriam-Webster's Collegiate® Dictionary, 11th Edition ©2014 by Merriam-Webster, Inc. (www.Merriam-Webster.com).

other's situation; though the self is personally experiencing pain as a result of the *other's* good fortune, when the self imagines itself to be in the *other's* situation it also imagines pleasure at having escaped harm.

Fitting Antipathy and Vicarious Mercy into the Equation: Antipathetic Mercy

The notion of antipathetic mercy is structured similarly to its counterpart sympathetic shame. Its use in the equation also encompasses the animosity from the definition of hate, but in addition to the notion of vicarious mercy. In order to maximize this conception of antipathetic mercy, an *other's* initial appraised value judgment of an entity, as conceived by the self, would have to have a harm variable that begins to approach zero and an efficacy variable approaching one. Consequently, for the *other's* subsequent appraised value judgment of an entity (as conceived by the self), the harm variables would have to continue to approach or remain at zero while the efficacy variables continue to move toward one. This means that the *other's* anxiety would head toward one. Alternatively, the *other's* appraised value judgment may head toward negative infinity if negative anxiety is at stake and further acquisition of the entity would lead away from a restoration of equilibrium.

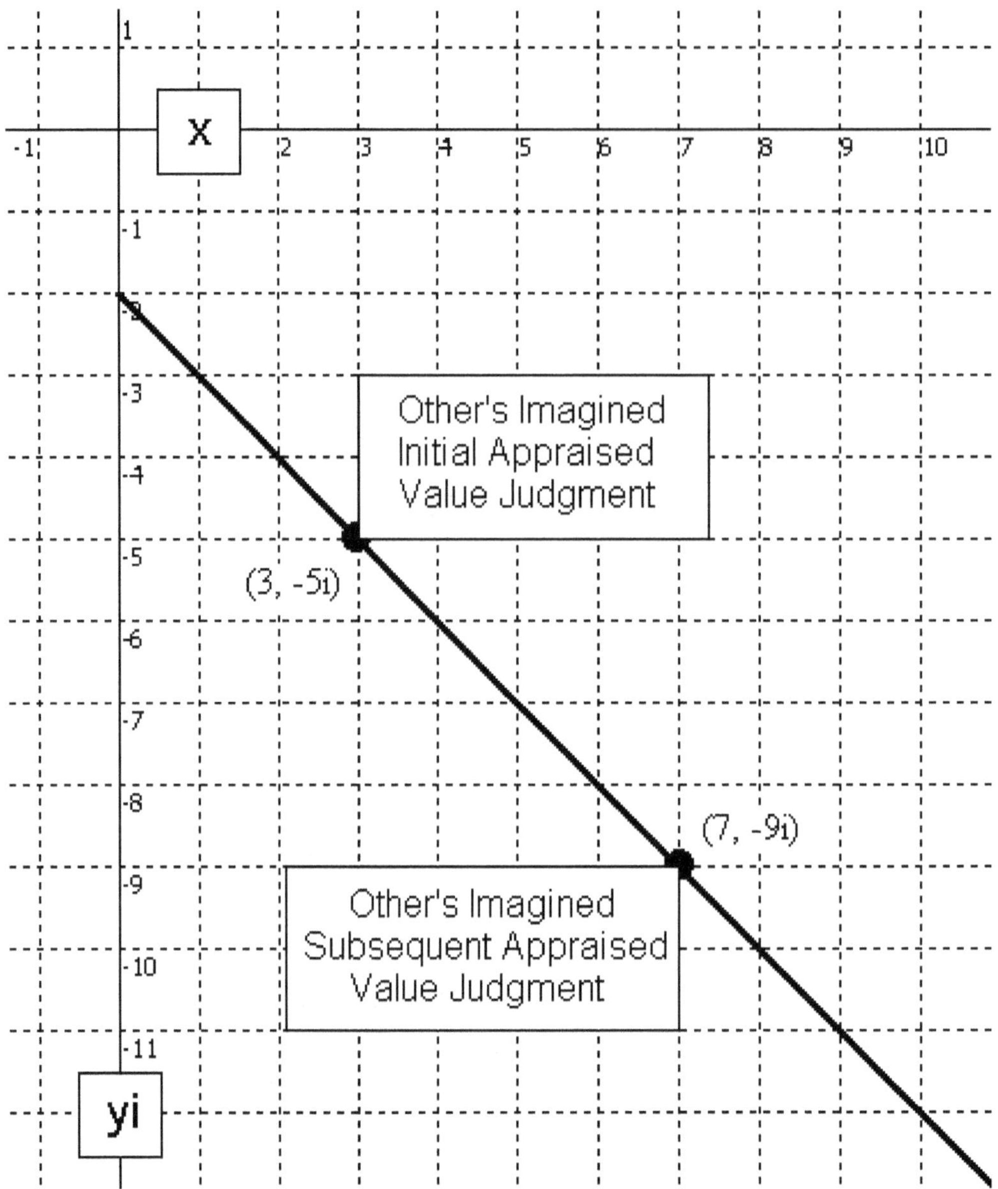

Figure 9.33 *Other's* Initial Appraised Value Judgment as conceived by the self and *other's* Subsequent Appraised Value Judgment as conceived by the self

Additionally, in order to ensure that this conception of antipathetic mercy is both being gauged and maximized, the self's initial appraised value judgment of the *other's* failure, based upon the *other's* valuation of an entity, as conceived by the self, would have to be approaching positive infinity if the *other* is successfully acquiring his or her desired entity and continues to acquire it. Therefore, the self's subsequent appraised value judgment of its desired entity, namely, the *other's* valuation as conceived by the self, would continue to approach positive infinity because the *other* is successfully acquiring his or her entity and this causes the self to feel pain. Alternatively, the self's appraised value judgment of its desired entity may head toward negative one if negative anxiety is at stake and the self's further acquisition of its desired entity, vicarious experience of the other's failure, would lead away from a restoration of equilibrium. This is an indirect relationship between the *other's* valuation of an entity, as conceived by the self, and the self's valuation of the *other's* nonfulfillment of the purpose.

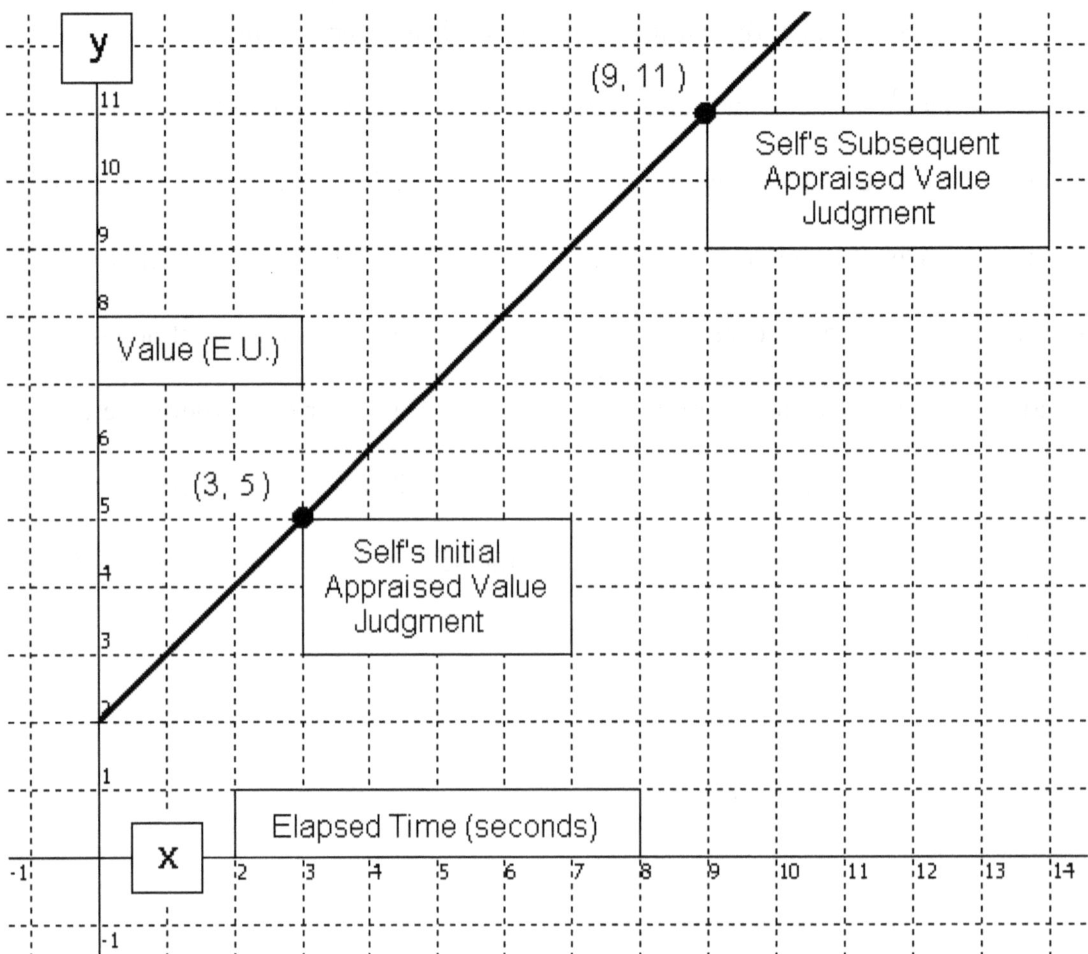

Figure 9.34 Self's initial Appraised Value Judgment of the *other's* valuation and self's subsequent Appraised Value Judgment of the other's valuation.

Valuation End Result for Antipathetic Mercy

The *other* is successful. The anxiety for the *other* lowers while the negative anxiety for the self decreases. Alternatively, the negative anxiety for the *other* may increase while the anxiety for the self elevates. Ultimately, this definition of antipathetic mercy consists in the self's ability to imagine another's pleasure and have a personal response to it that is the opposite, in this case of pain.

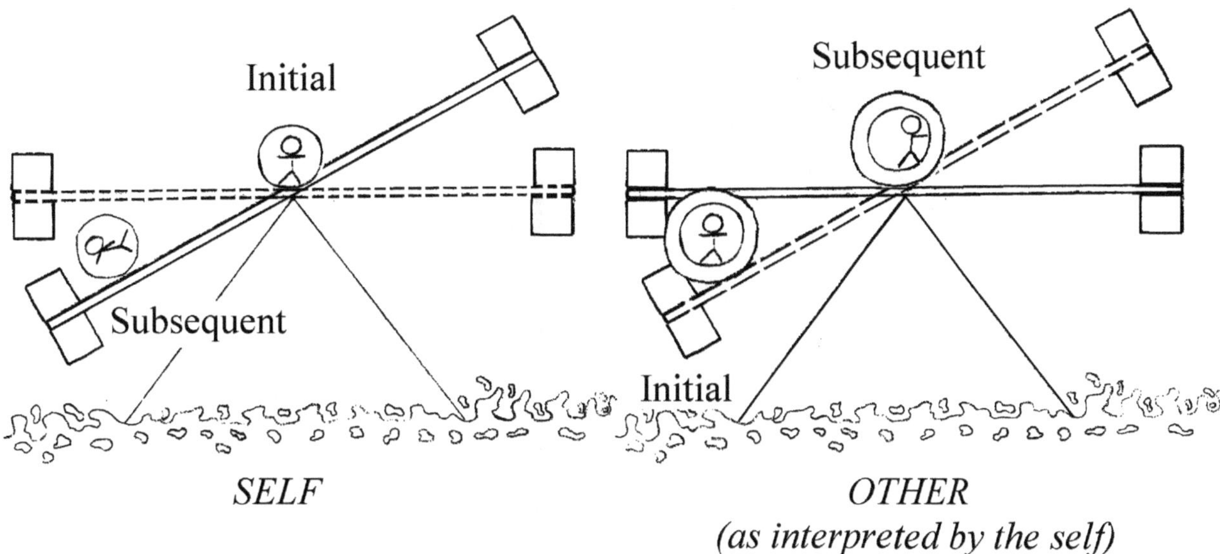

Figure 9.35 Sample End Result for Antipathetic Mercy (Roll Cage Theory of Drives Model). The self's desired entity is vicarious experience of the *other's* failure. First, the self interprets that the *other* is succeeding at his or her purpose. Thereafter, the self is not able to acquire the desired entity of vicarious experience of the *other's* failure. The *other* succeeds, and as a result the self does not acquire its desired entity, namely, vicarious experience of the *other's* failure. Success for the *other,* in this case, is the aim of achieving homeostasis.

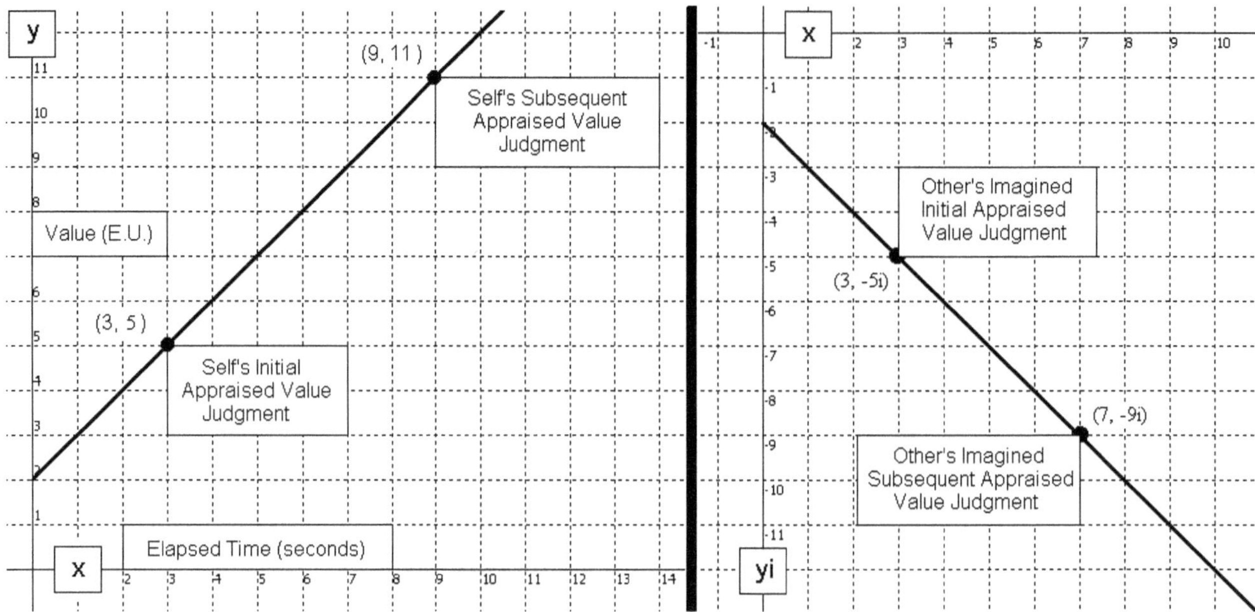

Figure 9.36 Sample End Result for Antipathetic Mercy (Cartesian Model).
The self's desired entity is vicarious experience of the *other's* failure. In this case, the *other* has acquired an entity in an amount that is greater than is required for achieving homeostasis, so negative anxiety is used with respect to a single purpose. Because the *other* is succeeding, the self feels pain, as the self's desired entity was vicarious experience of the *other's* failure. Regardless of whether pain or pleasure are concerned (e.g., anxiety or positive anxiety), the slope of the graph for the self is positive while the slope of the graph for the *other* is negative.

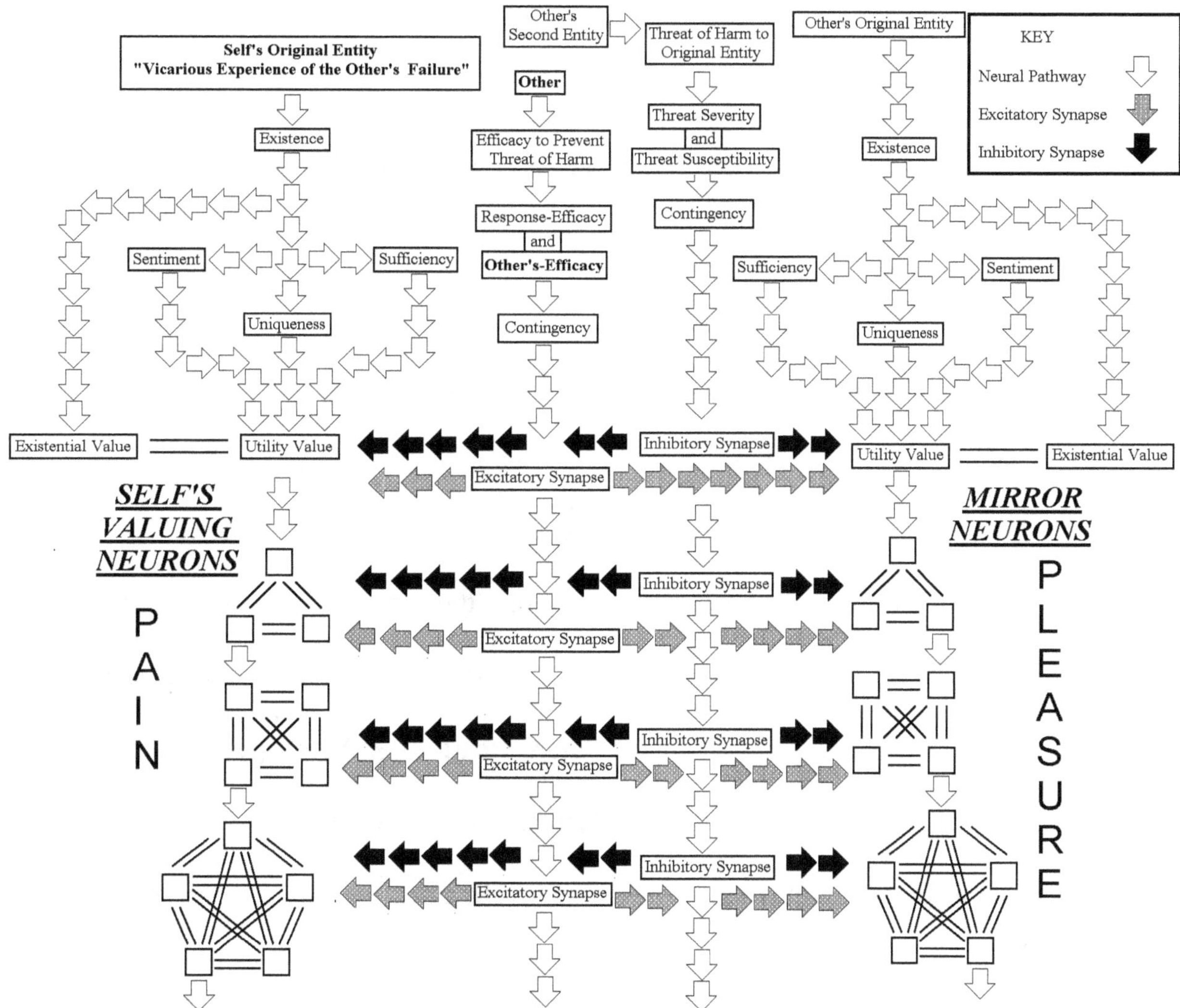

Figure 9.37 Sample End Result for Antipathetic Mercy (Neurological model): Threat < Efficacy. The self's desired entity is vicarious experience of the *other's* failure. Negative anxiety is depicted in the model. The self has no efficacy or threat component because it has no means for influencing the outcome of the *other's* situation. The self disapproves of the *other's* purpose and wants to witness and vicariously experience the *other's* failure. Necessarily, because the self disapproves of the *other's* purpose, the self would also feel vicarious mercy.

Breakdown of Antipathetic Mercy

The 1:1:1:1 ratio is maintained as a 2:2:2:2 ratio because the individual imagines himself or herself as two people. The expression:

√Self-Distinction = 1 represents the self;

√Self-Distinction = root of - 1 represents the individual's ability to imagine that he or she is valuing an entity with respect to a purpose while in another person's perspective (or animal or personified inanimate object). Hence, experience it vicariously.

1) The nonfulfillment of the *other's* purpose is the entity that the self is concerned with and wants to acquire. The self's desired entity can be thought of as an experience, namely, experiencing the *other's* failure vicariously. More specifically, this would be the elevation of anxiety or the decrease in negative anxiety that is vicariously felt for a specific purpose. If the self were feeling mercy for itself, then the graph and neurological model would look like the following (figure 9.38):

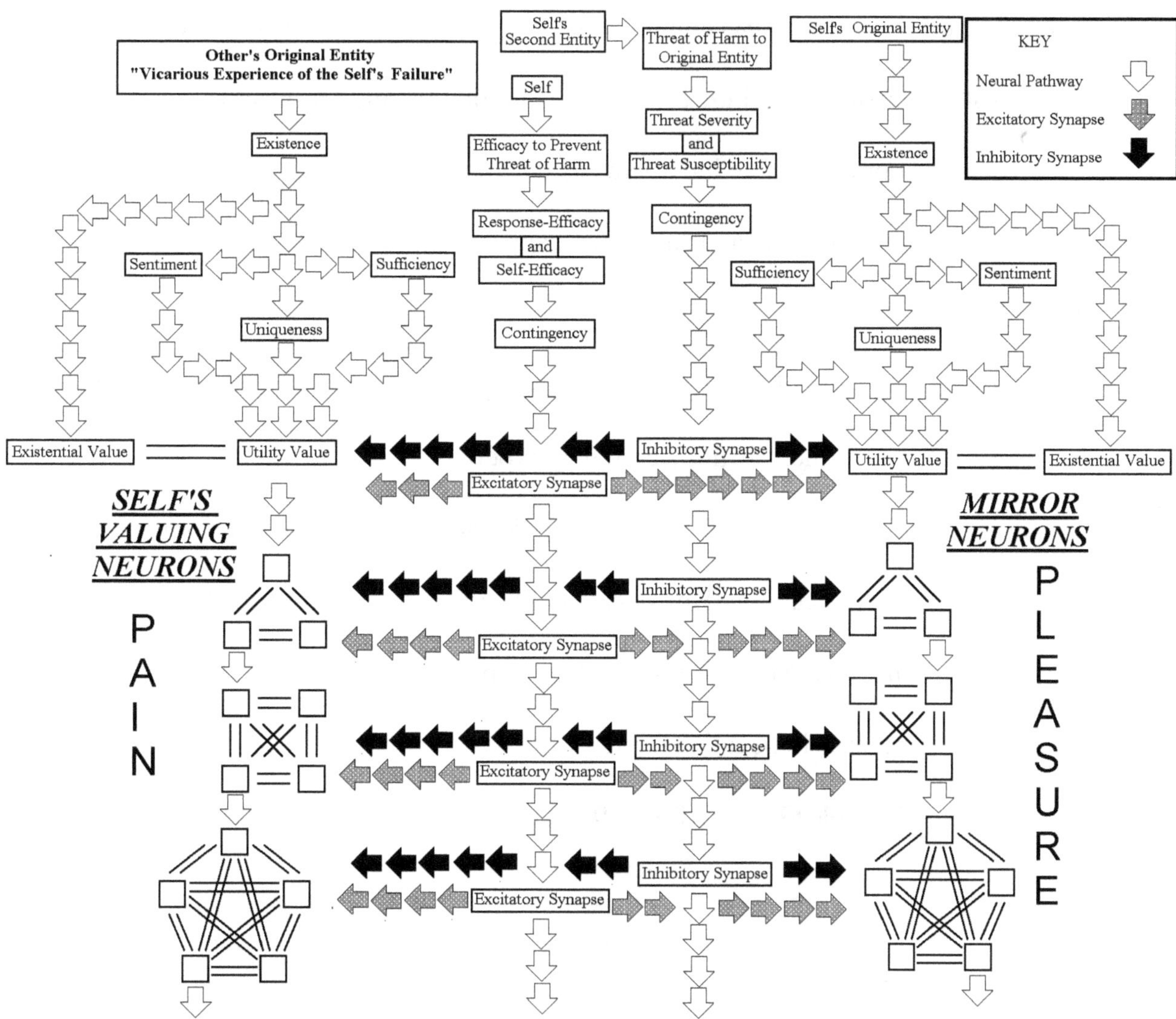

Figure 9.38 Sample End Result for Mercy (Neurological model): Threat < Efficacy. Neurological model of mercy felt by the self (right) and for acknowledging vicarious antipathy felt by an *other* (left). The self successfully achieves its purpose while being aware that an *other* disapproved of the self's purpose in the first place and wanted the self to fail or not succeed. Necessarily, vicarious antipathy is also felt by the self, as the self imagines itself as an *other* feeling pain as a result of the ability of the self to achieve its aims. Pleasure increases for the self, but pain increases for the *other* as imagined by the self. Efficacy is greater than Threat.

2) The self's appraised judgment value only has a mirrored and inflected value of the *other's* valuation, as conceived by the self.

3) The *other*, if active, has components for the Appraisal of the entity toward the restoration of equilibrium, drive reduction components, harm, and efficacy. Both the *other's* appraised value judgments as conceived by the self, and the self's mirrored values have an initial and subsequent appraisal.

4) Pattern

   a) The self's purpose, to vicariously experience the *other's* failure, outweighs its complement, to not vicariously experience the *other's* failure. The self desires to acquire the following entity: vicarious experience of the *other's* failure or not succeeding. However, the *other* successfully fulfills the purpose in question. Success, for the *other*, is measured as the lowering of anxiety or an increase in negative anxiety.

   b) Example Graphs (figure 9.39 and 9.40)

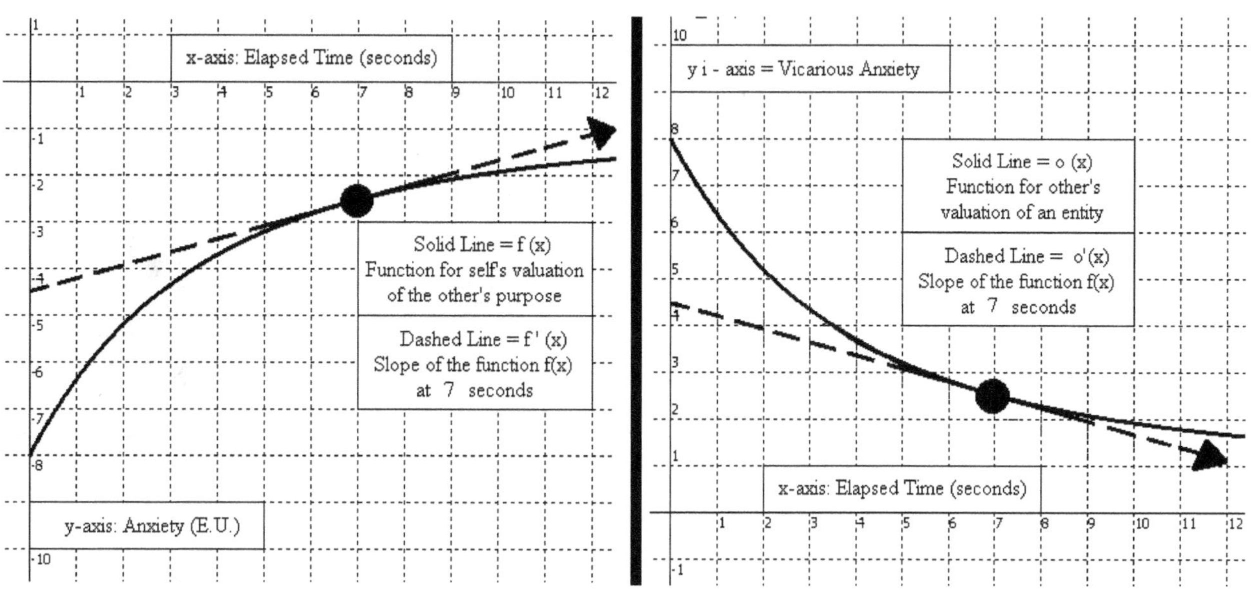

Figure 9.39 The above is a graph of antipathetic mercy with anxiety lowering for the *other* and negative anxiety decreasing for the self. The function f(x) represents the self's valuation of the *other's* purpose, on the left. The *other's* valuation of an entity is represented by the function o(x) on the right. The dashed lines f '(x) and o'(x) correspond to the slopes of each function at time 7 seconds and the magnitude of a specific emotion felt at time 7 seconds.

i) Anxiety lowering:

Graph of anxiety lowering for the *other* and negative anxiety decreasing for the self's estimation of the *other's* purpose. The predicted slope for the *other* is negative and for the self positive. The *other's* function is above the x-axis while the self's function is below the x-axis.

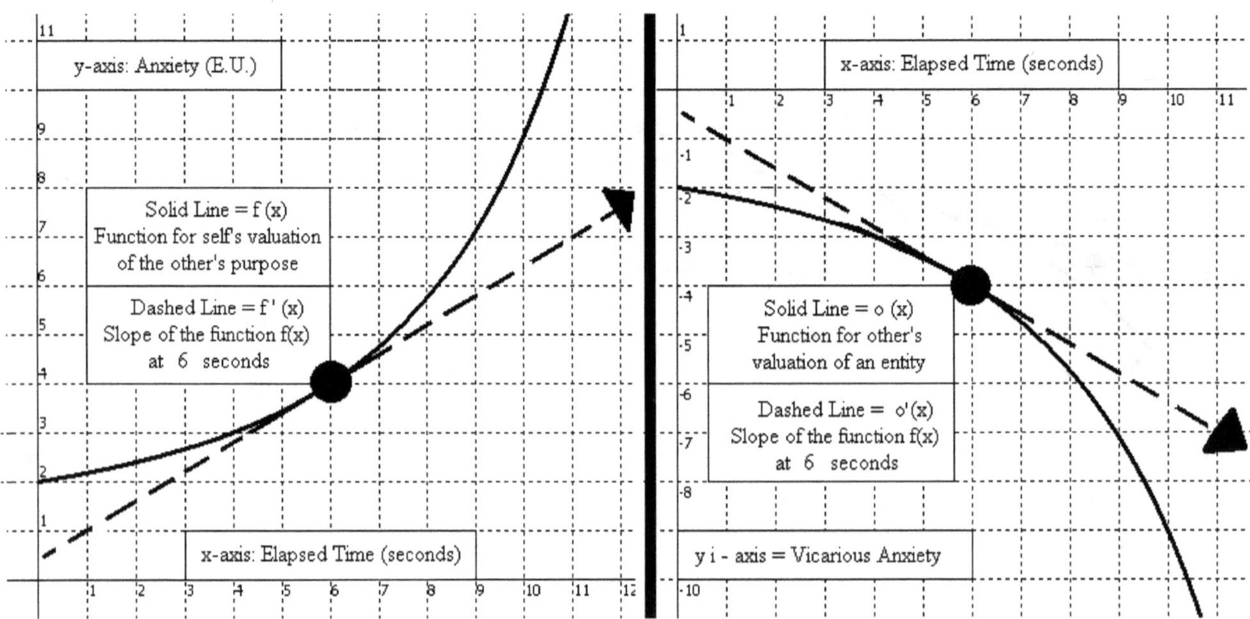

Figure 9.40 The above is a graph of antipathetic mercy with negative anxiety increasing for the *other* and anxiety increasing for the self. The function f(x) represents the self's valuation of the *other's* purpose, on the left. The *other's* valuation of an entity is represented by the function o(x) on the right. The dashed lines of f '(x) and o'(x) correspond to the slopes of each function at time 6 seconds and the magnitude of a specific emotion felt at time 6 seconds.

ii) Negative anxiety increasing:

Graph of negative anxiety increasing for the other and anxiety elevating for the self's estimation of the *other's* purpose. The predicted slope for the *other* is negative while the predicted slope for the self is positive. The *other's* function is below the x-axis while the self's function is above the x-axis.

c) Example for a situation where the self wants the *other's* anxiety to elevate or negative anxiety to decrease and the *other's* appraised judgment value of an entity for a purpose is the heavier of a complementary pair. Moreover, the *other* successfully acquires the entity in question.

i) A movie-goer at the same theater is cheering against the villain. The villain is trapped and doomed to fail as well with the hero closing in. However, the villain manages to escape and avoid capture (harm), meaning that the *other's* anxiety lowers while the self's negative anxiety decreases. What was once certain defeat for the villain has now become a draw (equilibrium).

ii) Alternatively, a moviegoer may be at a theater cheering against the villain and the villain has the clear upper hand, such as a doomsday device that is about to be set off. If the device does go off and the villain wins, then the *other's* negative anxiety will increase while the self's anxiety elevates and the effect is the same as in the previous scenario.

Loneliness, Neutrality, Vicarious Loneliness, Vicarious Neutrality, and Indifference

The final case considers scenarios where no empathy is felt, or a state of indifference is felt. Indifference refers to a state when one "has no preferences between alternative choices or courses of action."[26] Defining indifference in the equation is relatively straightforward compared to the four degrees of empathy. Whereas the self's valuation of the other's purpose fluctuated in a particular manner depending on whether or not the other was successfully acquiring an entity or not successfully acquiring it, for indifference the self's valuation of the other's purpose simply flat-lines or exhibits no response to the other's success or failure. So long as the self's valuation of the other's purpose does not veer from the x-axis, or alternately its projected course, while the self imagines itself in the *other's* situation, then a state of indifference is maintained. The *self*, due to the absence of any judgment from the self whatsoever, would feel vicarious loneliness in the *other*. Stated differently, the absence of either approval or disapproval from the self is what is felt by the self when it imagines itself to be the *other*. Not perceiving a reaction from the self, therefore, would lead an *other* to feel ignored or otherwise isolated and lonely due to the *self's* indifference.

Reversing the Direction of the Vicarious Emotions

As noted earlier, any of the vicarious emotions: pride; shame; humility; and mercy can be also be felt by the self if the passivity of the self is changed and the *other* ceases to be the active party. The interaction between the *other* and the self, then, would be slightly different. The vicarious emotions of pride, shame, humility, and mercy, when they are felt directly by the self, would parallel certain Category I emotions. Under such a scenario, the *other* would lack efficacy or threat (benefit) components and would simply want to vicariously experience the self either successfully or unsuccessfully acquire an entity.

Neurons, Empathy, and Category II Emotions (Inter-personal

CATEGORY II EMOTION	SELF IS PASSIVE		SELF IS ACTIVE		Category I Correspondence
	VICARIOUS EMOTION	SELF	VICARIOUS EMOTION	SELF	
Loving Pride	Vicarious Pride	Love	Vicarious Love	Pride	Happiness or Anger
Sympathetic Shame	Vicarious Shame	Sympathy	Vicarious Sympathy	Shame	Guilt or Sadness
Hateful Humiliation	Vicarious Humiliation	Hatred	Vicarious Hatred	Humiliation	Guilt or Sadness
Antipathetic Mercy	Vicarious Mercy	Antipathy	Vicarious Antipathy	Mercy	Happiness or Anger
Indifference	Vicarious Loneliness	Neutrality	Vicarious Neutrality	Loneliness	Content (flatlining)

Figure 9.41 The Category II Emotions: The Four Degrees of Empathy and Indifference. The Category I Correspondence refers to the Category I emotion that would be similar to the emotion felt by the self if it is active (hence, has efficacy and threat components). If the self is passive, then the Category I Correspondence would apply to the other.

Preview: Attention

It is generally accepted that emotions are of a fickle nature. What is vehemence one minute can turn into joviality the next. A joke misconstrued can turn a light-hearted encounter into a bloodbath. Although time may cause one to forget a slight, it can also permit a grudge to foment into vitriol. What is it then that enables this dynamic quality of emotion? Can it be expressed in the equation? More important, can it be harnessed?

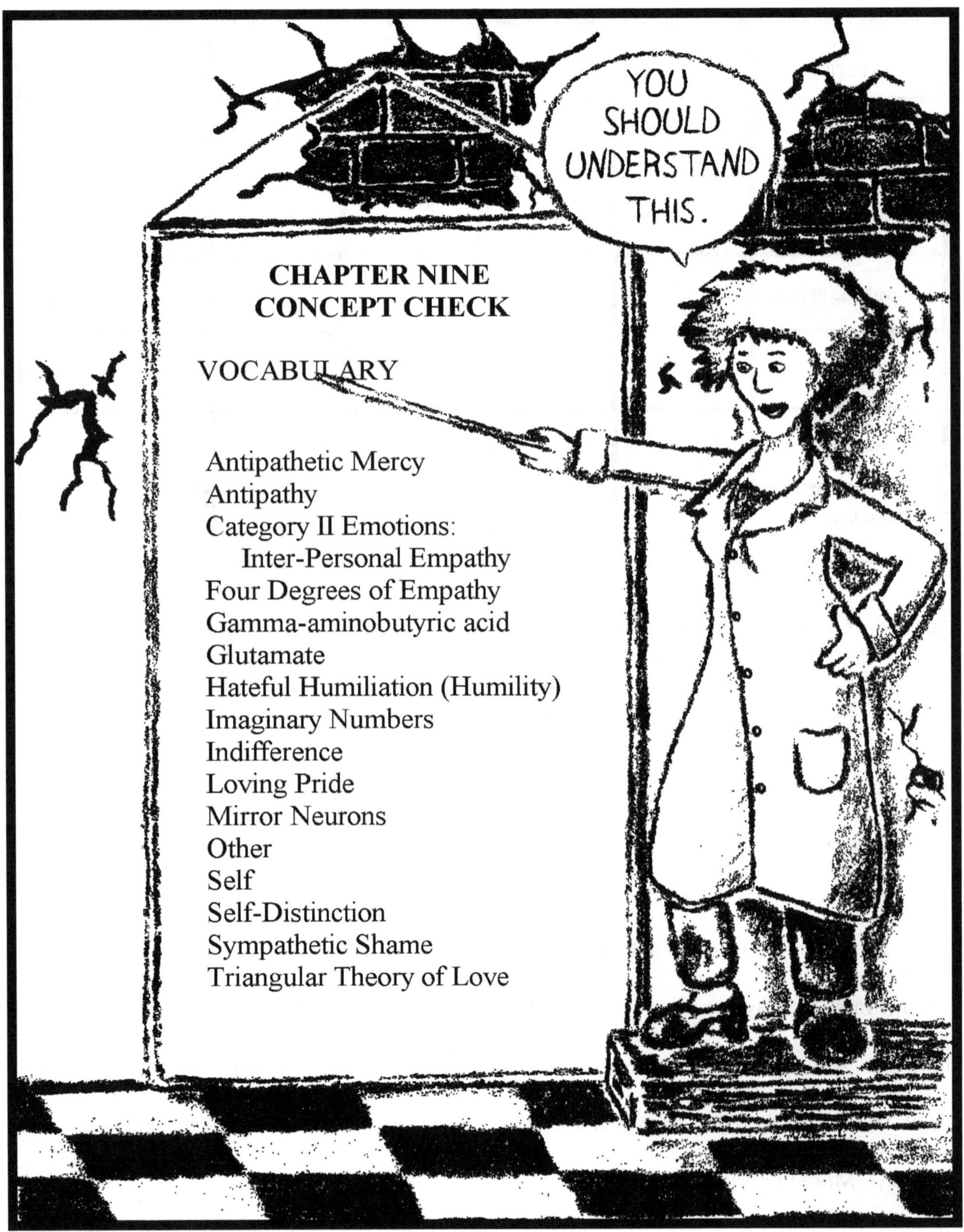
YOU SHOULD UNDERSTAND THIS.
CHAPTER NINE
CONCEPT CHECK
VOCABULARY
Antipathetic Mercy
Antipathy
Category II Emotions:
Inter-Personal Empathy
Four Degrees of Empathy
Gamma-aminobutyric acid
Glutamate
Hateful Humiliation (Humility)
Imaginary Numbers
Indifference
Loving Pride
Mirror Neurons
Other
Self
Self-Distinction
Sympathetic Shame
Triangular Theory of Love

Notes

1. Wilde, Oscar (2003). *The Picture of Dorian Gray*. New York: Barnes and Nobles Books. Print, p. 56. (Original work published 1890).

2. Reber, Arthur S., Rhianon Allen, and Emily S. Reber. *Penguin Dictionary of Psychology*. London. Penguin Books, 2009. Print, p. 488.

3. Reber, Arthur S., Rhianon Allen, and Emily S. Reber. *Penguin Dictionary of Psychology*. London. Penguin Books, 2009. Print, p. 727.

4. Reber, Arthur S., Rhianon Allen, and Emily S. Reber. *Penguin Dictionary of Psychology*. London. Penguin Books, 2009. Print, p. 395.

5. Reber, Arthur S., Rhianon Allen, and Emily S. Reber. *Penguin Dictionary of Psychology*. London. Penguin Books, 2009. Print, p. 601.

6. Reber, Arthur S., Rhianon Allen, and Emily S. Reber. *Penguin Dictionary of Psychology*. London. Penguin Books, 2009. Print, p. 601.

7. Reber, Arthur S., Rhianon Allen, and Emily S. Reber. *Penguin Dictionary of Psychology*. London. Penguin Books, 2009. Print, p. 329.

8. Reber, Arthur S., Rhianon Allen, and Emily S. Reber. *Penguin Dictionary of Psychology*. London. Penguin Books, 2009. Print, p. 601.

9. Reber, Arthur S., Rhianon Allen, and Emily S. Reber. *Penguin Dictionary of Psychology*. London. Penguin Books, 2009. Print, p. 319.

10. Reber, Arthur S., Rhianon Allen, and Emily S. Reber. *Penguin Dictionary of Psychology*. London. Penguin Books, 2009. Print, p. 259.

11. Reber, Arthur S., Rhianon Allen, and Emily S. Reber. *Penguin Dictionary of Psychology*. London. Penguin Books, 2009. Print, p. 259.

12. Reber, Arthur S., Rhianon Allen, and Emily S. Reber. *Penguin Dictionary of Psychology*. London. Penguin Books, 2009. Print, p. 477.

13. Reber, Arthur S., Rhianon Allen, and Emily S. Reber. *Penguin Dictionary of Psychology*. London. Penguin Books, 2009. Print, p. 543.

14. Reber, Arthur S., Rhianon Allen, and Emily S. Reber. *Penguin Dictionary of Psychology*. London. Penguin Books, 2009. Print, p. 436.

15. Reber, Arthur S., Rhianon Allen, and Emily S. Reber. *Penguin Dictionary of Psychology*. London. Penguin Books, 2009. Print, p. 436.

16. Reber, Arthur S., Rhianon Allen, and Emily S. Reber. *Penguin Dictionary of Psychology*. London. Penguin Books, 2009. Print, p. 437.

17. Reber, Arthur S., Rhianon Allen, and Emily S. Reber. *Penguin Dictionary of Psychology*. London. Penguin Books, 2009. Print, p. 796.

18. Reber, Arthur S., Rhianon Allen, and Emily S. Reber. *Penguin Dictionary of Psychology*. London. Penguin Books, 2009. Print, p 796.

19. Reber, Arthur S., Rhianon Allen, and Emily S. Reber. *Penguin Dictionary of Psychology*. London. Penguin Books, 2009. Print, p. 736.

20. Reber, Arthur S., Rhianon Allen, and Emily S. Reber. *Penguin Dictionary of Psychology*. London. Penguin Books, 2009. Print, p. 736.

21. Reber, Arthur S., Rhianon Allen, and Emily S. Reber. *Penguin Dictionary of Psychology*. London. Penguin Books, 2009. Print, p. 612.

22. Reber, Arthur S., Rhianon Allen, and Emily S. Reber. *Penguin Dictionary of Psychology*. London. Penguin Books, 2009. Print, p. 377.

23. Reber, Arthur S., Rhianon Allen, and Emily S. Reber. *Penguin Dictionary of Psychology*. London. Penguin Books, 2009. Print, p. 342.

24. Reber, Arthur S., Rhianon Allen, and Emily S. Reber. *Penguin Dictionary of Psychology*. London. Penguin Books, 2009. Print, p. 342.

25. Reber, Arthur S., Rhianon Allen, and Emily S. Reber. *Penguin Dictionary of Psychology*. London. Penguin Books, 2009. Print, p. 223.

26. Reber, Arthur S., Rhianon Allen, and Emily S. Reber. *Penguin Dictionary of Psychology*. London. Penguin Books, 2009. Print, p. 377.

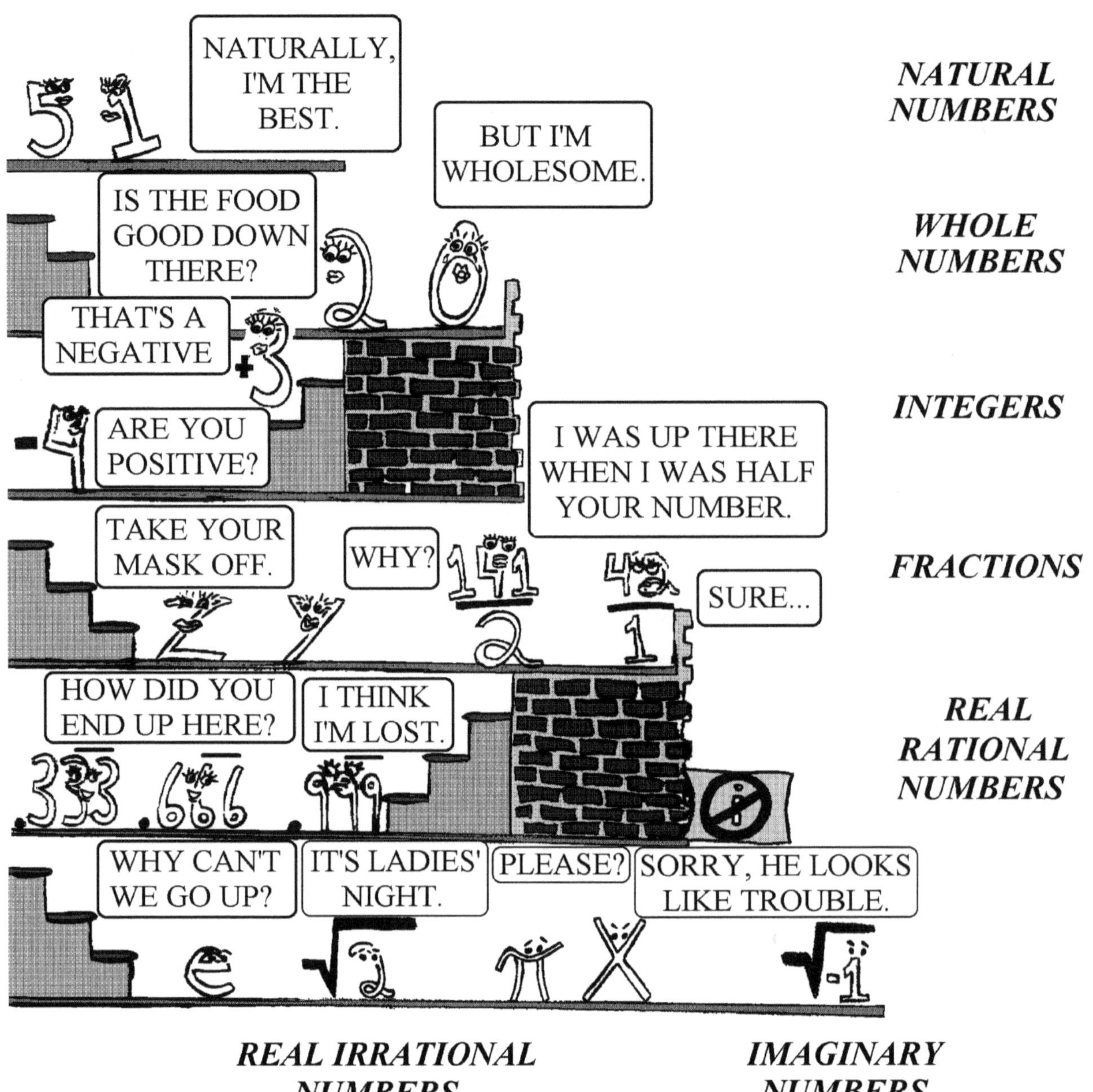

"I wonder who it was defined man as a rational animal. It was the most premature definition ever given. Man is many things, but he is not rational." - Oscar Wilde[1]

CHAPTER TEN

Motivation, Perception, Attention, and Reason

Front-loading: Associative Percept, Attenuation Theory, Attention, Attentional Decay, Blindsight, Blind Spot, Decay Theory, Deductive Reasoning, Desensitization, Executive Functioning, Extrinsic Motivation, Filter Theory, Forgetting, Gestalt Laws of Organization, Half-life of Attention with Error, Half-life of Attention without error, Inattentional Blindness, Inductive Reasoning, Inferred Percept, Intrinsic Motivation, Macrocosmic, Microcosmic, Motive, Motivation, Multi-tasking, Neglect, Organization, Percept, Perception, Positivism, Realism, Reasoning, Scotoma, Solipsism, Tabula Rasa, Template Matching, Volition, Will

Motivation

Two types of action may be said to occur; the first type carries a connotation of consciousness and purposefulness, such as in walking or talking.[2] The second type lacks the connotation of consciousness and is exemplified by autonomous activities such as one's heart beat, a neural action potential, or the action of a drug.[3] Action in the first type possesses a motive or aim, is under the control of volition, and the focus of the equation will fall on motivation. Volition is understood as the "voluntary selection of a particular action or choice from many potential choices or actions."[4] For all due purpose, volition may be considered synonymous with will here. Two types of motivation that will be of concern here are intrinsic motivation and extrinsic motivation.

Intrinsic motivation refers to feelings of satisfaction and fulfilment that are not derived from external rewards but are found in the task itself.[5] Wanting to learn to play the violin for its own sake would be one example. Conversely, extrinsic motivation originates outside the individual and is determined by rewards and/or threats from an outside force.[6] If someone desired to learn to play the violin in order to become famous or because of a fear of disappointing one's parents, then the motivation is extrinsic. Integrating these concepts into the equation only requires that one know which desired entity, and for what purpose, the highest Sentiment ranking exists. In the case of intrinsic motivation, the highest ranked purpose is learning to play the violin. In the case of extrinsic motivation, learning to play the violin becomes linked and essentially subsumed by the entity of fame or the entity of approval from one's parents. In other words, learning to play the violin would equal fame or it would equal approval from one's parents. In the case of extrinsic motivation, so long as learning to play the violin is fully sufficient, fully unique, and has a similar Sentiment to either fame or non-disapproval from one's parents, then the value of learning to play the violin will be indistinguishable from its external incentives. However, if alternatives to acquiring fame or avoiding disapproval from one's parents arise, or learning to play the violin no longer becomes sufficient to guarantee fame or avoid disapproval from one's parents, then the value of learning to play the violin will fall with respect to the external incentives.

Figure 10.1 If one plays the violin because one enjoys playing it, then the motivation is intrinsic. If one plays the violin to acquire fame or please a parent, then the motivation is extrinsic. If the motivation is extrinsic and something else becomes required or is offered as an alternative to achieving either of those ends, then the valuation of the violin will likely be lower and fluctuate more than its value if the motivation were intrinsic.

Ultimately, in order for an individual to be motivated by an entity, he or she has to know, or at least suspect, that the entity may physically exist. It would be unreasonable to hold that the inability to acquire an entity can cause an individual anxiety if the individual neither knows that the entity exists nor has any inkling of what it may or may not be. This distinction, concerning the absence of knowledge against acquired knowledge, leads to a few questions. How can individuals come to know that an entity exists or at least to suspect that it may exist? More important, how can the means by which an individual comes to know or suspect that an entity exists be represented in the equation?

Perception

Perception enables people to learn information, and consists of the collective processes giving "coherence and unity to sensory input."[7] This may also be extended to the awareness of organic processes, awareness of the truth of something, and the synthesis of sensory information.[8]

Most people are familiar with the five senses, but it is necessary to list them here to avoid becoming too presumptuous:

1) Audition: Audition concerns the mechanisms of hearing.[9]
2) Olfaction: Olfaction concerns the sense of smell, and receptors for olfaction are located in the olfactory epithelium in the nose.[10]
3) Tactile Perception and Kinesthesis: Tactile perception concerns the sense of touch.[11] Along with tactile perception goes kinesis, or "the study of the movements of the body and their communicative functions."[12]
4) Gustation: Gustation concerns the sense of taste.[13]
5) Vision: Vision concerns the process of seeing.[14]

An individual percept, however, concerns the "outcome of the process of perception."[15] Moreover, psychologists observe that a percept should not be confused with an actual object (i.e., a distal stimulus) or the energy acting on a receptor (i.e., a proximal stimulus), but thought of as "phenomenological or experiential."[16] A percept is "an impression of an object obtained by use of the senses."[1] In essence, the specific sense information gathered about a distal stimulus and proximal

---

[1] By permission. From Merriam-Webster's Collegiate® Dictionary, 11th Edition ©2014 by Merriam-Webster, Inc. (www.Merriam-Webster.com).

stimuli by the individual has to be reassembled in order to make sense to a person. Percepts, then, concern the manner by which this information is integrated by the individual.

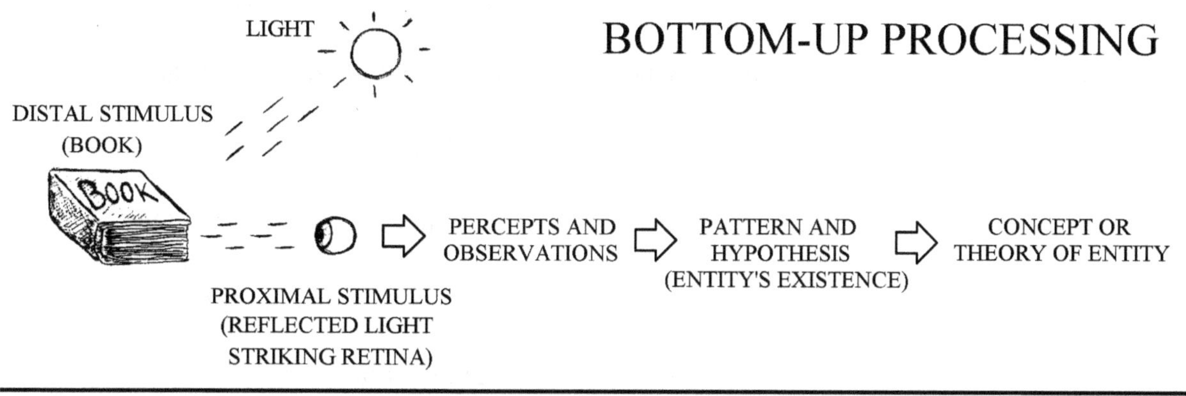

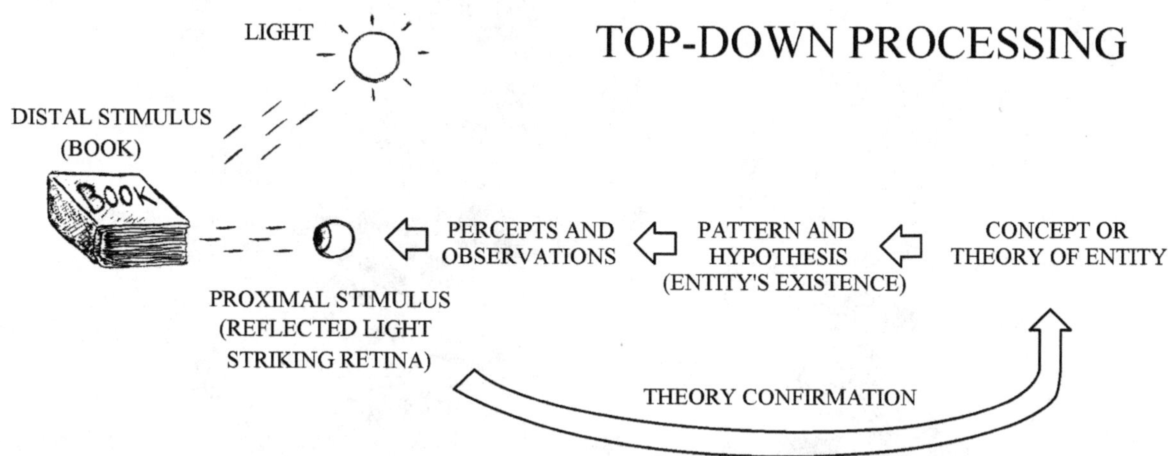

Figure 10.2 The above depicts the distal stimulus, proximal stimulus, percept, hypothesis of an entity's existence, and the concept of an entity with the example of a book and vision. The actual book itself is the distal stimulus, light energy reflected off of the book onto the eye represents the proximal stimulus, and then the individuals' organization of these sensory features forms the percept. The sensory features may be highly organized, for instance if the individual can classify the object as a book, or poorly organized if the sensory information alone is ambiguous and does not enable the individual to classify the object as a book, but only as a box-shaped object or a rectangular prism. The existence of the book may be acknowledged by either bottom-up or top-down processing.

Incorporation of Percepts

The next task involves incorporating perception and percepts into the equation. The five senses of perception have an influence on whether or not an individual believes an entity exists. However, one can suspect or believe that something exists without actually perceiving the entity per say, but rather by perceiving its influence or by analyzing evidence of its existence. One such entity is a black hole in outer space. Due to the theorized nature of a black hole, nothing, including light, may escape its gravitational pull once traversing beyond the event horizon.

Figure 10.3 A black hole with its accretion disk is an example of an inferred percept. Due to the inability of energy and matter to escape from its gravitational field, it is impossible to physically see a black hole, but its existence can be inferred by observing its effect on surrounding matter. Acknowledging the existence of an entity from an inference, hence an inferred percept, would be a top-down process.

Although no one has ever directly perceived a black hole using one of the five senses, or at least perceived one and lived to tell about it, the existence of black holes can be inferred using reasoning and logic from the effects that they have on other nearby objects. For instance, the accretion disk surrounding a black hole is one such source of evidence for a black hole. Thereafter, any time that an individual witnessed, galactic dust or a light wave being severely distorted by gravity, it would become analogous with either the presence of a black hole or some other celestial body. However, it should be remembered that the dust, gas, and radiation being influenced are not the black holes themselves. This phenomenon will be called an *inferred percept*, as the individual does not specifically perceive the entity itself, but perceives evidence suggesting its existence.

In addition to validating the hypothesis of an entity's existence, top-down processes may also be used to challenge a hypothesis that an individual has concerning the validity of an entity's existence. For instance, in the case of the four-headed giraffe from chapter five (Figure 5.17), it might have been the case that four giraffes merely looked like a four-headed giraffe from a particular vantage point and bottom-up processes. However, closer inspection could reveal otherwise.

Figure 10.4 Four giraffes, if lined up directly along the viewer's line of sight, might resemble a four headed giraffe (Figure 5.17**)**. A top-down process can spur an individual to re-examine the highly improbable, four-headed giraffe's existence to reveal otherwise.

Another example of a top-down percept in affect engineering is the *associative percept.* Lemons, pears, kiwis, cherries, strawberries, and blackberries are classified as fruits, which are "the usually edible reproductive body of a seed plant; especially : one having a sweet pulp associated with the seed . . ."[2] Although this definition of fruit makes sense to most people, fruit is not an actual object that exists in the world; rather it is a class of objects. One may, for instance, perceive a new food (e.g., tasting an orange) without knowing specifically what it is but may be able to classify it as a fruit. The individual knows it is not one of the six fruits he or she may already know of (figure 10.5) but suspects that it has something in common with them. He or she may even conceptualize an orange as being a combination of a some of their flavors due to its sweet and sour properties. Fruit then, does not represent a specific object that can be perceived in the world, but is a bridge

[2] By permission. From Merriam-Webster's Collegiate® Dictionary, 11th Edition ©2014 by Merriam-Webster, Inc. (www.Merriam-Webster.com).

between different entities that enables them to be associated under a single entity acting as an umbrella. It is associative in the sense that it concerns the association of ideas or images and occurs as a top-down process after the fact.

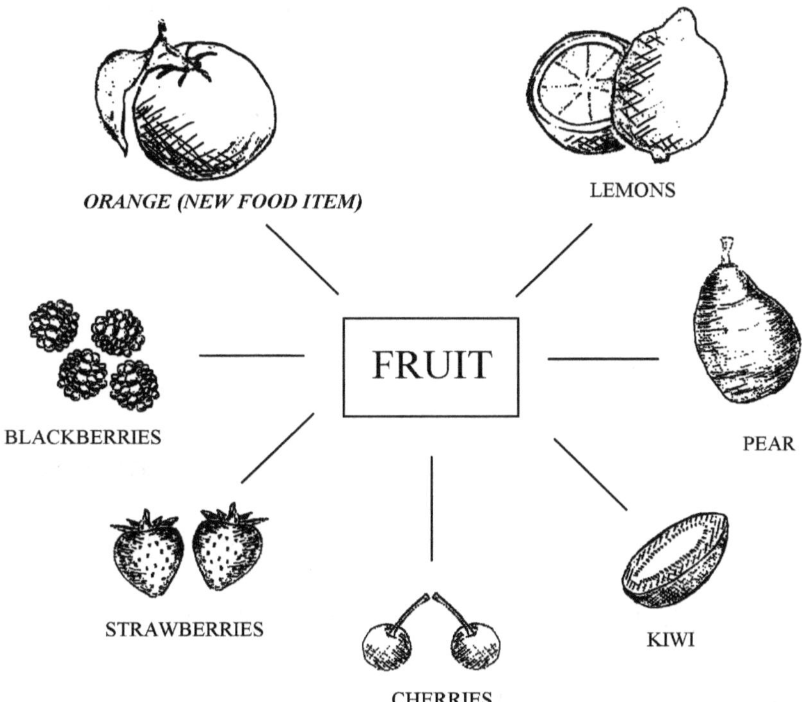

Figure 10.5 Specific entities are being classified under the associative entity of fruit, which is a high level and more abstract entity that does not correspond to a particular entity in the real world. Subsequent classifications at progressively higher and more abstract levels would include fruits, vegetables, grains, and legumes being classified under the associative entity of food.

How then can the equation both account for and represent these two routes to the conception of an entity's existence? One is being driven by bottom-up processing and a dependence on sensory perception, while the other is being driven by top-down processing, association, and inferences made from reasoning or belief. Frankly, affect engineering does not have to, as once an entity is conceived or thought of then it exists to the individual. The analysis of entities arising from bottom-up and top-down processes had begun with the assumption that the self starts with a tabula rasa. Tabula rasa literally means blank tablet and refers to a philosophical idea that humans come into the world without any innate ideas.[17] Of the two possibilities, it is the more difficult for which to genuinely account.

Existence: Bottom-Up Processing vs. Top-Down Processing

From chapter eight, bottom-up processes are those that proceed from simple and concrete elements toward "abstract and symbolic features" and incorporate inductive reasoning.[18] Inductive reasoning is a "process of reasoning in which general principles are inferred from specific cases."[19]

With regards to conceptualizing that an entity exists, a bottom-up process would start with the stimulus, perception of the stimulus, then move on toward percepts that are organized into a theory of the entity's Existence. Moreover, Existence, from chapter five, is never halfway; it is all or nothing. In the case of bottom-up processing, the acquisition of one percept related to an entity's Existence, regardless of modality (e.g., sight, touch, sound, taste, or smell), would be enough to trigger conception of a new entity or cognizance of an already known entity if the process is rerouted elsewhere. If all of the sensory information and related percepts are completely integrated into a specific entity, then each of these pathways would lead to the same entity with the same valuation.

Meanwhile, top-down processes proceed from complex and abstract representations toward the "analysis of inputs and lower level representations" and incorporate deductive reasoning.[20] Deductive reasoning is a form of reasoning that ". . . proceeds from the general to the particular," beginning with a set of assumptions from which conclusions are drawn and theorems derived.[21]

With regards to conceptualizing that an entity exists, a top-down process would begin with the theory of the entity's Existence, move outward to postulate a pattern of percepts for a hypothesis, and finally make observations of the percepts it is experiencing in order to confirm or refute the legitimacy of the original hypothesis. In the case of a top-down process, the hypothesis concerning the Existence of the entity is already established and only needs confirming. Conversely, in a bottom-up process, an entity's Existence may not yet be established, and creating a theory of its existence would be the end product (figure 10.2).

The circular nature of these two processes, inductive reasoning and deductive reasoning, complicates establishing a relationship between perception and existence but does not make it impossible. For now, affect engineering will err on the side of caution. It will be held that once an entity has been theorized to exist for the first time by an individual, via either a bottom-up process and inductive reasoning or a top-down process and deductive reasoning that is extrapolated associatively or from knowledge already in one's possession (e.g., a black hole's existence inferred from accretion disks), then the entity will exist indefinitely to the individual. The entity need not initially be directly perceivable so to speak, and may be imperceptible to the senses, but its Existence would have to at least be capable of being inferred from the percepts available to the individual or being established associatively.

BOTTOM-UP PROCESSES	TOP-DOWN PROCESSES
BEGIN WITH LOW LEVEL INFORMATION:	BEGIN WITH A HIGH LEVEL IDEA:
INDUCTIVE REASONING	***DEDUCTIVE REASONING***
VISUAL PERCEPTS AUDITORY PERCEPTS TACTILE OR KINESTHETIC PERCEPTS OLFACTORY PERCEPTS GUSTATORY PERCEPTS	INFERRED PERCEPTS ASSOCIATIVE PERCEPTS

Figure 10.6 In the case of bottom-up processing and inductive reasoning, percepts concerning sensory information such as sight, sound, touch, taste, and smell are activated first and this leads to inferences and associations made. In the case of top-down processing and deductive reasoning, inferred percepts and associative percepts are activated first and then sensory information is used to verify them.

Route to Existence: Four Steps

Nearly everyone has heard of the following philosophical question: "If a tree falls in the middle of a forest and no one is around to hear it, then does it make a sound?" From a solipsistic standpoint, one might hold that neither the existence of the forest nor the tree and the sound it might make from falling, can be known to exist. Solipsism is a philosophical position holding that one's own personal experience is the only thing of which one can be certain, and ". . . that one's experiences represent all of reality."[22] This line of thinking is along the more extreme lines of positivism, from chapter one. Positivism, to repeat, is the view that ". . . all knowledge is contained within the boundaries of science, and only those questions answerable from the application of the scientific method can be approached."[23] Logical positivism, which extends from this, holds that ". . . any nontautological proposition that could not, in principle, be verified by empirical, observational means was utterly devoid of meaning."[24] Alternatively, from a standpoint more in line with realism, one might hold that the tree and the sound it makes do exist regardless of whether or not it is witnessed. Realism is a philosophical position holding that "the physical world has a reality separate from perception and mind."[25] For the sake of empiricism, an approach will be taken between these two views that lies much closer to positivism and solipsism in the sense that an entity will be considered to exist to an individual if he or she can directly perceive its existence through one of the five senses or by using induction and what will be labeled inferred percepts and associative percepts. Hence, if an individual can only imagine or infer the entity's existence deductively and in combination with the perception of other entities, then the entity will still be considered to exist, even if only in the mind of the individual. An entity will only be considered to not exist to an individual if an individual has not perceived it, conceived it, imagined it, nor inferred its possible existence through other percepts or by association.

If a Tree Falls in a Forest and No One Is There to Hear It, Then Does it Make a Sound?

In accordance with the notion of a tabula rasa, before being asked this question a subject would have no knowledge of what a tree is, what a forest is, or what type of sound a tree would make if it fell in a forest. The following scenarios explain when a tree is considered to acquire existential value in the equation and highlights the levels of perception from the weakest amount of verification to the strongest amount of verification of the existence of the sound the tree made upon falling.

1) Scenario one (non-perception or non-existence of the tree):

   Before being asked the question the individual had no idea what a tree and a forest were, that they might exist, or that a tree might fall and make a sound (figure 10.7). The sound of the tree falling and any thought related to this is nonexistent to the individual, for the individual has no idea what a tree is or is not nor, and no inkling of the sound it might make upon falling.

Figure 10.7 Non-existence. The individual has no notion of what a tree is, a forest, or that the tree might make a sound if it fell. The tree has an existential value of zero.

2) Scenario two (suspected existence, weakest level of validation):

Existence of a tree falling is suspected (figure 10.8). After being asked the question, being shown what a tree is, and told what a forest is so that they are not simply unknown entities that exist, the individual can now imagine that a tree may have fallen in a forest. Even though the individual has neither seen a tree fall, he or she may have seen other objects fall and make a sound. Likewise, even though the individual has never seen a forest (a group of trees), he or she has probably seen groups of other objects. Utilizing the ability to associate, the individual can both imagine the possibility that a tree may have fallen in the middle of a group of other trees (a forest), and the individual can also imagine the sound its fall might make based upon its size. Thinking of an object is enough to bestow an existential value on an object in affect engineering. In essence, the mere suspicion or imagining of a tree falling is sufficient grounds for the Existence of a fallen tree being established for the individual. The idea only needs to be conceived in order to obtain an existential value of + 1 emotional unit. Henceforth, the theory of the tree's Existence is possessed by the individual, but the physical existence or nonexistence of the sound of an actual tree falling is unverifiable via sensory information.

Figure 10.8 Existence is weakly verified. The tree exists to the individual but the sound it might make is weakly verified. The tree and whatever sound it might make has an existential value of + 1, because it exists in the mind of the individual. Evidence for its Existence has neither been directly perceived through the senses nor inferred through evidence.

3) Scenario three (Inferred existence, moderate level of validation):

Existence of a tree falling is moderately inferred (10.9). If, after traipsing through a forest, the individual comes across a fallen tree in the forest, the sight of the fallen tree is sufficient evidence for the discoverer to reconstruct the tree's fall and the sound it likely would have made. Although the individual was not there to hear the tree fall, visual evidence of the fallen tree enables the reconstruction of a specific tree falling in a forest. The theory of the tree's Existence is possessed by the individual, and the actual existence of the sound of a tree falling can be moderately verified indirectly through inferred percepts, specifically, sight of the fallen tree.

Figure 10.9 Existence is moderately verified. The tree has an existential value of + 1 and the sound it might have made is moderately verified because the individual has found sensory evidence suggesting that the tree fell and made a sound in the past. The individual can infer that a particular tree fell in the forest and made a sound, but he or she was not there to witness it.

4) Scenario four (Direct perception, strongest level of validation):

If an individual is present to hear the tree fall, then there is no doubt of the tree making a sound because the individual perceives the sound directly and hears it first-hand (figure 10.10).

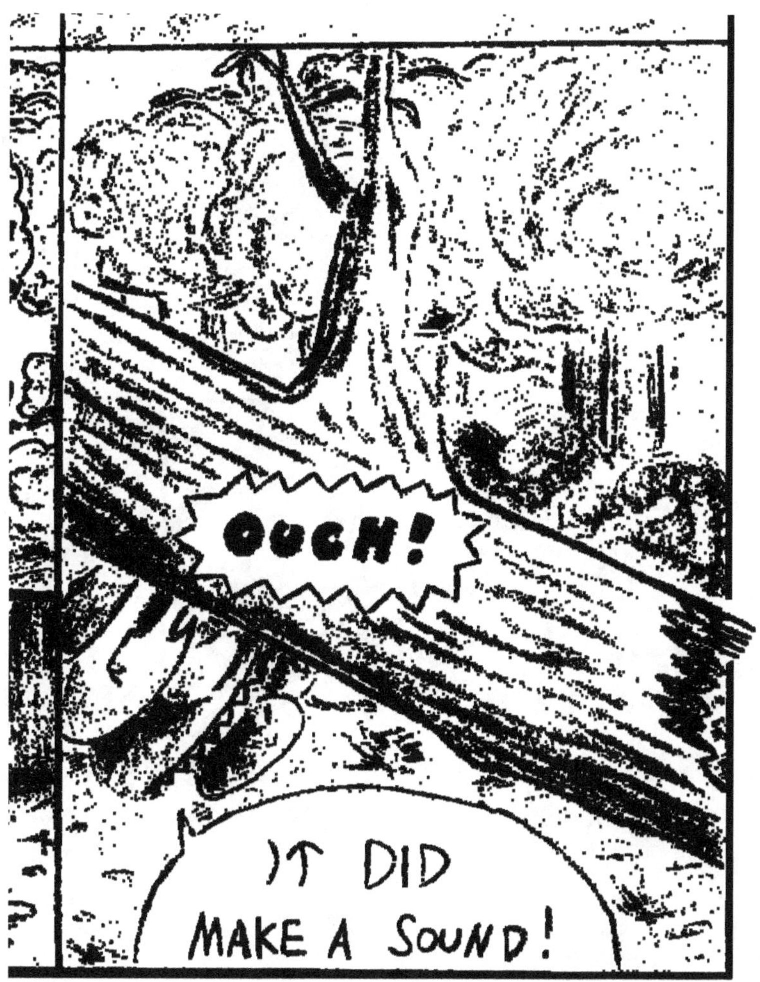

Figure 10.10 Existence is strongly verified. The tree has an existential value of + 1 because it exists in the mind of the individual and both its fall and sound were witnessed firsthand by the individual.

Figure 10.11 The levels of verification are: non-existence of the tree and the sound it made (top left); existence of the tree and sound it made from falling is weakly verified (top right); existence of the tree and sound it made from falling is moderately verified (bottom left); existence of the tree and sound it made from falling is strongly verified (bottom right).

Creating Something from Nothing: Coefficient Variable for Perception of an Entity's Existence

It is given:

1) Where Existence = 1, this means an entity exists to an individual;

2) Where Existence = 0, this means an entity does not exist to an individual;

3) If an entity has been perceived in one of the following seven ways: as a visual percept; a tactile percept; an auditory percept; a gustatory percept; an olfactory percept; as an inferred percept or as an associative percept, then Perception equals positive one for that route to an entity thereafter.

4) Although there are at least seven different routes that may lead to the creation of a percept, once one of them becomes linked to cognizance of an entity, then the Perception variable becomes positive one (+ 1) regardless of how many more routes become associated with the entity thereafter. Existence subsequently becomes + 1 as well.

The relationship between Perception of an entity and Existence may be broken down further and mathematically represented in a number of ways, such as in the following expression:

Existence = 0 ^ (1 - Perception), where 1 represents the Nonexistence presumed.

Normally, zero to the power of anything other than zero equals zero:

0 ^ 1 = 0,

0 ^ 2 = 0,

0 ^ 3 = 0.

While zero to the power of zero is sometimes considered to be equal to one:

0 ^ 0 = 1 for some mathematicians.

Why is this so and how did it come to be? Some mathematicians consider 0 ^ 0 = 1 simply because it is useful to do so. This can be discovered by checking the limit of the following function:

y = x ^ x

As x approaches zero . . .

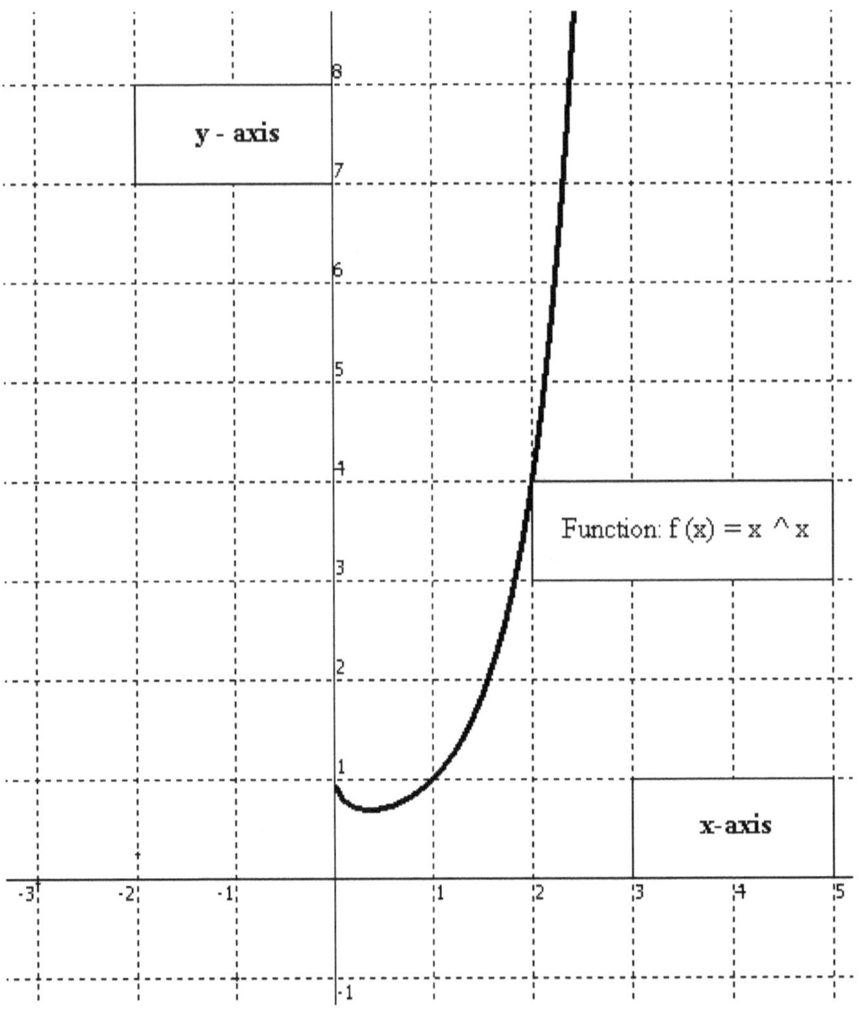

Figure 10.12 Limit of the function f(x) = x ^ x. As x goes to zero, f (x) approaches the limit of +1.

Here, if expressed in words, the above expression would be understood as follows:

Existence = Zero to the power of (the entity's Nonexistence minus Perception of the Entity via one of the seven percepts). Perception, in essence, would negate Nonexistence. This is only one of several ways one might choose to represent a percept leading to cognizance of an entity's Existence. Another would be to use a conditional *if*. . . *then* logical statement to simply hold that where the Perception of an entity is not equal to zero, then Existence equals + 1.

Establishing Existence in the Equation

In the equation, Existence necessarily concerns any entity considered by the individual. This includes the original entity, a second entity, the entity of the self, the entity of an *other*, and recognition of harm, benefit, or efficacy contingencies as they are events and objects as well.

The Perception of any entity's existence (e.g., original, second, contingency, etc.) would consist of the following: visual percepts; tactile percepts; auditory percepts; gustatory percepts; olfactory percepts. Also included are inferred percepts, for instances where the direct sensory perception of a stimulus is not possible but its Existence may be inferred by observing percepts related to other stimuli, such as in the case of a black hole. Finally, there are associative percepts, such as in the case of top-down processes concerning the category fruit. Perceived threat severity, for example, would refer to an individual's perception of a threat's intensity through one of the seven types of percepts established. The equation primarily addresses entities that the individual has cognized, has been made aware of, and has recognized to exist via one of the routes described above. Entities that influence physiology and whose Existence might go unrecognized make up entropy (discussed in chapter twelve).

Motivation, Perception, Attention, and Reason

Attention: Is Perception More than Sensation?

It is well known that the fact someone is perceiving an object does not mean that the individual is aware of the object's existence. A multitude of other factors aside from sensation alone play a role in determining what is recognized by an individual as existing and what is not recognized as existing. Psychologists have identified an array of concepts that can influence perception, some of which have already been considered and others that have not yet been addressed: attention; constancy; motivation; organization; set; learning; distortion; hallucination; and illusion.[26] Many of the concepts to be described exemplify top-down processes that screen, filter, arrange, distort, or bias incoming information coming from bottom up processing and the perceptive faculties.

With respect to attention, psychologists observe that for an "event to be perceived, it must be focused on or noticed," and moreover, that this process is selective.[27] Additionally, attending to one stimulus prevents or suppresses the ability of an individual to process other stimuli.[28]

Figure 10.13 The shock of witnessing a car accident has diverted Granny Smith's attention towards the spectacle before her and away from her apple. An apple thief has utilized the diversion to escape with her fruit in hand.

With respect to constancy, psychologists also note that one's "perceptual world tends to remain the same" in spite of drastic alterations in sensory input.[29] For instance, a book seen from an angle is still perceived as rectangular even though the "retinal image is distinctly trapezoidal."[30]

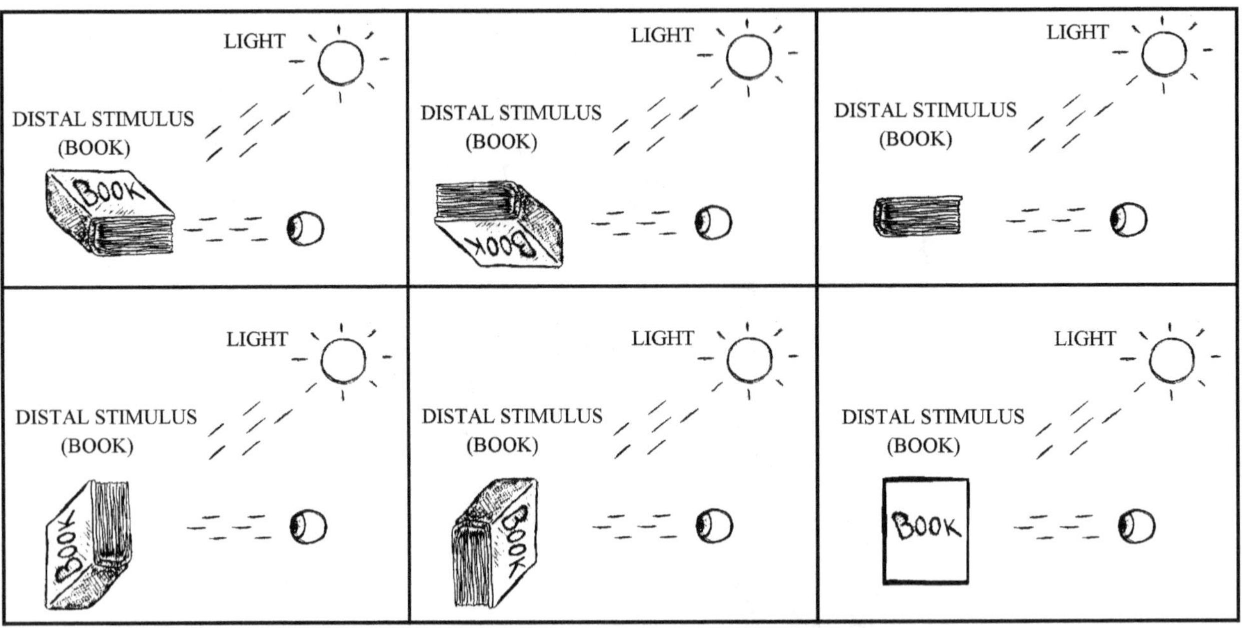

Figure 10.14 Despite the retinal image of a book being different when viewed from different angles and due to depth, the book is still perceived as being rectangular in shape.

With respect to motivation, one's motivational state can also influence what is perceived.[31] For example, a hungry person may see food objects in ambiguous stimuli that a sated person would ordinarily not see.[32] A brown pen might cause an individual to think about cinnamon sticks, a beige-colored rock might inspire thoughts of a freshly baked loaf of bread, or a frisbee might inspire a desire for pizza.

Additionally, with respect to organization, sensory elements are theorized to be organized into coherent wholes, and are not simply juxtaposed next to other sensory elements.[33] The tiles on a bathroom floor, for instance, may tend to be seen as a whole instead of as individual tiles. The bathroom floor may be sensed as a part of the bathroom, or the bathroom as a part of the house, or the house as part of a larger neighborhood.

Along the same lines, one's perception of what is perceived may also be influenced by the "cognitive and/or emotional stance taken toward a stimulus array" or set.[34] One may have a better aptitude or preference for processing visual percepts than auditory ones, or tactile percepts over gustatory ones. As a result, specific types of stimuli might be processed differently than other types even if they are delivering the same information concerning the same entity and same purpose.

With respect to learning, the problem of determining how much of one's perception is innate and how much is obtained through experience must be addressed. Another problem concerns how learning can "function to modify perception?"[35] If one wanted to hold that an individual's learning begins from a blank slate (a tabula rasa), this is fairly straightforward to accomplish in all three models: the equation, the roll cage model of drives, and the neurological model. An individual would have no initial conception of entities, meaning the Existence variable in the equation would start at zero, there would be no initial entity anxiety-mass in the roll cage model of drives, and neurons that correspond to the recognition of an entity would be unassigned. Alternatively, if one wanted to hold that an individual possessed innate knowledge, then this would be represented by the existence of entities in the individual's psyche prior to experience.

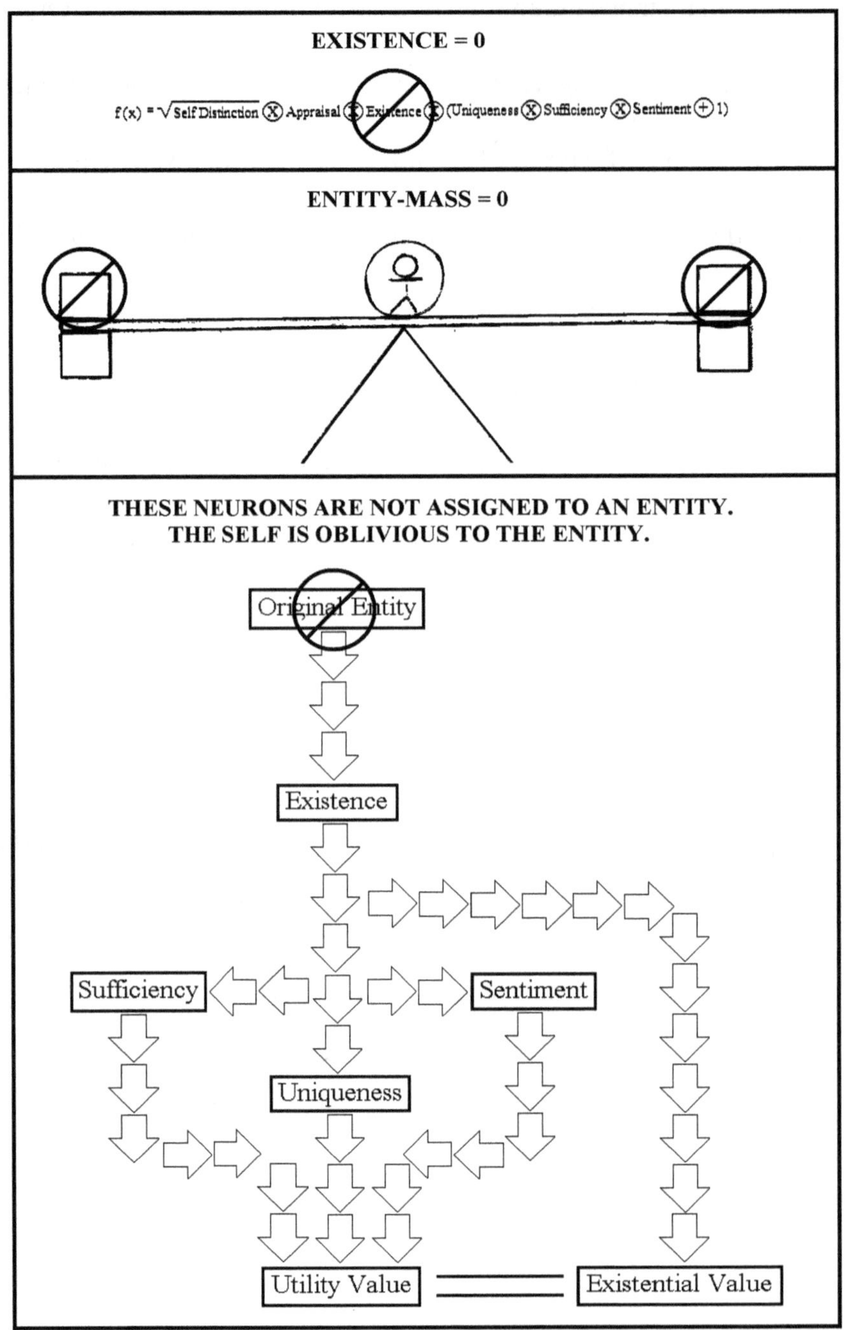

Figure 10.15 The equation, the roll cage model, and the neurological model depicting tabula rasa, where an entity has never been cognized.

The second question, concerning how the existence of knowledge already possessed influences the perception of new information, will require deeper analysis than what is currently being offered. It is a more open-ended and complex realm. Prior knowledge, for instance, may influence what an individual decides to attune to when perceiving new information in the future. Attention then will have to be incorporated into the equation before further progress can be made.

With respect to distortion and hallucination, the perception of information may be distorted by strong emotional feelings or hallucinations produced by a "variety of causes including drugs, lack of sleep, sensory deprivation, emotional stress, and psychosis."[36] The capability of the math equations, the Roll Cage Theory of Drives, and the neurological model to successfully illustrate distortion and hallucination, among other abnormalities, will also be crucial to achieve if defense mechanisms and psychological disorders are to be modeled later.

Finally, there are many instances where "what is perceived cannot be easily predicted from an analysis of the physical stimulus array."[37] An illusion, for instance, may be created when what is presented in the visual field (e.g., bottom-up processing) conflicts with the expectations of an individual (e.g., top-down processing). The flexibility of the models to account for the many things the mind can do will be put to the test in the coming chapters.

Attention as a Relationship of Awareness to Neglect

Figure 10.16 To pay attention is to distort reality. At bare minimum, any entity that exists to an individual has an existential value of one emotional unit, be it positive anxiety or negative anxiety that is concerned. Any additional value beyond that, such as Utility value (e.g., Sufficiency, Uniqueness, Sentiment), distorts the existential value of the entity. From a strictly technical standpoint, in the above image all the words may be said to simply exist on the page. To pay attention to any specific word or group of words elevates its value against every other word on the page. For instance, if one focuses his or her attention by investing more anxiety resources into the phrase "To pay attention is to distort reality," this elevates and distorts its value against other words on the page, such as the phrase "Everything exists and nothing more than that," in the smaller font and lower right corner.

Attention is generally considered to be the selective focusing on features of an environment relative to the exclusion of others.[38] Although attention may be conscious to the extent that an individual actively selects specific elements from a stimulus, for the most part people are not aware of the factors that influence why only a small part of a stimulus array is perceived.[39] In short, attention consists in the ability of an individual to include or exclude stimuli from awareness and this may be done either consciously or unintentionally.

One theory which is related to and useful in the study of attention is filter theory.[40] Filter theories are those that suggest neural mechanisms either "detect or block certain patterns of perceptual input."[41] An early filter theory, template matching, held that new stimuli were processed and compared to "various internal representations of objects stored in memory," or templates.[42] Although it could not account for how an individual could recognize that "A, **A,** and *a* are instances of the same letter," it did lead to the development of other filter theories.[43] Contemporary theories hold that neural cells or groups of neural cells are "maximally sensitive to particular patterns in the neural stimulation that they receive."[44] These cells would signal once a particular pattern occurred by increasing their rate of fire or alternatively, could display their filtering functions by decreasing their rate of fire.[45]

Attenuation theory is a version of filter theory that proposes that "unattended messages are attenuated (i.e., processed weakly) but not blocked entirely from further processing."[46] Moreover, under attenuation theory, different items in "unattended channels of information have different thresholds of recognition depending on their significance to the individual."[47]

Neglect

The type of neglect referenced here bears more similarity to the neurological disorder definition of neglect, where an individual is simply "not cognizant of (i.e., neglects) particular categories of information.[48] If asked to replicate a picture, for example, an individual with hemi-spatial neglect may only draw one side of an image; if eating, only food on the right or left side of a plate may be eaten. Typically, a person with hemi-spatial neglect may show no awareness of the "left half of space about him or her," yet, despite showing "no conscious awareness of the left-field stimuli, they will make appropriate, automated responses to stimuli in the neglected field."[49]

The Blind Spot and Blindsight

The blind spot is the spot on the retina where the neural fibers leave the eyeball to create the optic nerve and lack sensitivity to light.[50] Any partially or totally blind areas in the visual field are known as scotoma, with the blind spot sometimes being distinguished as a "physiological scotoma."[51] There are times, however, when an individual can be influenced by stimuli that fall within the blind areas.[52] Weiskrantz's term "blindsight" refers to circumstances where an individual exhibits the ability to respond to visual stimulation that he or she is not conscious of seeing.[53]

Inattentional Blindness

Similar to the notion of scotoma is the concept of inattentional blindness, hence, the "failure to perceive a presented stimulus when one's attention is focused on a different stimulus."[54] Hypothetically, an individual might be asked to watch a video and count the number of times that four balls being tossed back and forth are dropped. Meanwhile, a large man in a penguin suit could stroll across the screen and the individual might have no recollection of the man in the penguin suit if it were in a channel of information that was not being attended, despite being in the same visual field. Irvin Rock discovered that this phenomenon can occur when the "unattended stimulus is presented in the same location as the attended stimulus" and refers only to consciously perceived stimuli as opposed to information picked up implicitly.[55]

Defining Neglect in the Equation

The definition of Neglect utilized in the equation incorporates both the notions of lack of awareness and inattentional blindness while it excludes instances of blindsight. Blindsight might be construed as a sort of inferred percept in the sense that surrounding or other sensory stimulations (e.g., feeling the warmth of a light source) might be used to help fill in the blanks for scotoma to enable an individual to detect a change happening in a blind spot. Within the equation, Neglect will include any stimulus that activates sensory receptors but fails to be acknowledged or is excluded from being integrated into Awareness. Although Neglect will also extend to any entity that the person has no cognizance of, meaning those that have never been perceived by the individual, have never been conceived of indirectly via top-down processes (e.g., as an inferred percept or an associative percept), are for all practical concerns nonexistent if Existence equals zero.

Incorporating Awareness and Neglect into the Equation

At its core, the purpose of paying attention concerns emphasizing beliefs an individual may have concerning relationships between different entities. If an individual deemed that every entity in the universe had a zero contingency with every other entity in the universe, meaning that neither harmful nor beneficial relationships were theorized, then there would be no need to pay attention to the context (e.g., other entities) surrounding a particular entity for a purpose.

Every entity that an individual conceives of has an existential value of + 1 insofar that it exists, even if its existence is only limited to the individual's mind due to it being intangible.

However, paying attention to the fact that entity *A* has an existential value would be of little importance for decision making when every other entity in the alphabet *(B* through *Z*) that the individual has conceived also has this same existential value. Existential value alone provides no leverage for making a decision and in the absence of an existential value for an entity, it would be of no consequence at all because the individual would be oblivious to the very idea of it.

However, beyond the existential value of + 1, every other variable in the equation concerns a relationship. This may be between the original entity and a purpose (e.g., utility components). It may be between another entity and the original entity (e.g., a threat of harm or promise of benefit) or between the self and the threatening or beneficial entity (e.g., efficacy). Alternatively, it may be a relationship concerning the avoidance of pain or pursuit of pleasure (e.g., an Appraisal toward the restoration of equilibrium). It may finally concern a relationship between the self and an *other's* appraised value judgements (e.g., Self-Distinction).

In the equation, a more or less objective or base value of an entity has been established by the Existence, Sufficiency, and Uniqueness variables. Hence, the entity's objective value for a specific purpose is being described by most of the variables in the base of equation: Self-Distinction; Appraisal toward the restoration of equilibrium; the Utility components of Sufficiency; Uniqueness; and Existence. Meanwhile, a more or less subjective value of an entity has been established by observing the distorted effect on the objective value caused by Sentiment and the variables in the exponent of the equation: Threat Severity and Threat Susceptibility (or alternatively Benefit Intensity and Benefit Susceptibility), Response-Efficacy and Self-Efficacy. Along these lines, Attention will also be considered to have a distortionary effect on an entity's overall valuation. Although Attention would be most accurately modeled as a whole number (e.g., Sample Form IV of attention) to designate the attentional neurons that can activate threat, benefit, or efficacy components, it will be defined as a proportion or ratio of Awareness to Neglect in the examples throughout to avoid the inconvenience of trying to find an appropriate scale for the total units of attention available to value all entities.

Measurement for Attention with Error in the Equation

1) Attention for the Avoidance of Pain formula = Awareness ÷ Neglect.

2) Inattention for the Pursuit of Pleasure formula = Neglect ÷ Awareness.

3) Where Awareness + Neglect = one.

4) Awareness for two separate entities or contingencies can both never be more than .5 at the same time. Awareness here is a fraction of the individual's entire attentional resources (e.g., two thirds, one quarter) and cannot equal one in the Pursuit of Pleasure formula. Neglect can equal a value above .5 for more than one entity. However, Neglect cannot equal zero in the Avoidance of Pain formula; it may also be interpreted as the following:

   Neglect = one minus Awareness, or Awareness = one minus Neglect

   The only stipulation, is that neither Neglect nor Awareness can equal one or zero.

Earlier, in order to formulate an existence variable utilizing percepts, the possibility of extinction, such as in amnesia or Alzheimer's disease, was forsaken. Namely, once an entity has been conceived by an individual, its existential value in the equation and for the individual thereafter has always been + 1, meaning that it would always exist to the individual. While amnesia or memory loss can be represented in the equation, for instance, with the conception of an entity's existence being reduced to zero or oblivion, for the majority of examples remaining in the book no link between Attention and valuation extinction will be held. However, it is possible to do so with Sample Form I of attention. Generally, the forms of the equation used in examples will hold that not paying attention to an entity, after it has been cognized, will not reduce the entity to oblivion by eliminating it from consciousness; the entity's existential value will remain intact. If one desired, Attention could be linked to the extinction of an entity (e.g., loss of memory) by making it a coefficient outside the entire equation, thus giving it substantial influence on an entity's valuation.

The requirements for proper scaling might make the above model less user friendly without a calculator, but not impractical. In the next sections, four forms of the equation for modeling attention will be described, though only the second and third ones will be used predominantly in examples that follow. Although a multitude of variations are possible based upon the ones that will be mentioned here, the decision of which model to use depends primarily on the influence that one wants to give to Attention, if memory loss, forgetting, or extinction are things one wishes to account for, what types of memory are being considered, and what one wishes to measure.

Generally, two types of memory are considered by psychologists, short-term memory and long-term memory. Short-term memory generally refers to "memory for information that has received minimal processing or interpretation."[56] Long-term has generally been "well processed and integrated into one's general knowledge store."[57] Sample Form II or IV of attention would be best for modeling short-term memory, while Sample Form III would be best for modeling long-term memory. For all four models to be described, attention will be gauged at the onset of the perception of the stimulus and the percepts related to the perception of the stimulus. Thereafter, Attention will either not decline automatically if valuation extinction is being modeled but not attentional decay (Form I), or steadily decline if attentional decay theory is being modeled (form II and IV), or revert to normal after being distorted by attention in the case where valuation resilience is being modeled (Form III). If the stimulus is perceived again, via new sensory information, inferred percepts, or associative ones, then Attention changes. This new value for Attention, the second time or any number of additional times it is perceived, could be higher than the original value for Attention the first time it is perceived if more Awareness becomes available, for instance, by allowing only the values of other entities to decay (diminish over time) into forgetfulness in order to free up available anxiety. Alternatively, Attention could be lower if resources are being diverted away.

Sample Form One of Attention: Where Attention Contributes to Value Extinction, Value Distortion, but No Value Decay of an Entity's Valuation Automatically, with Error

Figure 10.17 illustrates how one might arrange Awareness and Neglect if one wanted to link Attention to the extinction of an entity from an individual's consciousness, memory loss, and distortion of an entity's value, but without attentional decay. In addition to Attention being in the exponent modifying Threat and Efficacy, it would also be a coefficient outside of the equation modifying the base value of the entity (i.e., alongside the root of Self-Distinction, Appraisal, and Existence), and would modify the original entity directly. Under this model, if an individual ceases to pay attention to an entity, then that entity also ceases to exist to the individual.

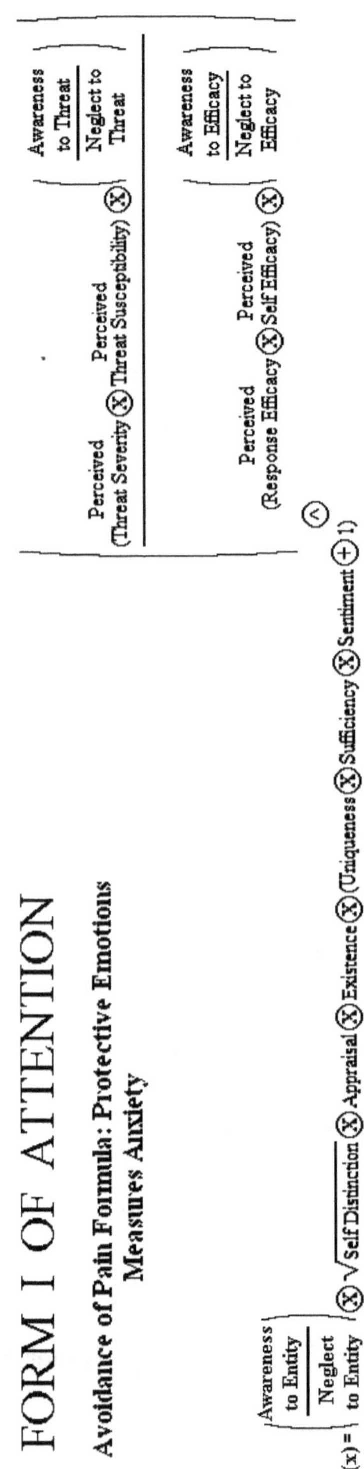

Figure 10.17 Sample Form I of Attention

The possibility for expressing forgetting and value deterioration is still available even if Attention is not linked to value extinction. It will be important to keep in mind that what will be described next are merely the chosen manners by which to represent Attention. All forms of the equation depict Attention as having a distortionary effect on an entity's value. The second, third, and fourth forms both will ascribe the *self's* Attention only to relationships between the original and the second entity (e.g., contingencies of harm, benefit, or efficacy). Unlike the example shown above in the first form, entities that exist to the individual will both have their existential value of plus one (+1) rendered untouchable by Attention, meaning that extinction of an entity will not occur.

In Sample Form II of the equation to be described next, Attention will have a distortionary effect that models forgetfulness and decay in memory. The term decay is used in the study of memory as a "biological metaphor to characterize the (presumed) gradual degradation and/or disintegration of neural traces."[58] The material modeled here, however, will only disappear up to a point, as it will leave the existential value of the entity intact. This form of the equation is best suited for modeling short-term memory.

In the third form of the equation to be described later, Attention will have a distortionary effect that models suppression, valuation resilience, and it will exclude both decay theory and extinction. This form of the equation will distort value in a similar manner to the first form, but will not model forgetfulness and decay. This means that all other variables such as the existential, utility, harm, benefit, and efficacy components, will initially be distorted by Attention, but will rebound back to their original value after a given amount of time. This form of the equation would best be suited to modeling long-term memory and information processing in technology, robots, or artificial intelligence, as machines do not simply forget information and data. People that do not forget information or who possess an exceptional aptitude for retaining knowledge might also be grouped in this category.

Sample Form Two of Attention: Attention Accounts for Value Distortion and Value Decay, but Does Not Contribute to Value Extinction

The decision to represent Attention as a ratio of Awareness to Neglect carries with it a number of considerations. Under this model of attention, paying attention is the process of holding constant the harm components, the efficacy components, or both components while allowing the valuations of other entities to decay to their existential or base value. If an individual is only considering the valuation of one entity for one purpose, then the harm components may be held constant while the efficacy components are permitted to fall into decay, vice-versa, both held constant, or neither one held constant.

The following images, Figure 10.18 and Figure 10.19, depict sample form two of Attention for the Avoidance of Pain equation and the Pursuit of Pleasure equation, but without attentional decay yet implemented.

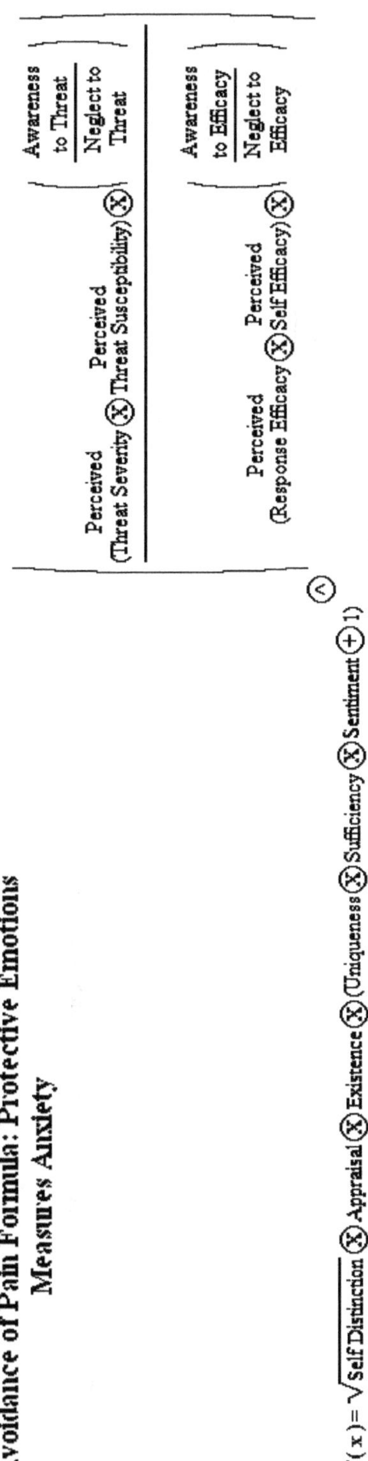

Figure 10.18 Sample Form II of Attention: Pain

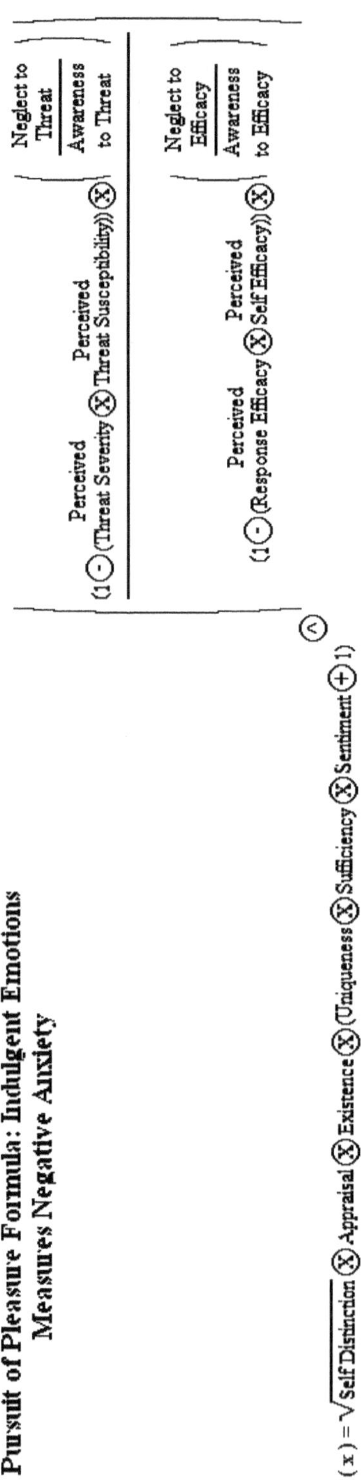

Figure 10.19 Sample Form II of Attention: Pleasure

For the Avoidance of Pain formula, Attention = Awareness ÷ Neglect

For the Pursuit of Pleasure formula, Inattention = Neglect ÷ Awareness

Where Awareness + Neglect = one,

Awareness directed toward two separate entities or contingencies can both never be more than .5 at the same time. Awareness is a fraction of the individual's total attentional resources (e.g., two thirds, one quarter). Neglect can be above .5 for more than one entity. However, if the entity exists, Neglect can never equal zero nor can Awareness equal one, as Neglect equals one minus Awareness. Awareness and Neglect are both percentages whose total must equal 100% for whatever component they are modifying. Neglect and Awareness are mutually exclusive and exhaustive:

Awareness to Threat + Neglect to Threat = one;

Awareness to Efficacy + Neglect to Efficacy = one.

If, for instance, someone is devoting 70% of his or her total attentional resources to threat components influencing an entity, then 30% of the resources have gone elsewhere. In other words, 30% of his or her attentional resources are neglecting the entity. The ratio of Awareness and Neglect would be the following:

Awareness to Threat ÷ Neglect to Threat = .7 / .3 = Attention

From this, the most Attention that could be devoted to valuing any other component or entity would only be 30% thereafter, or .3, and no other component could have an Awareness to Neglect ratio above .3 / .7. If, theoretically, seventy percent of an individual's assets were tied up in an investment, he or she could not also have forty percent of his or her assets invested elsewhere. One's attentional resources cannot be invested beyond what is available.

More information, however, is needed in the equation to mathematically express the concept of attention. Perception, if it were left unhindered and uninterrupted, would bombard the individual with so much information that thinking would be all but impossible. One's attentional resources would be overwhelmed by the influx of stimuli, sensations, and percepts. To combat neural paralysis by information overload, a number of tactics may be employed by the individual to reduce the perception of a stimulus. One such tactic, "desensitization," would be a decrease in sensitivity or reactivity to a stimulus.[59] For instance, a sudden noise may startle an individual, but after successive sudden noises within a short time the reaction will diminish and disappear.[60] An itch, if not scratched, will eventually diminish in intensity. The first jump into a pool of cold water may shock the system, but once in the pool the body would desensitize and the perception of the coldness diminishes.

Figure 10.20 After wearing a hat for an extended amount of time one might even forget that it is on one's head.

Thus, while the physical sensation of an object may still be present, with respect to a proximal stimulus, its perception by an individual does not necessarily remain constant. In this case it will be understood to decay, meaning that the attention paid to the perception of a stimulus will be held to diminish over time. In terms of the expression for Attention, this would mean that one's Awareness of a perception, from a specific point in time, will tend to decrease against the Neglect of that same perception over time until Neglect approaches 100% or one, and Awareness approaches 0% or zero. In essence, the loss of Awareness is equivalent to forgetting. A means of mathematically representing the concept of forgetting is needed. One method for representing forgetting is through the use of exponential decay, such as in the half-life of radioactive materials.

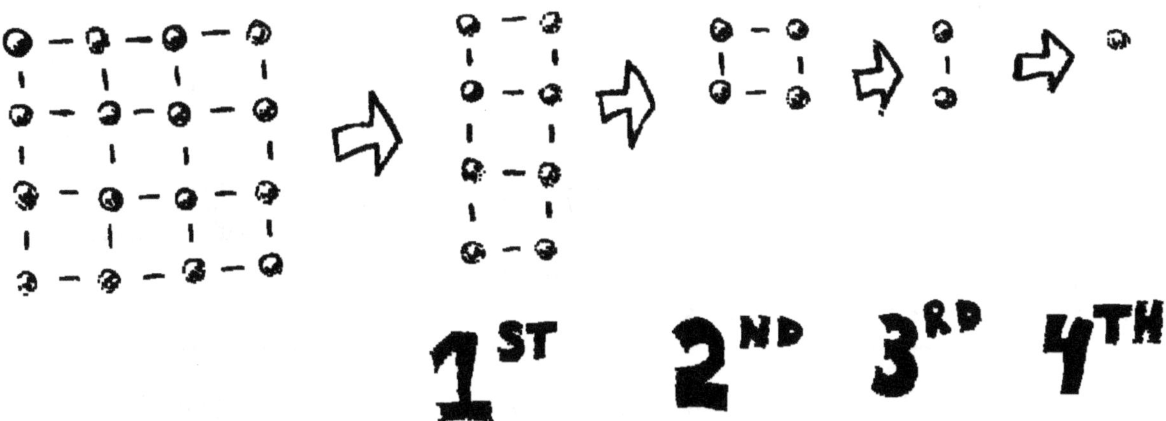

Figure 10.21 In nuclear physics, a half-life period is the amount of time it takes for half of a radioactive material to undergo radioactive decay. Attention, in the second form of the equation, is considered to decay exponentially. The application of half-lives to attention in the equation is similar to half-lives in nuclear physics. If an individual's attention has a half-life of 10 seconds and begins at a ratio of 16 (e.g., Awareness ÷ Neglect = 16), then after 10 seconds this would reduce to 8, after 20 seconds it would be 4, after 30 seconds 2, and after 40 seconds 1.

The combination of conscious awareness along with the notion of gradually receding impressions yields the concept of attentional decay. In essence, attentional decay will imply that one's Awareness is diminishing over time and can be likened to radioactive decay. The half-life of a radioactive element is the time required for half of it to disappear, and can be represented by the following math expression:

"Original Amount of Substance" × (1 / 2) ^ (["Original Time" – "Time at Second Point"] ÷ "Half-life of Substance)

If the variables are all known, then from this expression one can find out how much of the radioactive material will be left at a specific point later in time or how much of the substance was present before if one chose a point in time before the original time. The application of a half-life, then, to the definition of Attention that is being used in the equation would look like the following:

"Attention" × (1 / 2) ^ (["Original Time" – "Time at Second Point"] ÷ "Half-life of Attention")

For the Avoidance of Pain formula, the exponent's coefficient would look like figure 10.22.

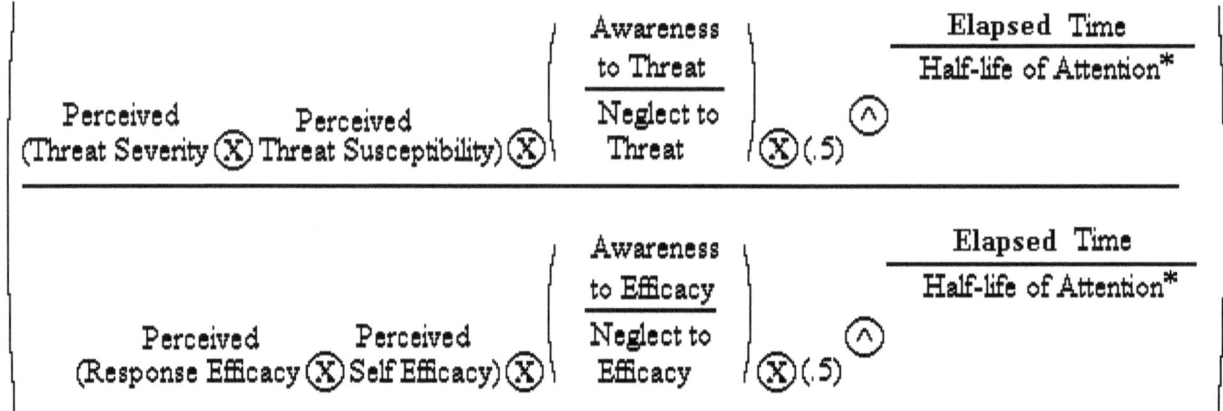

Figure 10.22"(Awareness / Neglect)" × (1 / 2) ^ (["Elapsed Time"] divided by "Half-life of Attention). The * indicates there is error within the expression used for the half-life of Attention and due to the measure used for Attention. Defining Attention as a ratio of Neglect to Awareness creates this error.

For the Pursuit of Pleasure formula, the exponent's coefficient would look like figure 10.23:

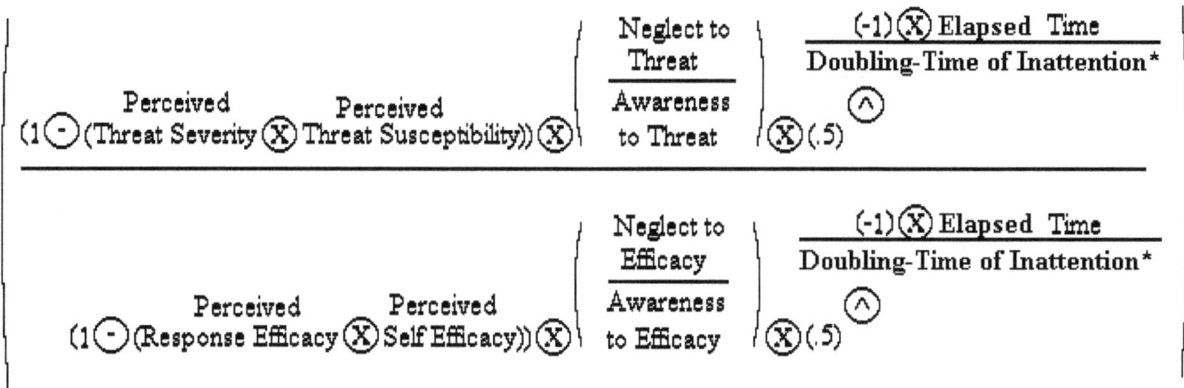

Figure 10.23 "(Neglect / Awareness)" × (1 / 2) ^ ((- 1) x ["Elapsed Time"] divided by the "Doubling Time of Inattention). The coefficient of (- 1) alongside the Elapsed Time accounts for the switching of Neglect and Awareness in the denominator and numerator in order to amplify or reduce negative anxiety in accordance with what is logical and intended. In the Pursuit of Pleasure equation, the *Doubling-Time of Inattention* is used. The Doubling-Time of Inattention is the length of time it takes the ratio of Neglect to Awareness to double. The * indicates there is error within the expression used for the Doubling-Time of Inattention if it is held constant here.

An Explanation of Error when Attention is Defined as a Ratio of Awareness to Neglect

If, for instance, one's Awareness is at .80 toward a specific component, 80% of his or her attentional resources are committed to it. Neglect of the entity is at 20% or .2. In the model of attention with error (for the Avoidance of Pain equation), Attention in this example would be represented by the number four, as eight-tenths divided by two tenths equals four. The next half-life of Attention would occur when Awareness is at 66.6% .666, and Neglect is at 33.3% or .333, as .666 divided by .333 yields two, which is half of four.

Awareness	*Neglect*	*Attention Level (Ratio of Awareness to Neglect)*	*Change in Awareness By Percentage Point*
.9411 (94.1%)	*.05882 (5.88%)*	*16*	
			5.23 Percentage Points
.888 (88.8%)	*.111 (11.1%)*	*8*	
			8.88 Percentage Points
.8 (80%)	*.2 (20%)*	*4*	
			13.3 Percentage Points
.666 (66.6%)	*.333 (33.3%)*	*2*	
			16.6 Percentage Points
.5 (50%)	*.5 (50%)*	*1*	
			16.6 Percentage Points
.333 (33.3%)	*.666 (66.6%)*	*.5*	
			13.3 Percentage Points
.2 (20%)	*.8 (80%)*	*.25*	
			8.88 Percentage Points
.111 (11.1%)	*.888 (88.8%)*	*.125*	
			5.23 Percentage Points
.05882 (5.88%)	*.9411 (94.1%)*	*.0625*	

Figure 10.24 Chart for Half-Life of Attention with error: Attention is being measured as the ratio between Awareness and Neglect in the Avoidance of Pain equation. The Doubling-Time of Inattention would be the mirror to this, where it would equal Neglect divided by Awareness.

A look at the chart reveals that this distribution resembles a bell curve that is close to, but not exactly that, of a standard deviation. Based upon the chosen expression for Attention, if an individual's Awareness level changes then the percentage point change in Awareness is most likely to occur when the Attention value is between one half and one, or between one and two; henceforth, Awareness would be between 33.3% and 50% or between 50% and 66.6%.

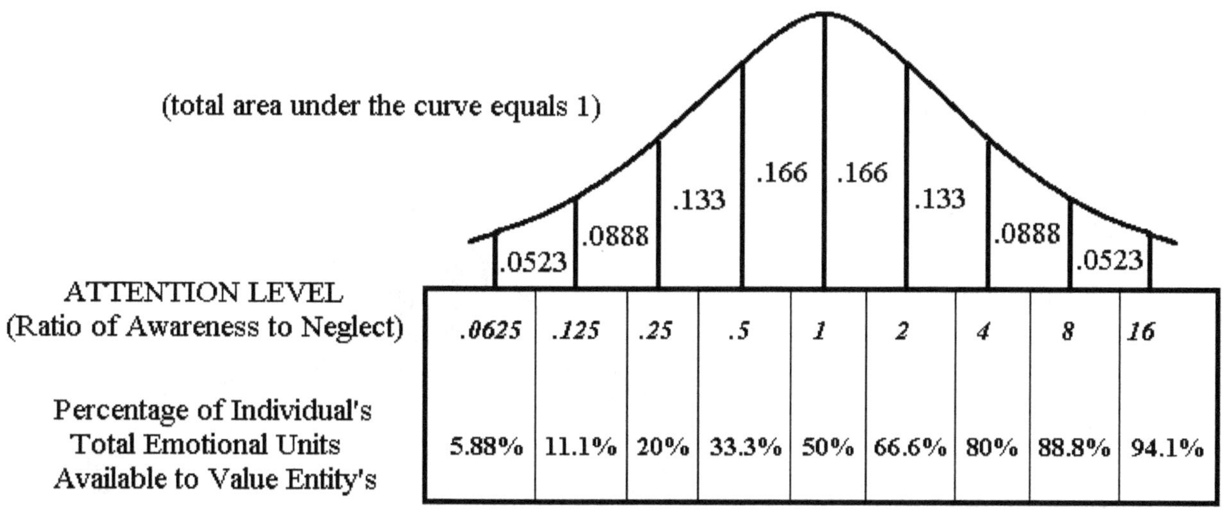

Figure 10.25 Distribution curve for the Half-life of Attention with Error, where the deviation equals one Half-Life of Attention, where Attention is defined as Awareness divided by Neglect.

Further analysis of the chart and graph reveals that the number of percentage points from Awareness required to change the Attention level by a factor of one-half decreases toward the poles and is highest in the middle. For instance, if it is given that the Attention levels of three separate people will decrease by a half-life two times, and each one's starting Awareness levels were different, such that the following is true:

Person *A*: starting Awareness = 20.0%,

Person *B*: starting Awareness = 66.6%,

Person *C*: starting Awareness = 94.1%,

then after two half-lives person *A*'s Awareness will have dropped approximately 14.1 percentage points to 5.88%, person *B*'s Awareness will have dropped approximately 33.3 percentage points to 33.3%, and person *C*'s Awareness will have dropped approximately 14.1 percentage points to 80%.

Henceforth, the half-life of attention with error would hold that person A and person C had an equal change in their Awareness, while person B's total change in Awareness was almost twice that of the other two; this is the source of the error.

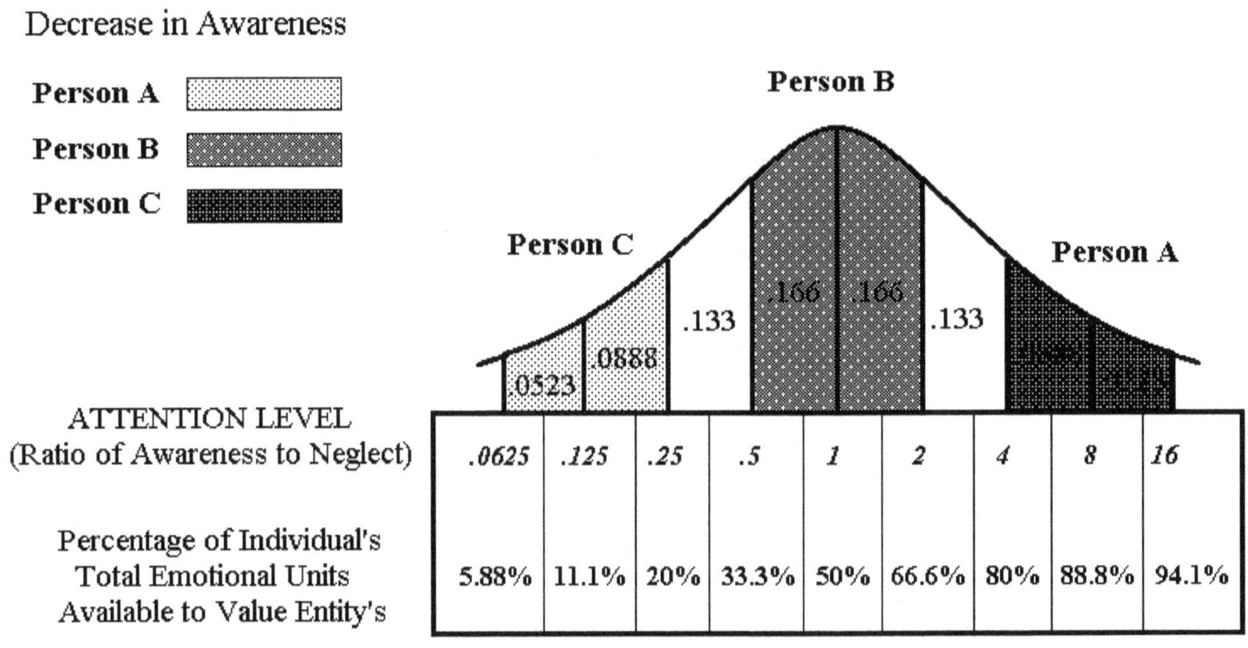

Figure 10.26 The Half-life of Attention with Error (i.e., a drop in two half-lives of attention).

Implications of Utilizing a Half-life of Attention with Error

The implications of this model are as follows. Attentional decay is being measured by using half-lives. However, as a consequence of defining Attention as Awareness divided by Neglect, changes in Awareness at the extreme ends of the spectrum are given more significance and weight than changes in Awareness in the middle of the spectrum per half-life, as they affect the ratio of Awareness to Neglect more so at the poles than when Awareness and Neglect are roughly equivalent.

For instance, if an individual were allowing an entity to fall into attentional decay, and his or her Awareness was initially at 99.9%, strongly outweighing Neglect, using Sample Form II of Attention with the function would hold the following as true:

1) Over a specified number of half-lives (e.g., twenty), Attention would drop off very slowly during the first four or five half-lives, drop off quickly toward the middle, and then drop of very slowly again once Attention's value goes further below one, meaning Awareness decreases beneath 50% or .5.

2) The model would also hold the reverse to be true. If Attention started off on a very low level, it would not take a large increase in Awareness to double the Attention level initially, then it would take a significantly larger amount Awareness to double toward the middle of the distribution before beginning to taper off again at the upper bound once Attention's value goes well above one, meaning Awareness increases above 50% or .5.

While the above assessments are not necessarily improbable, they are error nonetheless. Error is a "general term used for any difference between the true value of a statistic within a population and the estimated value of that statistic derived from a sample of the population."[61] Although this measure for Attention, a ratio of Awareness to Neglect, is perhaps the most practical as it circumvents the need to know specifically how many units of attention are being directed to amplify or reduce anxiety invested in an entity via threat components, benefit components, or efficacy components, this expression for the Half-Life of Attention also becomes a variable that is subject to error, and will be designated as the Half-Life of Attention with error. The Doubling-Time of Inattention is subject to the same error.

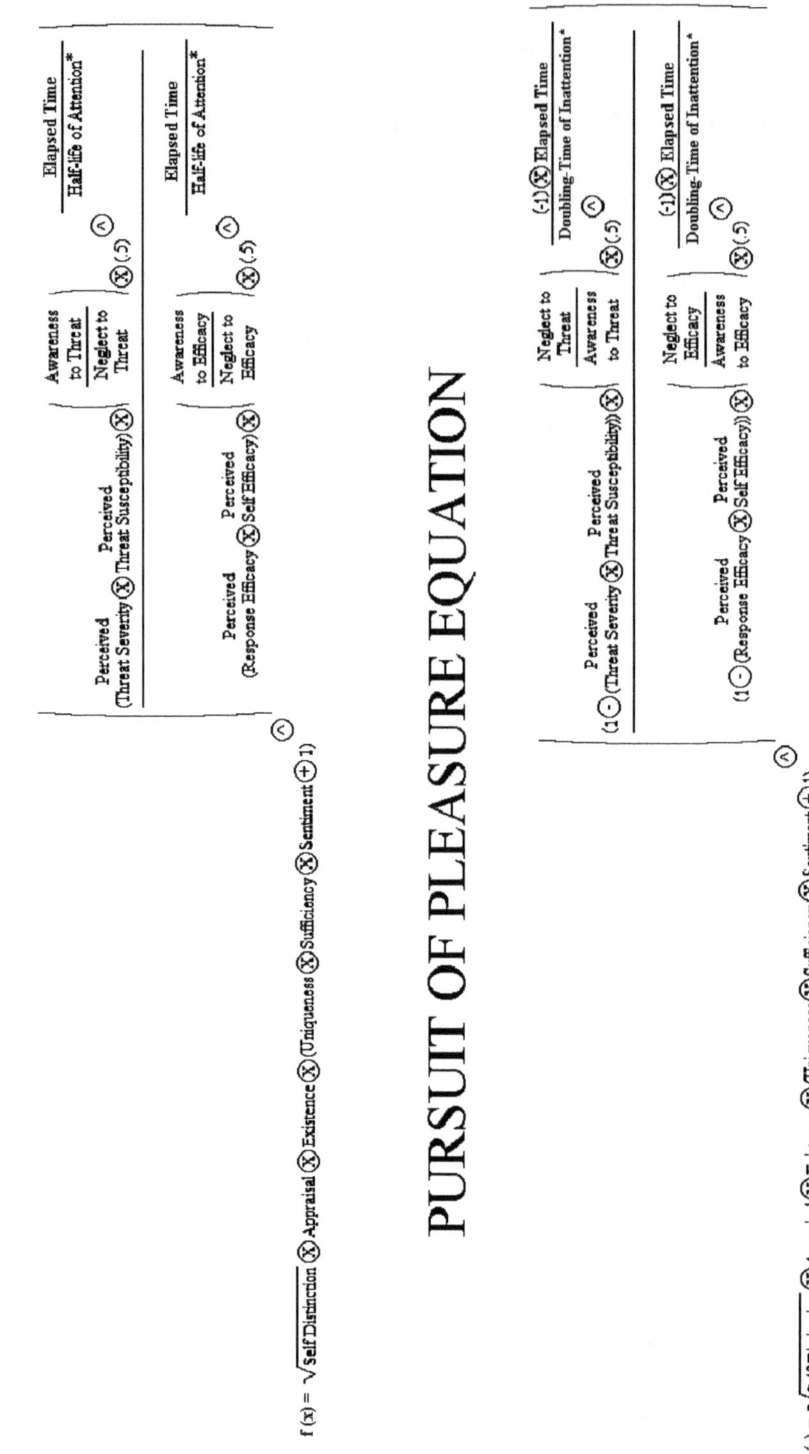

Figure 10.27 Half-Life of Attention and Doubling-Time of Inattention with error in the Avoidance of Pain and Pursuit of Pleasure functions.

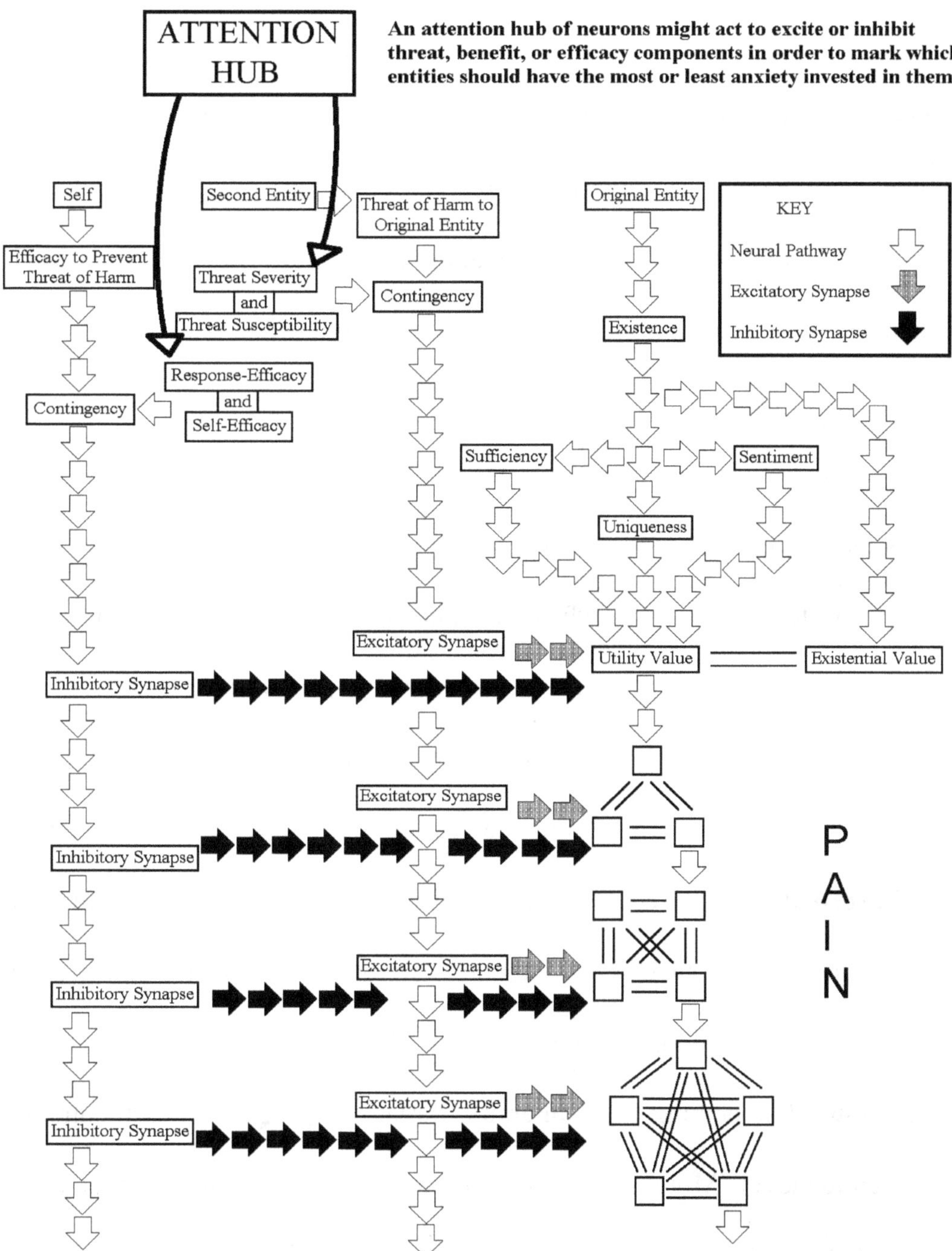

Figure 10.28 A neurological model of attention where an attention hub of neurons exerts influence on an entity's value by exciting or inhibiting threat, benefit, or efficacy components. If attention were linked to value extinction (e.g., in Sample Form I of Attention), then it would also exert influence on the original entity itself.

Alternative Measure for Attention: Half-life of Attention Without Error or Sample Form IV

Alternatively, if one wanted to use a half-life of attention value without this error, then Attention would have to be assessed slightly differently. For instance, instead of including Neglect in the denominator for the Avoidance of Pain equation, Attention would simply be described as a positive real number. One's maximum level of Attention would be the total number of neurons or groups of neurons (e.g., attentional units) in an attention hub that are capable of exciting or inhibiting threat components, benefit components, or efficacy components. This in turn would modify the valuing neurons that gauge the emotional unit. Specifically, one attentional unit would be the minimum quantity of attentional neurons whose activation is required to maintain the base firing-rate of a specific component at the component's existential value over a given time frame. Attention levels that fall below one would indicate valuation extinction of whichever component is concerned.

In this form, Attention would be construed as the total number of attentional units being deployed to mobilize or demobilize components that direct how anxiety is invested in an entity. For instance, if an individual had a total of 1,000 attentional units (e.g., neuron groups) in an attentional hub available to mobilize or demobilize all components, then whatever number of attentional units is being directed toward something (e.g., an entity, threat component, efficacy component, etc.) would be the Attention level. Moreover, that number would have to be less than one thousand, as one thousand would be 100% of his or her attentional resources. One attentional unit would be the minimum number of attentional neurons required to maintain the base firing-rate of a specific component (e.g., Threat, Benefit, Efficacy). Fifteen attentional units directed to a component would imply that the component's firing-rate is amplified fifteen times above its base or existential level.

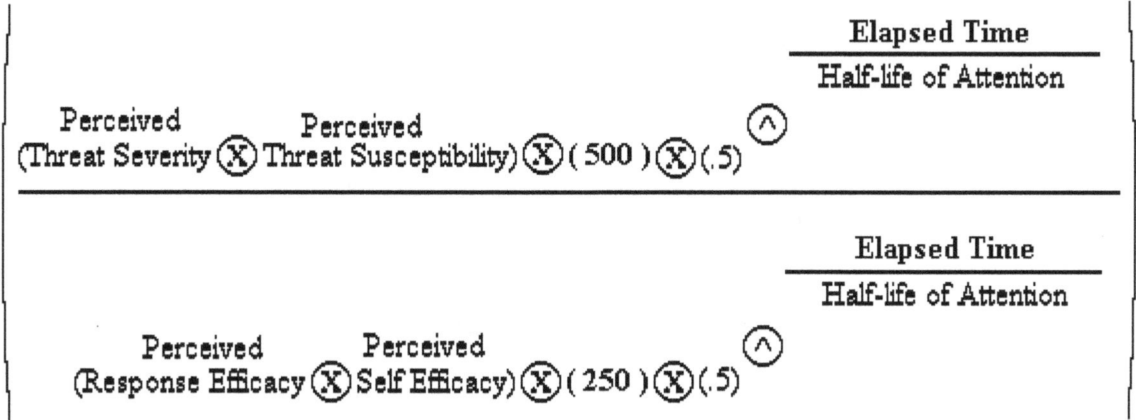

Figure 10.29 Sample form IV of the Half-Life of Attention with *no error*. Where 1,000 attentional units is an individual's maximum amount of attention, the above expression in the exponent of the formula depicts 50% of an individual's attentional resources being directed toward threat components, and 25% of an individual's attentional resources being directed toward efficacy components. The Attention level is being measured by the attentional units being directed toward an entity, threat, benefit, or efficacy component to mobilize or demobilize anxiety. The above would correspond to an Attention level of *1* in the numerator and *1 / 3* in the denominator using sample form II of attention.

One, however, runs into the problem of juggling large numbers using this more precise measure for Attention. The number of such attentional units would likely be much greater than 1,000 and perhaps in the tens of billions. For example, one variable might have 10 attentional units directed toward it whereas another might have 10,000,000,000 or more depending on how much total attention is available. This would stretch the graph vertically along the y-axis and, although more accurate, it would not be practical to do this without the aid of powerful computational tools and the ability to actually gather that information concerning an individual's attention directed toward a specific component of the equation as a whole number. Though cleaner, this form of Attention will not be used in the examples throughout the book, as the numbers would necessarily lead to incredibly large values. However, for those interested in running simulations, an example is shown.

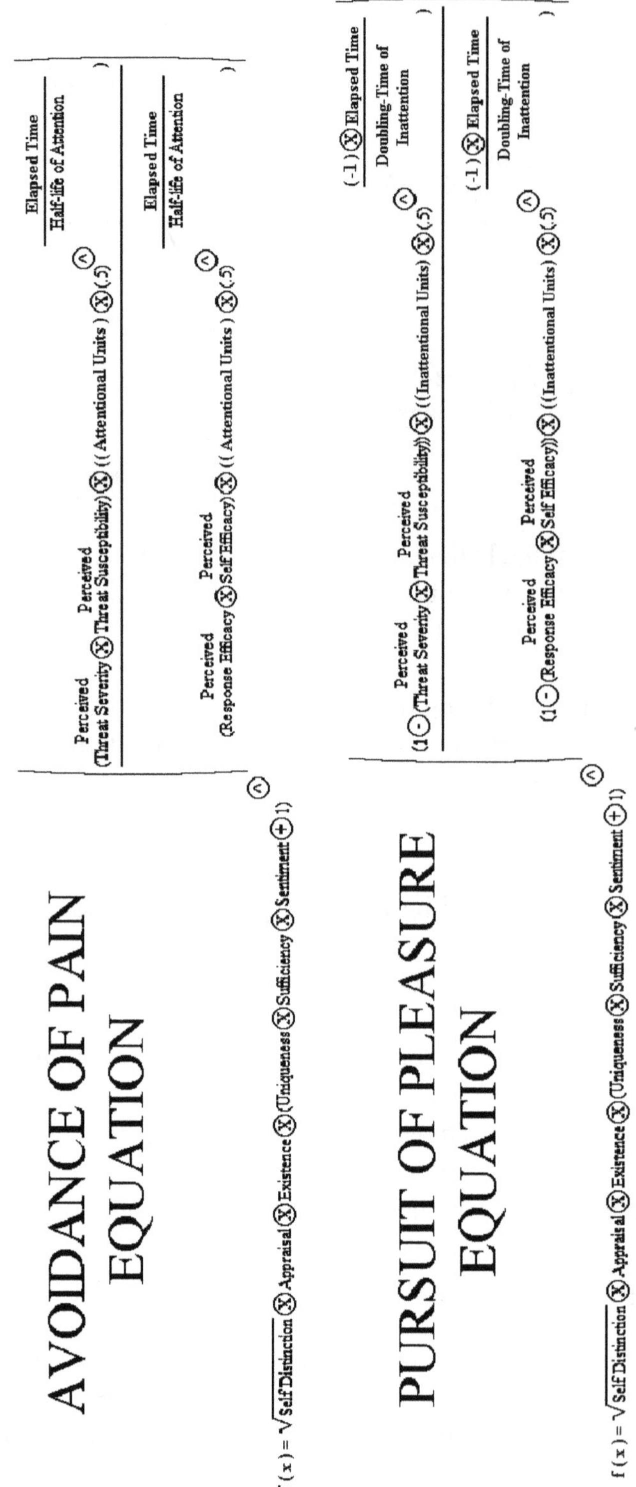

Figure 10.30"Half-Life of Attention" and"Doubling-Time of Inattention" without Error.

Of note, for the pursuit of pleasure equation formula the variables are the slightly different. The inattentional unit and the Doubling-Time of Inattention would be used, but the set up for the function is similar. In the numerator, (1 - Threat components) leaves the likelihood that the entity will be left intact, which is ideal, and inattentional units would be the number of attentional units that are neglecting Threat. In the denominator, (1 - Efficacy components) leaves the inability to perform a recommended behavior, which is not ideal, and inattentional units would be the number of attentional units neglecting Efficacy components. The Pursuit of Pleasure function also requires negative one (- 1) as a coefficient in the exponent next to Elapsed Time.

Controlling Attention

With the introduction of the half-life into the equation comes an added bonus. The variable of time may finally be implemented into the equation. After wandering around without a resting place for five chapters, the x-variable will finally be able to find a home in the equation!

Understandably, in most cases Attention does not simply operate on its own accord. Most people are capable of using their mental faculties or executive functioning (e.g., command center) to hold attention on an entity or to exclude an entity from being attuned to. Therefore, a means of representing reasoning's ability to direct attentional resources will have to be implemented into the equation. Reasoning, then, will act to choose what is attuned to and what is not attuned to. In essence, it will selectively distort the aspects of some relationships between entities by either holding them constant or by permitting them to fall into attentional decay. Moreover, as reasoning necessarily presumes that the individual is thinking about the entity, it must leave the existential value of the entity intact. Finally, reasoning will have to be able to account for action as well, given that one of the stated aims in writing this book was to establish a method by which to mathematically express how reasoning, affect, and emotions interact to lead to behavior.

Reasoning and Constancy

Reasoning typically indicates that an individual is thinking, either inductively or deductively. Reason, a synonym of logical thought, may also be viewed as a justification for one's actions and is sometimes expressed "in terms of one's motivations."[62] Moreover, the term reasoning is used with the sense that it concerns cognitive processes, "not whether the correct outcome is achieved" as the wrong solution can be deduced from perfectly logical reasoning if the original assumptions are at fault.[63] Given the importance that Attention currently has in the equation, the process of implementing a variable for Reasoning would have to be such that it is capable of subduing Attention in order to direct an individual's anxiety resources toward a course of action. For instance, if it is generally assumed that an individual wants to achieve homeostasis by lowering the value of an entity that is useful to a purpose, there are a number of options available for which reasoning must account. One would concern the individual acquiring the entity, in which case he or she may simply hold the value of an entity constant or at an artificially elevated level until it is obtained. This could be achieved by selectively paying attention to the threat components of a relationship while ignoring or not paying attention to the efficacy components (figure 10.31). Oppositely, if the individual chose to not attend to a valuable entity immediately, reasoning would need to be able determine how the entity's value could be reduced, such as by selectively paying attention to the efficacy components while not attuning to the threat components (figure 10.32).

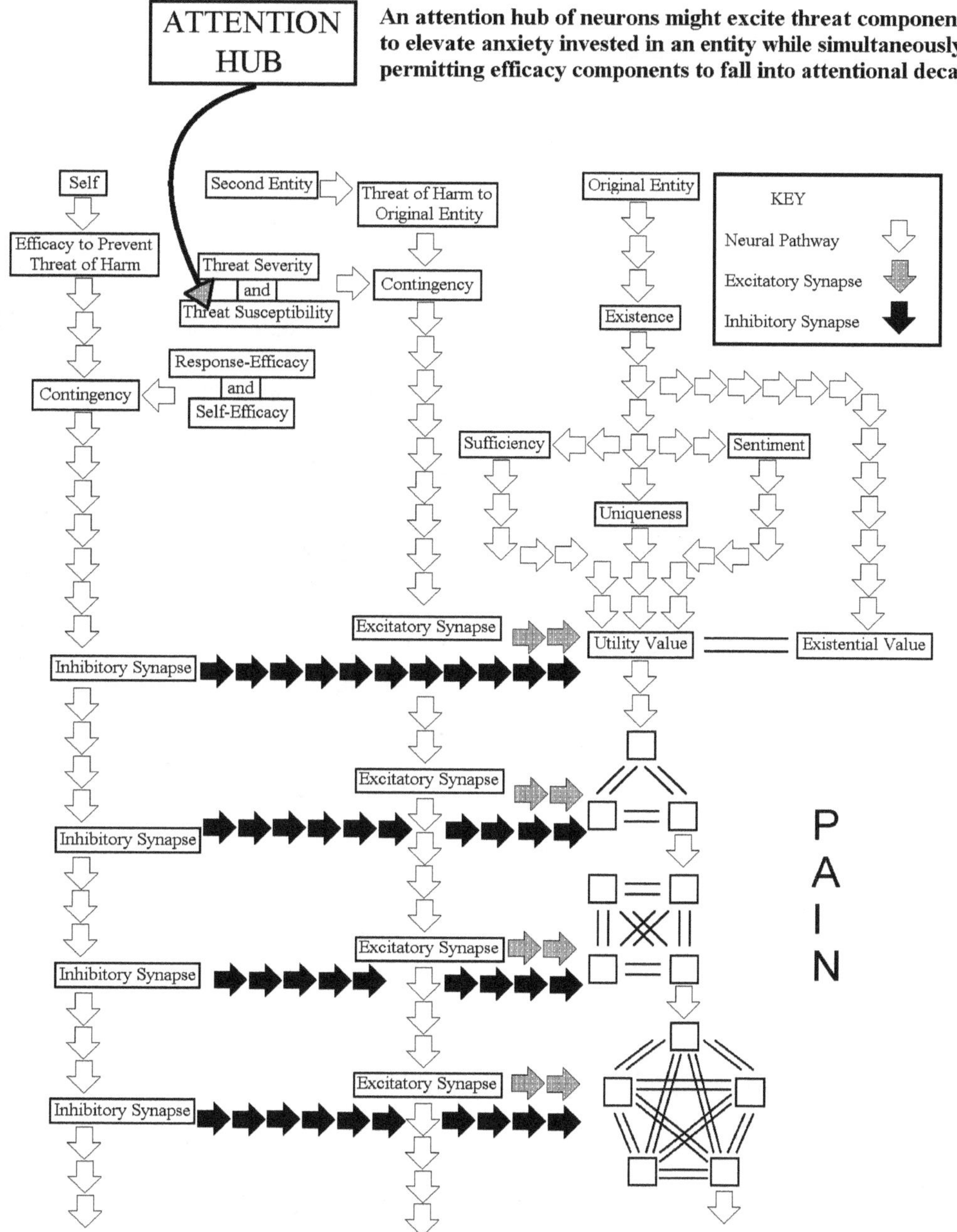

Figure 10.31 Selectively paying attention to the threat components to elevate an entity's value

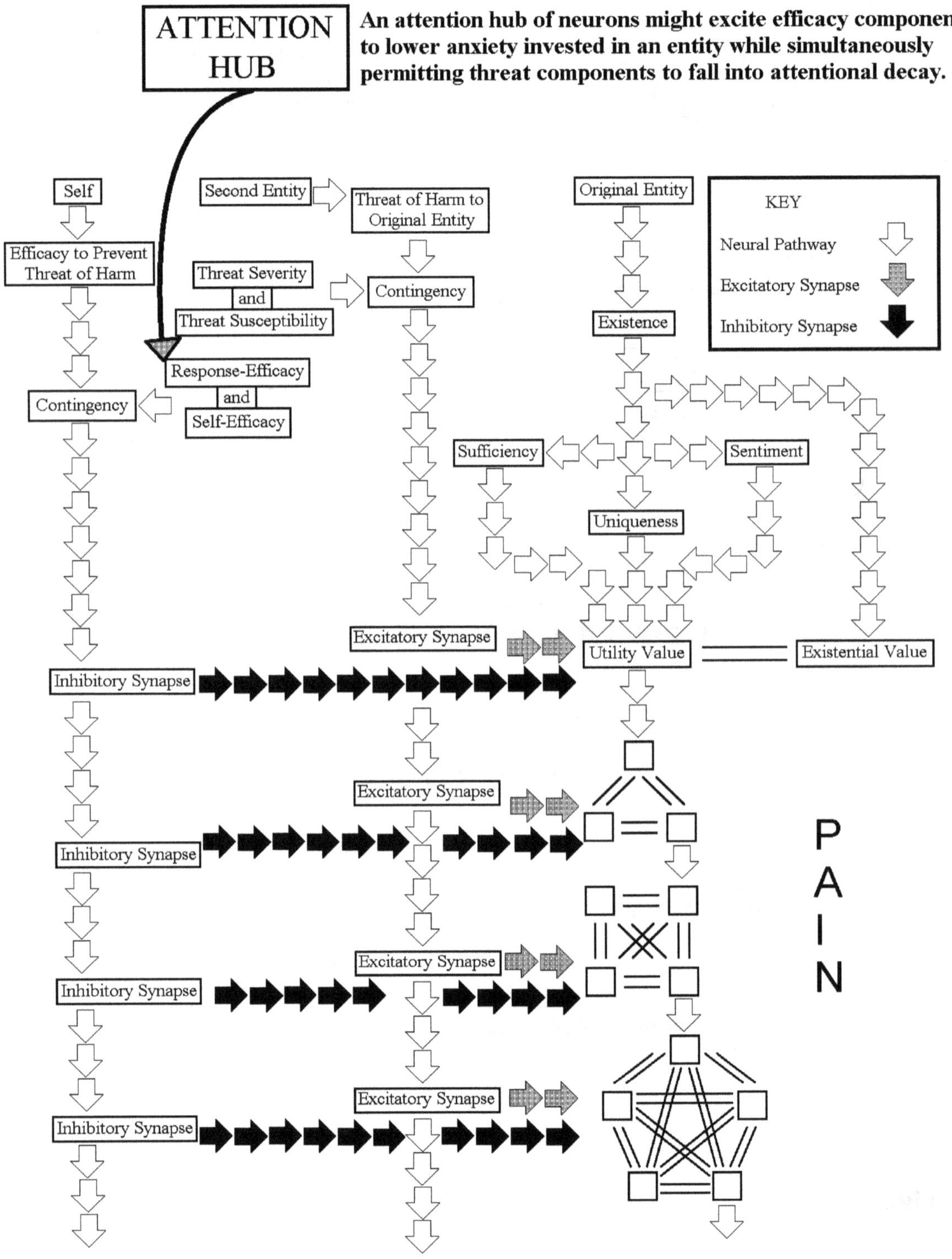

Figure 10.32 Selectively paying attention to the efficacy components to lower an entity's value

At its core, reasoning can be likened "problem-solving."[64] The values of entities, as they relate to purposes, can be distorted by reasoning so as to selectively direct the individual's anxiety investment toward or away from entities which require immediate action by giving them the most Attention or none at all. Although reasoning acts to hold components constant and steady, if all components are not held constant then further distortion results.

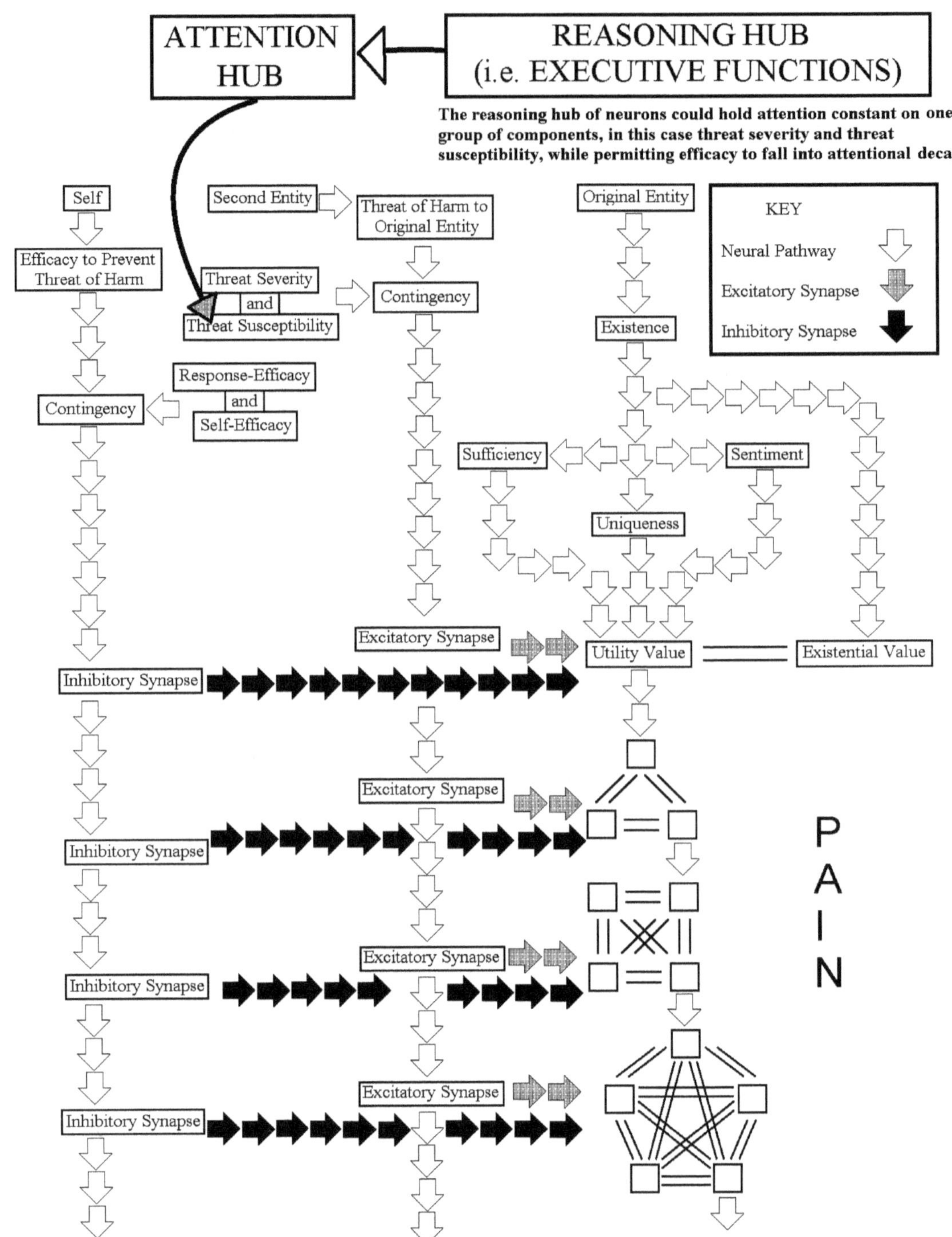

Figure 10.33 Neurological Model illustrating how Reasoning and/or executive functions might act to influence Attention, for instance, by holding Attention constant on threat components to prevent attentional decay.

A return to the idea of constancy will also help further illuminate the role of reasoning. The ability of the mind to hold the perception of something constant, even while features of it appear to change, must be considered. Reasoning, for instance, could hold the conception of an entity constant even as the stimulus properties of an entity are changing, such as its size on the retinal image.

Exponent in the Avoidance of Pain Equation

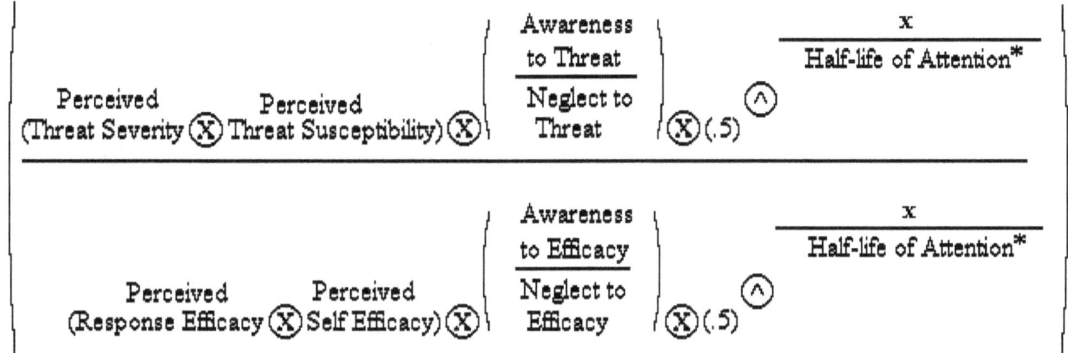

Exponent in the Pursuit of Pleasure Equation

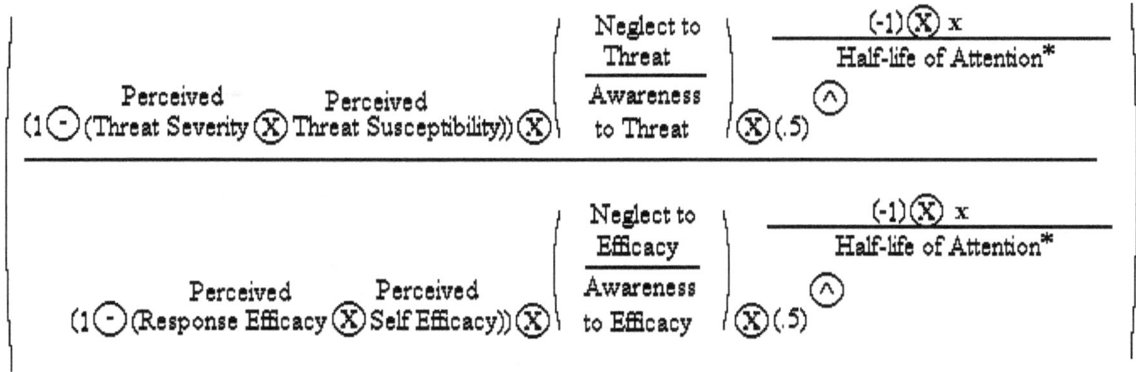

Figure 10.34 Time is officially implemented into the exponent of each equation as the **x** variable. The word **Time** will be used to indicate the variable *x* in subsequent images to avoid confusion with the multiplication sign.

Reasoning in the equation would act as a modifier of Attention. Reasoning is the decision to uphold or not uphold the legitimacy of Attention's distorting effect by maintaining Attention on variables in the exponent, such as harm (Perceived Threat Severity and Perceived Threat Susceptibility), efficacy (Perceived Self-Efficacy and Perceived Response Efficacy), or benefit (Perceived Benefit Intensity and Susceptibility) and by indirectly modifying them through Attention.

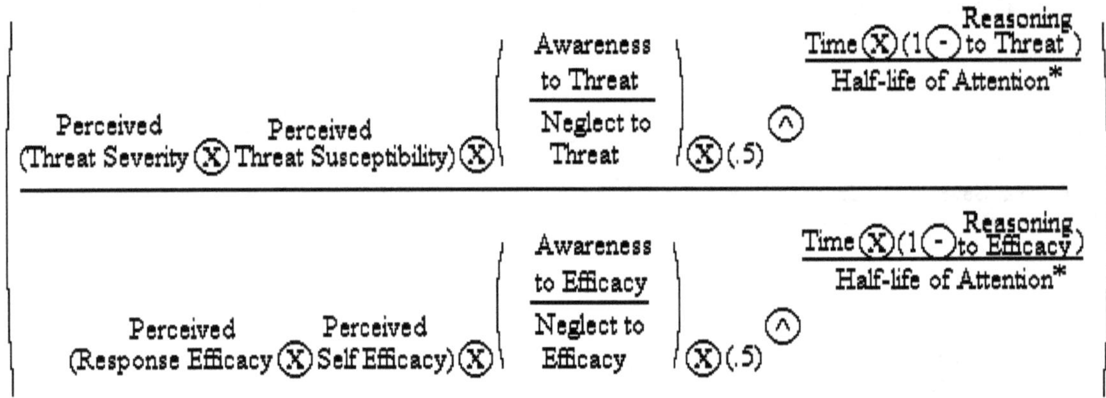

Figure 10.35 Reasoning implemented in the exponent for the Avoidance of Pain Equation

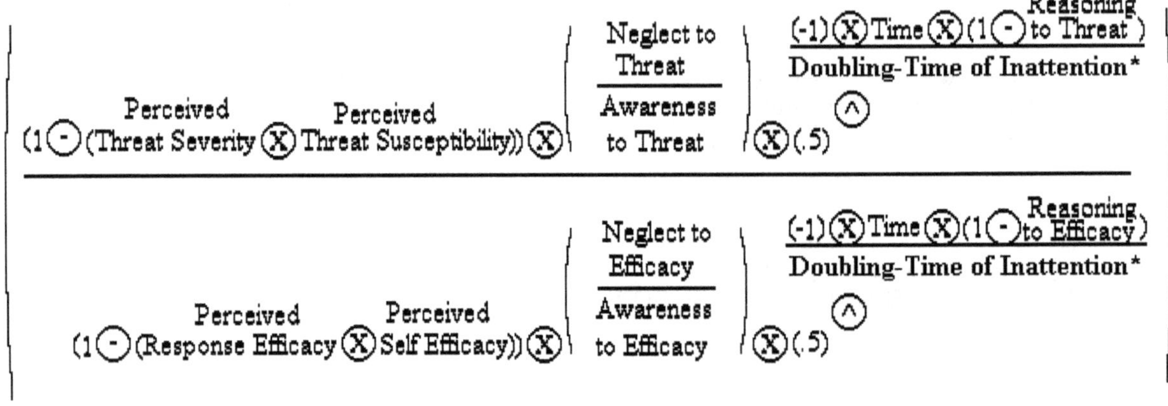

Figure 10.36 Reasoning implemented in the exponent for the Pursuit of Pleasure Equation

Maintaining Attention

Instances where Reasoning is used to prevent a component from falling into attentional decay would be those where Awareness is above zero, Neglect is below one but not equal to zero, and an individual is using his or her Reasoning resources to maintain an elevated value for the entity. This can be accomplished by using one's Reasoning to direct Awareness toward the harm components of Perceived Threat Severity and Perceived Threat Susceptibility while withdrawing Reasoning in the denominator to direct Awareness away from the efficacy components of Perceived Response-efficacy and Perceived Self-efficacy. As Reasoning goes to + 1, Attention's half-life grows longer until it eventually ceases to decay. As Reasoning goes to zero, Attention's half-life decreases to its normal rate and the attuned entity falls into attentional decay. If both the harm and efficacy are held constant (e.g., Reasoning is at +1 for both) then an entity's value is held constant. In this manner Reasoning would act as a gatekeeper to determine which entities will be valued highest and for which purpose.

Although Reasoning in a single equation is a value between zero and one, an individual's Reasoning capacity is simply a positive real number and is a finite resource. Hence, the maximum number of entities that an individual can use his or her Reasoning to hold Attention constant would be representative of the individual's rational power. For instance, if an individual can maintain his or her Reasoning faculties on no more than seven different items at once, then the Reasoning stockpile would be seven. If two units of Reasoning power are used to maintain Attention on a specific entity's threat and efficacy components, then only five units would be available thereafter.

However, if no Attention is being directed to an entity because all of one's reasoning resources are being used elsewhere, then the model holds that nothing can be done to influence the flow of Attention until Reasoning resources becomes available. This can, however, be accomplished by dropping Reasoning resources in another location to free them for use elsewhere.

Moreover, as exponents, Attention and Reasoning cannot modify the entity's base or existential value of + 1, but they can elevate the entity's worth above this value in Sample Forms II and IV of attention. When Reasoning diminishes, attentional decay resumes and the entity's overall value diminishes to its existential value of + 1 when the exponent's value approaches zero. Figure 10.37 depicts Reasoning being used to direct Attentional resources toward efficacy components while permitting harm components to fall into attentional decay. This would be exemplified by an individual trying to avoid feeling overwhelmed from a seemingly insurmountable threat of harm.

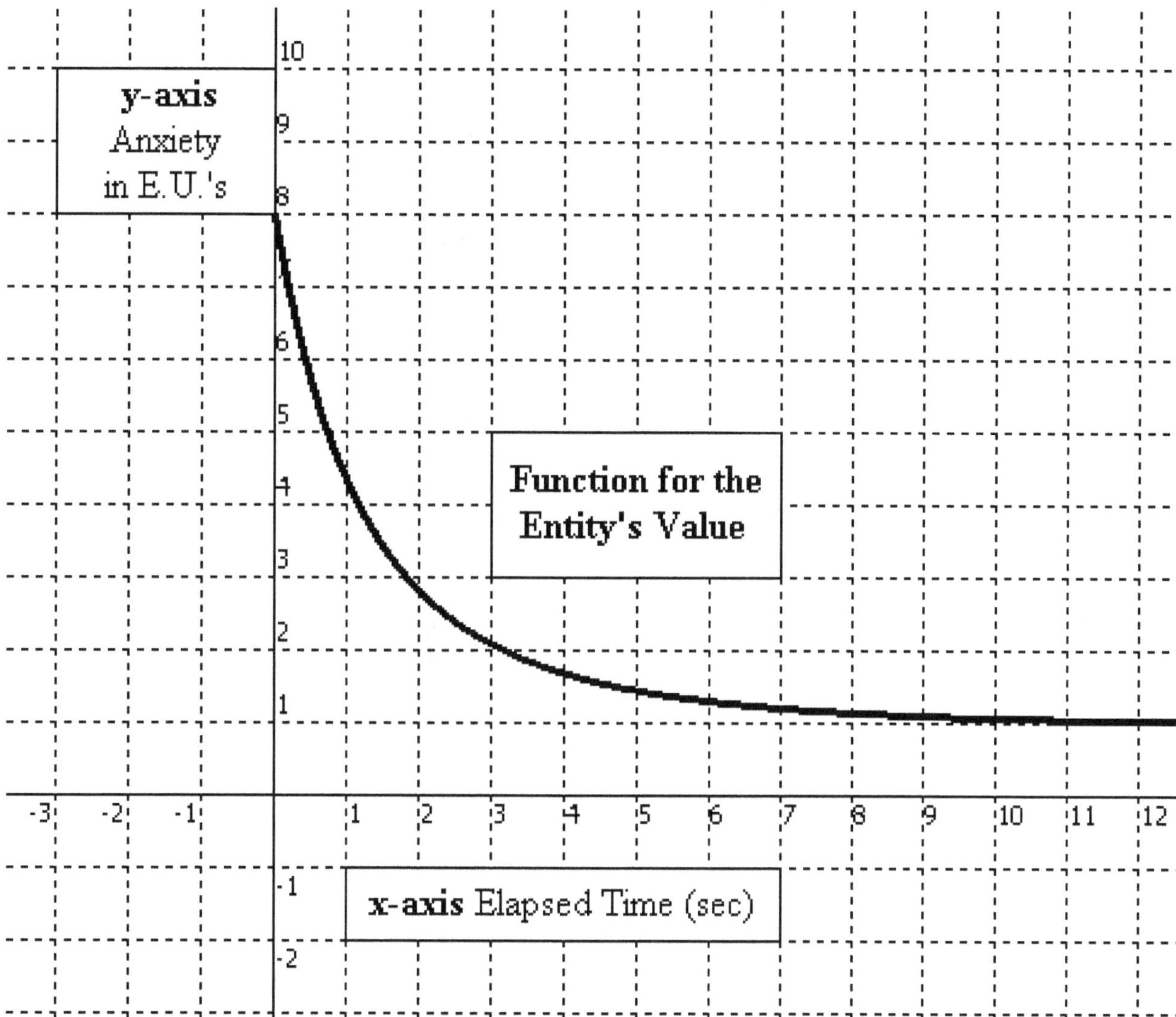

Figure 10.37 Graph where Attention has been held constant for the efficacy components while the threat components have been permitted to fall into attentional decay. Threat Susceptibility and Threat Severity were each at .75 and 1, meaning there was a 75% likelihood that 100% of an entity would be damaged, destroyed, or access to it would be denied. Response Efficacy and Self Efficacy were both set to .5, meaning that a recommended behavior had a 50% chance of being successful with a 50% likelihood that the self would be able to perform it. Awareness and Neglect were at .5 (50%) in both the denominator and the numerator. Reasoning to Threat was set at zero, and so the threat components were allowed to fall into attentional decay. Reasoning to Efficacy was set at one, and so the efficacy components were prevented from falling into attentional decay. The Half-Life of Attention with error was set to two seconds. All other variables equaled one. Time zero is the onset of the contingency's conception.

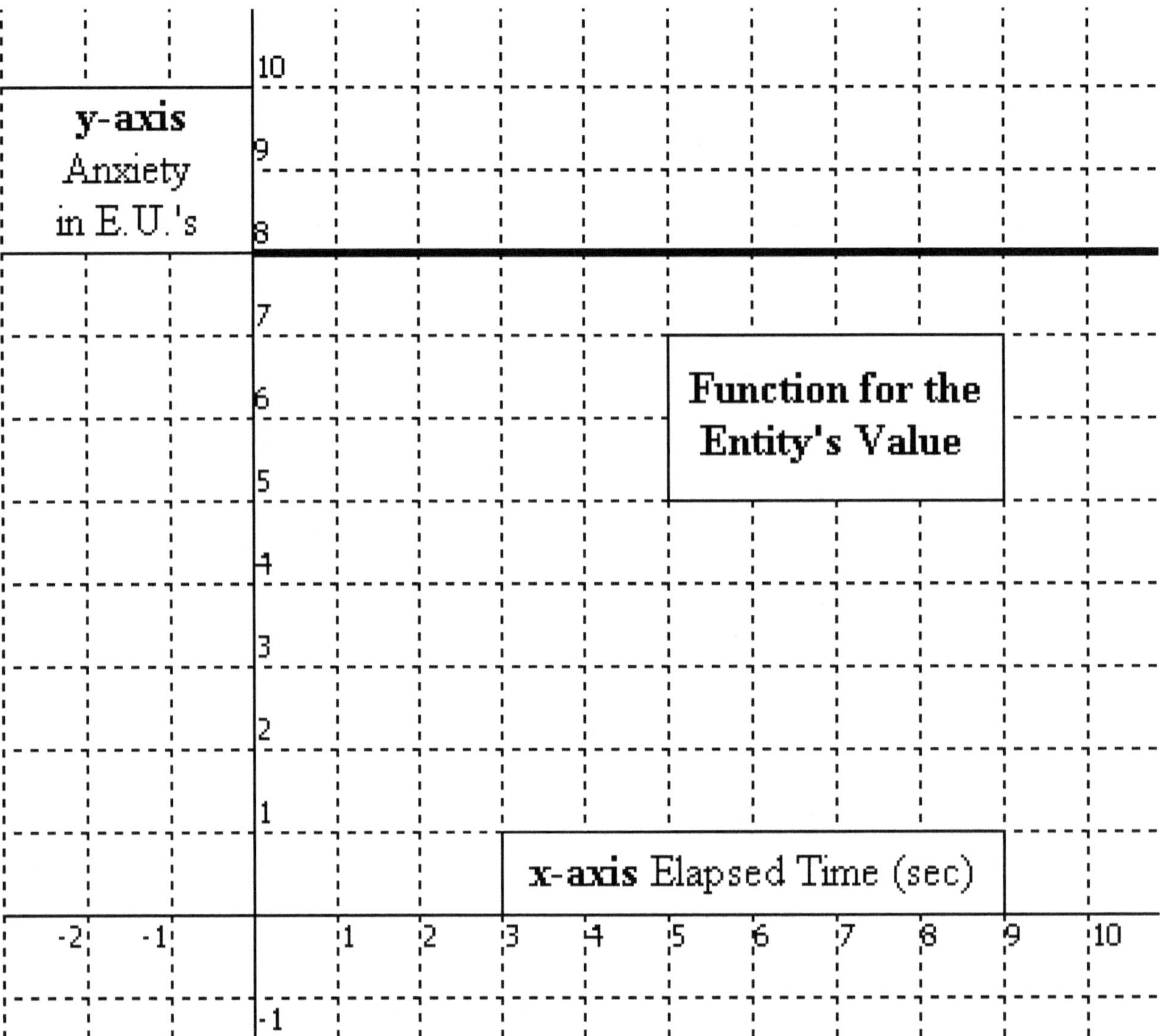

Figure 10.38 Graph of the same equation in Figure 10.37 except that Reasoning to Threat is also held at one, meaning that the individual prevents the threat components from falling into attentional decay and prevents the efficacy components from falling into attentional decay. Both functions start at eight emotional units of anxiety invested in the entity. In the case where the threat components are allowed to fall into attentional decay and the efficacy components do not, the total anxiety invested in the entity drops to its existential or base value. However, in the case where neither the threat nor efficacy components are allowed to fall into attentional decay, the level of anxiety invested in the entity stays constant at eight emotional units of anxiety. The individual stays vigilant with respect to valuing the entity.

Figure 10.38 depicts Reasoning being used to direct attentional resources toward Efficacy components and Harm components so that neither are permitted to fall into attentional decay. This might be exemplified by an individual trying to keep the valuation of the entity stable while he or she is comparing its value to another entity that is useful toward fulfillment of the same purpose. Thirdly, when Awareness approaches one and Neglect approaches zero, the Attention variable in the Avoidance of Pain equation approaches positive infinity and becomes undefined at the point in time where Neglect would equal zero, a vertical asymptote. However, because zero cannot be a denominator in math, Neglect never reaches zero percent, which also means that Awareness will never reach one hundred percent.

If Awareness does approach one and Neglect approaches zero, typically it would be an instance where one would want to thereafter use Reasoning to direct Attentional resources away from the entity. For instance, if proportionally large amounts of Attention are being directed toward harm components (e.g., Perceived Threat Severity and Perceived Threat Susceptibility), to minimize the influence created by this threat the individual might use Reasoning to direct Attentional resources away from harm and toward efficacy components. Notwithstanding, there are instances where one might wish to elevate the anxiety invested in an entity. If one was wary of the Attention being given to efficacy components, then an individual could raise his or her guard by directing Attention away from efficacy components and toward harm components. Figure 10.39 depicts Reasoning being used to direct attentional resources toward threat components while permitting efficacy components to fall into attentional decay. This might be exemplified in an individual trying to avoid feeling complacent from a possible false sense of security, or from a distrust of heightened estimations of one's efficacy to perform a recommended behavior (e.g., to avoid taking it for granted).

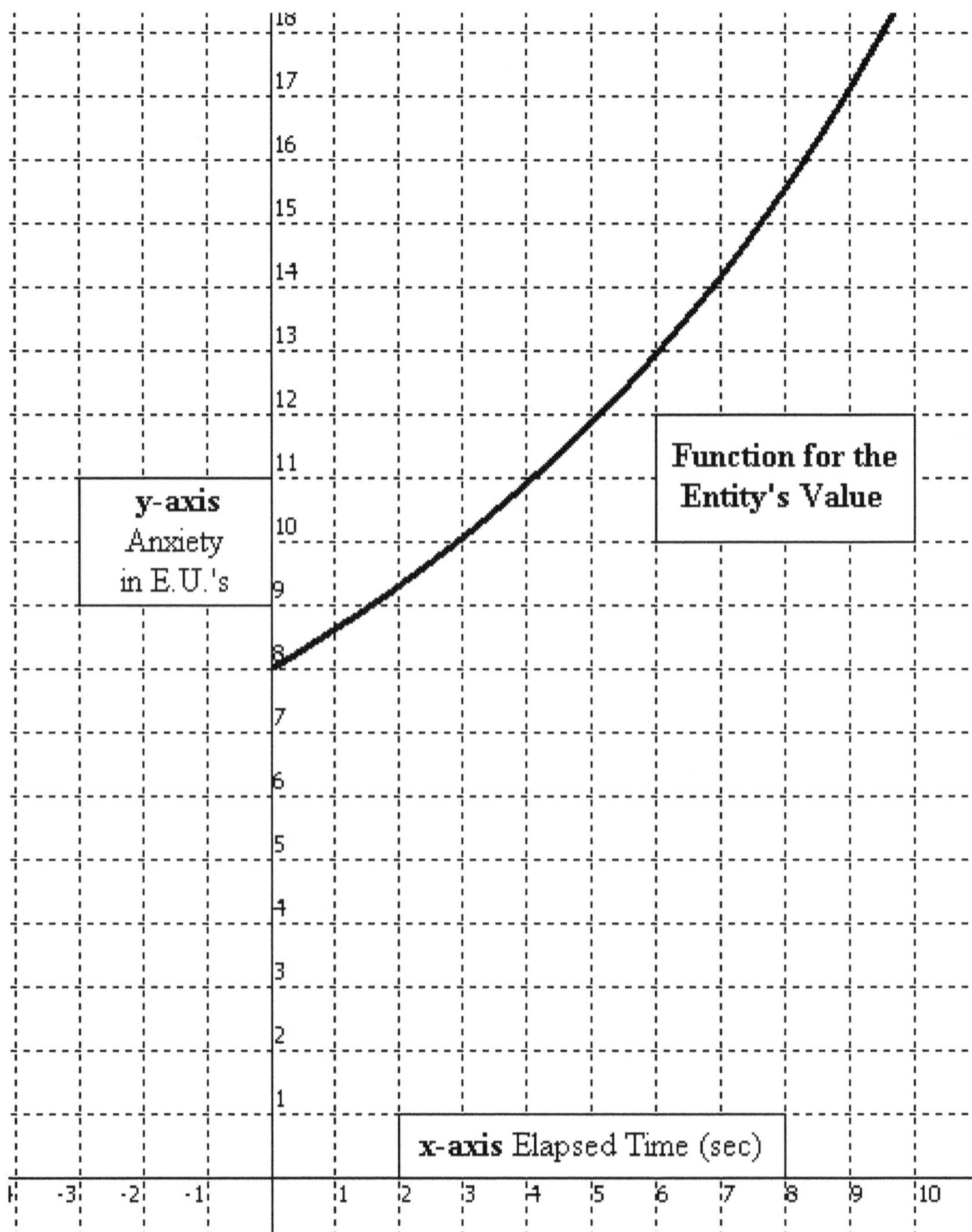

Figure 10.39 Graph of the same function from Figure 10.37 and Figure 10.38. Reasoning to Threat was set to one, meaning that it was not permitted to fall into attentional decay. Reasoning to Efficacy was set to .9, meaning that it was gradually permitted to fall into attentional decay.

The Tipping Point of Awareness against Neglect: the Notion of Multitasking

Multi-tasking, from the prefix multi- for many, implies that an individual is capable of performing multiple tasks simultaneously. In order to perform multiple tasks at once, a person's attention must be split between two different tasks or "acts that one must accomplish."[65] The equation permits for Attention to be split. However, as more tasks are added, the mental effort that it is required to prevent an entity from falling into attentional decay increases as Reasoning and one's executive functioning must possess sufficient potential to manage them all. In Sample Form II of Attention is structured, no more than one entity can be held with an Attention level above one through the use of Reasoning to direct Attention. Two entities could have an Attention level of one if they were the only two entities that the individual held to exist. For example, if Awareness is evenly distributed over a specific number of entities, and if Attention is defined as the following:

Awareness ÷ Neglect, hence . . . Awareness ÷ (1 - Awareness)

1) For one entity, Awareness to Neglect would approach positive infinity, or the vertical asymptote at 1 / 0.
2) For a total of two entities, Awareness to Neglect would be .5 / .5 for each.
3) For a total of three entities, Awareness to Neglect would be approximately .33 / .66 for each entity.
4) For a total of four entities, Awareness to Neglect would be .25 / .75 for each entity.
5) For a total of five entities, Awareness to Neglect would be .2 / .8 for each entity.

Conclusion to Sample Form II and Attentional Decay

With Reasoning held at zero, the tendency of an entity's value is to return to its existential value of + 1 if attentional decay is considered to be in the form of a half-life. Awareness and Neglect, like the secondary components of harm and efficacy, distort an individual's appraised judgment value of an entity. Harm, hence Perceived Threat Severity and Perceived Threat Susceptibility, is modeled to elevate an entity's value upward toward positive infinity for anxiety and toward negative one for negative anxiety. Efficacy, both Perceived Response-efficacy and Perceived Self-efficacy, is modeled to decrease an entity's value downward, toward negative infinity for negative anxiety and toward positive one for anxiety. Attention, depending on whether Awareness is diverted toward harm components and away from efficacy components, or toward efficacy components and away from harm components, can ultimately amplify an entity's value toward positive infinity, toward negative infinity, toward + 1, or toward - 1. The only case where attentional decay would not reduce the harm components and the efficacy components would be when Reasoning is equal to one for both the numerator and the denominator in the exponent. This would mean that an individual would be using his or her Reasoning resources to hold constant all the components concerning an entity's value for a single purpose. Reasoning, and its role as explained here, will be helpful to keep in mind when coping is examined, and vital to keep in mind once defense mechanisms are taken into consideration. Reasoning's role in Sample Form III of attention will be similar.

Sample Form Three of Attention: Attention Accounts for Value Distortion but with Valuation Resilience. Value Decay and Value Extinction Are Not Occurring.

A third and alternate method for measuring Attention, instead of using exponential decay, incorporates the idea of valuation resilience. In this final sample model of Attention, Reasoning forcibly directs Attention to alter one's valuation of an entity. This would mean that Attention is always under the control of Reasoning and the distinction between the two would be noticeably less than in the first two models of attention.

This form of attention is ideal if one did not wish to consider forgetting in the equation. This would be the case if emotions are modeled on someone or something that does not forget (e.g., a computer), or if long-term memory is under consideration. Instead of finding a half-life of Attention, hence, the time it takes for the ratio of Awareness to Neglect to change by a factor of one half, the *Root-Life of Attention* would be used. The Root-Life of Attention would be the time it takes Attention (as a ratio of Awareness to Neglect) to revert back toward one by a factor of one square root from its original value at the time it was conceived. For the Avoidance of Pain function, this would change the variables in the exponent for Attention and Reasoning to the following:

(Awareness ÷ Neglect) ∧ ((.5) ∧ { [Time (1 - Reasoning to Threat)] ÷ Root-life })

in the numerator, and

(Awareness ÷ Neglect) ∧ ((.5) ∧ { [Time (1 - Reasoning to Efficacy)] ÷ Root-life })

in the denominator.

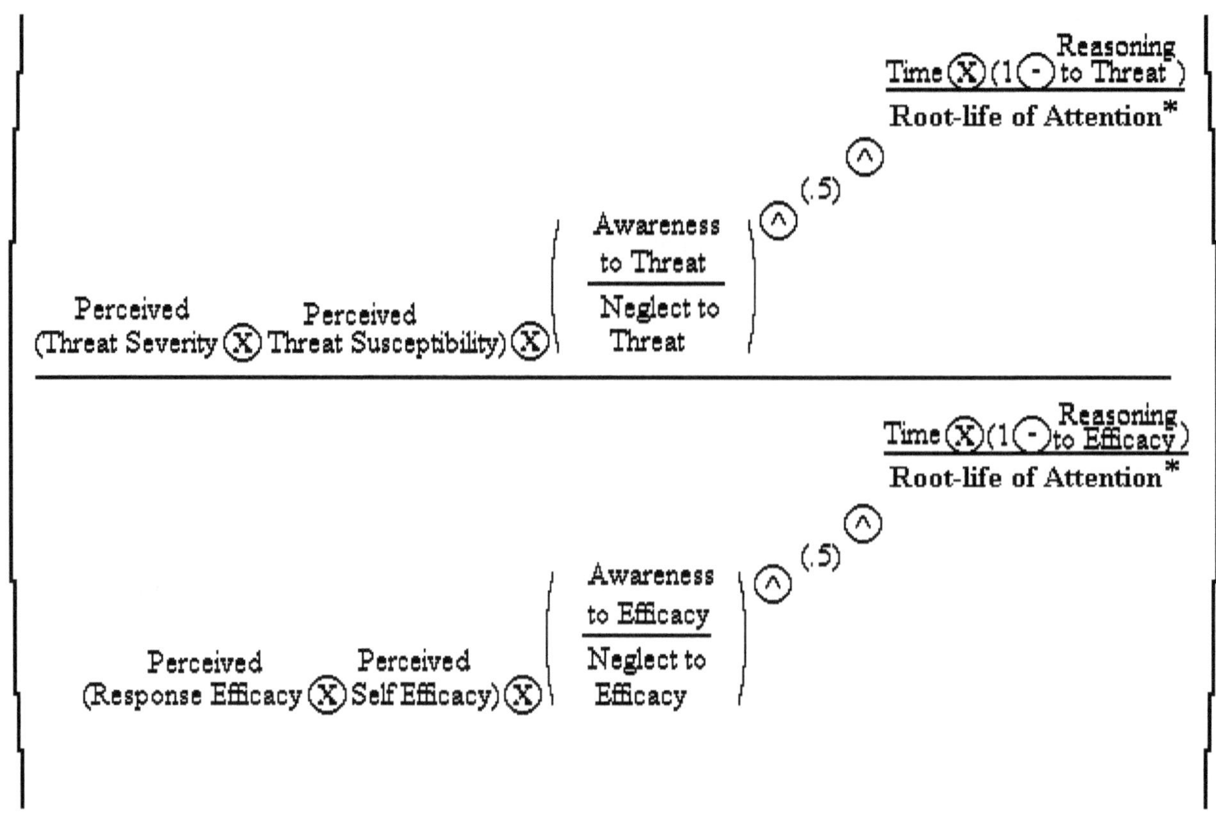

Figure 10.40 Sample Form III of Attention, Valuation Resilience
The above depicts the exponent of the Avoidance of Pain Function when Attention is modeled with valuation resilience. Instead of calculating the half-life of attention, a root-life of attention is used. The root-life of attention can be thought of as the time it takes for Attention to recoil towards one (e.g., where Awareness and Neglect both equal .5) by a factor of one root.

As the root-life here is the amount of time it takes the Awareness to Neglect ratio to recoil back toward 1:1 by a factor of one square. Increasing one's level of Awareness for one entity necessarily suppresses the Awareness of all other entities. The Root-life of Attention is exemplified in the following graph. The Attention level to the threat components is two, meaning that Awareness equals approximately .666 and Neglect equals approximately .333 in the numerator, or 2/3 and 1/3 respectively. The Attention level to the efficacy components is .5, meaning that Awareness equals approximately .333 and Neglect equals approximately .666 in the denominator, or 1/3 and 2/3 respectively. The root-life of Attention is set to two seconds, and Reasoning to both Threat and Efficacy are maximized at one; all other variables equal one as well.

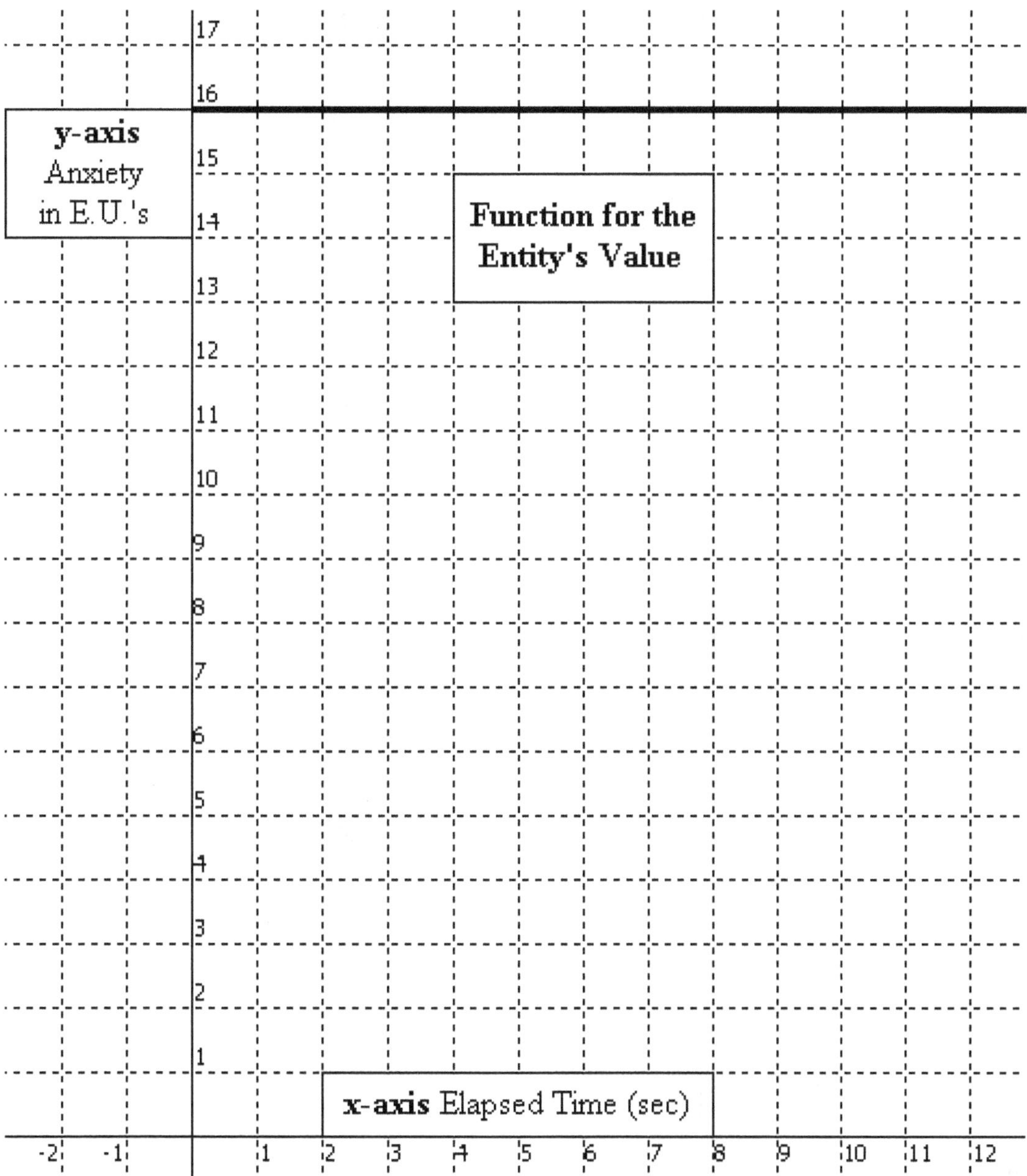

Figure 10.41 Valuation Resilience modeled (Sample Form III of Attention).The Root-life of Attention is set to 2 seconds. Attention in the numerator equals two (Awareness : Neglect = 2/3 : 1/3). Attention in the denominator equals one half (Awareness : Neglect = 1/3 : 2/3). All other variables equal one. Attention in the numerator is amplified while Attention in the denominator it is suppressed.

If the individual ceased to amplify Attention to the threat components in the numerator (e.g., Reasoning to Threat equals zero or is anything less than one) while continuing to suppress the efficacy components, then the graph would change to look like figure 10.42.

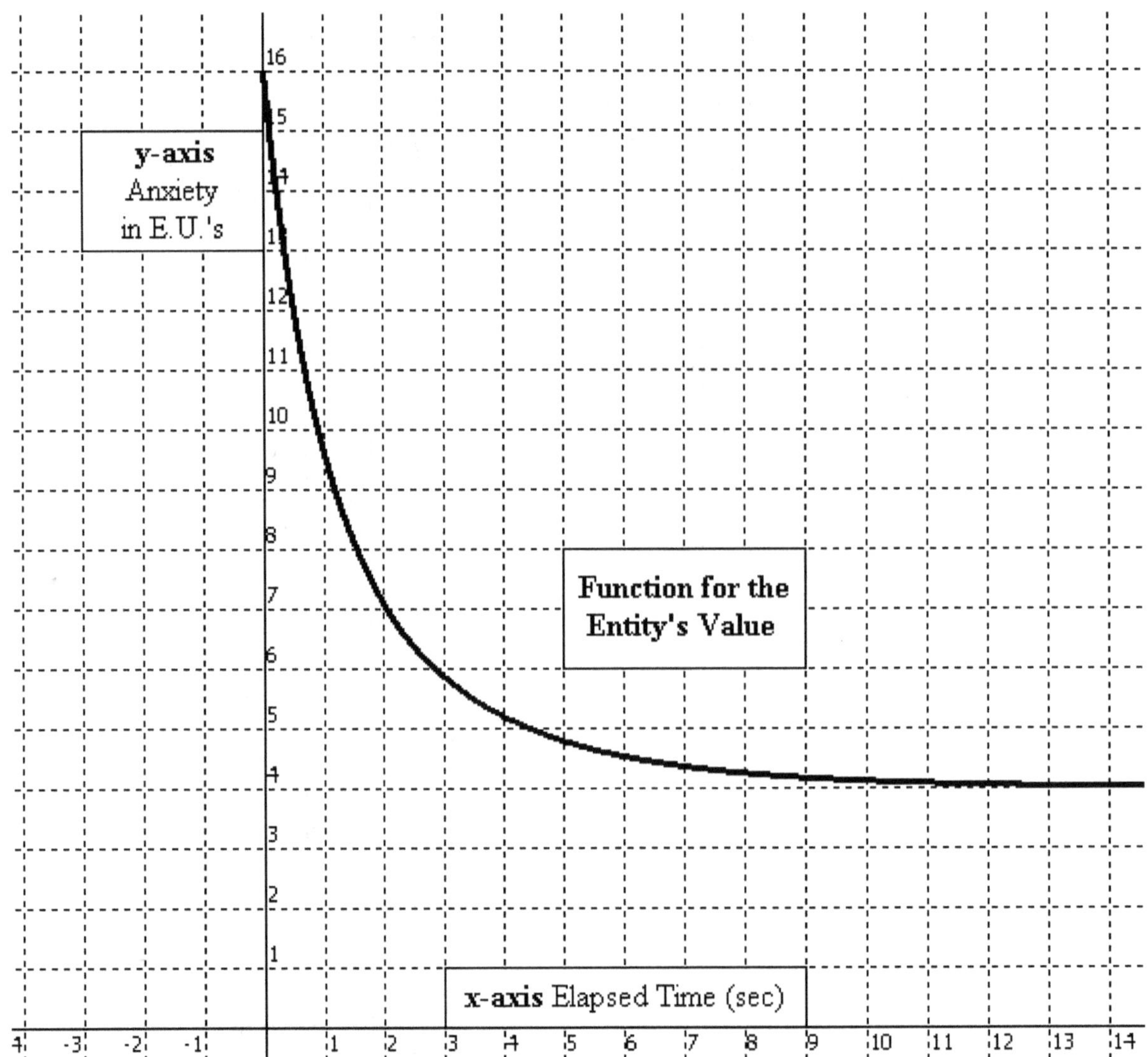

Figure 10.42 Valuation resilience is modeled (Sample Form III of Attention). Like Figure 10.41, the Root-life of Attention is set to 2 seconds. Attention in the numerator equals two (Awareness : Neglect = 2/3 : 1/3). Attention in the denominator equals one half (Awareness to Neglect = 1/3 : 2/3). However, Reasoning to Threat equals zero in this instance, meaning that the amplified Attention to the threat components is permitted to recoil back to one. All other variables equal one.

If the individual stopped suppressing efficacy components while maintaining the threat components at an elevated level, then anxiety would also diminish as Attention to efficacy recoiled upwards toward one. The graph would change to look like figure 10.43.

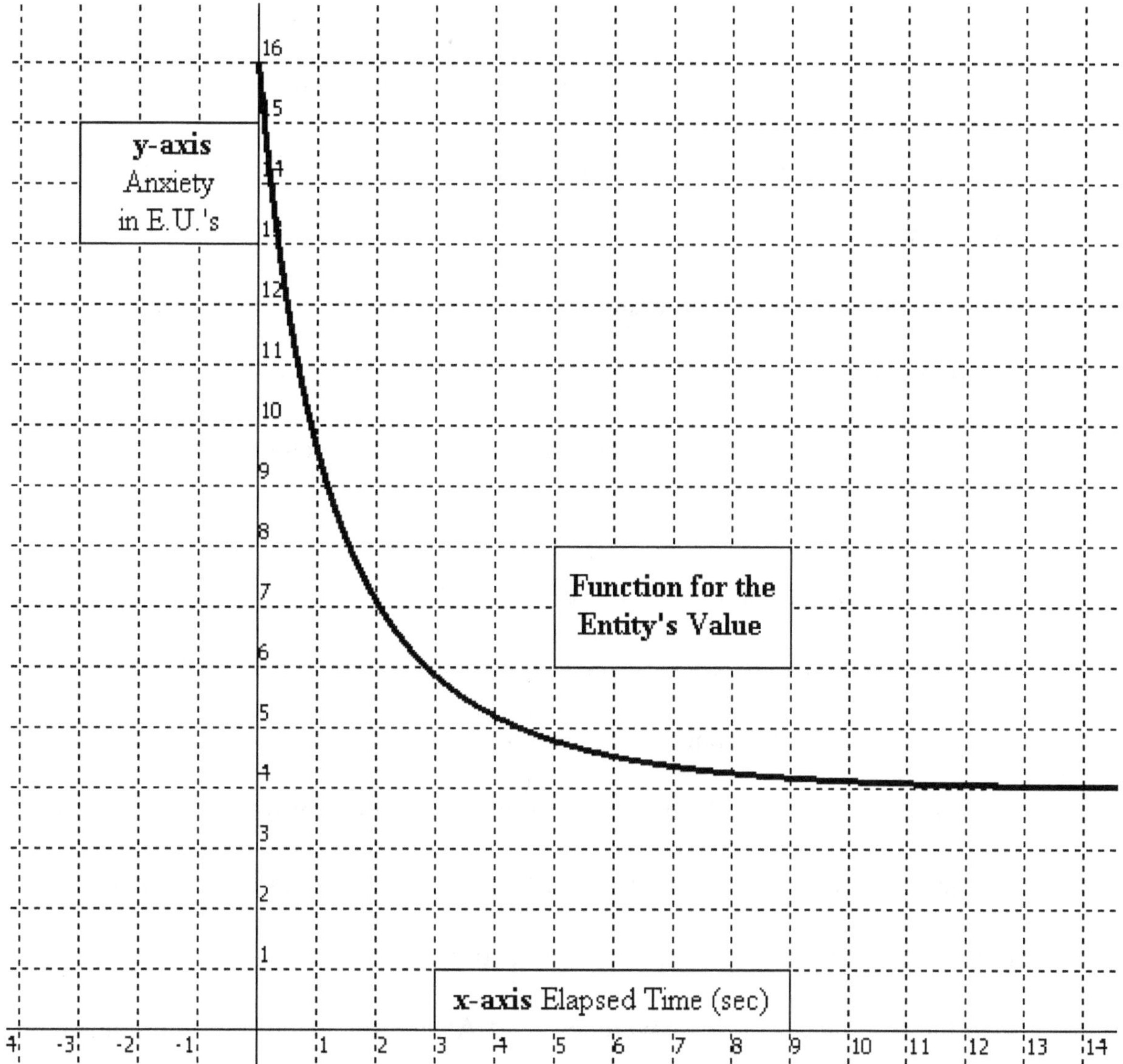

Figure 10.43 Valuation resilience is modeled (Sample Form III of Attention). Like Figure 10.41 and Figure 10.42, the Root-life of Attention is set to 2 seconds. Attention in the numerator equals two (Awareness : Neglect = 2/3 : 1/3). Attention in the denominator equals one half (Awareness : Neglect = 1/3 : 2/3). However, Reasoning to Efficacy equals zero in this instance, meaning that its suppressed Attention is permitted to recoil back to one. All other variables equal one. In this case, the decline in anxiety happens to be identical to that of Figure 10.42.

Worth noting, the graph in figure 10.43 is flawed only in the sense that it cannot be possible for Awareness to be greater than .5 for more than two components. Something would have to give first, such as the suppression of the numerator to make way for the denominator. Awareness must be reduced and suppressed elsewhere (e.g., another component in the equation or from a different entity entirely) to free up Awareness for the efficacy components reemerging from suppression. This can only be done for one component of one entity, as only one Attention level can be elevated above one at any given time, hence, where Awareness is greater than Neglect. However, the 4:1 ratio between the threat components in the numerator and the efficacy components in the denominator can be replicated with much smaller amounts of Awareness invested into each. The graph in figure 10.43, therefore, could be modeled while staying true to the parameters that have been set for Awareness.

In the case where both Attention to Threat components and Attention to Efficacy components are both less than one, elevating anxiety invested is a simpler process. If one wanted to increase the Attention level devoted to it, then one would simply cease to suppress it. For instance, in the equation used for figures 10.41, 10.42, and 10.43, if the Attention level were at one-third for both Threat and Efficacy components (e.g., Awareness : Neglect = 1/4 : 3/4), then setting Reasoning to Threat to zero would permit the Awareness to Neglect ratio (Attention level) to recoil toward one while staying in the parameters set for Awareness. If Awareness is still not in ample supply to accomplish this, then Awareness would have to be forcibly reduced elsewhere to accommodate. Total Awareness toward all entities or components of entities (threat, benefit, efficacy) cannot exceed 100% or 1.

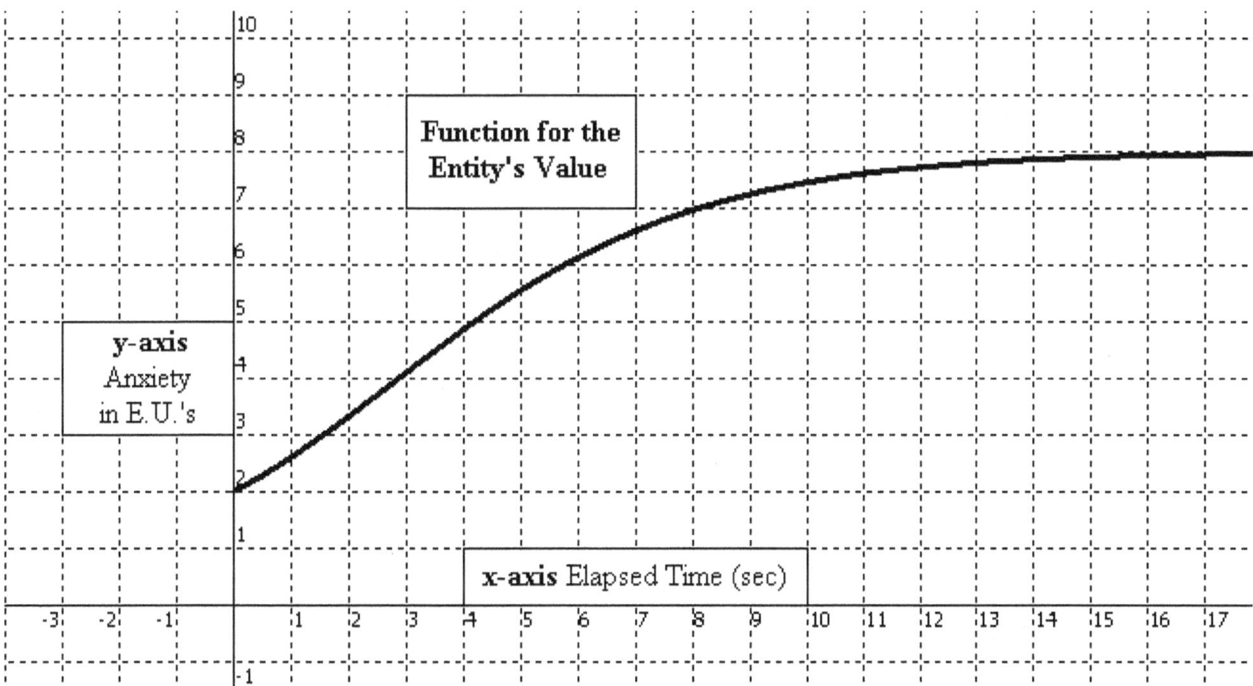

Figure 10.44 Valuation resilience is modeled (Sample Form III of Attention). Like figure 10.41 and figure 10.42, the Root-life of Attention is set to 2 seconds. However. Attention in both the numerator and denominator is less than one, equaling 1/3 in both cases (Awareness : Neglect equals .25 : .75). Reasoning to Threat equals zero in this instance, meaning that Attention to the threat components is permitted to recoil from being suppressed back towards one. All other variables equal one. In this case, anxiety tapers off at approximately 8 emotional units and would only be able to go above this if Awareness is reduced elsewhere in order to amplify it.

When Reasoning ceases to be used to direct Awareness toward an entity, then Attention will diminish back toward one if it was amplified above one or surge upwards toward one if it was suppressed below one. The vast majority of unattended entities that an individual has in his or her psyche would have Attention toward them suppressed constantly. Those entities and components that are attended to would be selectively permitted to surge toward one if none have an Attention value above one. If one component is actively attended to, meaning its Attention value is above one, then Attention directed toward it would be amplified while Attention to all else would be suppressed.

In the valuation resilience model of attention, any Attention value above one, where Awareness is greater than Neglect, has been amplified. Any Attention value between zero and one, where Awareness is less than Neglect, has been suppressed. Although only one entity can have an Attention value above one, meaning whatever component in the equation with an Attention value that is more than one has been amplified, it is still possible to model fluctuations in anxiety investment for other entities. For instance, to model a case where an entity's valuation has become elevated or has decreased and both the Threat and Efficacy components have Attention levels that are less than one, one component will continue to be suppressed while the other permitted a resurgence.

In short, instead of an entity always being reduced to its existential value over time, entities would always tend to revert back to their original appraised judgment value after the harm and efficacy components had been considered. This would be the case regardless of whether or not an entity was given substantial Attention at the onset. Conversely, all of the other entities would have a smaller ratio of Awareness to Neglect and deflated valuations as only one entity would be able to have more than 50% of the individual's total attentional resources devoted to it. Attention, then, would act to actively suppress all other values to prevent them from re-surging while maintaining the attuned component (e.g., Threat or Efficacy) at an elevated value if Awareness to it is greater than Neglect, or a constant one if Awareness and Neglect are the same, at .5. It follows that Reasoning would be used to hold an individual's Attention on an entity constant at a specific level while suppressing Attention on all other entities in order to budget anxiety resources. This form of Attention in the equation would also be the most useful for modeling trauma and psychological disorders where an entity or event has to be continuously suppressed to prevent it from unconsciously re-entering into prominence (e.g., if Reasoning slips up). Consequently, the demands and expectations placed on Reasoning are greater in Sample Form III of attention than other forms.

Valuation Resilience and the Root-life of Attention in the Pursuit of Pleasure Equation

If one wished to model the Pursuit of Pleasure with the root-life of attention, then the exponent's transformation would be close to the Avoidance of Pain function and Sample Form II, save that the negative one coefficient adjacent to Time (the **x** variable) would not be used.

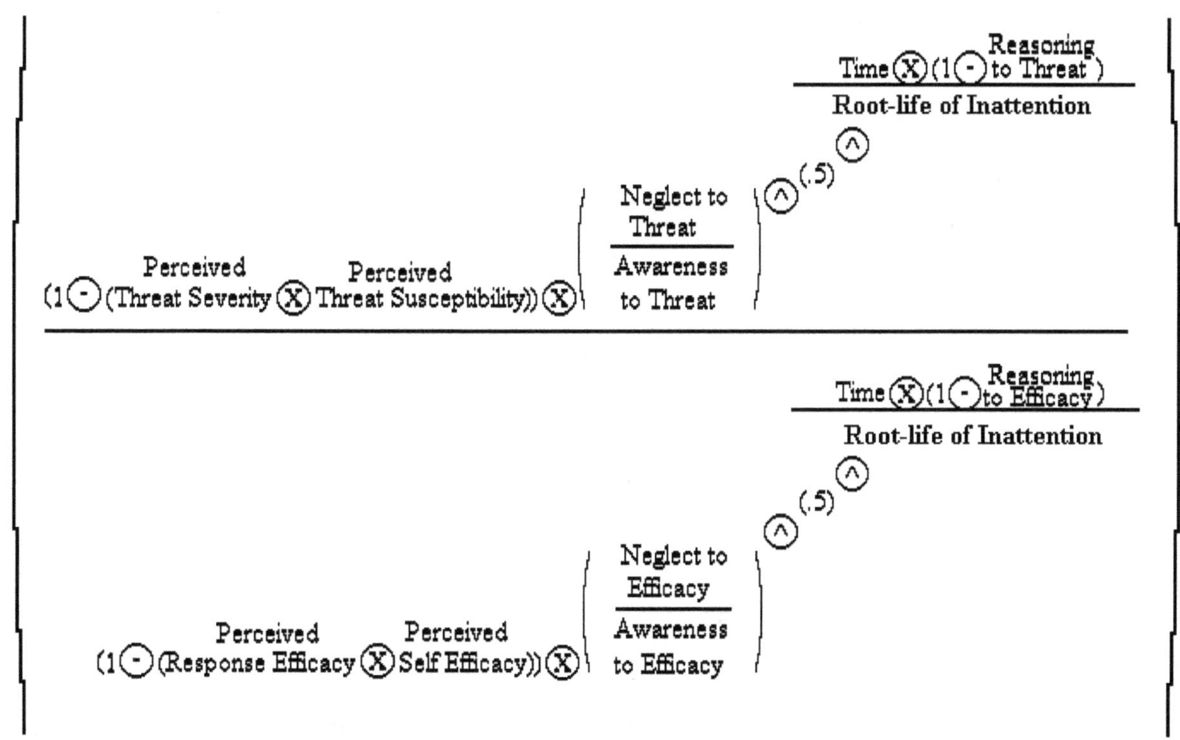

Figure 10.45 Valuation resilience: Exponent in the Pursuit of Pleasure Equation
The above depicts the exponent of the Pursuit of Pleasure equation when Inattention is modeled with valuation resilience. Instead of calculating the Half-life of Attention, a Root-life of Inattention is used. The (-1) coefficient is removed from its place beside Time and is not needed as it was in Sample Form II of Attention, as .5 five is now an exponent modifying the Inattention level.

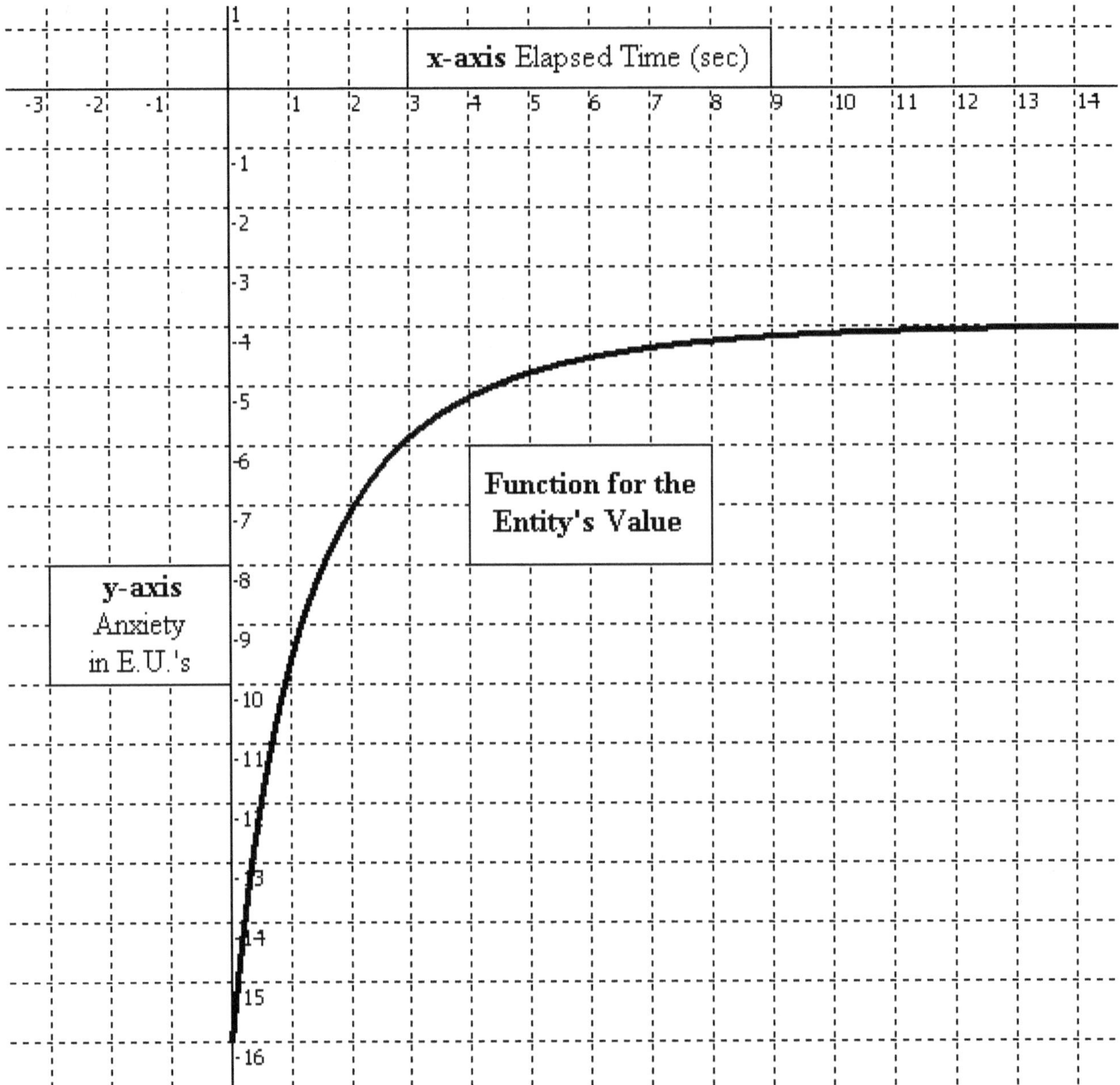

Figure 10.47 Valuation resilience is modeled in the Pursuit of Pleasure Function. Like figure 10.46, the Root-life of Inattention is set to 2 seconds. Inattention in the numerator equals two (Neglect : Awareness = 2/3 : 1/3). Inattention in the denominator equals one half (Neglect : Awareness = 1/3 : 2/3). Threat Susceptibility and Self Efficacy are set to .5, meaning a 50% likelihood of the threat occurring and a 50% likelihood the recommended behavior will be performed. However, Reasoning to Efficacy is set to zero in this instance, meaning that its suppressed level of Neglect is permitted to recoil back to one over time. All other variables equal one. A similar graph could be achieved while maintaining the parameters set for Awareness if the Inattention levels were 1 / 10 in the numerator and 4 / 10 in the denominator.

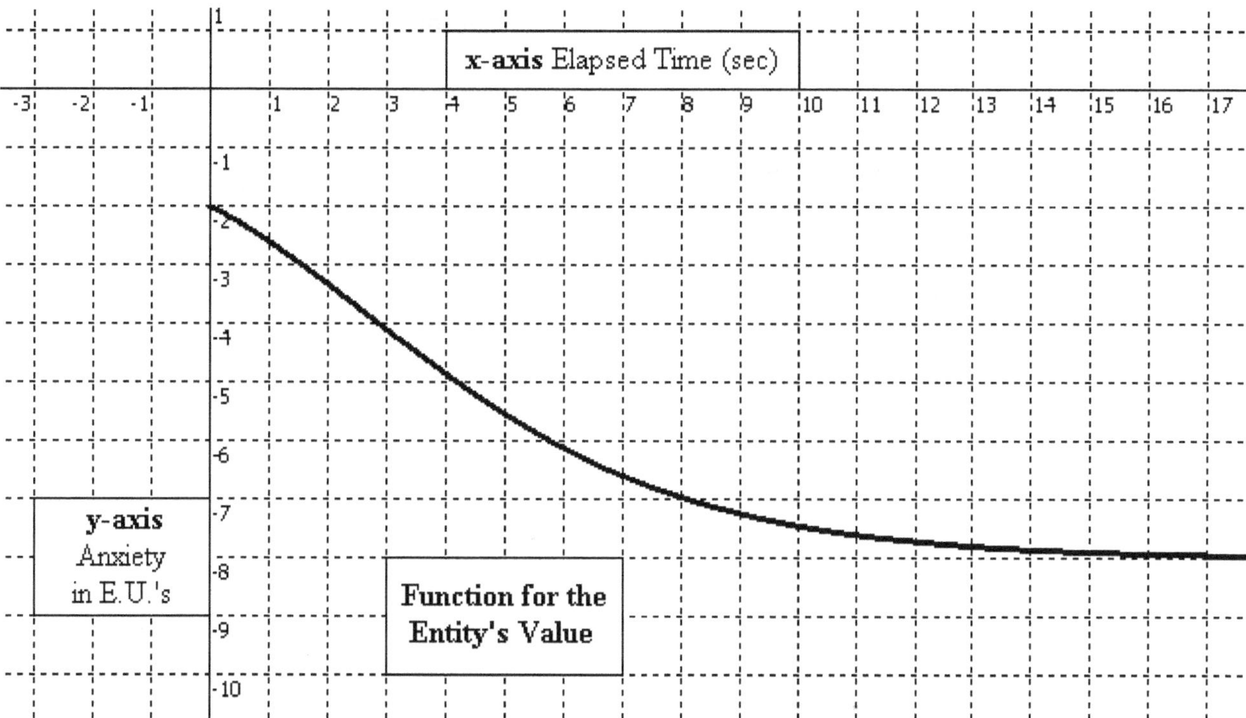

Figure 10.48 Valuation resilience is modeled in the Pursuit of Pleasure equation. Like figure 10.46 and figure 10.47, the Root-life of Inattention is set to 2 seconds. Threat Susceptibility and Self Efficacy are set to .5, meaning there is a 50% likelihood of the threat occurring and a 50% likelihood the recommended behavior will be performed. The Neglect to Awareness ratio in the numerator and the denominator is equal to 3 in both cases (Neglect : Awareness = .75 : .25). Reasoning to Efficacy equals zero in this instance, meaning that the Awareness to Neglect ratio is permitted to recoil from being suppressed back towards one. All other variables equal one. In this case, negative anxiety tapers off at approximately 8 emotional units and would only be able to decrease further if Awareness is reduced elsewhere in order to amplify it.

In sum, Sample Form III's ability to model how an unwanted valuation of an entity may force its way back into the forefront (e.g., from being suppressed in the unconscious) distinguishes it from Sample Form II of attention. Ultimately, it will depend on what one is looking to model. If one wished to use the equation to model dissociative disorders for example, Sample Form III of attention might prove more useful for explaining certain concepts than Sample Forms I and Form II of attention. Alternatively, if one were modeling something else, such as Attention Deficit Disorder, then Sample Form II of Attention might prove more useful than Sample Forms I and III of attention. Going further still, if one wished to model neural atrophy, such as in Alzheimer's disease, then Sample Form I of attention may prove more useful than forms II or III. Finally, if one wished to model Attention on a grand scale with the aid of powerful computing devices, the model of Attention without error, Sample Form IV, would be preferred to the models of Attention with error such as forms I and II.

Motivation Revisited: Organization, Sets, and Learning

Motivation energizes behavior that impels or "drives an organism to action" and one's motivational state can sometimes influence what is perceived, making certain items more or less likely to be recognized.[66] In perception, a perceptual defense "occurs whenever the recognition threshold for a stimulus is raised."[67] One such instance of a perceptual defense might occur where a subject is prevented from "seeing a taboo word," or reporting that the inappropriate word was presented to the subject by an experimenter.[68] On the other hand, "perceptual vigilance," or "perceptual sensitization," refers to instances where a recognition threshold is lowered.[69] An individual becomes hypersensitive to the perception of a specific stimuli. If, for example, someone is terrified of insects, the sight of a table with six legs may be sufficient to induce the perception of a bug and cause fright.

Figure 10.49 A six-legged table may resemble an insect enough to induce fear in someone who is afraid of bugs.

Accounting for Motivation

The equation can incorporate the lowering or the raising of a recognition threshold via the linking of multiple or fewer percepts and perceptual routes to an entity's existence. Hence, some percepts may be more likely to make it pass an attention filter because the number of links to a specific entity might occupy a greater percentage out of the total number of routes to entities. For instance, an entity with ten different perceptual pathways would be more likely to be triggered than an entity with only one. If the whole numbers from zero to nineteen are given as stimuli, and entity

A consists of odd natural numbers, then entity *A* has a 50% chance of being activated by a perceptual route from bottom-up processes. Entity *B* consists of even natural numbers. If entity *C* consists of everything else, then it only has one route leading to it and would have a five percent chance of being activated by a perceptual route if there is no motivational interference. Top-down, motivational influence in this case could occur if the individual decides to divert Attention away from even numbers, hence, those that end in 0, 2, 4, 6, or 8 beforehand. While this might prime the individual to respond to odd numbers more readily to identify instances of entity *A*, it would have the opposite effect on entity *C*. Though ending in 0, zero itself is not a natural number.

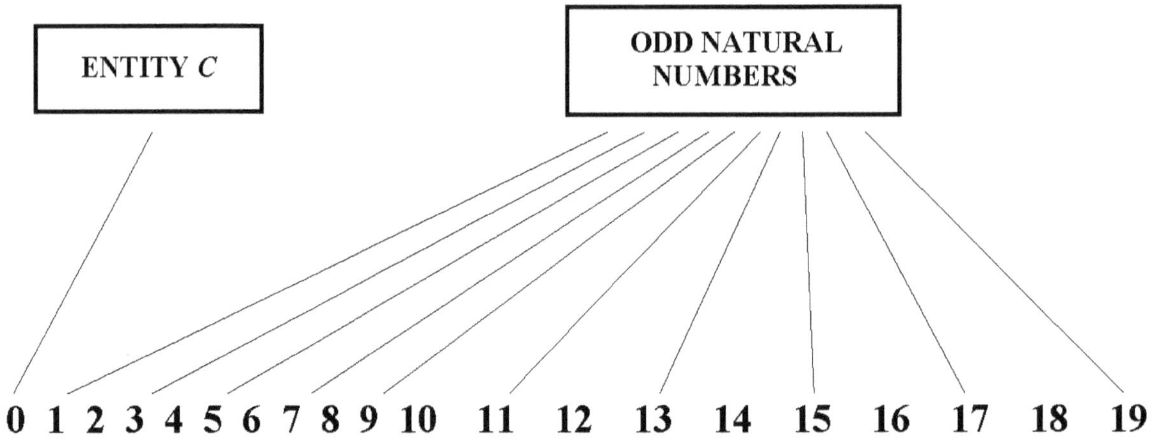

Figure 10.50 An entity with ten different perceptual pathways is more likely to be triggered than an entity with only one perceptual pathway if a total of twenty are possible (50% vs. 5% likelihood). The odd natural numbers are more likely to be activated than instances of *B* (natural even numbers) or *C* (everything else). If subjects in a study were asked to distinguish between the three types in a random trial, and they utilized a top-down process to focus Attention on numbers ending in 1, 3, 5, 7, and 9 and to exclude numbers ending in 0, 2, 4, 6, and 8, then they might be expected to respond more readily and accurately to instances of entity *A* or *B* than instances of entity *C*.

Organization

Organization, in the Gestalt sense, is the principle that the separate elements are not equally important in determining what is perceived, but rather it is the "integration of the whole that is deemed critical."[70] A few of these laws include "continuation," "similarity," "closure," "proximity," and "good shape or form."[71]

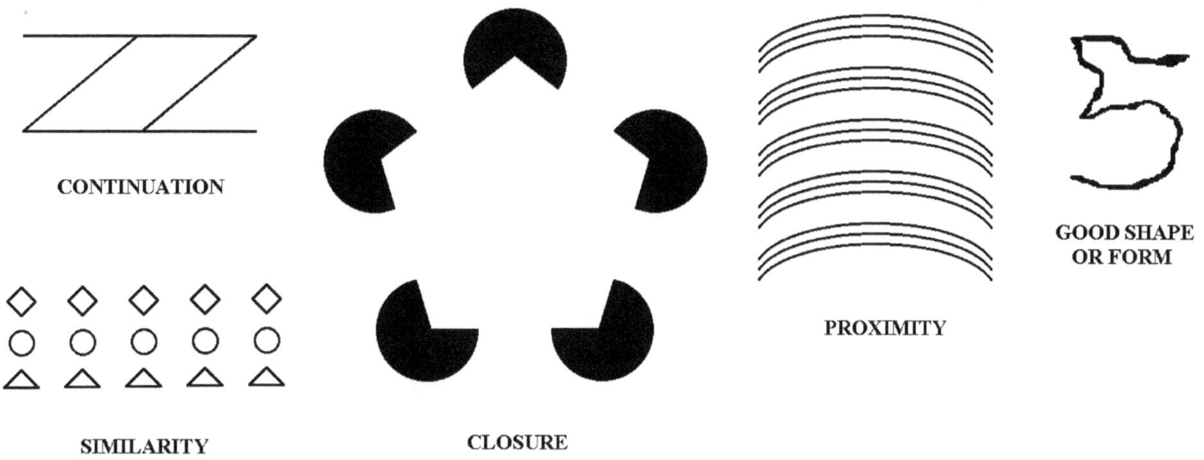

Figure 10.51 Images demonstrating a few Gestalt principles:

1) Continuation is "the tendency to perceive a line as maintaining its established direction."[72] One may be more likely to see the figure above (top left) as a diamond wedged between two parallel bars, as opposed to two Z's.
2) Similarity is when parts of a stimulus that are physically similar are "perceived as belonging together."[73] One tends to see the figure above (bottom left) as a row of five diamonds, five circles, and five triangles instead of seeing them as fifteen objects.

3) Closure is "the tendency to perceive incomplete objects as complete."[74] The white pentagon above (middle left), though incomplete, tends to be seen as a complete one.

4) Proximity is when "events or stimuli that are close to one another spatially or temporally are perceived as belonging together."[75] One may see the figure above (middle right) as five groups of three curved lines each instead of fifteen separate curved lines.

5) Good shape or form is the "tendency to perceive figures in their symmetric, uniform, and stable way."[76] Despite being deformed, the shape above (top right) is still recognizable as the number five.

Organization and Gestalt Principles

Gestalt principles necessarily concern the relationship between the part and a whole. People tend to look at the whole of something, the parts that make it up, or they jump back and forth. If the whole is ambiguous based upon the arrangement of the parts, then an individual might pick one over another depending on which principles are utilized. The two terms that will be used here to describe this process concern the macrocosm and the microcosm. Things that are macrocosmic generally concern universal concepts that can be applied to individual cases. People that utilize a macrocosmic approach to the world might say things like:

1) *I like to see the big picture.*
2) *I never pay enough attention to detail.*
3) *One-size-fits-all.*

They tend to look at the whole scenario and utilize inductive reasoning along with top-down processing. Things that are microcosmic generally concern smaller cases that can be applied to larger or more general principles. People that utilize a microcosmic approach to the world might say things like:

1) *I like to analyze things closely.*

2) *I am too in the middle of things.*

3) *I can't see the forest for the trees.*

They tend to look at the parts and utilize deductive reasoning along with bottom-up processing.

Accounting for Organization in the Equation

People that tend to look at the whole, when confronted with a stimulus, might have more perceptual routes devoted to the activation of a symbol or something that represents all of the entities rather than the individual entities themselves. These individual entities may, for instance, feed into the overarching concept. Conversely, people that tend to look at the parts, when confronted with a stimulus, might have more perceptual routes devoted to the activation of individual entities concerning the stimulus rather than a symbol that represents a unified conception of these entities (figure 10.52).

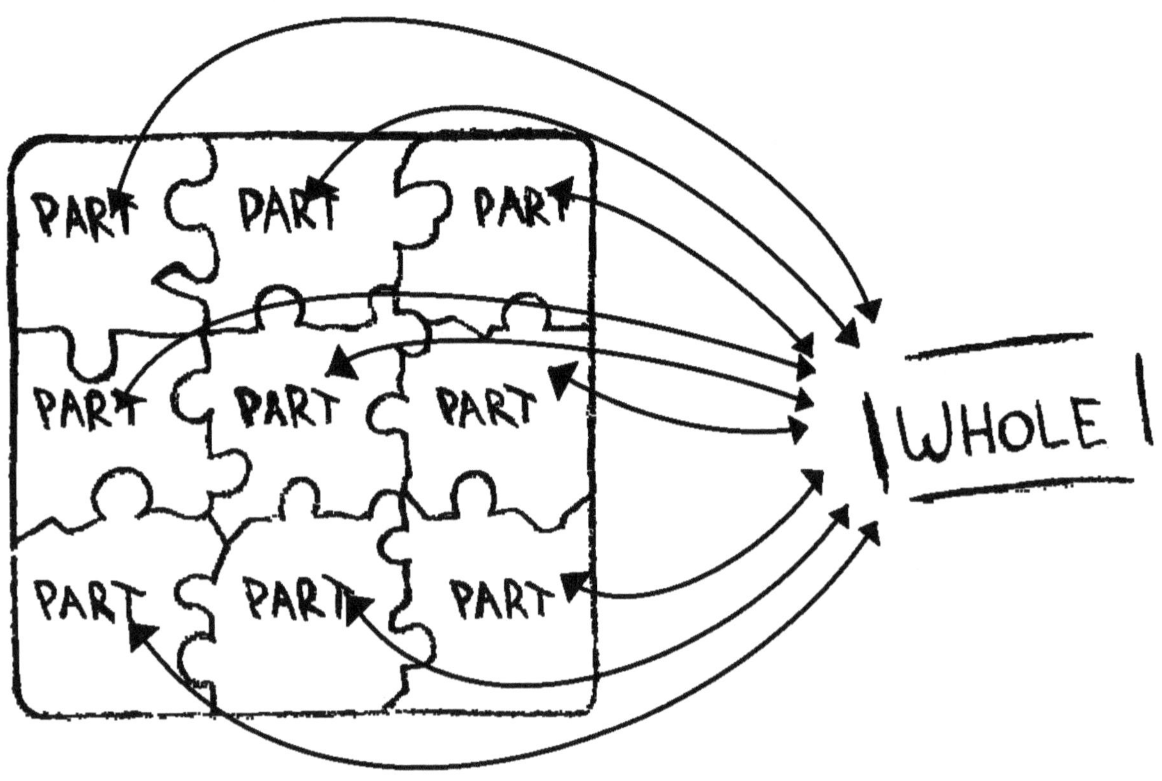

Figure 10.52 In the puzzle above, one's Attention may either be directed toward the entity of the whole or directed to one of the nine entities that make up the whole.

Sets

A set is an "aggregate or series of things sharing some defining property or properties such that they can be regarded collectively."[77] Readiness to perceive particular stimuli or classes of stimuli is denoted by the term "Einstellung."[78] One type of set may include stimuli that are all auditory and the presentation of them to an individual would prepare him or her for those stimuli to the exclusion of visual stimuli.[79] This bears much in common with the concept of priming. Priming is the presenting of an event or episode that "prepares a system for functioning."[80] Priming and perceptual sets prepare the system to act.

Sets and Priming

If part of a perceptual route to the cognizance of an entity is already active, then the activation of the remaining part of the route should not be expected to take longer than the activation of it in its entirety if it were completely at rest.

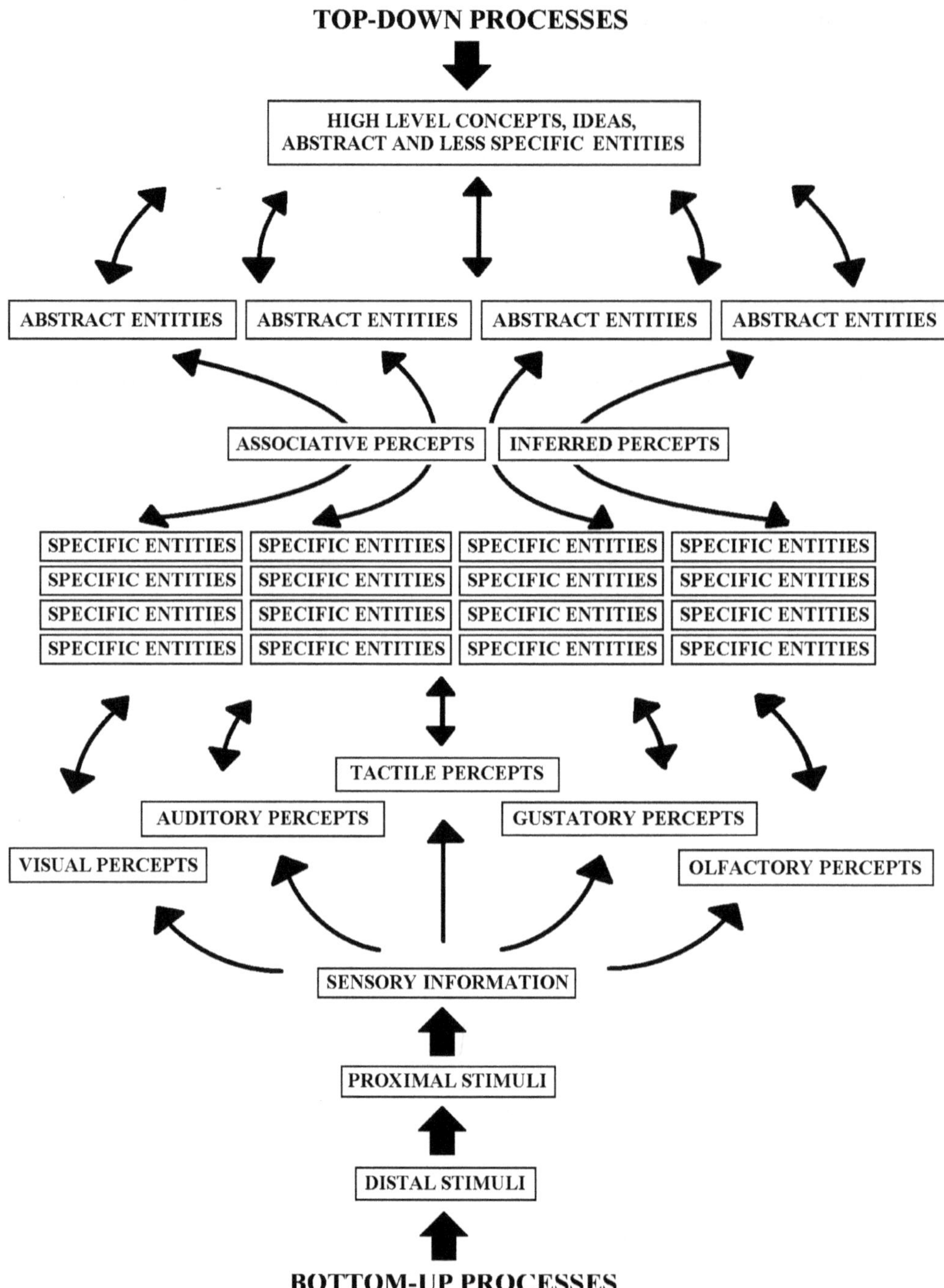

Figure 10.53 Bottom-up Processes, Top-down Processes, Sets, and Priming
If one perceptual pathway is already active from an individual being presented certain types of stimuli, for instance auditory, then the cognizance of new auditory stimuli may be facilitated if the previous experiences have lowered the activation threshold or readied certain neurons to fire. Conversely, previous experiences might be expected to raise the activation threshold, for instance, if fatigue or boredom (e.g., a diminishment in Attention) have resulted.

Learning

Learning is the "process of acquiring knowledge or the actual possession of such."[81] The learning of new information, in the equation, would be represented by the cognition and addition of new entities or new functions for valuations of entities. In the neurological model this would be new pathways or links established between one entity and other entities. In the Roll Cage Theory of Drives, it would be additional anxiety-mass. Learning may occur if additional perceptual pathways become associated with the same entity. For instance, if someone who has only seen snow in movies and pictures suddenly winds up in the middle of a blizzard, then the entity of snow gains a tactile perceptual pathway as the individual can now link the physical sensation of touch to snow along with the bellowing of the storm. If he or she also eats the snow, then gustatory and olfactory pathways are established as well. What was originally one pathway has suddenly become five.

Distortion

In nontechnical terms, distortion refers to any "twisting or contorting that alters the shape of something" into something other than what it was originally intended to represent.[82] This may be evidenced by systematic errors in the recollection of information.[83] These systematic errors can be precipitated by "noise," which generally refers to any "unwanted, interfering stimulus."[84]

Accounting for Distortion

Distortion can be represented in the equation by interference from percepts unrelated to the stimulus at hand. For example, this might occur if another entity's valuation is triggered when a separate entity is being considered by an individual. If bananas become associated with bad luck because something tragic always seems to happen after one eats a banana (e.g., losing one's wallet, dog running away, getting fired from a job, etc.), then the valuation of bananas may become distorted by its context in past experiences. As any number of abnormalities might occur from mistakenly associating one entity with another entity when high level concepts, inferred percepts, and associative percepts are entered into the mix, using the equation to model distortion might only prove speculative unless the inferences and associations could also be verified by the individual. A competent independent observer, over time, might be able to identify them as well.

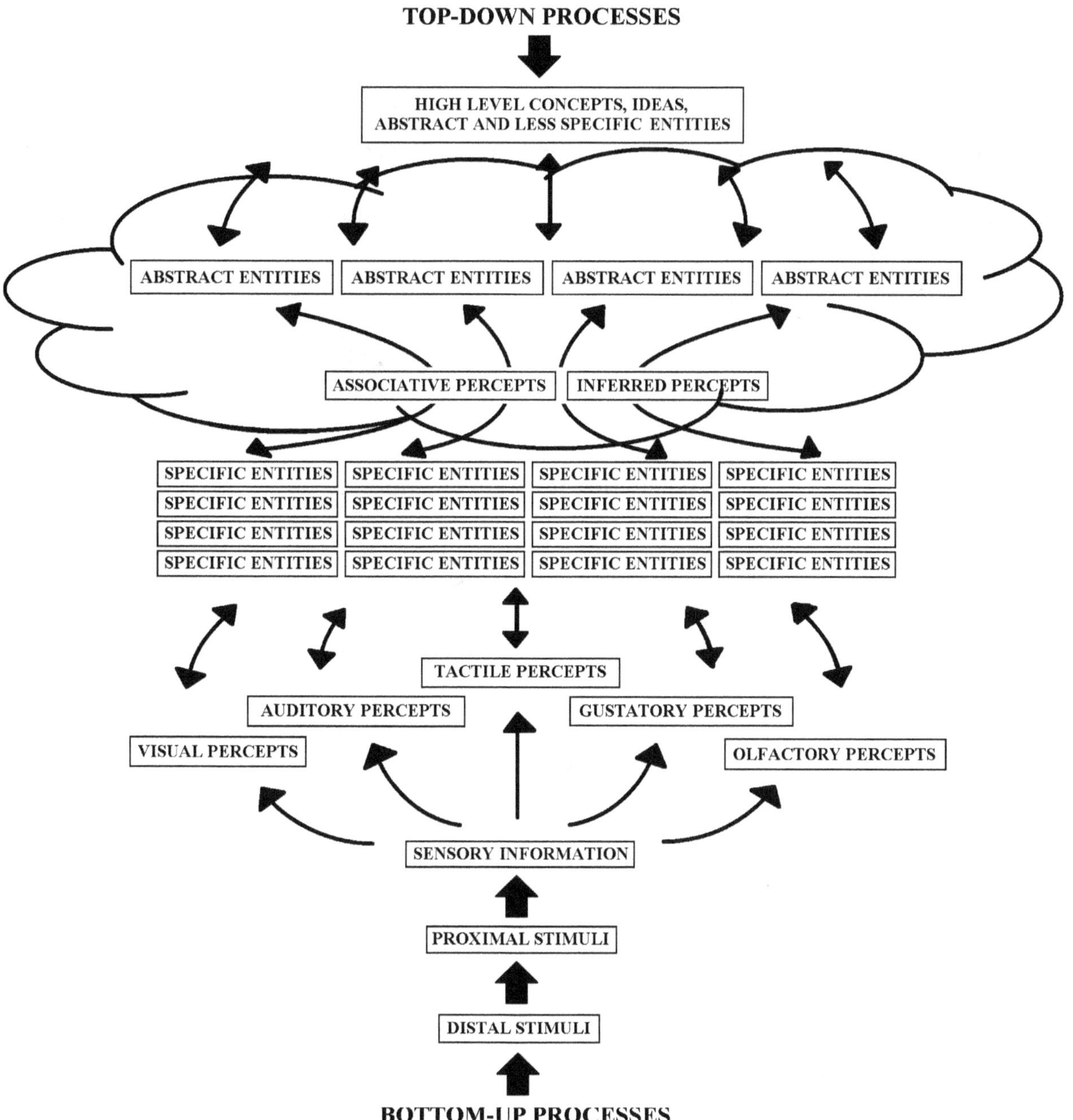

Figure 10.54 The above is a model of where distortion might be expected to originate (inferred and associative percepts in the cloudy region).

Hallucination

A hallucination is "a perceptual experience with all the compelling subjective properties of a real sensory impression but without the normal physical stimulus for that modality."[85] One may hear voices where there are none or see images when they are not present. Hallucinations may be the result of psychotic disorders, but may also be induced by drug use.

Accounting for Hallucination

Like distortion, hallucination would be represented by interference, albeit at a much stronger level. This would be one capable of mimicking the sensation of the stimulus itself and any relevant percepts that would follow. Similar to distortion, successfully using the equation to model hallucinations would require knowledge of what sensory percepts the individual is falsely perceiving. Moreover, any abnormalities in the collection of information would likely be found at the perceptual level.

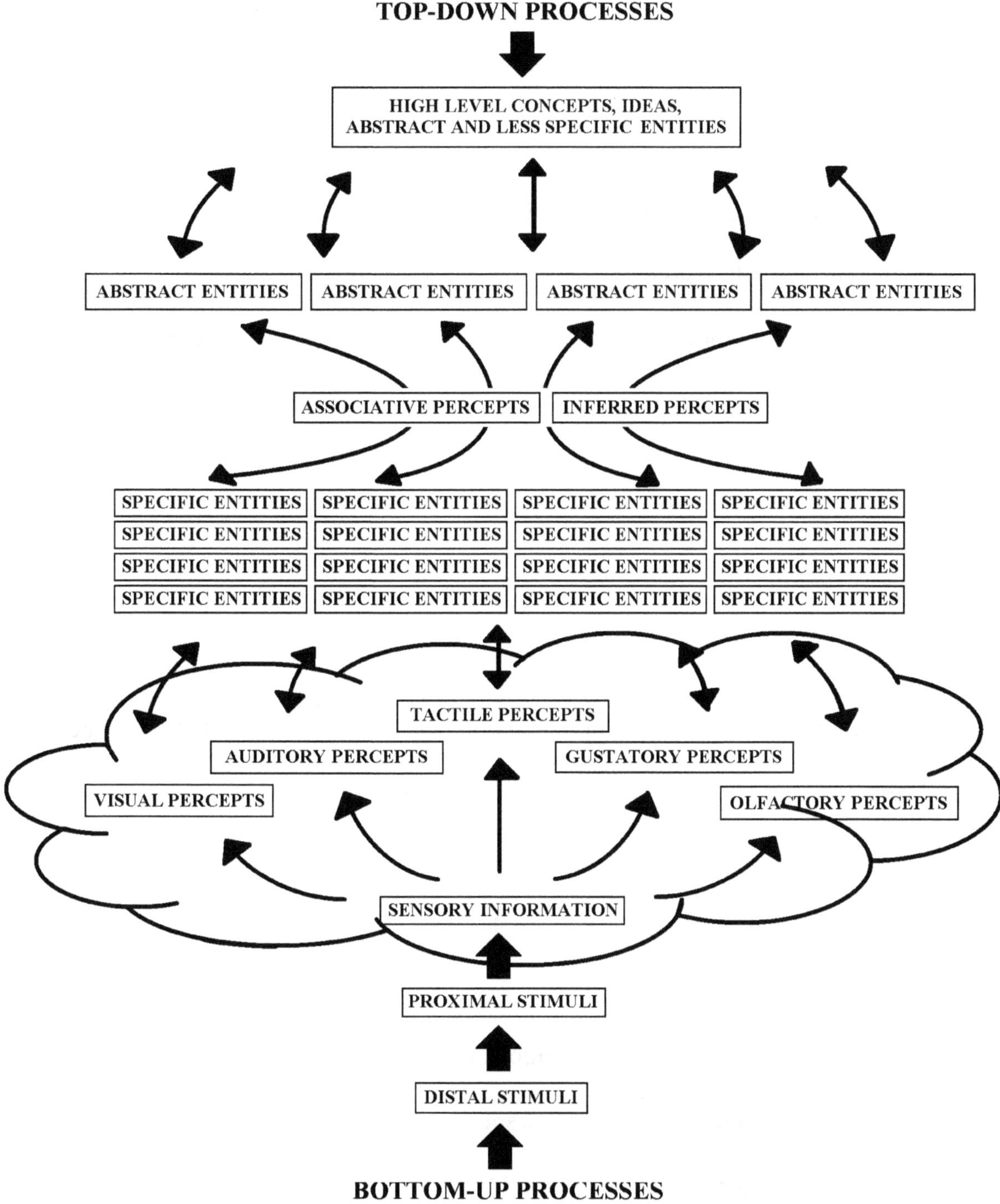

Figure 10.55 The above is a model of where hallucinations might be expected to originate (cloudy region), which would be the sensory percepts leading up to the entity's existence.

Illusion

An illusion is "any stimulus situation in which that which is perceived cannot be predicted, prima facie, by a simple analysis of the physical stimulus."[86] Unlike a hallucination, the stimulus is actually present, but it is misinterpreted. A possible explanation for this is that bottom-up processing may interact with top-down processing and what is expected does not necessarily correlate with what is physically perceived. For instance, the white pentagon from Figure 10.51 (middle left) is an illusory contour.

Accounting for Illusion

Illusion could be understood as an interaction between higher-level thought processes and lower-level sensory perception. Conflict between top-down processes and bottom-up processes might lead to ambiguity and mixed interpretations of particular stimuli (figure 10.52). An expectation from top-level processes, for example, may not correspond to sensory information, leading an individual to question both.

In sum, with the incorporation of Attention and Reasoning, most of the essential features of equations have been described. However, before specific defense mechanisms can be brought under inspection, a number of other ideas must be addressed first. Among them are two additional categories of emotion, biological rhythms, and the expansion of the equation into multiple dimensions.

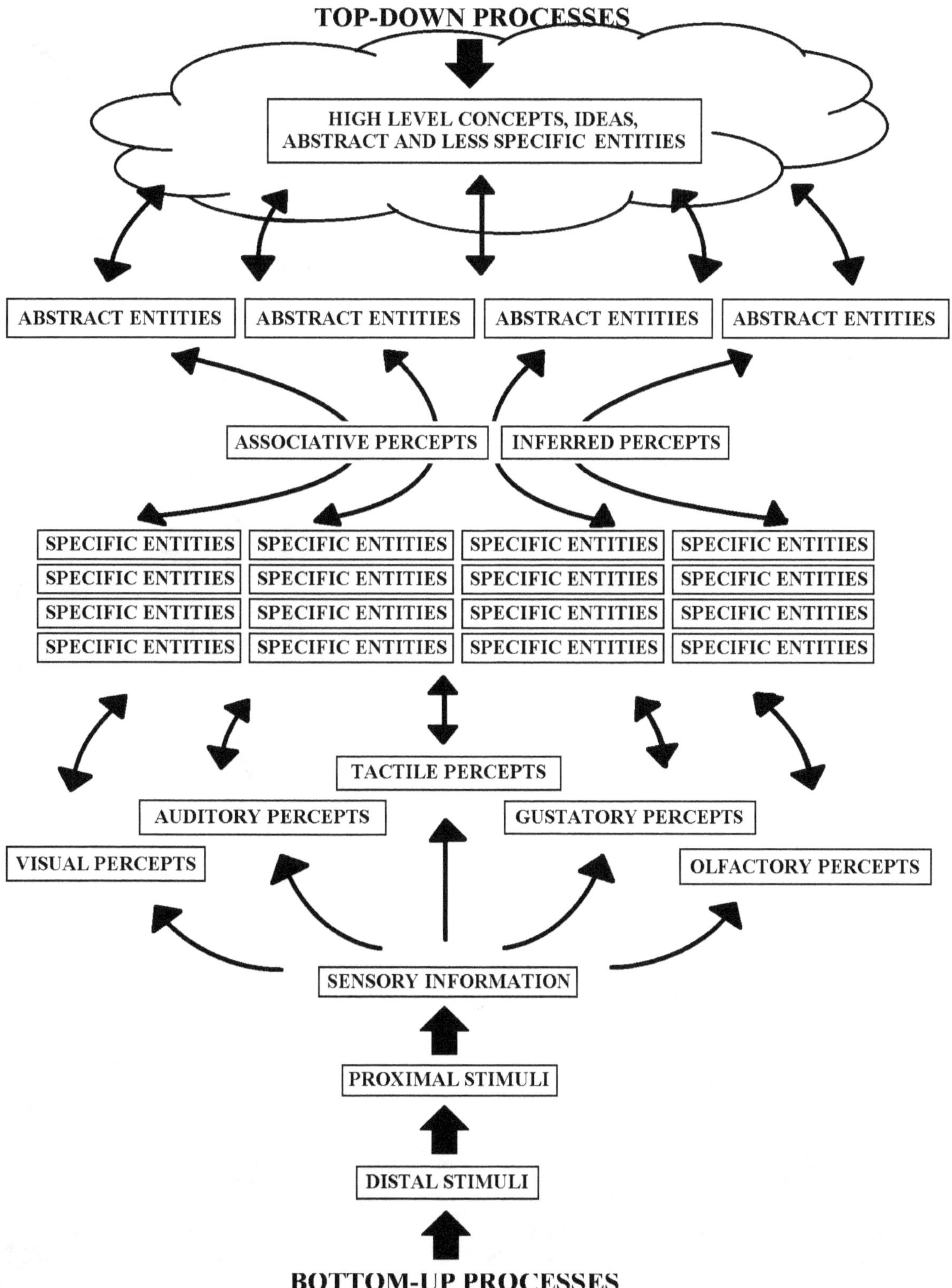

Figure 10.56 The cloudy region above is where illusions might be expected to originate.

YOU SHOULD UNDERSTAND THIS.
CHAPTER TEN
CONCEPT CHECK
VOCABULARY
Associative Percept
Attenuation Theory
Attention
Attentional Decay
Decay Theory
Deductive Reasoning
Executive Functioning
Filter Theory
Forgetting
Half-life of Attention with Error
Half-life of Attention without error
Inductive Reasoning
Inferred Percept
Multi-tasking, Neglect
Percept (Five Senses)
Perception
Reasoning
Tabula Rasa

Notes

1. Wilde, Oscar (2003). *The Picture of Dorian Gray.* New York: Barnes and Nobles Books. Print, p. 30. (Original work published 1890).

2. Reber, Arthur S., Rhianon Allen, and Emily S. Reber. *Penguin Dictionary of Psychology*. London. Penguin Books, 2009. Print, p. 10.

3. Reber, Arthur S., Rhianon Allen, and Emily S. Reber. *Penguin Dictionary of Psychology*. London. Penguin Books, 2009. Print, p. 10.

4. Reber, Arthur S., Rhianon Allen, and Emily S. Reber. *Penguin Dictionary of Psychology*. London. Penguin Books, 2009. Print, p. 868.

5. Reber, Arthur S., Rhianon Allen, and Emily S. Reber. *Penguin Dictionary of Psychology*. London. Penguin Books, 2009. Print, p. 399.

6. Reber, Arthur S., Rhianon Allen, and Emily S. Reber. *Penguin Dictionary of Psychology*. London. Penguin Books, 2009. Print, p. 283.

7. Reber, Arthur S., Rhianon Allen, and Emily S. Reber. *Penguin Dictionary of Psychology*. London. Penguin Books, 2009. Print, p. 566.

8. Reber, Arthur S., Rhianon Allen, and Emily S. Reber. *Penguin Dictionary of Psychology*. London. Penguin Books, 2009. Print, p. 566.

9. Reber, Arthur S., Rhianon Allen, and Emily S. Reber. *Penguin Dictionary of Psychology*. London. Penguin Books, 2009. Print, p. 74.

10. Reber, Arthur S., Rhianon Allen, and Emily S. Reber. *Penguin Dictionary of Psychology*. London. Penguin Books, 2009. Print, p. 527.

11. Reber, Arthur S., Rhianon Allen, and Emily S. Reber. *Penguin Dictionary of Psychology*. London. Penguin Books, 2009. Print, p. 800.

12. Reber, Arthur S., Rhianon Allen, and Emily S. Reber. *Penguin Dictionary of Psychology*. London. Penguin Books, 2009. Print, p. 411.

13. Reber, Arthur S., Rhianon Allen, and Emily S. Reber. *Penguin Dictionary of Psychology*. London. Penguin Books, 2009. Print, p. 338.

14. Reber, Arthur S., Rhianon Allen, and Emily S. Reber. *Penguin Dictionary of Psychology*. London. Penguin Books, 2009. Print, p. 863.

15. Reber, Arthur S., Rhianon Allen, and Emily S. Reber. *Penguin Dictionary of Psychology*. London. Penguin Books, 2009. Print, p. 566.

16. Reber, Arthur S., Rhianon Allen, and Emily S. Reber. *Penguin Dictionary of Psychology*. London. Penguin Books, 2009. Print, p. 566.

17. Reber, Arthur S., Rhianon Allen, and Emily S. Reber. *Penguin Dictionary of Psychology*. London. Penguin Books, 2009. Print, p. 800.

18. Reber, Arthur S., Rhianon Allen, and Emily S. Reber. *Penguin Dictionary of Psychology*. London. Penguin Books, 2009. Print, p. 107.

19. Reber, Arthur S., Rhianon Allen, and Emily S. Reber. *Penguin Dictionary of Psychology*. London. Penguin Books, 2009. Print, p. 378.

20. Reber, Arthur S., Rhianon Allen, and Emily S. Reber. *Penguin Dictionary of Psychology*. London. Penguin Books, 2009. Print, p. 823.

21. Reber, Arthur S., Rhianon Allen, and Emily S. Reber. *Penguin Dictionary of Psychology*. London. Penguin Books, 2009. Print, p. 193.

22. Reber, Arthur S., Rhianon Allen, and Emily S. Reber. *Penguin Dictionary of Psychology*. London. Penguin Books, 2009. Print, p. 755.

23. Reber, Arthur S., Rhianon Allen, and Emily S. Reber. *Penguin Dictionary of Psychology*. London. Penguin Books, 2009. Print, p. 599.

24. Reber, Arthur S., Rhianon Allen, and Emily S. Reber. *Penguin Dictionary of Psychology*. London. Penguin Books, 2009. Print, p. 599.

25. Reber, Arthur S., Rhianon Allen, and Emily S. Reber. *Penguin Dictionary of Psychology*. London. Penguin Books, 2009. Print, p. 657.

26. Reber, Arthur S., Rhianon Allen, and Emily S. Reber. *Penguin Dictionary of Psychology*. London. Penguin Books, 2009. Print, p. 566-567.

27. Reber, Arthur S., Rhianon Allen, and Emily S. Reber. *Penguin Dictionary of Psychology*. London. Penguin Books, 2009. Print, p. 566.

28. Reber, Arthur S., Rhianon Allen, and Emily S. Reber. *Penguin Dictionary of Psychology*. London. Penguin Books, 2009. Print, p. 566.

29. Reber, Arthur S., Rhianon Allen, and Emily S. Reber. *Penguin Dictionary of Psychology*. London. Penguin Books, 2009. Print, p. 567.

30. Reber, Arthur S., Rhianon Allen, and Emily S. Reber. *Penguin Dictionary of Psychology*. London. Penguin Books, 2009. Print, p. 567.

31. Reber, Arthur S., Rhianon Allen, and Emily S. Reber. *Penguin Dictionary of Psychology*. London. Penguin Books, 2009. Print, p. 567.

32. Reber, Arthur S., Rhianon Allen, and Emily S. Reber. *Penguin Dictionary of Psychology*. London. Penguin Books, 2009. Print, p. 567.

33. Reber, Arthur S., Rhianon Allen, and Emily S. Reber. *Penguin Dictionary of Psychology*. London. Penguin Books, 2009. Print, p. 567.

34. Reber, Arthur S., Rhianon Allen, and Emily S. Reber. *Penguin Dictionary of Psychology*. London. Penguin Books, 2009. Print, p. 567.

35. Reber, Arthur S., Rhianon Allen, and Emily S. Reber. *Penguin Dictionary of Psychology*. London. Penguin Books, 2009. Print, p. 567.

36. Reber, Arthur S., Rhianon Allen, and Emily S. Reber. *Penguin Dictionary of Psychology*. London. Penguin Books, 2009. Print, p. 567.

37. Reber, Arthur S., Rhianon Allen, and Emily S. Reber. *Penguin Dictionary of Psychology*. London. Penguin Books, 2009. Print, p. 567.

38. Reber, Arthur S., Rhianon Allen, and Emily S. Reber. *Penguin Dictionary of Psychology*. London. Penguin Books, 2009. Print, p. 69.

39. Reber, Arthur S., Rhianon Allen, and Emily S. Reber. *Penguin Dictionary of Psychology*. London. Penguin Books, 2009. Print, p. 69.

40. Reber, Arthur S., Rhianon Allen, and Emily S. Reber. *Penguin Dictionary of Psychology*. London. Penguin Books, 2009. Print, p. 298.

41. Reber, Arthur S., Rhianon Allen, and Emily S. Reber. *Penguin Dictionary of Psychology*. London. Penguin Books, 2009. Print, p. 298.

42. Reber, Arthur S., Rhianon Allen, and Emily S. Reber. *Penguin Dictionary of Psychology*. London. Penguin Books, 2009. Print, p. 806.

43. Reber, Arthur S., Rhianon Allen, and Emily S. Reber. *Penguin Dictionary of Psychology*. London. Penguin Books, 2009. Print, p. 806.

44. Reber, Arthur S., Rhianon Allen, and Emily S. Reber. *Penguin Dictionary of Psychology*. London. Penguin Books, 2009. Print, p. 299.

45. Reber, Arthur S., Rhianon Allen, and Emily S. Reber. *Penguin Dictionary of Psychology*. London. Penguin Books, 2009. Print, p. 299.

46. Reber, Arthur S., Rhianon Allen, and Emily S. Reber. *Penguin Dictionary of Psychology*. London. Penguin Books, 2009. Print, p. 32.

47. Reber, Arthur S., Rhianon Allen, and Emily S. Reber. *Penguin Dictionary of Psychology*. London. Penguin Books, 2009. Print, p. 32.

48. Reber, Arthur S., Rhianon Allen, and Emily S. Reber. *Penguin Dictionary of Psychology*. London. Penguin Books, 2009. Print, p. 501.

49. Reber, Arthur S., Rhianon Allen, and Emily S. Reber. *Penguin Dictionary of Psychology*. London. Penguin Books, 2009. Print, p. 501.

50. Reber, Arthur S., Rhianon Allen, and Emily S. Reber. *Penguin Dictionary of Psychology*. London. Penguin Books, 2009. Print, p. 104.

51. Reber, Arthur S., Rhianon Allen, and Emily S. Reber. *Penguin Dictionary of Psychology*. London. Penguin Books, 2009. Print, p. 711.

52. Reber, Arthur S., Rhianon Allen, and Emily S. Reber. *Penguin Dictionary of Psychology*. London. Penguin Books, 2009. Print, p. 102.

53. Reber, Arthur S., Rhianon Allen, and Emily S. Reber. *Penguin Dictionary of Psychology*. London. Penguin Books, 2009. Print, p. 104.

54. Reber, Arthur S., Rhianon Allen, and Emily S. Reber. *Penguin Dictionary of Psychology*. London. Penguin Books, 2009. Print, p. 375.

55. Reber, Arthur S., Rhianon Allen, and Emily S. Reber. *Penguin Dictionary of Psychology*. London. Penguin Books, 2009. Print, p. 375.

56. Reber, Arthur S., Rhianon Allen, and Emily S. Reber. *Penguin Dictionary of Psychology*. London. Penguin Books, 2009. Print, p. 461.

57. Reber, Arthur S., Rhianon Allen, and Emily S. Reber. *Penguin Dictionary of Psychology*. London. Penguin Books, 2009. Print, p. 459.

58. Reber, Arthur S., Rhianon Allen, and Emily S. Reber. *Penguin Dictionary of Psychology*. London. Penguin Books, 2009. Print, p. 192.

59. Reber, Arthur S., Rhianon Allen, and Emily S. Reber. *Penguin Dictionary of Psychology*. London. Penguin Books, 2009. Print, p. 208.

60. Reber, Arthur S., Rhianon Allen, and Emily S. Reber. *Penguin Dictionary of Psychology*. London. Penguin Books, 2009. Print, p. 208.

61. Reber, Arthur S., Rhianon Allen, and Emily S. Reber. *Penguin Dictionary of Psychology*. London. Penguin Books, 2009. Print, p. 271.

62. Reber, Arthur S., Rhianon Allen, and Emily S. Reber. *Penguin Dictionary of Psychology*. London. Penguin Books, 2009. Print, p. 658.

63. Reber, Arthur S., Rhianon Allen, and Emily S. Reber. *Penguin Dictionary of Psychology*. London. Penguin Books, 2009. Print, p. 658.

64. Reber, Arthur S., Rhianon Allen, and Emily S. Reber. *Penguin Dictionary of Psychology*. London. Penguin Books, 2009. Print, p. 658.

65. Reber, Arthur S., Rhianon Allen, and Emily S. Reber. *Penguin Dictionary of Psychology*. London. Penguin Books, 2009. Print, p. 802.

66. Reber, Arthur S., Rhianon Allen, and Emily S. Reber. *Penguin Dictionary of Psychology*. London. Penguin Books, 2009. Print, p. 487.

67. Reber, Arthur S., Rhianon Allen, and Emily S. Reber. *Penguin Dictionary of Psychology*. London. Penguin Books, 2009. Print, p. 586.

68. Reber, Arthur S., Rhianon Allen, and Emily S. Reber. *Penguin Dictionary of Psychology*. London. Penguin Books, 2009. Print, p. 568.

69. Reber, Arthur S., Rhianon Allen, and Emily S. Reber. *Penguin Dictionary of Psychology*. London. Penguin Books, 2009. Print, p. 568.

70. Reber, Arthur S., Rhianon Allen, and Emily S. Reber. *Penguin Dictionary of Psychology*. London. Penguin Books, 2009. Print, p. 540.

71. Reber, Arthur S., Rhianon Allen, and Emily S. Reber. *Penguin Dictionary of Psychology*. London. Penguin Books, 2009. Print, p. 326.

72. Reber, Arthur S., Rhianon Allen, and Emily S. Reber. *Penguin Dictionary of Psychology*. London. Penguin Books, 2009. Print, p. 326.

73. Reber, Arthur S., Rhianon Allen, and Emily S. Reber. *Penguin Dictionary of Psychology*. London. Penguin Books, 2009. Print, p. 326.

74. Reber, Arthur S., Rhianon Allen, and Emily S. Reber. *Penguin Dictionary of Psychology*. London. Penguin Books, 2009. Print, p. 326.

75. Reber, Arthur S., Rhianon Allen, and Emily S. Reber. *Penguin Dictionary of Psychology*. London. Penguin Books, 2009. Print, p. 326.

76. Reber, Arthur S., Rhianon Allen, and Emily S. Reber. *Penguin Dictionary of Psychology*. London. Penguin Books, 2009. Print, p. 326.

77. Reber, Arthur S., Rhianon Allen, and Emily S. Reber. *Penguin Dictionary of Psychology*. London. Penguin Books, 2009. Print, p. 730.

78. Reber, Arthur S., Rhianon Allen, and Emily S. Reber. *Penguin Dictionary of Psychology*. London. Penguin Books, 2009. Print, p. 251.

79. Reber, Arthur S., Rhianon Allen, and Emily S. Reber. *Penguin Dictionary of Psychology*. London. Penguin Books, 2009. Print, p. 730.

80. Reber, Arthur S., Rhianon Allen, and Emily S. Reber. *Penguin Dictionary of Psychology*. London. Penguin Books, 2009. Print, p. 614.

81. Reber, Arthur S., Rhianon Allen, and Emily S. Reber. *Penguin Dictionary of Psychology*. London. Penguin Books, 2009. Print, p. 422.

82. Reber, Arthur S., Rhianon Allen, and Emily S. Reber. *Penguin Dictionary of Psychology*. London. Penguin Books, 2009. Print, p. 227.

83. Reber, Arthur S., Rhianon Allen, and Emily S. Reber. *Penguin Dictionary of Psychology*. London. Penguin Books, 2009. Print, p. 227.

84. Reber, Arthur S., Rhianon Allen, and Emily S. Reber. *Penguin Dictionary of Psychology*. London. Penguin Books, 2009. Print, p. 512.

85. Reber, Arthur S., Rhianon Allen, and Emily S. Reber. *Penguin Dictionary of Psychology*. London. Penguin Books, 2009. Print, p. 340.

86. Reber, Arthur S., Rhianon Allen, and Emily S. Reber. *Penguin Dictionary of Psychology*. London. Penguin Books, 2009. Print, p. 369.

IN A FLATLAND SOMEWHERE

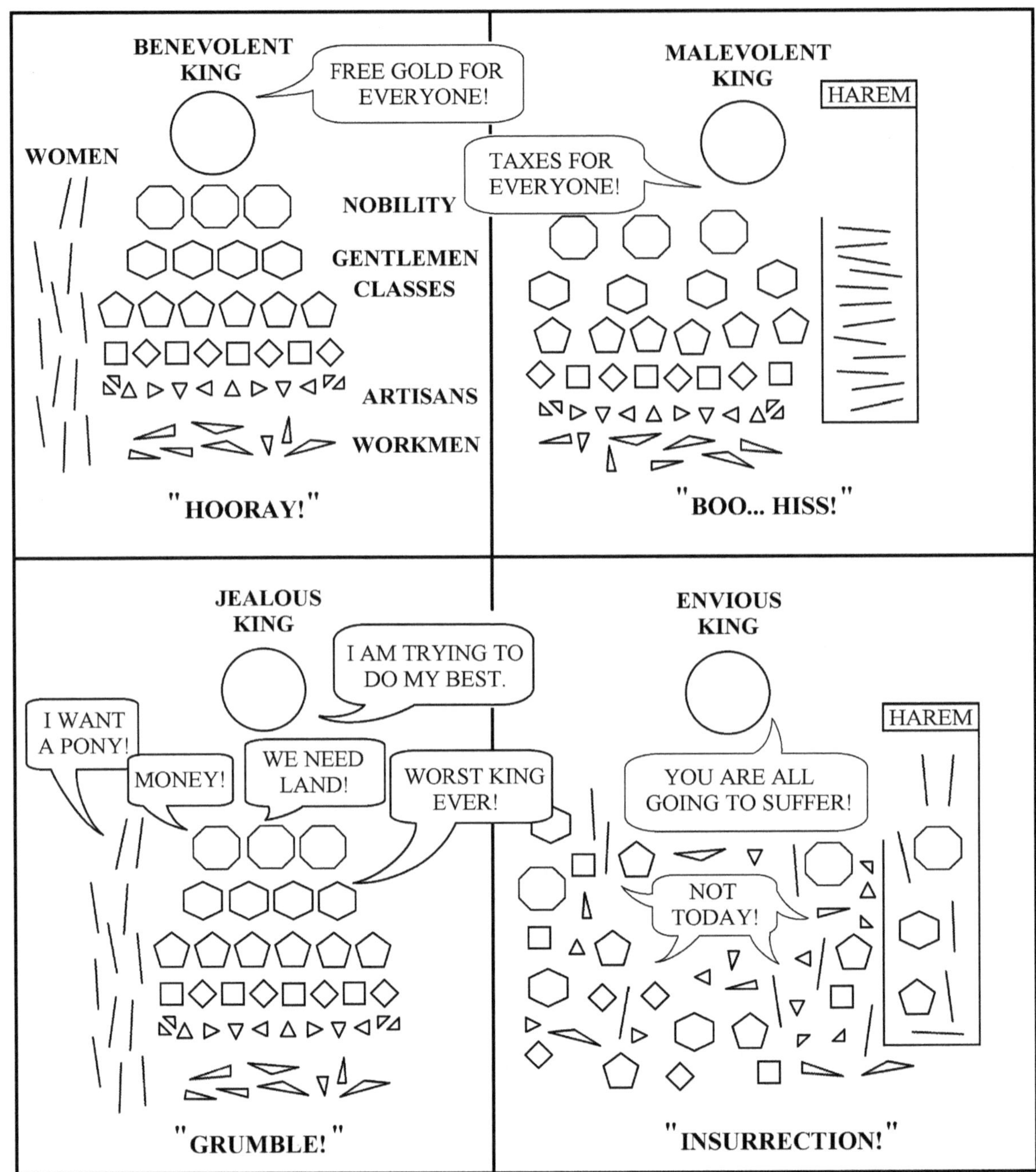

"To be good is to be in harmony with one's self . . . Discord is to be forced to be in harmony with others." - Oscar Wilde[1]

CHAPTER ELEVEN

Category Three Emotions (Compound Interactive)

Front-loading: Benevolence, Category III Emotions: Compound Interactive Emotions, Envy, Indulgent Type, Jealousy, Malevolence, Primary Emotions, Protective Type, Secondary Emotions

It can be observed that the equations from figures 1.1 and 1.2 have been addressed in their entirety, but not everything has yet been considered. A return to Plutchik's model will reveal some useful things to take into account. In Plutchik's Psychoevolutionary theory, on one level emotions are arrange in terms of their least and most intense manifestation.[2] On a second level in Plutchik's model, emotions vary in terms of their similarity.[3] On a third level, some emotions are considered to be the opposite of others, and are ordered along their polarity to another emotion.[4] Finally, some emotions are considered to be "primary emotions," while others are considered to be mixed or "secondary emotions" because they are derived from combinations of the primary emotions.[5] Although the emotions in the equations will not be organized in a fashion similar to Plutchik's, these principles will be accounted for into a third category of emotions are considered. They are the Category III, or Compound Interactive Emotions.

Category Three Emotions (Compound Interactive)

Accounting for Plutchik's Grouping Scheme:

Manifestations of Intensity

In the equation, manifestations of intensity can be accounted for under Category I or Intra-Personal Emotions depending on the slope of a function and its acceleration. For instance, whereas sadness might be indicated by a positive slope, fear might be indicated by a slope that is initially zero but begins to increase before tapering off. Grief, likely being a more intense form of sadness, might have a slope that is positive, accelerating, and approaching a vertical asymptote. Anxiety or negative anxiety invested over a given amount of time could be found using integration to uncover the area between curve and x-axis. On the other side, whereas happiness might be indicated by a slope that is negative, courage might be indicated by a slope that is initially zero and then accelerates negatively before tapering off. Euphoria, likely being more intense than happiness, might be indicated by a negative slope that approaches a vertical asymptote.

Category Three Emotions (Compound Interactive)

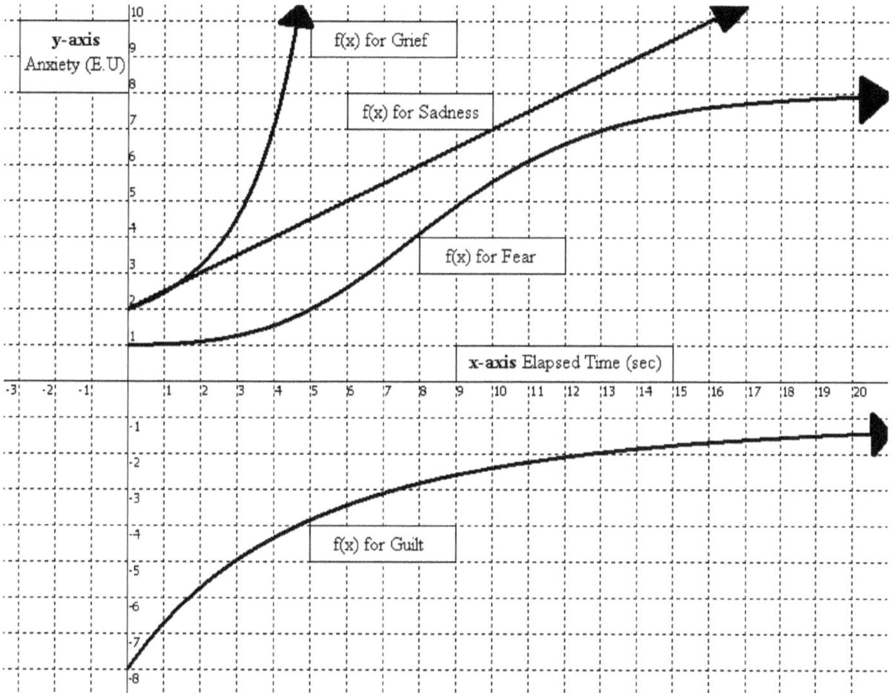

The Derivative, f ' (x) = (Magnitude of an emotion)

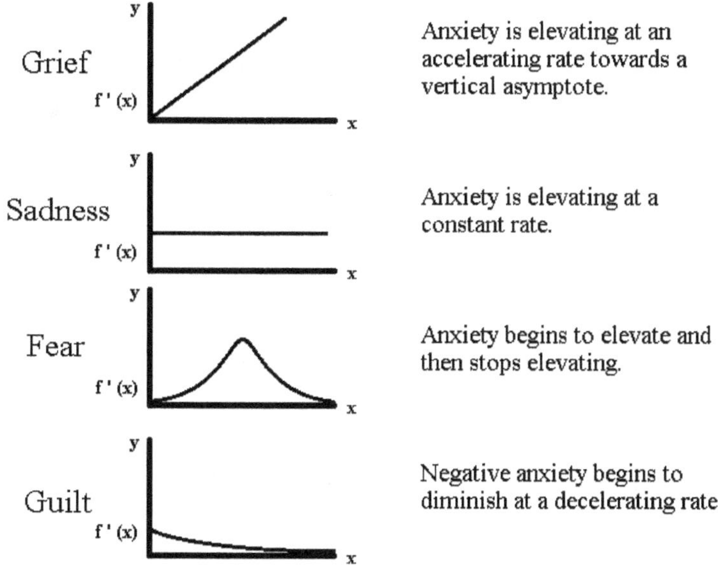

Figure 11.1 Category I Emotions are depicted. Valuations that might be expected to lead to grief, sadness, fear, and guilt are shown on the Cartesian plane above. The slope of the functions, represented by f '(x), would be indicative of the type of emotion involved.

Category Three Emotions (Compound Interactive)

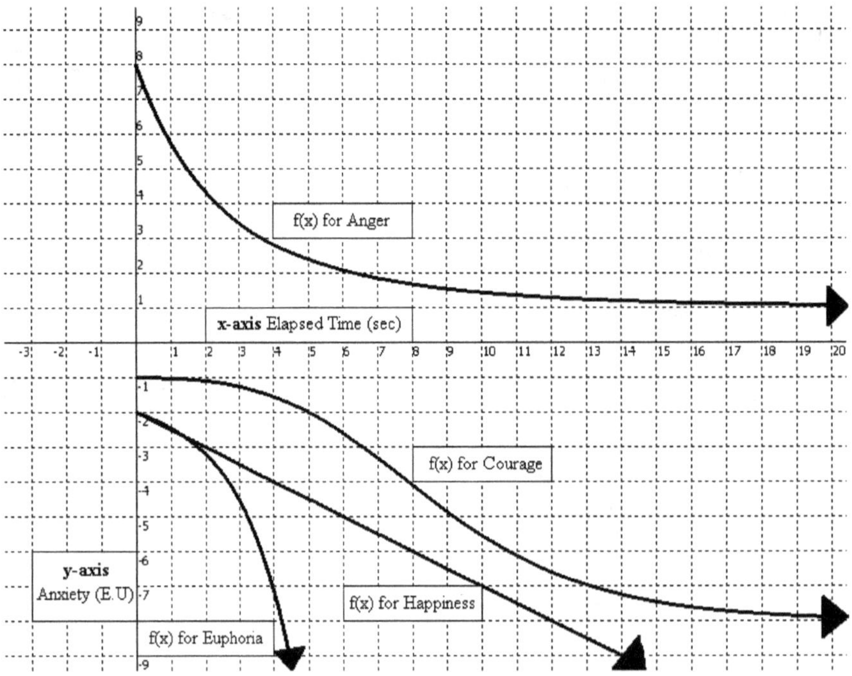

The Derivative, f ' (x) = (Magnitude of an emotion)

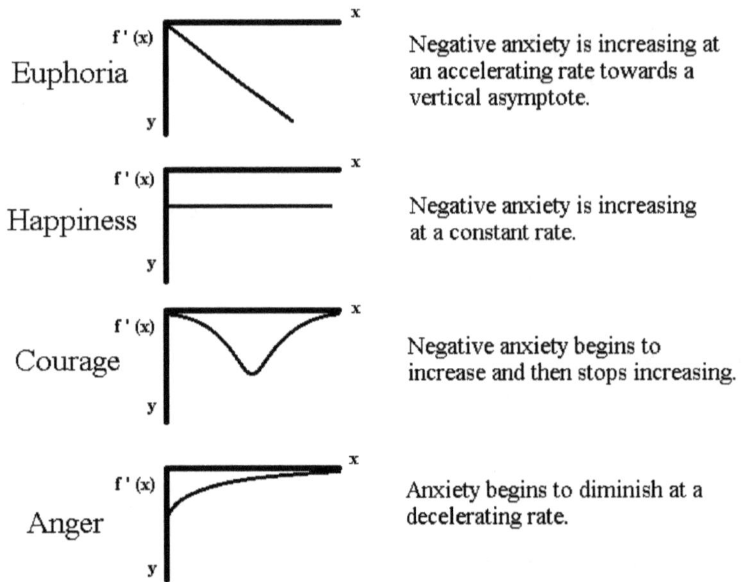

Figure 11.2 Category I Emotions are depicted. Valuations that might be expected to lead to euphoria, happiness, courage, and anger are shown on the Cartesian Plane above. The slope of the functions, indicated by f '(x), would be indicative of the type of emotion involved.

Category Three Emotions (Compound Interactive)

Similarities

Although anger is listed as an Avoidance of Pain emotion and guilt is listed as a Pursuit of Pleasure emotion, each bears more in common with particular emotions from the opposite side of the x-axis that are used in the alternate equation. For instance, anger is modeled to have more in common with courage and happiness, as a lowering of anxiety would be similar to an increase in negative anxiety, while guilt is modeled to have more in common with fear and sadness because the elevation of anxiety is similar to a decrease in negative anxiety. Fear, sadness, grief, and guilt would be similar because they are modeled to result in an elevation of anxiety or a decrease in negative anxiety. Happiness, courage, euphoria, and anger are grouped together because they result in a lowering of anxiety or an increase in negative anxiety. Fear, sadness, grief, and guilt all have positive slopes. Courage, happiness, euphoria, and anger all have negative slopes.

Polarities

Based upon the slopes of the sample scenarios modeled above, the following would be established:

1) Euphoria and grief are mirror opposites because of their slopes.
2) Happiness and sadness are mirror opposites because of their slopes.
3) Courage and fear a mirror opposites because of their slopes.
4) Anger and guilt are mirror opposites because of their slopes.

Category Three Emotions (Compound Interactive)

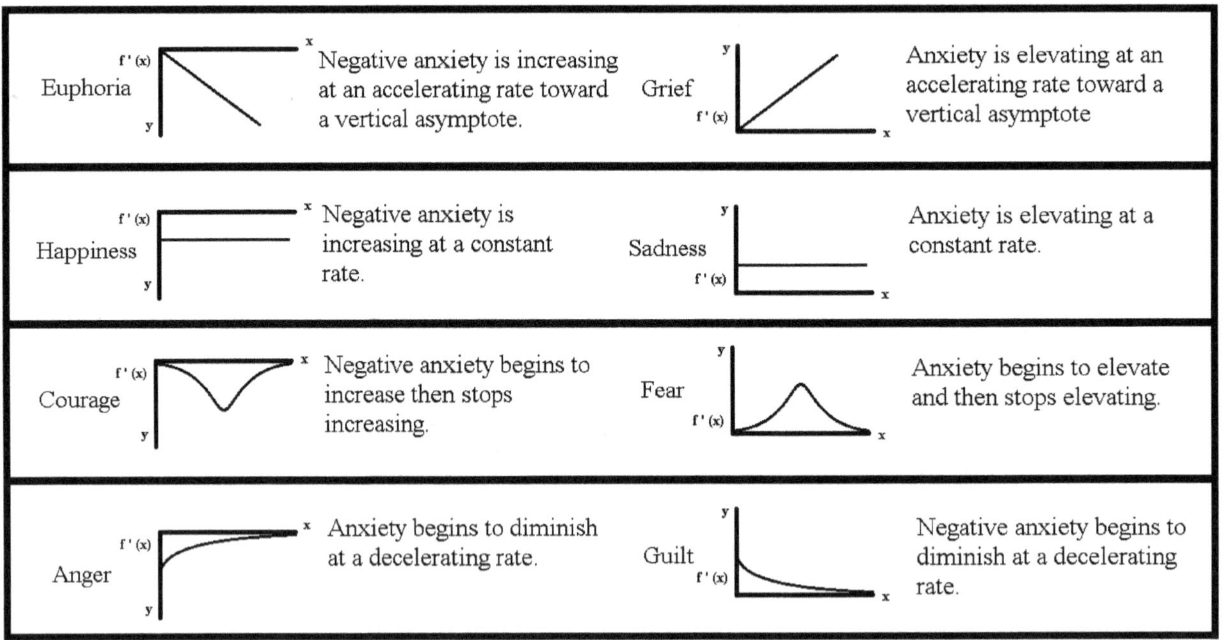

Figure 11.3 Modeled slopes of Category I Emotions alongside their mirror opposites

Similarities and Polarities Among Category II Emotions: Inter-Personal Emotions or the Four Degrees of Empathy

Among the Category II Emotions, a similar array of distinctions can be made among the four emotions of loving pride, sympathetic shame, hateful humiliation, and antipathetic mercy. Loving pride and sympathetic shame mirror one another across the x-axis, as the self wants the *other* to succeed, while hateful humiliation and antipathetic mercy mirror one another across the x-axis, as the self wants the *other* to not succeed. Moreover, the distinguishing feature of hateful humiliation from loving pride is that in the case of the former the self's anxiety lowers when the *other's* anxiety elevates. The distinguishing feature of antipathetic mercy from sympathetic shame is that in the case of the former the self's anxiety elevates when the *other's* anxiety lowers.

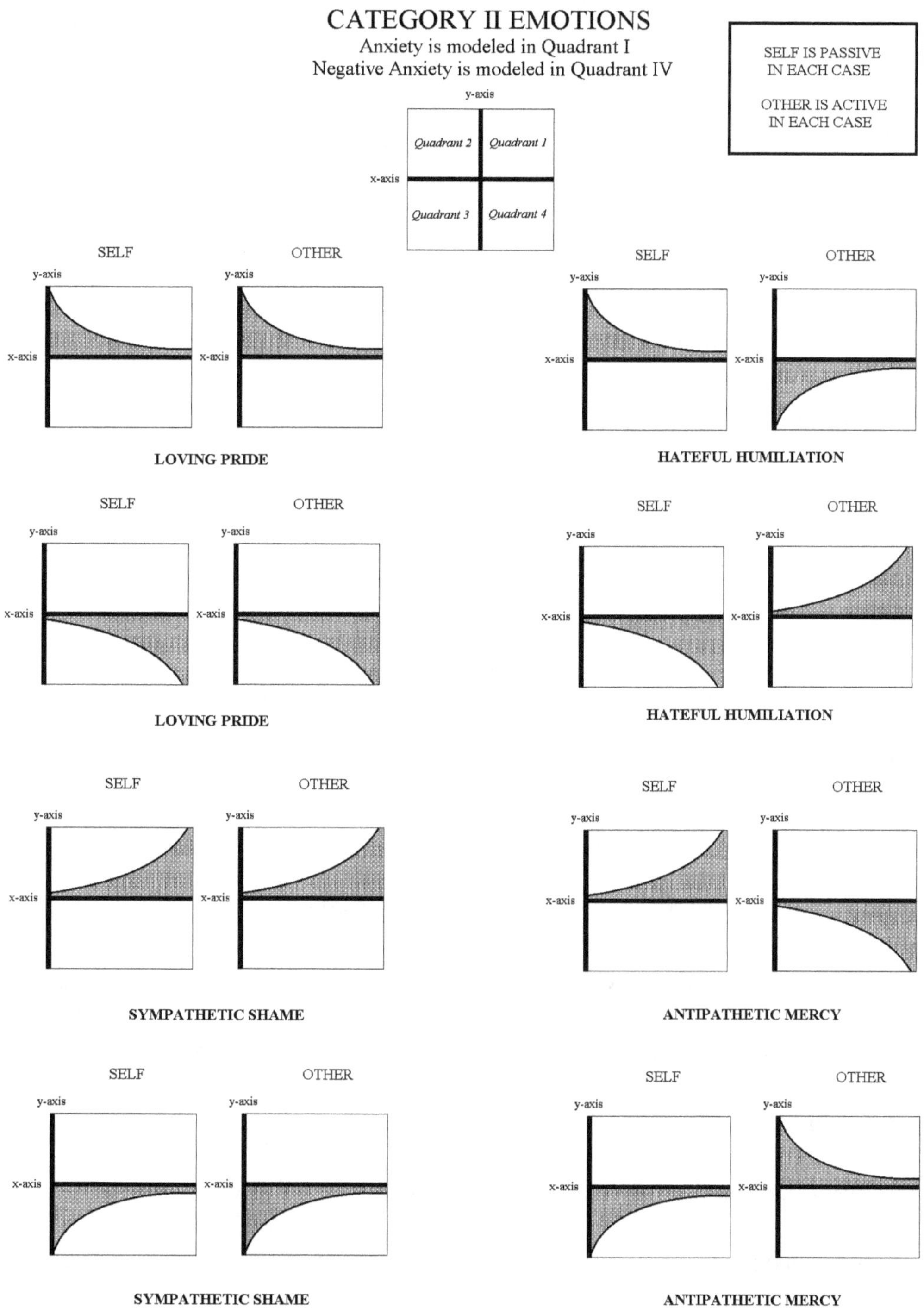

Figure 11.4 The Category II Emotions: The Four Degrees of Empathy

Category Three Emotions (Compound Interactive)

Primary Emotions in the Equation

Primary emotions are established as those that are not mixed emotions. In order to maintain the 1:1:1:1 ratio, it was held that a single entity could evoke no more than one value for an individual for a single purpose. All Category I Emotions, the Intra-Personal Emotions or emotions of the self, are primary emotions. This includes grief, sadness, fear, anger, indifference, guilt, courage, happiness, and euphoria, (figure 11.5).

Secondary Emotions of the Equation

Loving pride, sympathetic shame, hateful humiliation, and antipathetic mercy are mixed or secondary emotions in Plutchik's sense of the term.[6] They combine a valuation of the self with a valuation of an *other* to formulate a mixed emotion. The 1:1:1:1 ratio is maintained because the self imagines itself as an *other* as well, thus seeing itself as two people with two separate valuations for two separate entities. Most important, in the case of Category II Emotions, either the self is passive or the *other* is passive. If the self is passive, then the self does nothing to influence the outcome of the *other's* situation. In such a case, the *other's* success or failure is the desired entity for the self and the self would lack both efficacy and harm components while the *other* would have harm or benefit and efficacy components concerning a contingency (figure 11.5).

Category Three Emotions (Compound Interactive)

CATEGORY I EMOTION					
AVOIDANCE OF PAIN	Grief	Sadness	Fear	Anger	Content
PURSUIT OF PLEASURE	Euphoria	Happiness	Courage	Guilt	Content

CATEGORY II EMOTION	SELF IS PASSIVE		SELF IS ACTIVE	
	VICARIOUS EMOTION	SELF	VICARIOUS EMOTION	SELF
Loving Pride	Vicarious Pride	Love	Vicarious Love	Pride
Sympathetic Shame	Vicarious Shame	Sympathy	Vicarious Sympathy	Shame
Hateful Humiliation	Vicarious Humiliation	Hatred	Vicarious Hatred	Humiliation
Antipathetic Mercy	Vicarious Mercy	Antipathy	Vicarious Antipathy	Mercy
Indifference	Vicarious Loneliness	Neutrality	Vicarious Neutrality	Loneliness

Figure 11.5 All Category I Emotions would be considered primary, in Plutchik's sense of the term. All Category II Emotions would be considered mixed, as they are the result of an interaction between two purposes, two valuations, two entities, and two people (the self and the *other* imagined by the self).

Category Three Emotions (Compound Interactive)

Pride, shame, humiliation, and mercy would also have the distinction of being a combination of more than one emotion. In the equation, if felt by the self then they would take into account a response from an *other*, as conceived by the self. Contrarily, vicarious pride, vicarious shame, vicarious humiliation, and vicarious mercy would involve the self's conception of an *other's* valuation that takes into account a response from the self. Both are thought of as mixed.

Recap

Thus far, Category I Emotions have been established as those belonging to the individual and they are primary in the sense that they are not a mixture of two emotions. Secondly, Category II Emotions have been established as combined emotions where an individual imagines himself or herself to be someone else in addition to the self and the 1:1:1:1 ratio remains upheld. In effect, a single individual conceives itself to be two individuals and the self values the entity of vicariously experiencing the success or failure of an *other* or vice versa. Category II Emotions would be secondary in the sense they are mixtures of primary emotions concerning the self and an *other*.

Unlike a Category II Emotion where one party is passive, what is proposed next is a type of emotion where an individual may actively influence the outcome of the other's situation. In essence, they would be like Category II Emotions except that the self would possess both efficacy and harm components because it can take action to influence the other's situation. Namely, the individual acts on a purpose belonging to itself in order to influence the vicariously appraised value judgment of an entity that corresponds to a purpose belonging to an *other* rather than the self. These scenarios will be distinguished by the title of Category III Emotions, or Compound Interactive Emotions.

One such example will be the combining of love, vicarious pride, and an active form of happiness to create the emotion of Benevolence. Like loving pride, benevolence would be characterized by the individual imagining himself or herself in an *other's* place valuing an entity and wanting the *other* to succeed in order to vicariously experience the *other's* success. Ordinarily, for a Category II Emotion, the *other's* appraised value judgment would have both drive reduction, harm, and efficacy components while the self's appraised value judgment would only have drive-reduction components because the self desires an outcome that it cannot influence. However, in the case of a Category III, Compound Interactive emotion, the self is aware of a means by which its own actions can influence the outcome of an *other's* appraised judgment value of an entity, so the relationship between the purposes takes on a new form. The individual, acting as if the *other's* purpose were an extension of the self, moves to influence the outcome of the vicariously appraised value judgment (where the coefficient is root negative one for the other). There are still only two valuations present, one for the entity valued by the *other* as conceived by the self, and one concerning the self's vicarious experience of the *other's* success or failure. The difference, however, would lie in the role that the self plays. Specifically, both the self and the *other* would have harm or benefit and efficacy components.

Four Category III Emotions are proposed, of two types each. These four emotions are benevolence, malevolence, jealousy, and envy. They involve an interplay between Category I and Category II emotions and may be of two types each: 1) an indulgent type; 2) a protective type. This makes for a total of eight separate distinctions.

Category Three Emotions (Compound Interactive)

Category III Emotions: Compound-Interactive Emotions

Category Three Emotions are divided into two types depending on whether or not they concern anxiety or negative anxiety for the self and four separate emotions based upon particular patterns of appraised values. First, Compound-Interactive Emotions, where negative anxiety is at stake for the self, will be labeled indulgent, while those where anxiety is at stake for the self will be labeled protective. Secondly, the emotions of benevolence and jealousy are considered similar to the extent that the self desires for the *other* to succeed and the self takes action in order to attempt to make this so; likewise, the emotions of malevolence and envy are considered similar to the extent that the self desires for the *other* to fail and the self takes action in order to attempt to make this so. Thirdly, the emotions of benevolence and malevolence are considered to be opposite in nature, while those of jealousy and envy are considered to be opposite in nature. Fourthly, although they are opposite in nature, benevolence and malevolence are considered to be similar to the degree that the self achieves its desired outcome; likewise, jealousy and envy are considered similar to the degree that the self fails to achieve its desired outcome. Fifthly, benevolence and envy are considered similar to the degree that the other achieves its desired outcome; malevolence and jealousy are considered similar to the degree that the other fails to achieve its desired outcome.

Category Three Emotions (Compound Interactive)

Indulgent Forms of Benevolence, Jealousy, Envy, and Malevolence

Indulgent Benevolence

Benevolence literally means an "act of kindness."[1] Mathematically, it is represented by the core theme of desiring to vicariously experience an *other's* success and the self successfully accomplishes a purpose (e.g., preventing a threat of harm) in order to facilitate this. The self's negative anxiety increases because it was successful in its endeavors. The *other's* anxiety may lower or alternatively negative anxiety may increase. Indulgent Benevolence is a combination of the following Category I and Category II emotions where f(x) is the self's valuation of vicariously experiencing the *other's* success, and o(x) is the *other's* valuation of an entity as conceived by the self:

---

[1] By permission. From Merriam-Webster's Collegiate® Dictionary, 11th Edition ©2014 by Merriam-Webster, Inc. (www.Merriam-Webster.com).

Category Three Emotions (Compound Interactive)

Loving pride (love and vicarious pride) + happiness

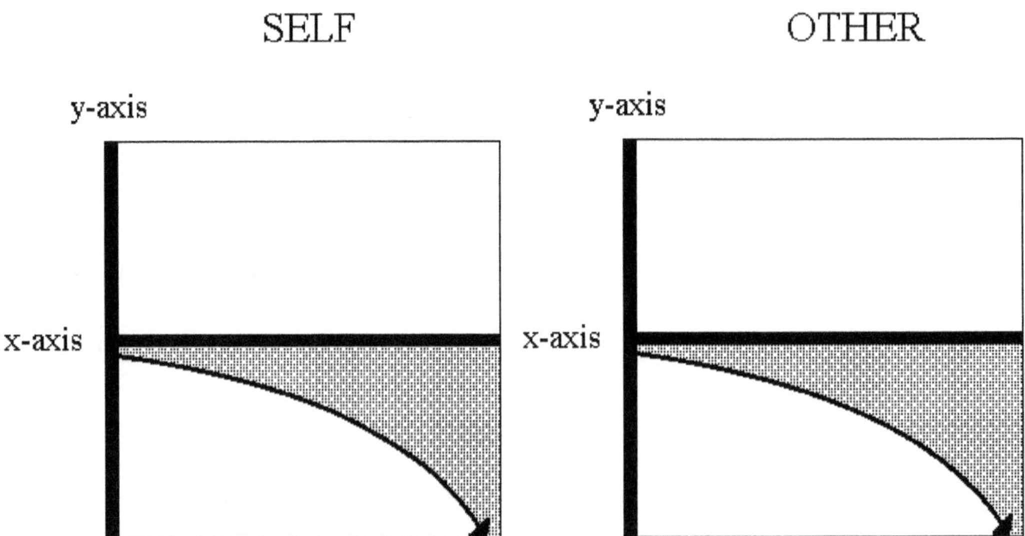

HAPPINESS + LOVING PRIDE

Both the *self* and the *other* are active, meaning they each have harm/benefit components and efficacy components.

Figure 11.6 Graph of Indulgent Benevolence: Loving Pride (Love + Vicarious Pride), and Happiness are all felt by the Self

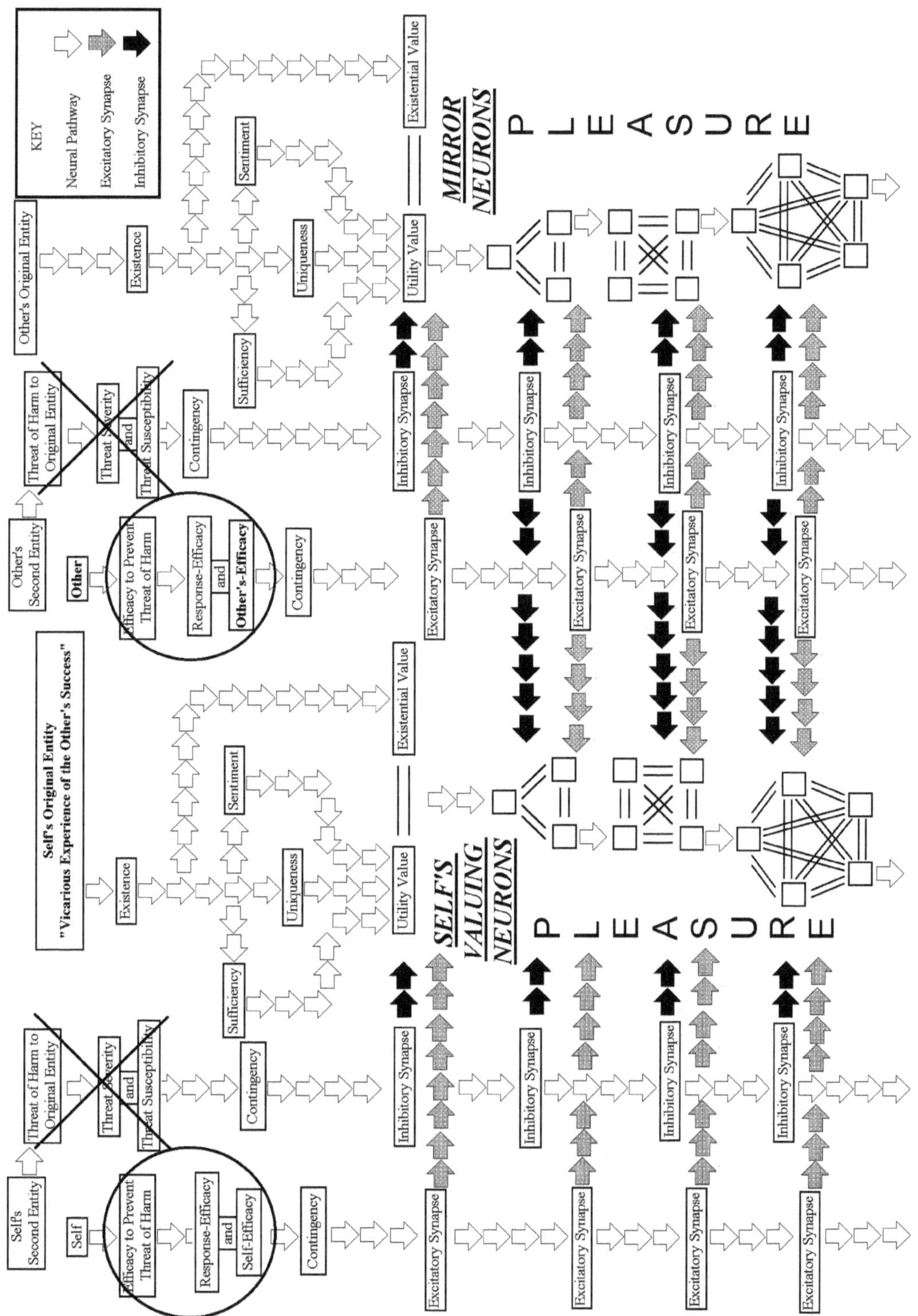

Figure 11.7 Neurological model of Indulgent Benevolence (similar to figure 9.13). The self's efficacy and the other's efficacy components (O) are increasing against the threat components (X).

Category Three Emotions (Compound Interactive)

Indulgent Jealousy

Jealousy is generally considered to be a state where a sense of anxiety arises from the "lack of a sense of security in the affections of one who is loved."[7] In psychology, the jealousy is "directed toward a third party, the rival who is perceived as garnering the affections of the object of love."[8] Although it is only used in the romantic sense in psychology, its use in the equation will be made slightly broader and extended to also encompass individuals to whom the individual feels companionate love, but will be less specific than its contemporary usage that often incorporates coveting into its definition.

At its core then, jealousy would be an arousal of anxiety in the self based upon the assessment that an *other* is dissatisfied with the self's failure to help the *other* successfully achieve a purpose. If the other is satisfied and successfully achieves his or her purposes as a result of the actions of the self, then the threat from a potential rival is nil. If the *other* is dissatisfied and does not successfully achieve his or her purpose as a result of the actions of the self, anything that has the potential to alleviate the *other's* dissatisfaction would become a rival. This rival, from the self's point of view, would not necessarily have to be a rival lover or living creature, and may even be an inanimate entity, such as an activity that offers something the self or another person cannot provide.

Mathematically, the core, distinguishing feature of indulgent jealousy in the equation is that the self desires to vicariously experience the success of an *other*, but the self fails to perform an action that would ensure that the *other* succeeds. Because the self is unsuccessful in its attempt, its negative anxiety decreases. The *other's* anxiety elevates, or alternative, negative anxiety may decrease. Indulgent jealousy is a combination of the following Category I and Category II emotions where f(x) is the self's valuation of vicariously experiencing the *other's* success, and o(x) is the *other's* valuation of an entity as conceived by the self:

Sympathetic Shame (Sympathy and Vicarious Shame) + Guilt

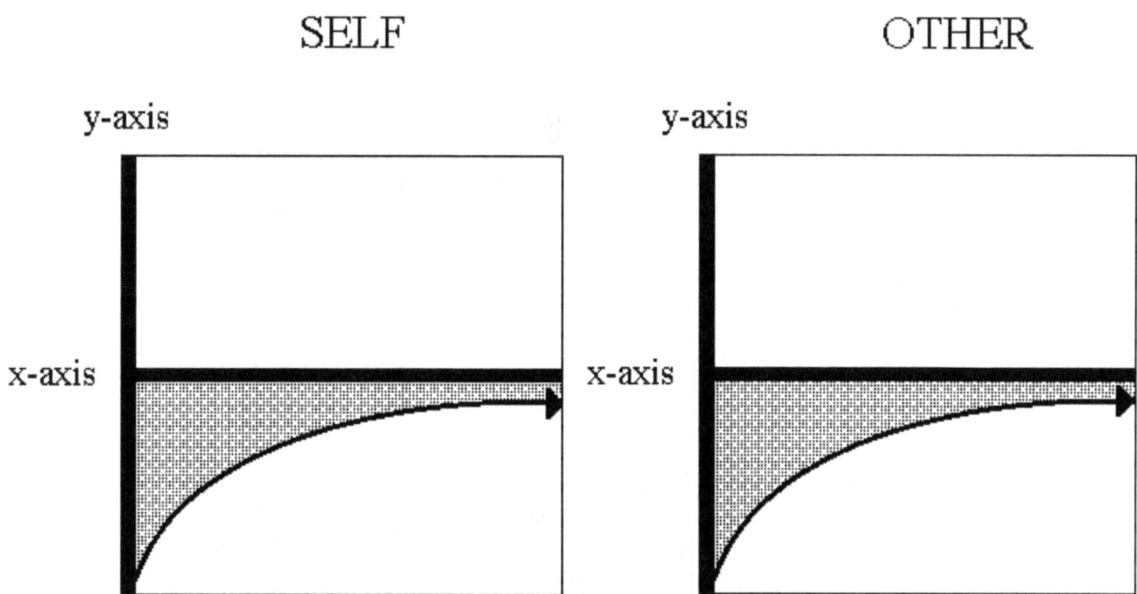

GUILT + SYMPATHETIC SHAME

Both the *self* and the *other* are active, meaning they each have harm/benefit components and efficacy components.

Figure 11.8 Graph of Indulgent Jealousy: Sympathetic Shame (Sympathy + Vicarious Shame), and Guilt are all felt by the Self.

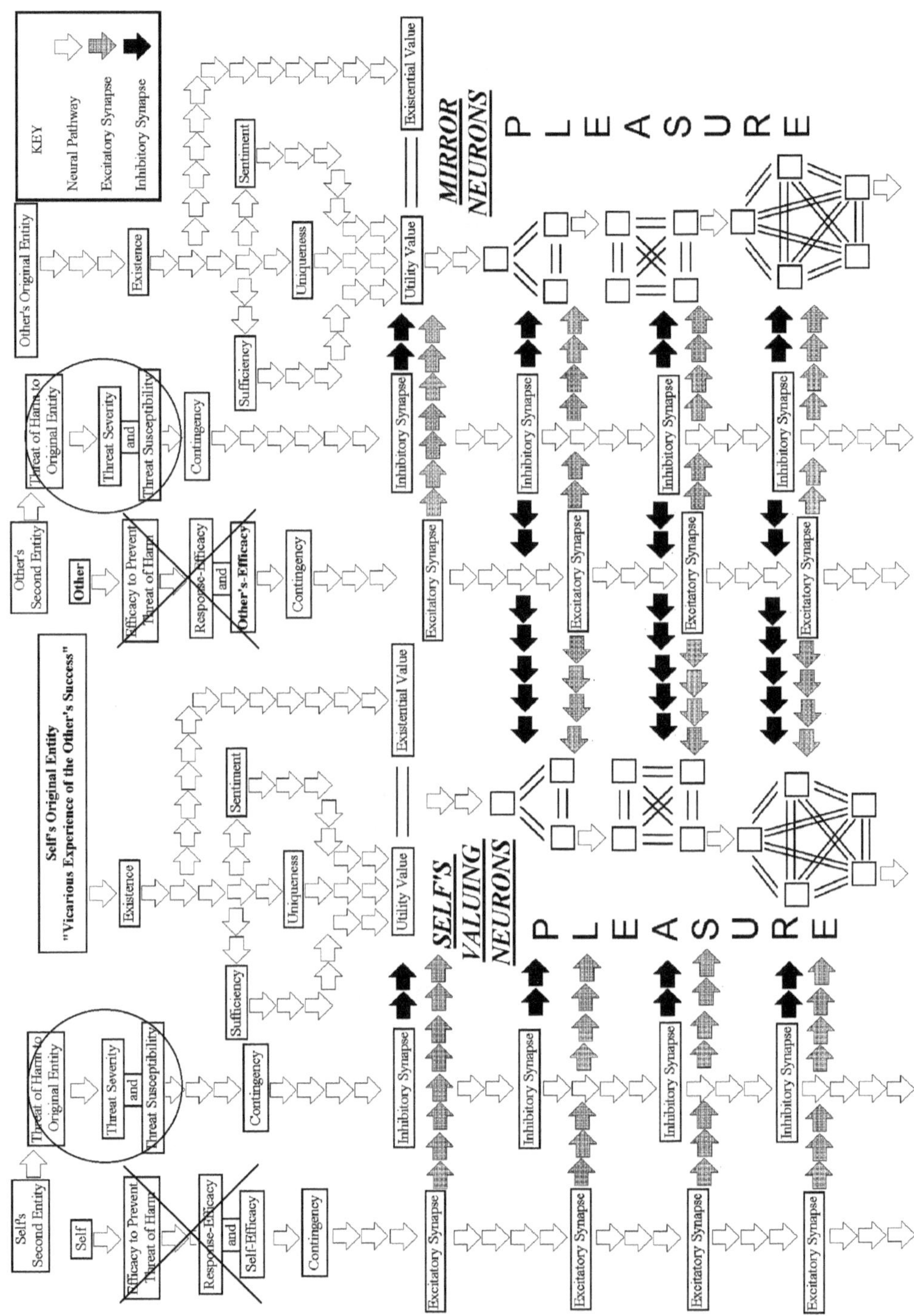

Figure 11.9 Neurological model of Indulgent Jealousy. The efficacy of the self and other (X) are decreasing against the threat components (O).

Category Three Emotions (Compound Interactive)

Indulgent Malevolence

Someone who is malevolent is "having or showing a desire to cause harm to another person."[2] Mathematically, the core distinguishing feature of indulgent malevolence in the equation is that the self desires to vicariously experience the failure of an *other* and the self is successful in its endeavors to make this happen. Because the self is successful, its negative anxiety increases. The *other*, having failed, will experience an elevation of anxiety or a decrease in negative anxiety. Indulgent malevolence is a combination of the following Category I and Category II Emotions where f(x) is the self's valuation of vicariously experiencing the other's failure, and o(x) is the *other's* valuation of an entity as conceived by the self:

[2] By permission. From Merriam-Webster's Collegiate® Dictionary, 11th Edition ©2014 by Merriam-Webster, Inc. (www.Merriam-Webster.com).

Category Three Emotions (Compound Interactive)

Hateful Humiliation (Hatred and Vicarious Humiliation) + Happiness

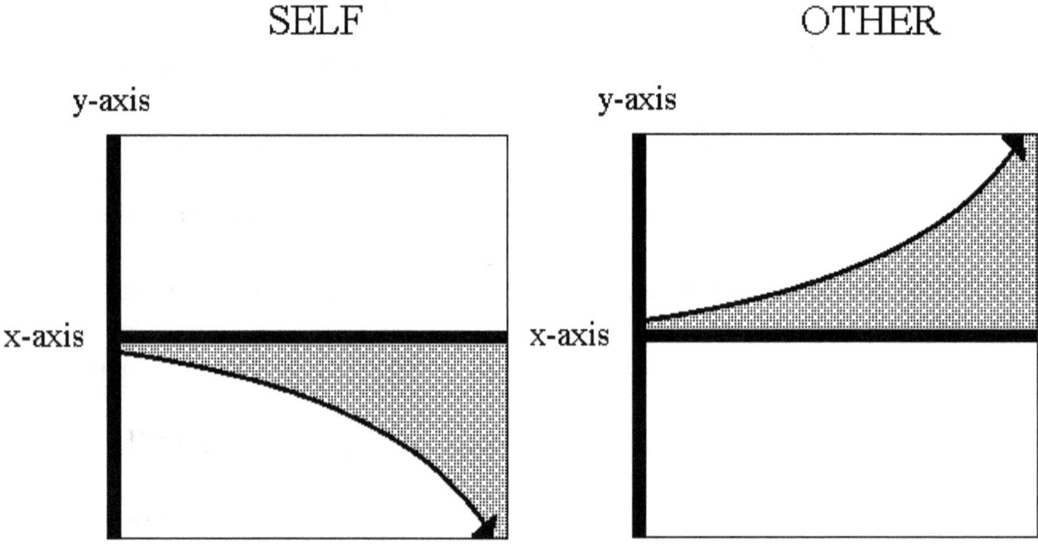

HAPPINESS + HATEFUL HUMILIATION

Both the *self* and the *other* are active, meaning they each have harm/benefit components and efficacy components.

Figure 11.10 Graph of Indulgent Malevolence: Hateful Humiliation (Hatred + Vicarious Humiliation), and Happiness are all felt by the Self.

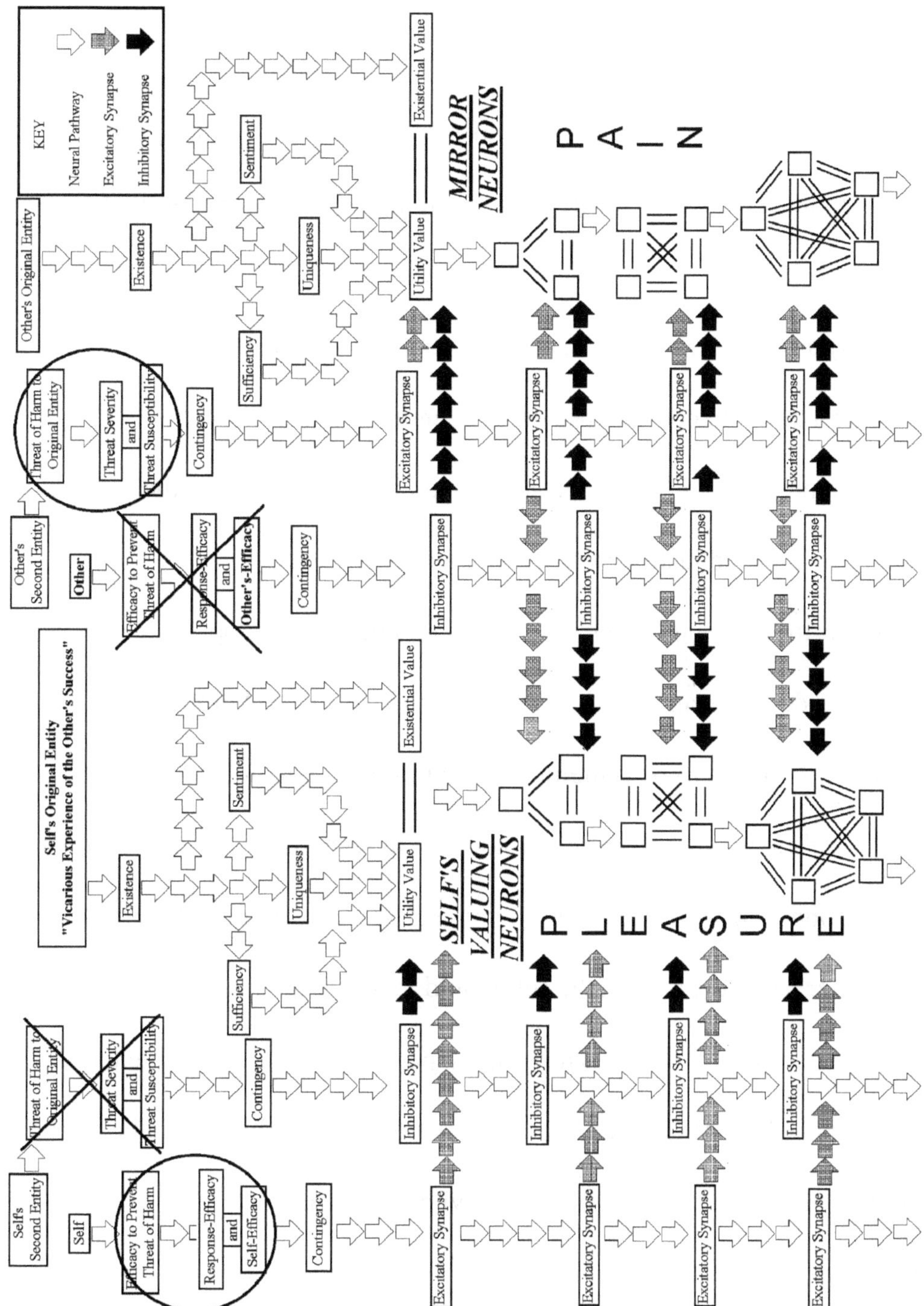

Figure 11.11 Neurological model of Indulgent Malevolence. The self's efficacy components (O) increase against the threat components (X) while the other's efficacy components (X) decrease against threat (O)

Indulgent Envy

Envy is generally "classified as a special form of anxiety," where, according to William McDougall, an individual possesses a "'grudging contemplation of more fortunate persons.'"[9] Envy is distinguished from jealousy by the absence of a necessity for a loved one, "merely a desire for things possessed by a rival."[10] Envy is "painful or resentful awareness of an advantage enjoyed by another joined with a desire to possess the same advantage."[3] In the equation, envy will be thought of as the self's contempt for an *other's* gain, stemming from the belief that the *other* did not deserve the gain or somehow escaped justice by acquiring an entity.

With regards to the ambiguity between jealousy and envy in popular usage, if one wished to use the equation to describe an emotion felt in response to suspicion that a lover was being unfaithful, then it only needs to be remembered that one must uphold the 1:1:1:1 ratio. Envy or contempt for another's gain is what the self would feel in response to the successful ambitions of a rival, for instance, one who is suspected of having a tryst with the self's lover. Jealousy, hence, a fear of losing another's loyalty, is what the self would feel in response to the ambitions of the self's lover not being met because of the self's inability to perform actions to enable them. Thereafter, if the self comes to loathe his or her lover for being unfaithful, then this would be a separate purpose and envy would be felt toward the scorned lover who sought and acquired something that the self did not want him or her to acquire.

Mathematically, the core distinguishing feature of indulgent envy in the equation is that the self desires to vicariously experience the failure of an *other*, but the self is unsuccessful in its

---

[3] By permission. From Merriam-Webster's Collegiate® Dictionary, 11th Edition ©2014 by Merriam-Webster, Inc. (www.Merriam-Webster.com).

endeavors. Because the self fails, its negative anxiety decreases. The *other*, having succeeded, will experience a lowering of anxiety or an increase in negative anxiety. Indulgent envy is a combination of the following Category I and Category II Emotions where f(x) is the self's valuation of the *other's* purpose, and o(x) is the *other's* valuation of an entity as it is conceived by the self:

Antipathetic Mercy (Antipathy and Vicarious Mercy) + Guilt

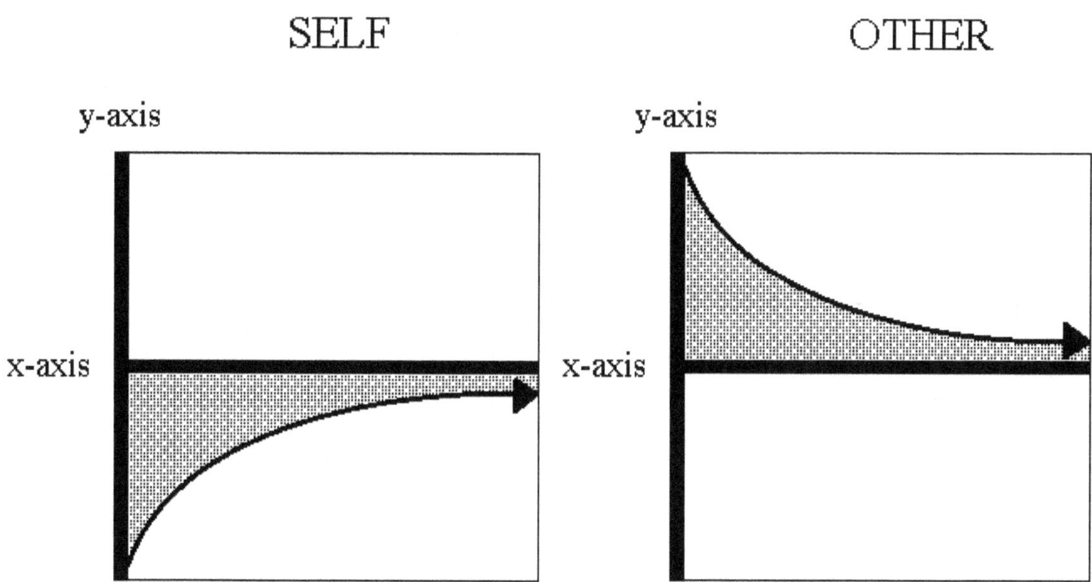

GUILT + ANTIPATHETIC MERCY

Both the *self* and the *other* are active, meaning they each have harm/benefit components and efficacy components.

Figure 11.12 Graph of Indulgent Envy: Antipathetic Mercy (Antipathy + Vicarious Mercy), and Guilt are all felt by the Self

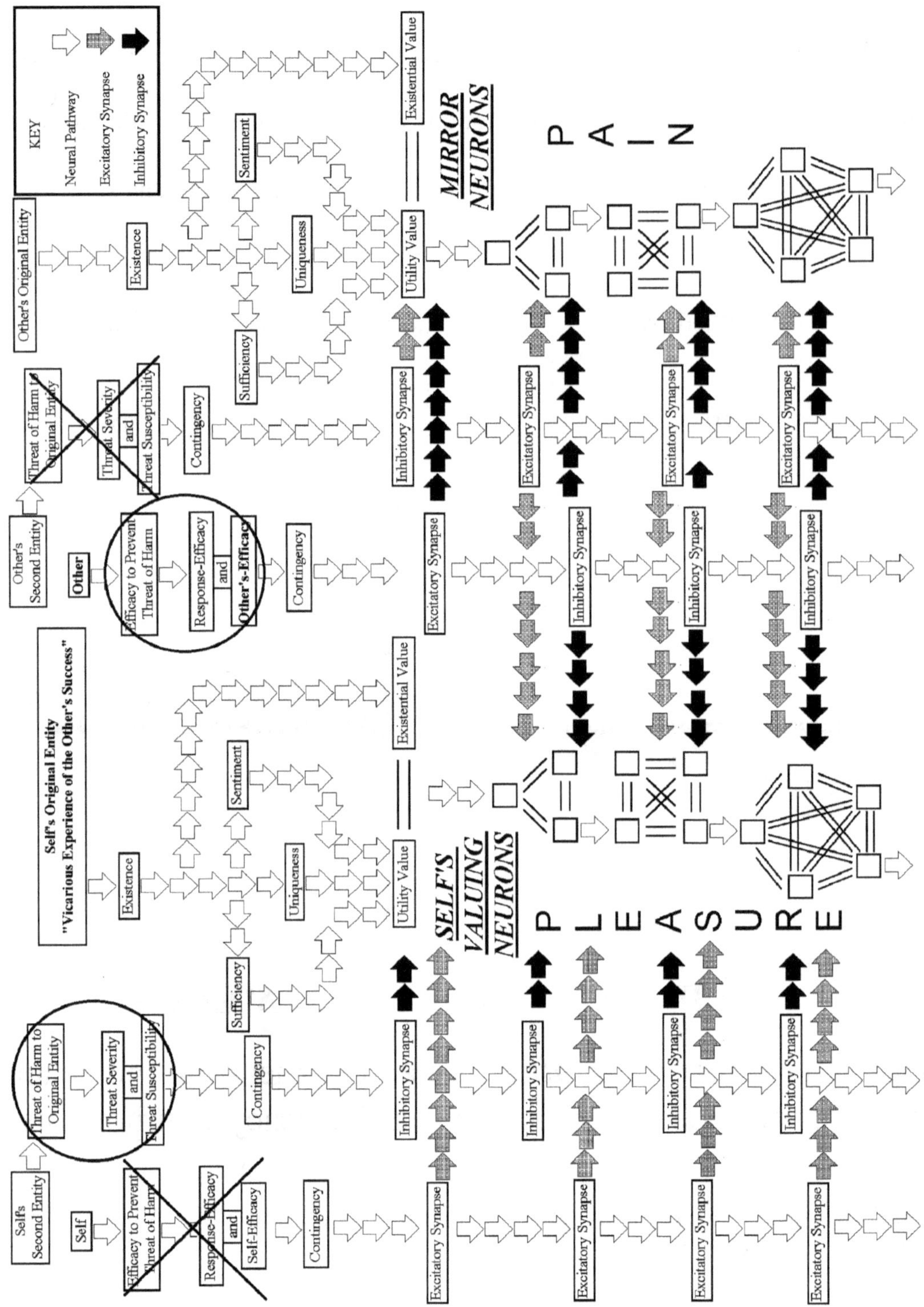

Figure 11.13 Neurological model of Indulgent Envy. The self's efficacy components (X) decrease against the threat components (O) while other's efficacy components (O) increase against threat (X)

Category Three Emotions (Compound Interactive)

Protective Forms of Benevolence, Jealousy, Envy, and Malevolence

The protective forms of the four Compound-Interactive Emotions just described are similar in nearly all manners except those concerning the emotion felt by the self. In the case of Protective Benevolence, the self feels anger rather than happiness. For Protective Jealousy sadness is felt instead of guilt. For Protective Malevolence anger instead of happiness is felt. For Protective Envy sadness is felt instead of guilt.

Protective Benevolence

Mathematically, protective benevolence is represented by the core theme of desiring to vicariously experience another's success and the self successfully accomplishes a purpose (i.e., acquire an entity) in order to facilitate this. The self's anxiety decreases because it is successful in its endeavors. The *other's* anxiety may lower or alternatively negative anxiety may increase. Protective Benevolence is a combination of the following Category I and Category II Emotions, where f(x) is the self's valuation of vicariously experiencing the other's success and o(x) is the *other's* valuation of an entity as it is conceived by the self:

Category Three Emotions (Compound Interactive)

Loving Pride (Love and Vicarious Pride) + Anger

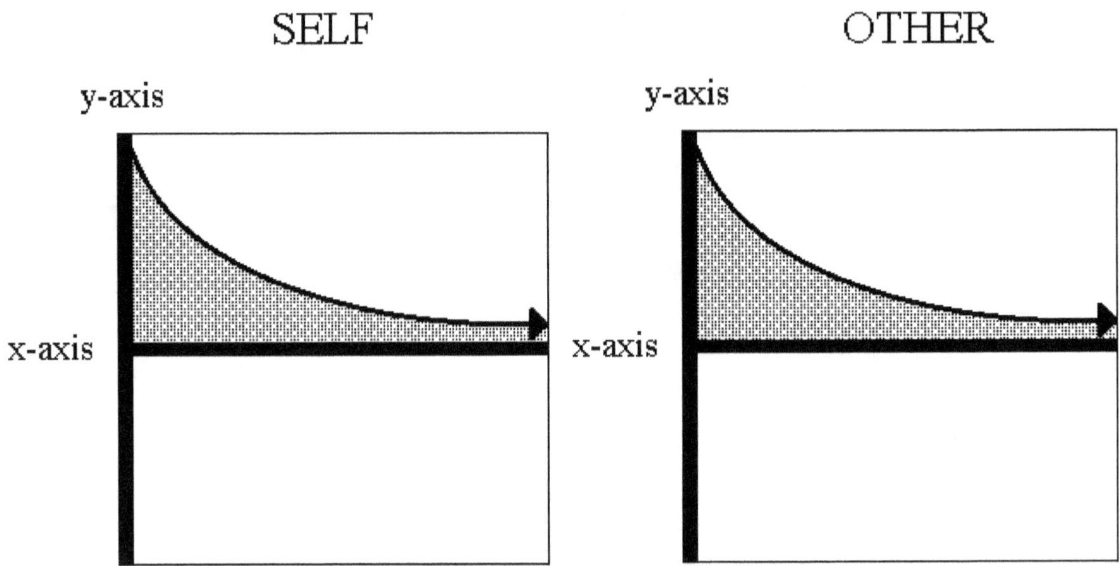

ANGER + LOVING PRIDE

Both the *self* and the *other* are active, meaning they each have harm/benefit components and efficacy components.

Figure 11.14 Graph of Protective Benevolence: Loving Pride (Love + Vicarious Pride), and Anger are all felt by the Self

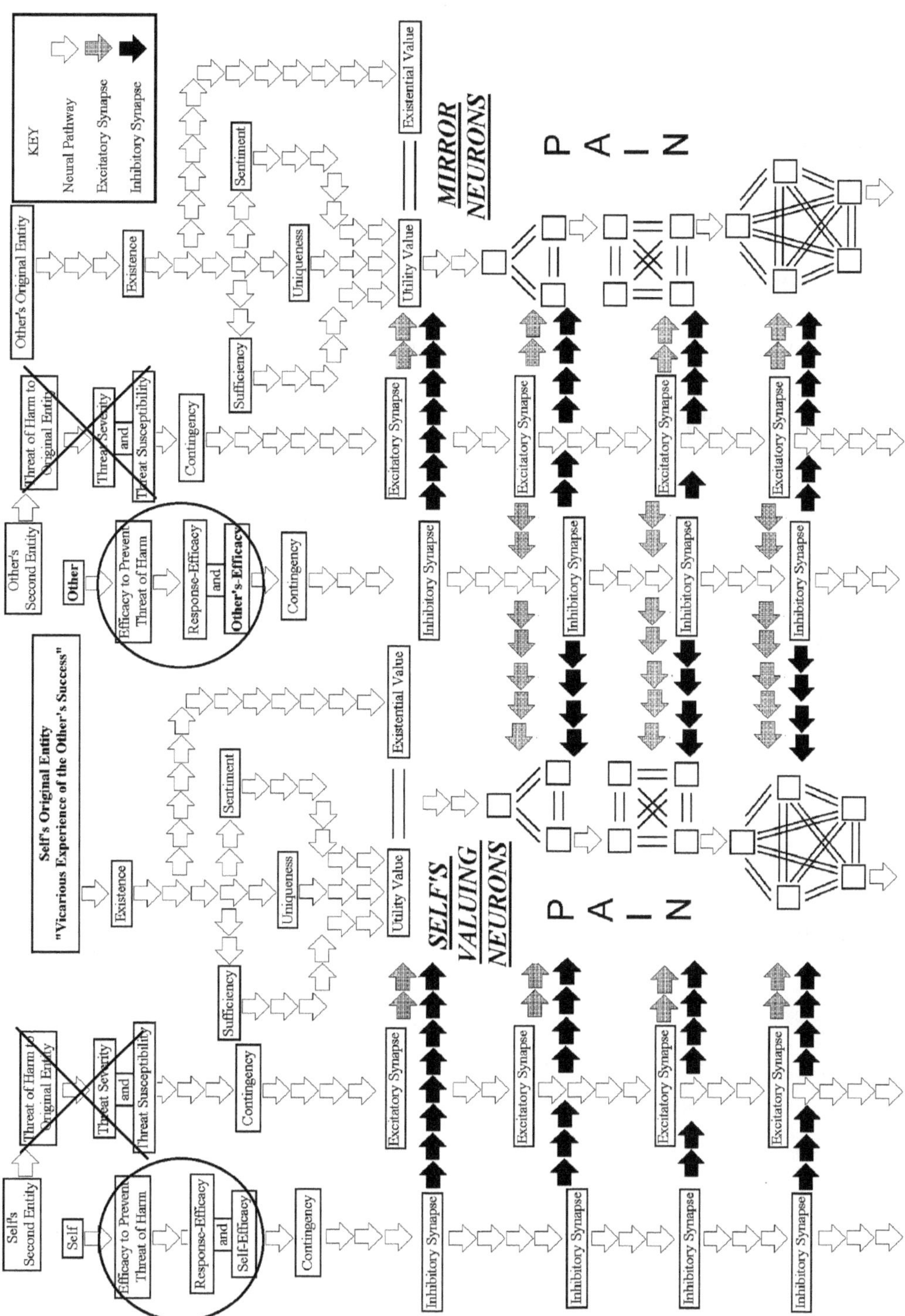

Figure 11.15 Neurological model of Protective Benevolence. The self's and the other's efficacy components (O) are increasing against threat components (X).

Category Three Emotions (Compound Interactive)

Protective Jealousy

The core, distinguishing feature of protective jealousy in the equation is that the self desires to vicariously experience the success of an *other* but fails to do so. Because the self is unsuccessful in its attempt to enable the *other* to succeed, then its anxiety increases. The *other's* anxiety elevates, or alternatively, negative anxiety may decrease. Protective jealousy is a combination of the following Category I and Category II Emotions, where f(x) is the self's valuation of vicarious experience of the other's success and o(x) is the other's valuation of an entity as it is conceived by the self:

Category Three Emotions (Compound Interactive)

Sympathetic Shame (Sympathy and Vicarious Shame) + Sadness

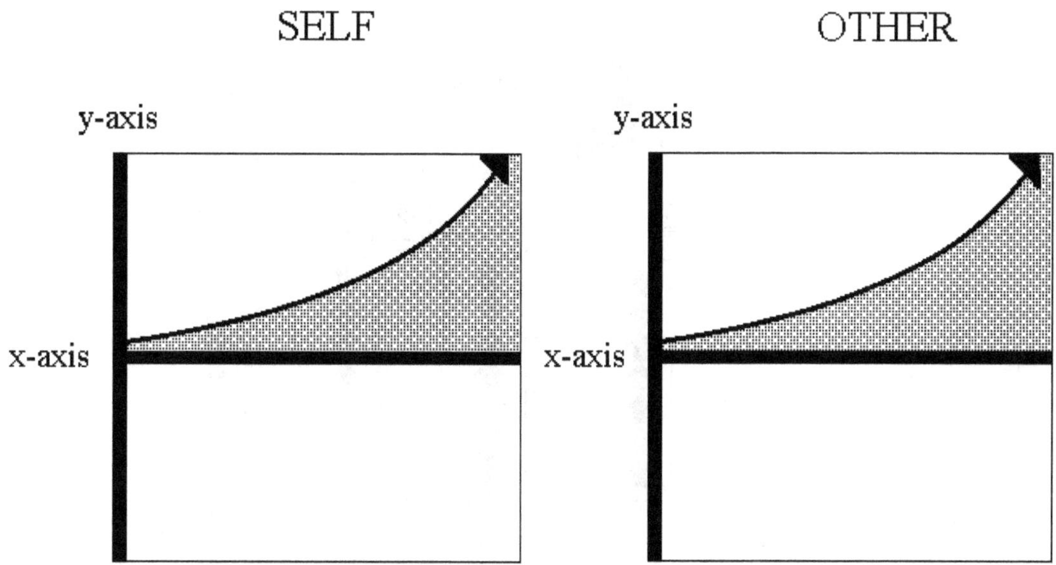

SADNESS + SYMPATHETIC SHAME

Both the *self* and the *other* are active, meaning they each have harm/benefit components and efficacy components.

Figure 11.16 Graph of Protective Jealousy: Sympathetic Shame (Sympathy + Vicarious Shame), and Sadness are all felt by the Self.

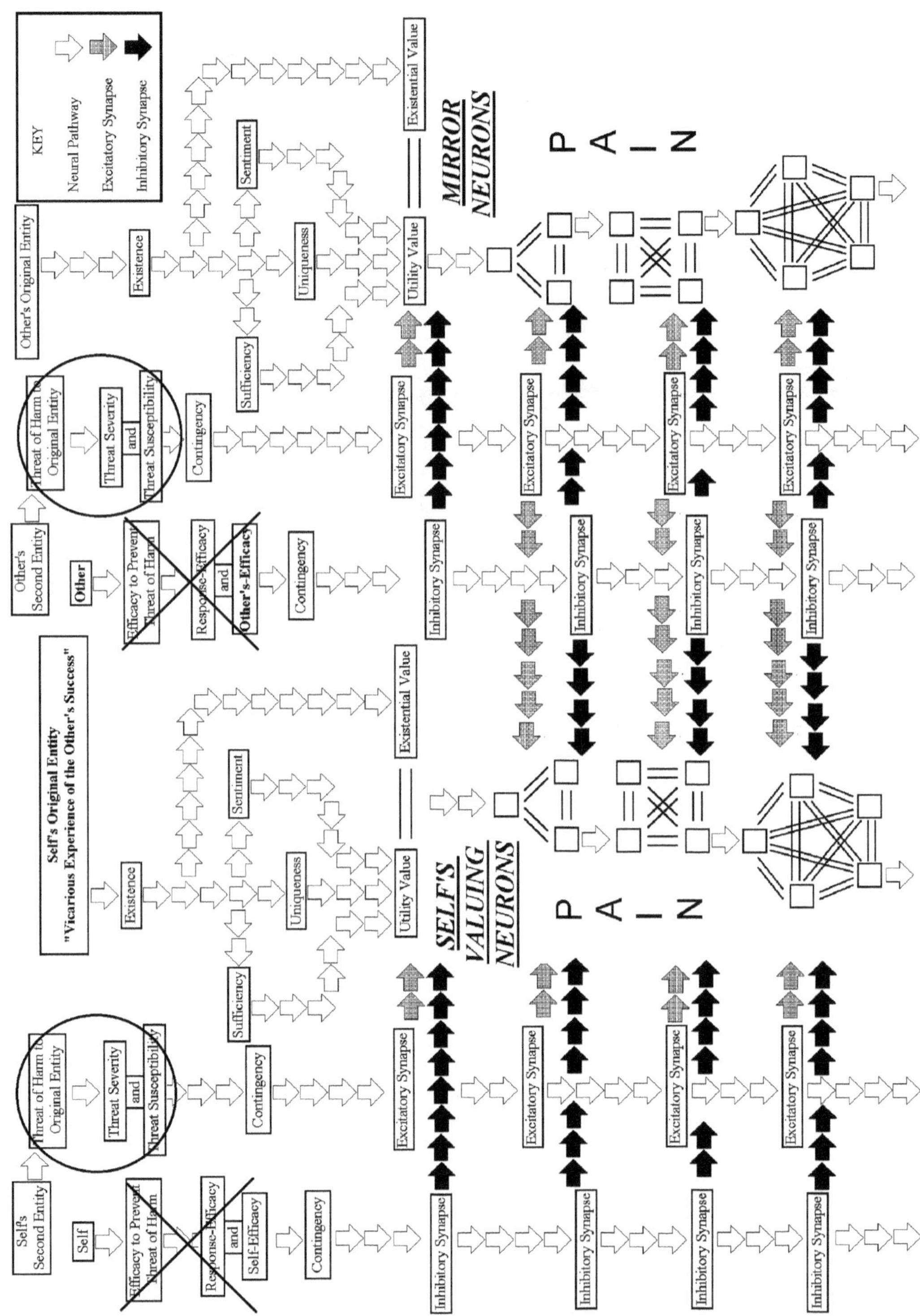

Figure 11.17 Neurological model of Protective Jealousy. The self's and the other's efficacy components (X) are decreasing against threat (O).

Category Three Emotions (Compound Interactive)

Protective Malevolence

The core distinguishing feature of protective malevolence in the equation is that the self desires to vicariously experience the failure of an *other* and the self is successful in its endeavors. Because the self is successful, its anxiety decreases. The *other*, having failed, will experience an elevation of anxiety or a decrease in negative anxiety. Protective malevolence is a combination of the following Category I and Category II Emotions where f(x) is the self's valuation of the vicarious experience of the *other's* failure, and o(x) is the *other's* valuation of an entity as it is conceived by the self:

Category Three Emotions (Compound Interactive)

Hateful Humiliation (Hatred and Vicarious Humiliation) + Anger

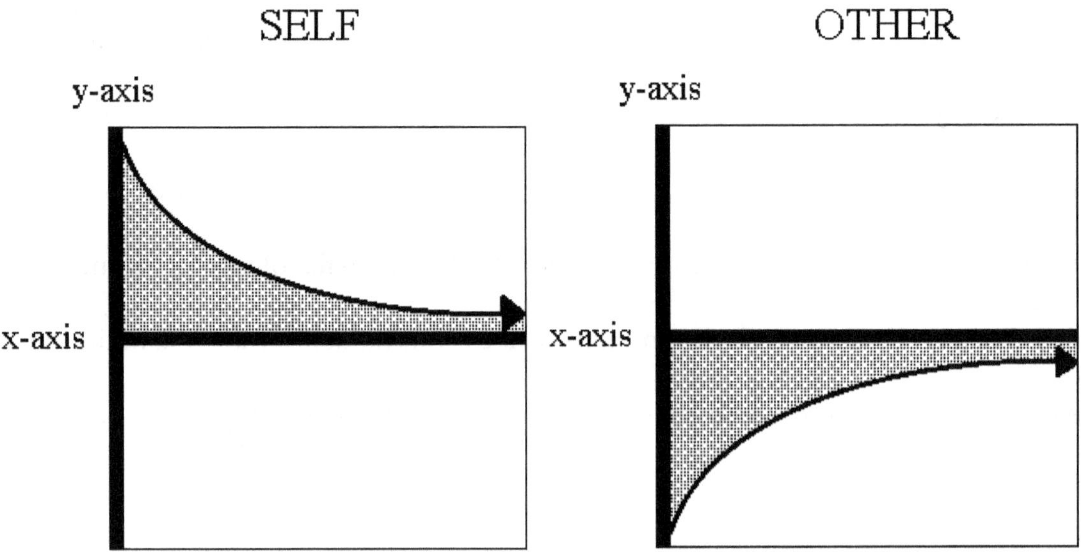

ANGER + HATEFUL HUMILIATION

Both the *self* and the *other* are active, meaning they each have harm/benefit components and efficacy components.

Figure 11.18 Graph of Protective Malevolence: Hateful Humiliation (Hatred + Vicarious Humiliation), and Anger are all felt by the Self

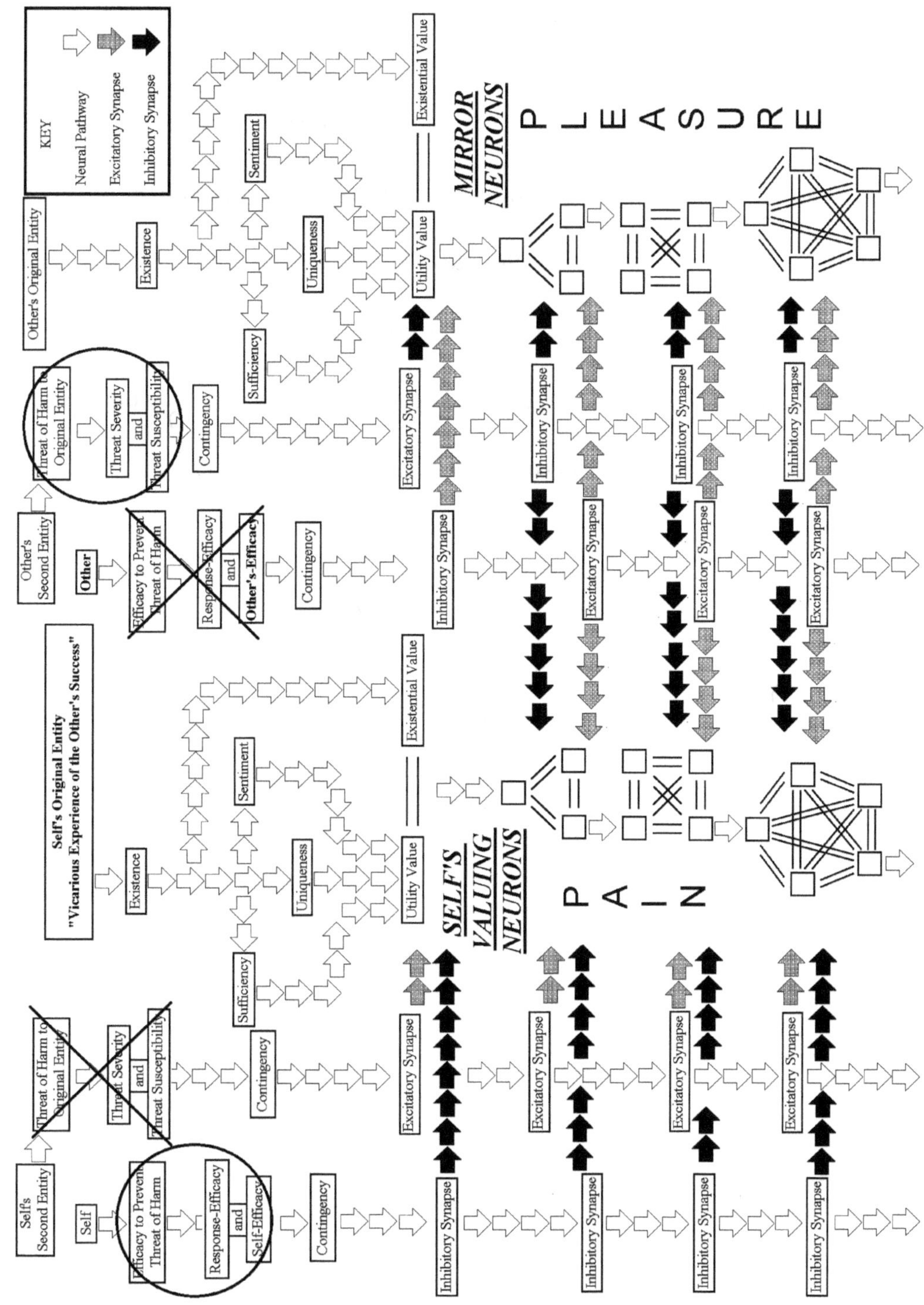

Figure 11.19 Neurological model of Protective Malevolence. Self's efficacy components (O) are increasing against threat (X), while other's efficacy (X) decreases against threat (O).

Category Three Emotions (Compound Interactive)

Protective Envy

Mathematically, the core distinguishing feature of protective envy in the equation is that the self desires to vicariously experience the failure of an *other* but the self is unsuccessful in its endeavors. Because the self fails, its anxiety elevates. The *other*, having succeeded, will experience a lowering of anxiety or an increase in negative anxiety. Protective envy is a combination of the following Category I and Category II emotions, where f(x) is the self's valuation of the *other's* purpose, and o(x) is the other's valuation of an entity as it is conceived by the self:

Category Three Emotions (Compound Interactive)

Antipathetic Mercy (Antipathy and Vicarious Mercy) + Sadness

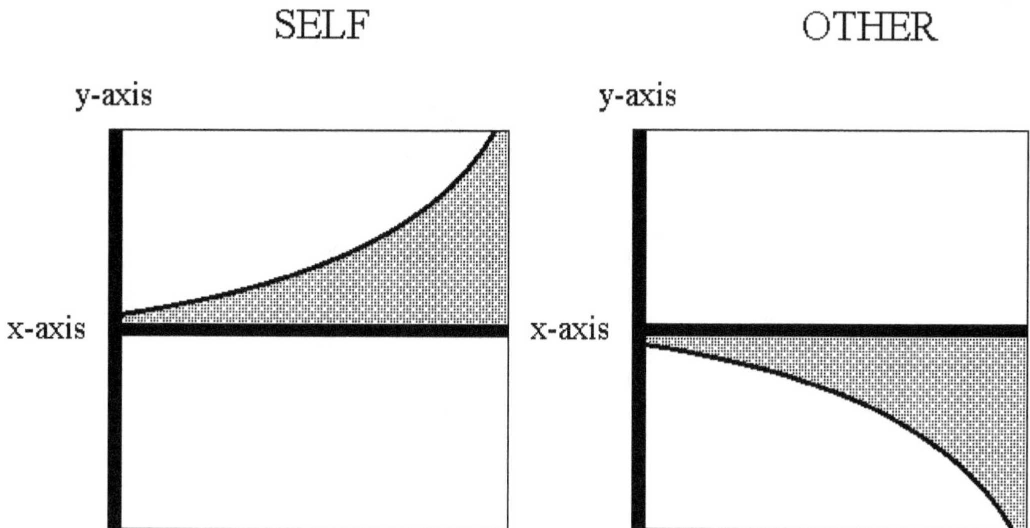

SADNESS + ANTIPATHETIC MERCY

Both the *self* and the *other* are active, meaning they each have harm/benefit components and efficacy components.

Figure 11.20 Graph of Protective Envy: Antipathetic Mercy (Antipathy + Vicarious Mercy), and Sadness are all felt by the Self

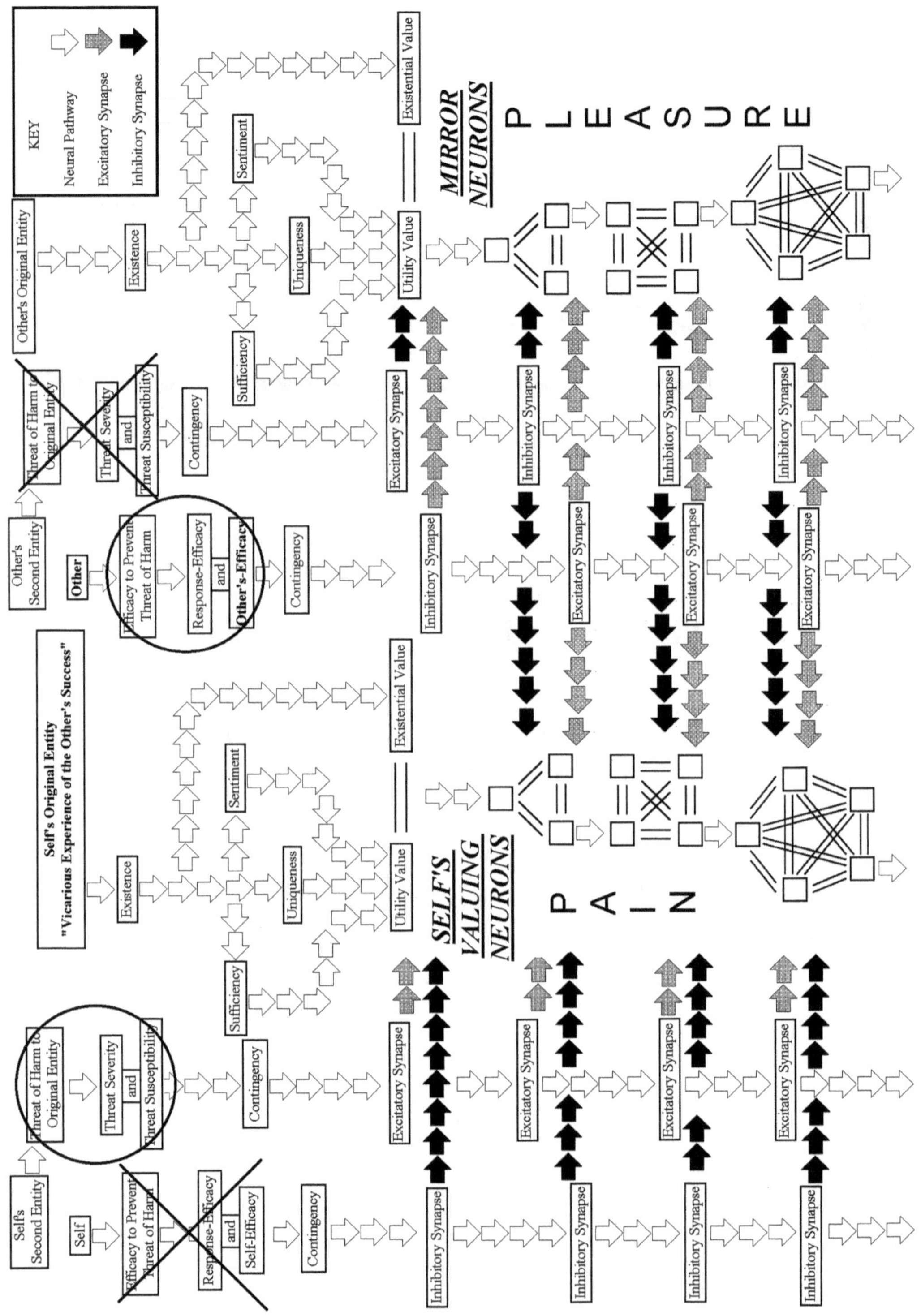

Figure 11.21 Neurological model of Protective Envy. Self's efficacy components (X) are decreasing against threat (O), while other's efficacy (O) increases against threat (X).

Category Three Emotions (Compound Interactive)

Logic Check

From the above graphs and models, four emotions of two types each were added and described by the equation as the following:

1) Indulgent Benevolence and Protective Benevolence - These refer to the desire to vicariously experience an *other's* success and the self is successful in the attempt to personally do something to ensure this.

2) Indulgent Jealousy and Protective Jealousy - These refer to the desire to vicariously experience an *other's* success and the self is unsuccessful in the attempt to personally do something to ensure this.

3) Indulgent Malevolence and Protective Malevolence - These refer to the desire to vicariously experience an *other's* failure and the self is successful in the attempt to personally do something to ensure this.

4) Indulgent Envy and Protective Envy - These refer to the desire to vicariously experience an *other's* failure and the self is unsuccessful in the attempt to personally do something to ensure this.

Category Three Emotions (Compound Interactive)

CATEGORY III EMOTION	INDULGENT TYPE		PROTECTIVE TYPE	
	CAT. I EMOTION +	CAT. II EMOTION	CAT. I EMOTION +	CAT. II EMOTION
BENEVOLENCE	Happiness	Loving Pride	Anger	Loving Pride
JEALOUSY	Guilt	Sympathetic Shame	Sadness	Sympathetic Shame
MALEVOLENCE	Happiness	Hateful Humiliation	Anger	Hateful Humiliation
ENVY	Guilt	Antipathetic Mercy	Sadness	Antipathetic Mercy

Figure 11.22 Category III Emotions: Compound Interactive Emotions

Consistency with Logic: Check of Limits and Definitions

The following definitions appear to be in accordance with what is modeled by the equation:

1) Benevolence - Benevolence in the equation is understood as the self's desire to help an *other* to succeed.

2) Jealousy - Jealousy is described as "a lack of sense of security in the affections of one who is loved."[11] In the equation, this is understood as an elevation of anxiety precipitating from the possibility of losing an *other's* loyalty, for instance, if the *other's* purposes are not fulfilled on account of the self's failure to perform a

recommended behavior.

3) Malevolence - Malevolent in the equation is understood as the desire by the self to ensure that an *other* fails and does not succeed at whatever his or her aim was.

4) Envy - Envy is described as "a grudging contemplation of more fortunate persons."[12] In the equation, this can be understood as contempt for an *other's* gain. Hence, this would consist of cases where the self was unable to prevent an *other* from succeeding when the self wished to do so.

Preview

Observations here have revealed some useful ways by which the equation may be used to isolate instances of specific emotions. However, the possibility that other emotions or moods may still exist and have not been accounted for is a very real one. Moreover, factors that may influence the perception of affect, such as time and its measurement, have not been fully considered. A fourth category of emotions, Emotive States, will be described in the next chapter along with different conceptions of the notion of time, biological rhythms, and the modeling of affect during dream states.

YOU SHOULD UNDERSTAND THIS.
CHAPTER ELEVEN
CONCEPT CHECK
VOCABULARY
Benevolence
Category III Emotions:
Compound Interactive Emotions
Envy
Indulgent Type
Jealousy
Malevolence
Primary Emotions
Protective Type
Secondary Emotions

Notes

1. Wilde, Oscar (2003). *The Picture of Dorian Gray.* New York: Barnes and Nobles Books. Print, p. 82. (Original work published 1890).

2. Cornelius, R. R. *The Science of Emotion: Research and Tradition in the Psychology of Emotion.* Upper Saddle River, NJ. Prentice-Hall, Inc. 1996. Print, p. 46.

3. Cornelius, R. R. *The Science of Emotion: Research and Tradition in the Psychology of Emotion.* Upper Saddle River, NJ. Prentice-Hall, Inc. 1996. Print, p. 46.

4. Cornelius, R. R. *The Science of Emotion: Research and Tradition in the Psychology of Emotion.* Upper Saddle River, NJ. Prentice-Hall, Inc. 1996. Print, p. 46.

5. Cornelius, R. R. *The Science of Emotion: Research and Tradition in the Psychology of Emotion.* Upper Saddle River, NJ. Prentice-Hall, Inc. 1996. Print, p. 46.

6. Cornelius, R. R. *The Science of Emotion: Research and Tradition in the Psychology of Emotion.* Upper Saddle River, NJ. Prentice-Hall, Inc. 1996. Print, p. 46.

7. Reber, Arthur S., Rhianon Allen, and Emily S. Reber. *Penguin Dictionary of Psychology*. London. Penguin Books, 2009. Print, p. 406.

8. Reber, Arthur S., Rhianon Allen, and Emily S. Reber. *Penguin Dictionary of Psychology*. London. Penguin Books, 2009. Print, p. 406.

9. Reber, Arthur S., Rhianon Allen, and Emily S. Reber. *Penguin Dictionary of Psychology*. London. Penguin Books, 2009. Print, p. 264.

10. Reber, Arthur S., Rhianon Allen, and Emily S. Reber. *Penguin Dictionary of Psychology*. London. Penguin Books, 2009. Print, p. 406.

11. Reber, Arthur S., Rhianon Allen, and Emily S. Reber. *Penguin Dictionary of Psychology*. London. Penguin Books, 2009. Print, p. 406.

12. Reber, Arthur S., Rhianon Allen, and Emily S. Reber. *Penguin Dictionary of Psychology*. London. Penguin Books, 2009. Print, p. 264.

"I cannot repeat an emotion. No one can, except sentimentalists." - Oscar Wilde[1]

CHAPTER TWELVE

Emotive States (Category IV Emotions) and Time

Front-loading: Biological Rhythm, Category IV Emotions: Emotive States, Confusion, Delirium, Entropy, Fatigue, Greed, Joyfulness, Helplessness, Infradian Rhythm, Learned Helplessness, Mood, Post-Traumatic Stress Disorder (PTSD), Restlessness, Surprise, Ultradian Rhythm

The formal implementation of the variable of Time into the equation enables the analysis of less specific affective states such as moods. Generally, moods are emotions that have relatively short durations. Fortunately, accounting for moods in the equation is a straightforward process. Moods can be understood as trends or anomalies in a function or group of functions. For instance, the value of a specific entity may change, it may be unfamiliar, or the values of a collection of entities may change unbeknownst to the individual. Moods will be referred to as Category IV Emotions, or Emotive States due to strongly being determined by the variable of time. At least six emotions will be classified as emotive states in the next section, though the possibility for classifying others is possible. One of these emotions, indifference, has already been identified, but it will be given more redress here. The other emotive states specifically identified are surprise, restlessness, joyfulness, helplessness, and confusion.

Emotive States (Category IV Emotions) and Time

Category IV Emotions: Emotive States

Surprise

To surprise is "to strike with wonder or amazement especially because unexpected."[1] Someone who is surprised might be left in astonishment or amazement at the circumstances surrounding a situation. Identifying specific things that surprise individuals, however, is a difficult task, as different people tend to be surprised by different things. However, at the core of this definition of surprise lies the notion that the outcome of an event or situation was drastically different from what was expected by the individual. For instance, this might be indicated by a sharp increase in an individual's appraised valuation of an entity when no such change was expected. Alternatively, it might be a dynamic reduction in an individual's appraised valuation of an entity if an individual expected its appraised value to continue to fluctuate wildly.

At its core, the Category IV Emotion of surprise would be indicated by a drastic change of an individual's appraised valuation of an entity. For example, this may be precipitated by new information concerning a threat of harm to an entity that causes the individual's appraised value judgement of it to spike in one direction or another. A sudden change in an entity's Appraisal variable, for instance, would be a good indicative of the emotive state of surprise.

---

[1] By permission. From Merriam-Webster's Collegiate® Dictionary, 11th Edition ©2014 by Merriam-Webster, Inc. (www.Merriam-Webster.com).

Emotive States (Category IV Emotions) and Time

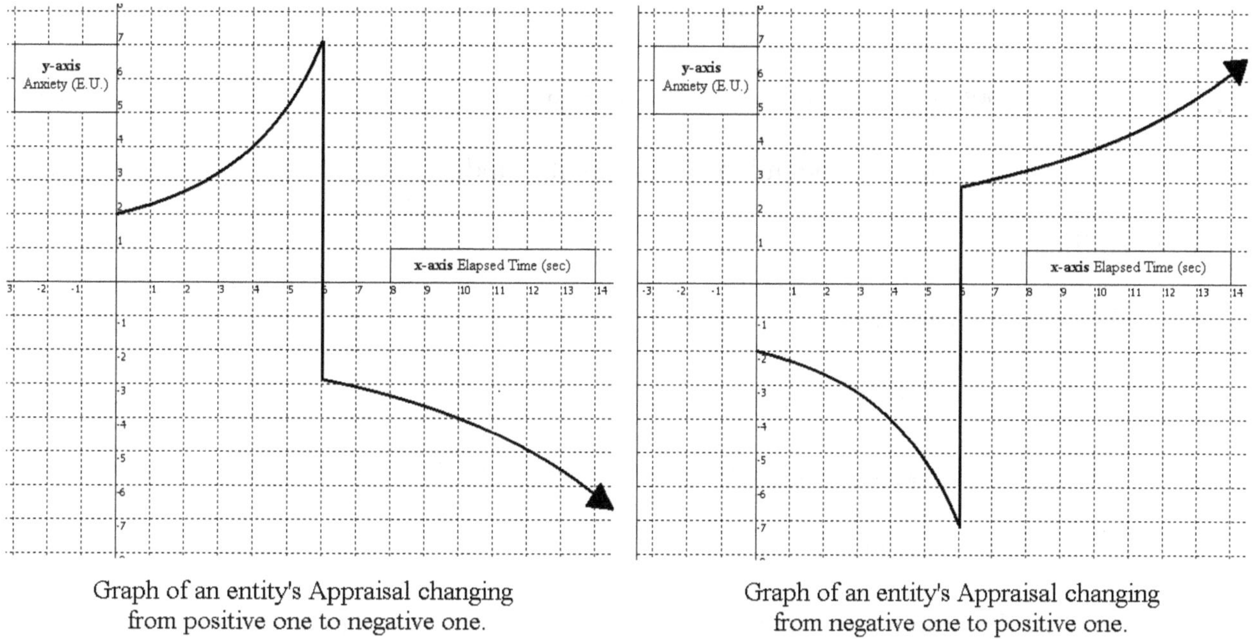

Figure 12.1 The above depicts a graph of an entity's Appraisal changing from +1 to -1 (left) and a graph of an entity's Appraisal changing from -1 to +1 (right). Additionally, Reasoning to Threat is initially greater than Reasoning to Efficacy in the image on the left, but is lesser following the Appraisal change. In the image on the right, Reasoning to Efficacy is initially greater than Reasoning to Threat, but is lesser following the Appraisal change.

Restlessness

Someone who is restless is "characterized by or manifesting unrest especially of mind."[2] Unrest suggests that an individual is in an agitated condition. These descriptions imply that a person afflicted with restlessness is dealing with his or her issues of the day at a frenetic pace. This may develop, for instance, if two or several purposes and recommended courses of action conflict with

---

[2] By permission. From Merriam-Webster's Collegiate® Dictionary, 11th Edition ©2014 by Merriam-Webster, Inc. (www.Merriam-Webster.com).

one another in such a manner that all of their values become elevated. Hypothetically, if a C.E.O. has ten different corporate negotiations to attend to on top of running a corporation itself, conflicts between different obligations are bound to arise, notwithstanding any that may be present in his or her personal life besides.

At its core, the Category IV Emotion of restlessness would be indicated by an abnormal elevation of one or several of an individual's appraised valuations of entities. This might be precipitated by the inability to marshal enough resources (e.g., Reasoning power) because they are spread too thinly between different purposes, the purposes were not prioritized in the most efficient way possible (e.g., Sentiment), or too many purposes have been brought to the fore. Trying to tackle ten purposes at once, for example, instead of giving priority to time sensitive purposes or prioritizing them in a way that will optimize one’s anxiety investment can lead to an elevation of all of their appraised valuations and an overinvestment of anxiety. Also, the creation of additional purposes, or setting as one's purpose the purpose of having more purposes, can spread an individual's anxiety resources too far and make him or her begin to feel as if there is not enough time to do everything.

Emotive States (Category IV Emotions) and Time

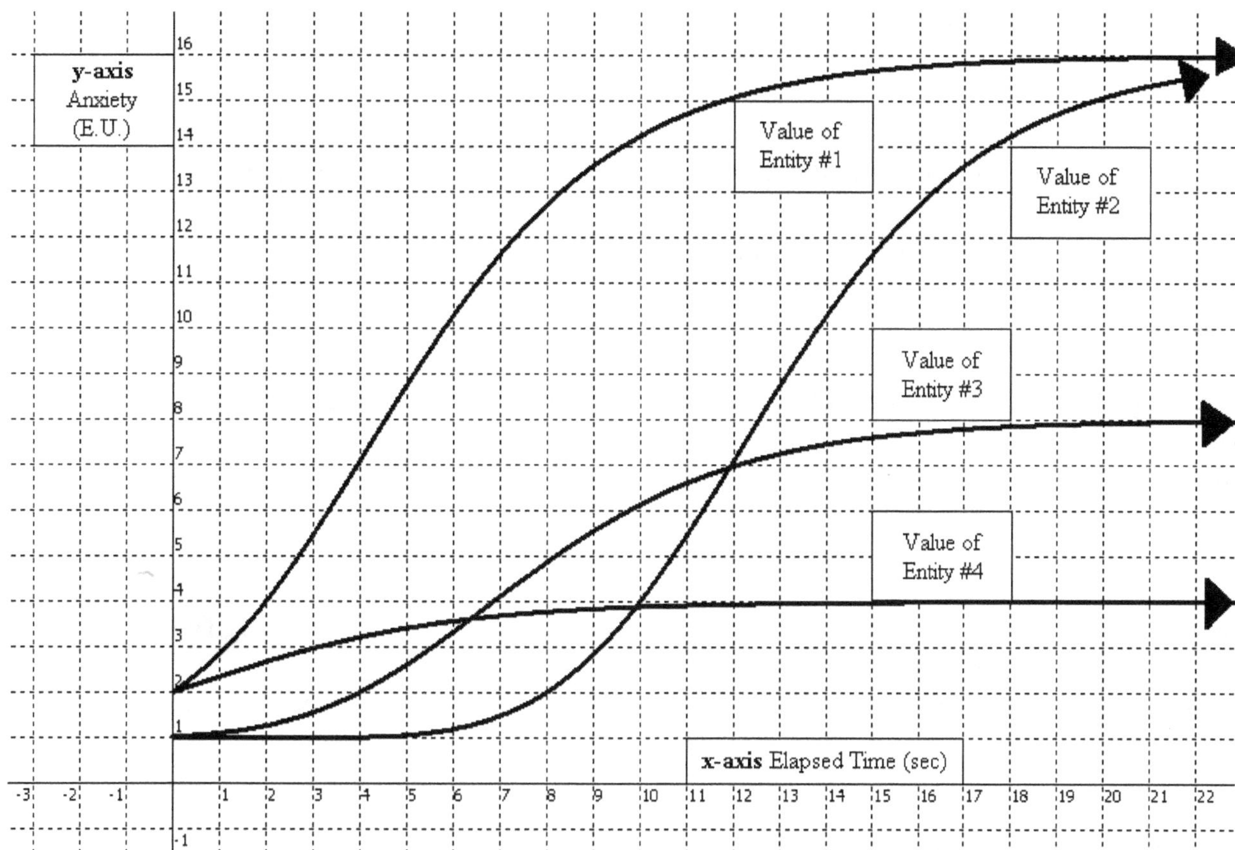

Figure 12.2 The above is a graph of multiple entities whose valuations and the anxiety invested in them have become abnormally high. The individual may have several purposes to which he or she is attending and various entities to acquire. A feeling of being overwhelmed with how much is yet to be done (i.e., entities that need to be acquired) precipitates a sense of urgency to accomplish everything at once.

Emotive States (Category IV Emotions) and Time

Joyfulness

Joy is generally thought of as synonymous with happy, though for a considerable amount of time. An individual experiencing joyfulness is successfully settling his or her issues of the day and has gone above and beyond handling them and need not worry about them for a given time. This is exemplified in hoarding or caching behavior. A farmer who has stockpiled enough grain to last the winter and has extra grain available for security or trade may feel a sense of joyfulness.

At its core, the Category IV Emotion of joyfulness would be indicated by an abnormal lowering of one or several of an individual's appraised valuations of entities. This might be, for instance, precipitated by a surplus in one's anxiety resources available to value entities and subsequently the anxiety invested in the currently valued entities may not seem as taxing. In the case of joyfulness, because the values of entities concerning specific purposes are negative due to being acquired in excess, an individual's energies can be devoted elsewhere if he or she chooses or nowhere at all. It is plausible that one might set the purpose of having no purposes to be one's purpose, thus preventing one's anxiety resources from being overinvested and spread too thin.

Emotive States (Category IV Emotions) and Time

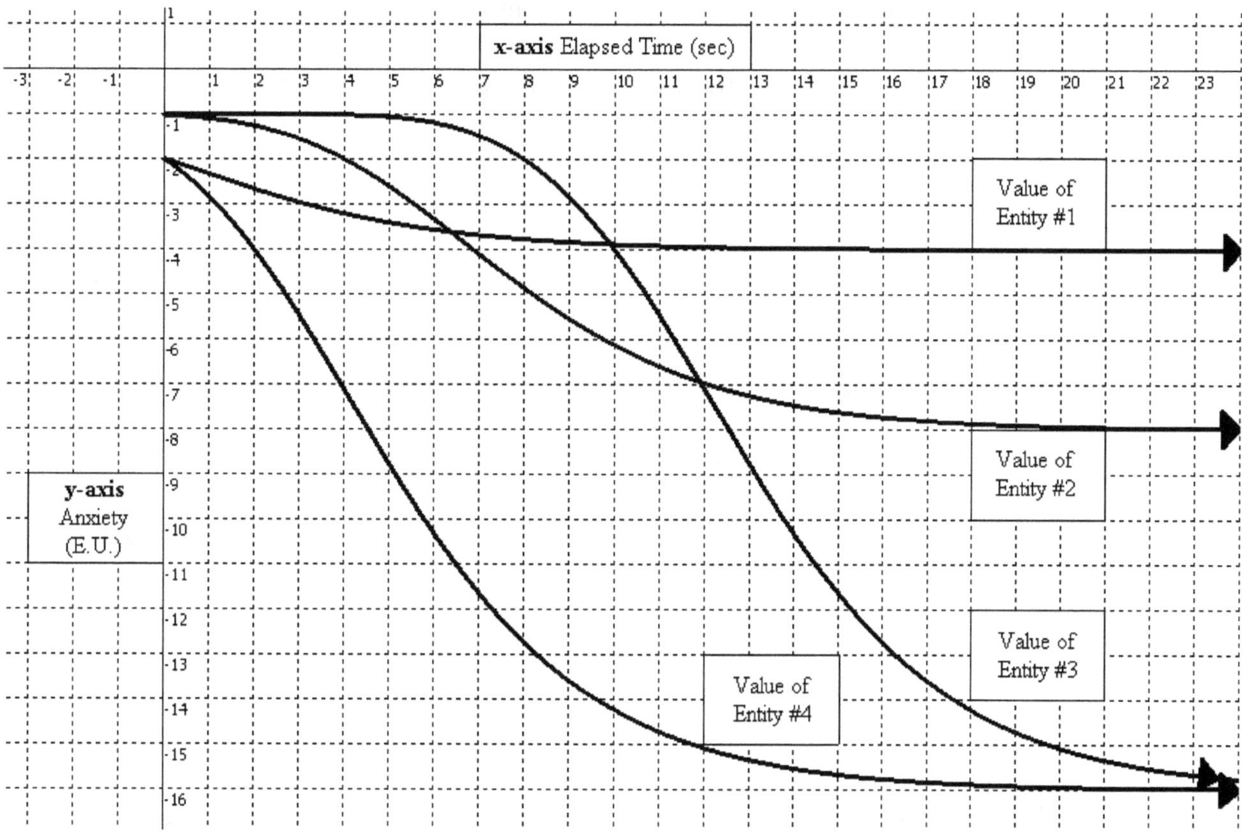

Figure 12.3 The above is a graph of multiple entities whose valuations and the anxiety invested in them have become abnormally low. The individual may have several purposes to which he or she is attending and various entities that have already been acquired well beyond what is required for homeostasis. A feeling of being at ease with how much has already been accomplished might precipitate a sense of elation and/or the desire to invest anxiety elsewhere in new ventures.

Indifference

Indifference, if it is recalled, refers to a state where one has no preference between alternative choices.[2] This suggests that an individual who is feeling indifferent would have an absolute valuation of entities that is at or near their minimum of one. At its core, the Category IV Emotion of indifference would be indicated by a flat-lining of the values of one or more entities where the value hovers near positive one or negative one depending on the Appraisal of the entity. Necessarily, the utility value of entities would be de-emphasized to the point where they are negligible. This might be precipitated by a person already being at homeostasis and not wishing to concern himself or herself with the usefulness of entities that surround. An individual, who is at equilibrium or homeostasis with respect to a purpose and its complementary purpose, would have no immediate preference as to whether or not this entity became acquired or inaccessible. In a sense, he or she might be content with simply acknowledging its existence and no more (e.g., apathy, disinterest).

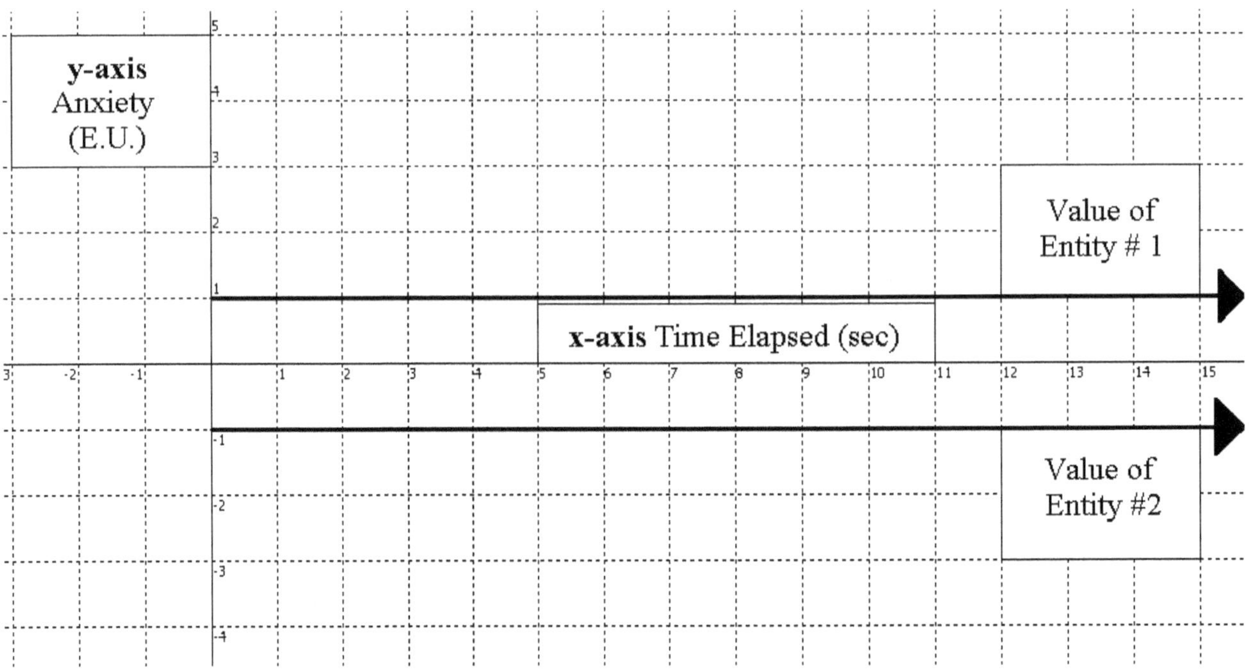

Figure 12.4 The above depicts a graph of two entities whose valuation is at its minimum or existential value. Although the Appraisal value is still being considered, the individual is indifferent to any utility value in the entity.

Emotive States (Category IV Emotions) and Time

Confusion

One who is confused is generally unclear about things. Confusion is sometimes likened to being in a state of delirium or clouded consciousness where "hallucinations, illusions, and misinterpretations of events" are prevalent.[3] This description suggests that an individual may be unclear about the factors surrounding a decision or choice to be made.

At its core, the Category IV Emotion of confusion would be indicated by one or more variables in an equation being unknown, being known only within a specific range, or being mixed up with other variables and values from other entities, such as in the case of hallucinations or misinterpretations. In general, there is a quality of distortion about the function in the sense that an entity's value may fluctuate due to a variable only being known to an estimated range. Alternatively, the fluctuation may be due to a variable being completely misplaced. This may happen in the case where interference is accepted from another entity's valuation or from a different purpose.

Emotive States (Category IV Emotions) and Time

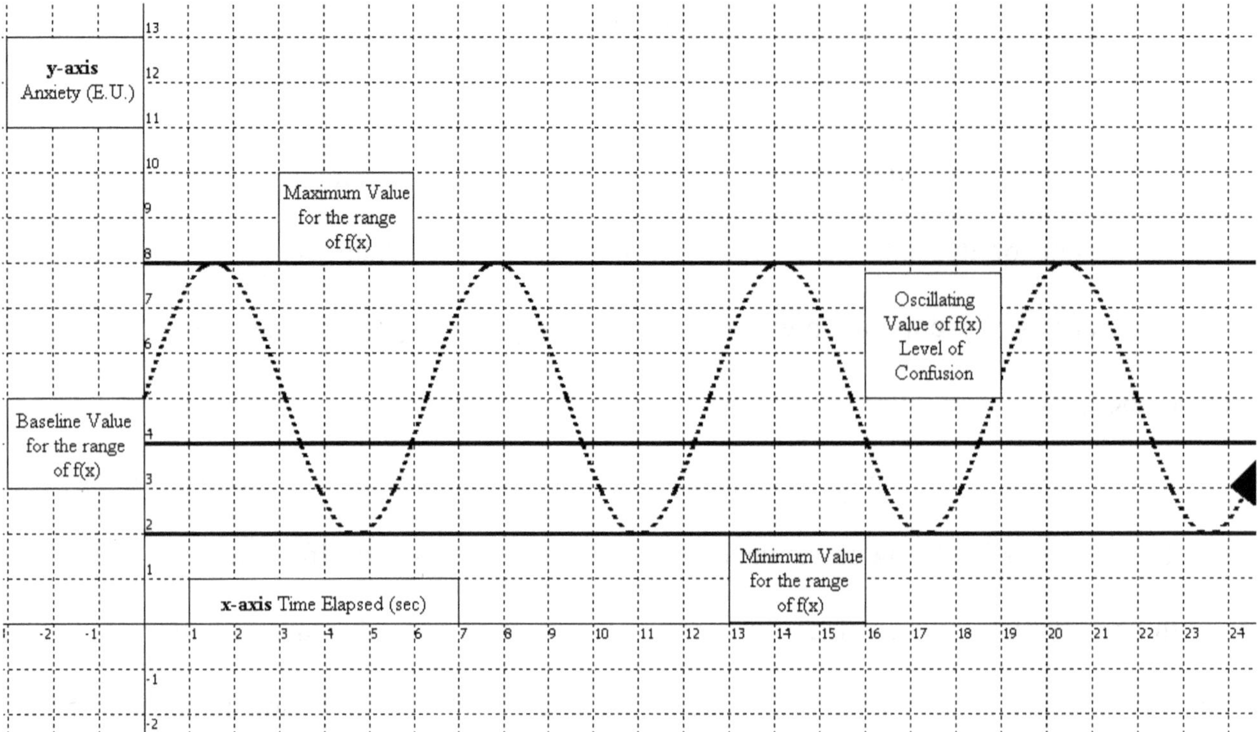

Figure 12.5 The graph (dotted line) is of an entity whose valuation is uncertain. The solid lines at y = 4 (baseline value), y = 2 (minimum value), and y = 8 (maximum value), correspond to the valuation of an entity where Attention is balanced in the denominator and numerator, and all other variables equal one except for Self-Efficacy and Threat susceptibility. Self-Efficacy is held constant at 25% (.25). Threat Susceptibility initially equals 50% (.5), but fluctuates between 75% (.75) and 25% (.25) due to conflicting information the self receives regarding the likelihood of the threat. A Threat Susceptibility of 75% (.75) would correspond to the maximum level of anxiety invested in the entity, where y = 8. A Threat Susceptibility of .25% would correspond to the minimum level of anxiety invested in the entity, where y = 2. Uncertainty with respect the entity's value due to the oscillation of Threat Susceptibility (.25 Threat Susceptibility < .75) is modeled on the sinusoidal function y = 3 (sin (x)) + 5 where an individual assesses Threat Susceptibility as 25%, then 75%, and back to 25% again approximately every six seconds (2 pi seconds). The individual is confused with respect to deciding whether he or she should invest or divest anxiety in the entity.

Emotive States (Category IV Emotions) and Time

Helplessness

Learned helplessness is a term that was coined by Seligman and characterizes a state produced by ". . . exposure to noxious, unpleasant situations in which there is no possibility of escape or avoidance."[4] It would be reasonable to say that a helpless individual might feel as if he or she lacked sufficient efficacy (response or self) in order to acquire entities to adequately fulfill his or her purposes. Moreover, the individual's Reasoning resources might also be unable to adequately direct Attention away from the inability to prevent a threat of harm, thus exacerbating the sense of powerlessness against an obstacle. Many cases of learned helplessness, a term coined by Martin Seligman, involve trauma and are produced by "exposure to noxious, unpleasant situations in which there is no possibility of escape or avoidance."[5] At the extreme end of this are stressful or traumatic situations, such as a "natural disaster, a bad accident, war or rape" that may lead to post-traumatic stress disorder.[6]

At its core, the Category IV Emotion of helplessness would be indicated by the efficacy variables being chronically low against threat. Moreover, one's Reasoning ability would be unable to adequately redirect Attention and anxiety resources away from the source of angst or to efficacy. In general, the individual's anxiety would be at the beck and call of whatever threats to entities are present in an environment; this might lead to an individual taking a passive role in the allocation of anxiety, negative anxiety resources, and being completely at the mercy of his or her emotions. In the equation, an elevation of anxiety (pain) and a decline in negative anxiety (pleasure) are nearly guaranteed when the efficacy components approach zero, the exception being the case where the threat components are also zero.

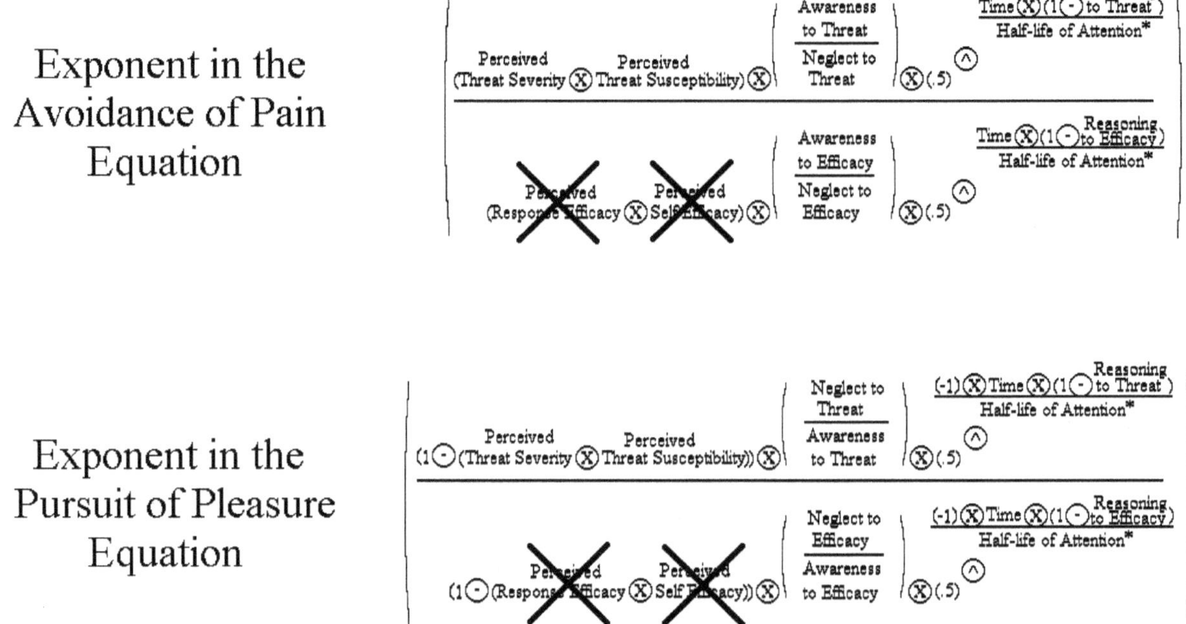

Figure 12.6 When the efficacy components approach zero in both the Avoidance of Pain function and the Pursuit of Pleasure function, overall anxiety (pain) may be expected to increase and overall negative anxiety (pleasure) may be expected to decrease. An individual who feels helpless would be expected to lack efficacy.

Modeling Miscellaneous Emotions and Affective States: Greed and Limerence

Greed is "a selfish and excessive desire for more of something (as money) than is needed."[3] Someone who is greedy or feeling greedy would be in a state where they feel as if they are desiring

[3] By permission. From Merriam-Webster's Collegiate® Dictionary, 11th Edition ©2014 by Merriam-Webster, Inc. (www.Merriam-Webster.com).

something beyond reason. This definition suggests that an individual who feels greedy has determined that his or her desire and the corresponding purpose to possess an entity conflicts with another purpose the individual holds (e.g., one that is not the complement of the main purpose). The original purpose and compliment may, for instance, might be the one of eating food to eliminate hunger and its complement of not eating food. It may also be the case that the self's success in the attempt to acquire the entity of food has directly lead to an *other* being unsuccessful at an endeavor where the self wanted them to be successful, for instance, if the *other* was unable to eat anything because the self ate all of the food. Using the equation, the emotion that the self would feel with respect to the *other's* purpose is jealousy. Complementary purposes aside, greed would then be modeled as a type of conflict that arises primarily between three different purposes: the self's purpose of acquiring food to eat; the self's purpose of vicariously experiencing an *other's* success; and the *other's* purpose of acquiring food to eat, as it conceived by the self. Greed, then, would be construed as an instance where the self's perseverance to fulfill a purpose (e.g., eating) comes at the expense of being able to vicariously experience an *other's* success. The 1:1:1:1 ratio, of course, must always be upheld.

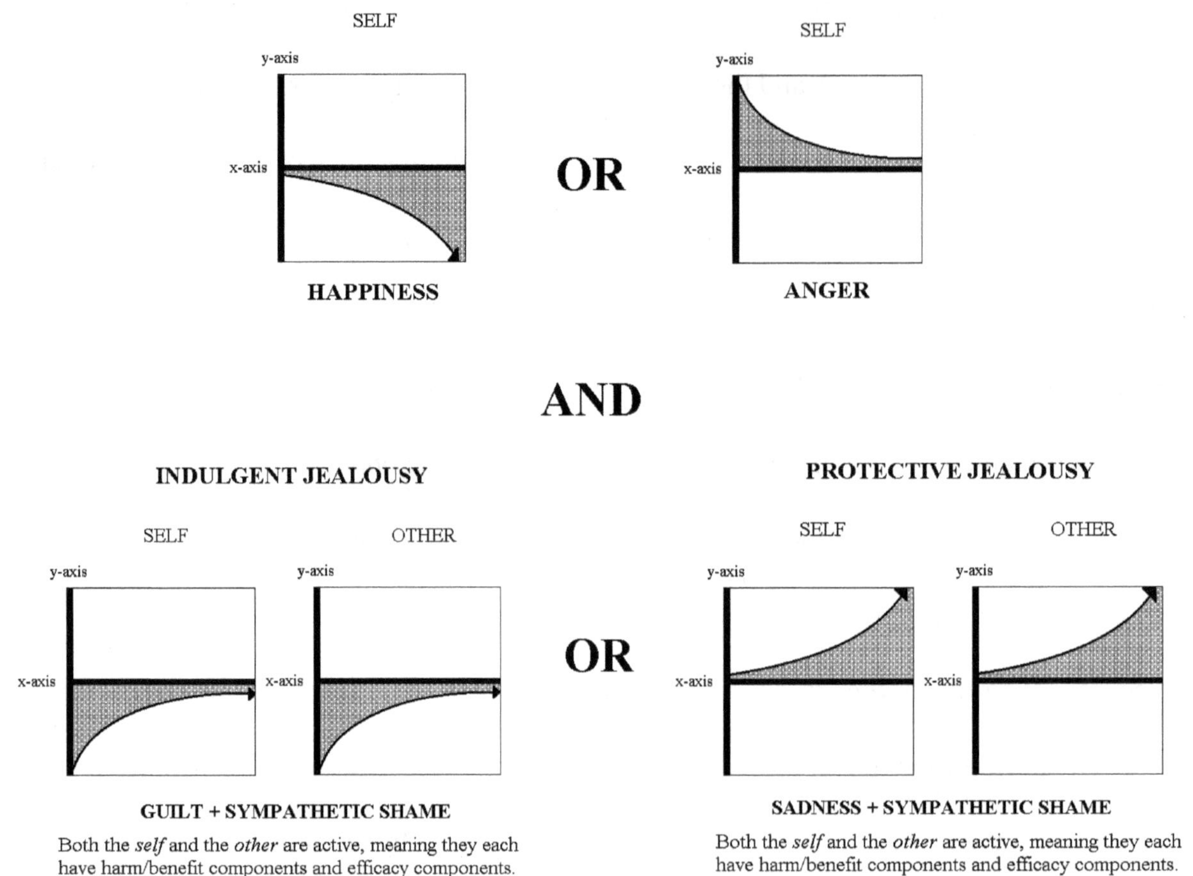

Figure 12.7 Greed in the equation is modeled as conflict between at least three purposes. The self, by fulfilling one purpose for himself or herself, inhibits an *other* from fulfilling a purpose. Moreover, the self wanted the *other* to successfully achieve an aim. Similarly, charitableness, the opposite of greed, would also be modeled with at least three purposes and slopes opposite those of greed above.

Limerence, as noted from chapter nine, is a cognitive and affective state where an individual strongly desires for a return of feeling from another person. This desire for a return of feeling may border on obsession or infatuation and is involuntary.

Like greed, limerence involves an interplay between different purposes. With respect to limerence and due to the uncertainty (e.g., confusion) harbored by the self, no less than four purposes and their four complements would be involved in the self's analysis, along with a multitude of Category I, II, and III emotions. The four purposes assessed by the self would be the following:

1) The self desires the entity of a romantic relationship with an *other*.
2) The self desires the entity of a return of feeling (e.g., vicarious experience of the *other* successfully acquiring a romantic relationship with the self).
3) The *other's* desire to acquire the entity of a romantic relationship with the self, as conceived by the self.
4) The *other's* desire for the entity of a return of feeling, as conceived by the self. This would necessarily entail the self mirroring the *other's* mirror neurons and empathy. For instance, the self imagines the *other* to want vicarious experience of the self successfully acquiring a romantic relationship with the *other*.

The four purposes above would be judged against whichever purposes are linked to their complements (e.g., the self not desiring the entity of a romantic relationship with an *other*, the self not desiring a return of feeling) Additionally, two things would be certain. The self desires the limerent object's affections (e.g., romantic relationship) and the self wants a return of feeling, which is a specific purpose from the *other*. However, a return of feeling, where mirror neurons and empathy are concerned, becomes complex. It is a short walk down a slippery slope from the self wanting a return of feeling to the self wanting to successfully experience the *other* vicariously experiencing and wanting the self to succeed. The result in such a scenario might be the equivalent of someone looking at his or her own reflection in two mirrors bounce back and forth eternally.

Finally, the self would simultaneously consider that an *other,* the limerent object, is returning a feeling while weighing it against the possibility that the *other* is not returning a feeling, as limerence entails reason to doubt. Hence, if the self desired a return of the feeling of love from an *other*, then the self would simultaneously consider the possibility that the *other* has the purpose of loving the self held above not loving the self as its purpose and vice versa. The effect would be dizzying to say the least.

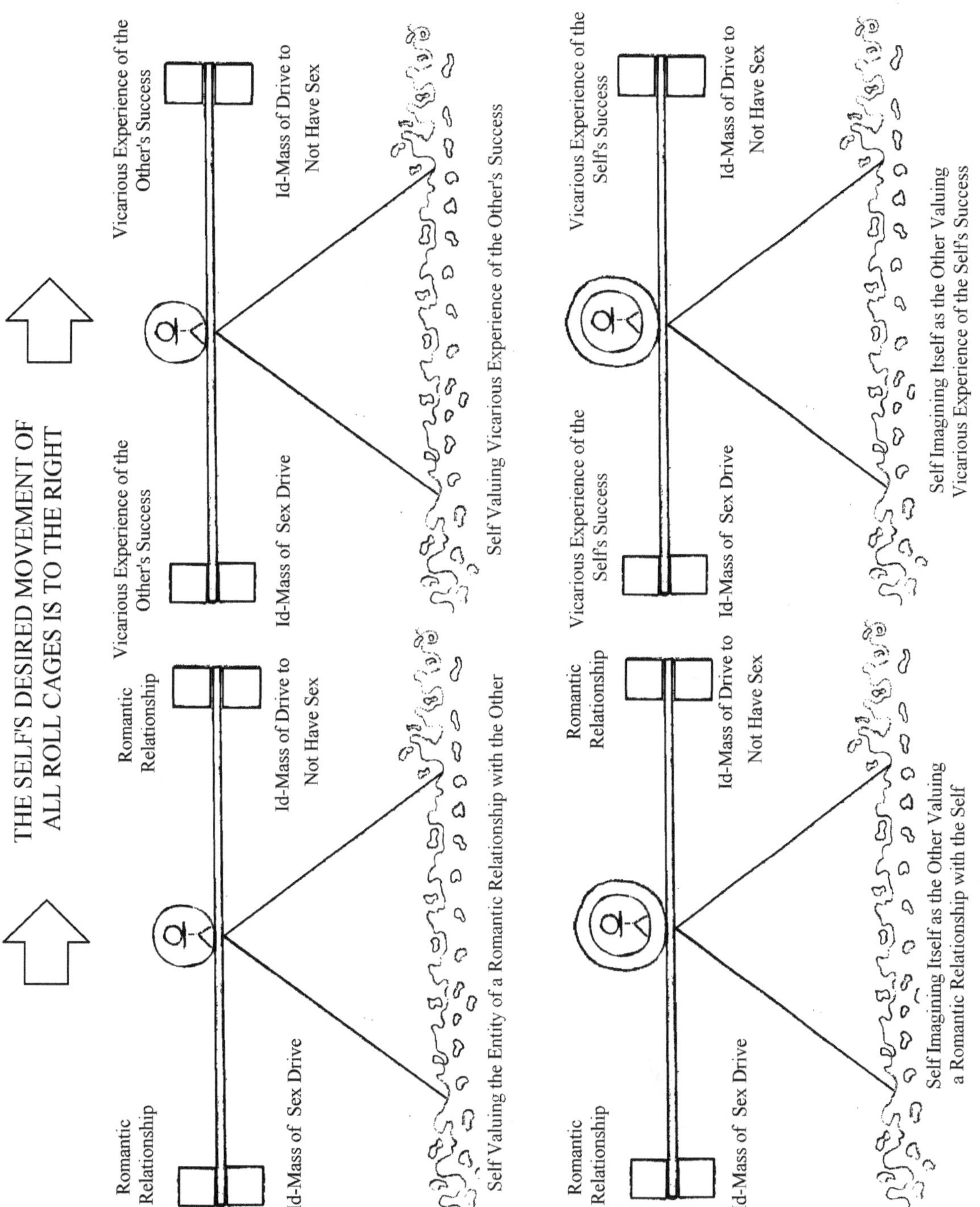

Figure 12.8 Limerence is depicted with the Roll Cage Theory of Drives. The self desires each roll cage in the above model to move to the right to acquire a romantic relationship (top left), vicarious experience of the *other* acquiring one (top right). Interpretations of the *other's* acquisition of a relationship and vicarious experience are on the bottom.

Emotive States (Category IV Emotions) and Time

Time and Special Considerations:

Absolute Time vs. Relative Time

Usually, time is understood in the absolute sense (e.g., an atomic clock). However, in psychology, time and the perception of it are distinguished from each other. In the literal sense, time is “the measured or measurable period during which an action, process, or condition exists or continues.”[4] However, "from a psychological point of view, time is always dealt with relativistically."[7] The perception of time's passage concerns the ability of an individual to be aware of durations of events.[8] Experienced, or subjective time, then, is a sense of duration by an individual that is "independent of external markers like clocks, calendars, and day/night cycles."[9] This sense of time is dependent upon "internal, endogenous events," such as cognitive markers, circadian rhythms, and biological clocks.[10] Fortunately, the equation can account for each of these interpretations of time by inserting an appropriate x-value. However, the notion of biological rhythms introduces other factors into the equation that will need to be taken into consideration as well.

Absolute Time: Strengths and Weaknesses

For practical purposes, using time in a more objective sense is helpful for modeling emotions

---

[4] By permission. From Merriam-Webster’s Collegiate® Dictionary, 11th Edition ©2014 by Merriam-Webster, Inc. (www.Merriam-Webster.com).

mathematically and can be used to attempt to account for affect and emotions that are felt at all times. However, in this strength lies its weakness. Most people may not assess their emotions and affective states regularly, let alone around the clock. Using time in an absolute sense imparts a greater sense of flow to a function and a greater sense of an individual's ability to perceive his or her own affective state than may actually be warranted.

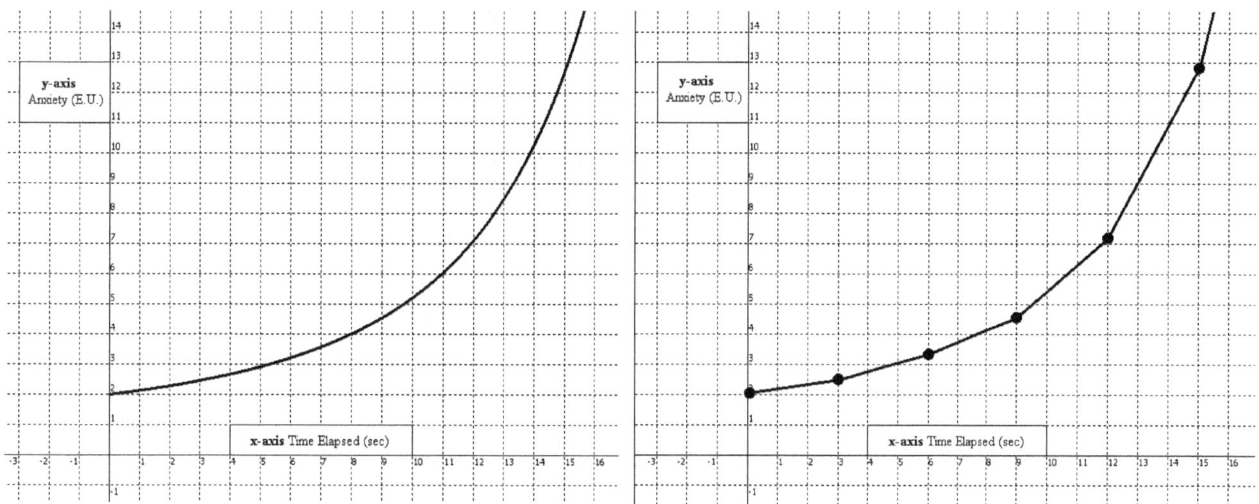

Figure 12.9 An individual is assessing an entity's value, holding Attention constant on threat components while permitting efficacy components to fall into attentional decay. The graph on the left models the rise in value constantly and as if the individual is always reassessing an entity's value. The graph on the right models the rise in value at specific intervals and as if the individual were reassessing an entity's value periodically (e.g., every three seconds).

Cognitive Markers: Strengths and Weakness

A cognitive marker is Robert Ornstein's term for a "representation of an individual mental event such as a thought, an image, or an impression" among other things.[11] Ornstein held that the passage of subjective time was related to the number of cognitive markers; he offered this as an explanation for why it "seems to take longer to travel to a new location than to an old one," given that one "notices more specific details" due to them being novel.[12] Although this may be a more realistic model of how an individual perceives time, the difficulty of formulating values of entities when more than one person is involved would make it less elegant to use, as cognitive markers are specific to an individual.

The use of cognitive markers by an individual might suggest something different to the individual than what may actually be the case. For example, the person of Gares, an uncle, might give a gift to his niece Sophia every twelve months to commemorate the passing away of her father (his brother) to cancer. He sees her on a daily basis and knows that his niece is struggling with grief after losing her father. Gares observes that each time she receives a gift she seems a lot happier than before, but her anxiety surges during the eleven months between when she receives a gift. Gares, in this case, has a near absolute sense of his niece's emotional state.

Terry, another uncle, gives a gift to Sophia at the same time as Gares. However, Terry is only able to visit her one month during the year, as he lives halfway around the world. Each time he visits she is a lot happier than she normally is the rest of the year. However, from Terry's vantage point of only having a cognitive marker every eleven months instead of each day, this would not be evident as her anxiety surges a specific amount while he is away and then diminishes before he sees her again.

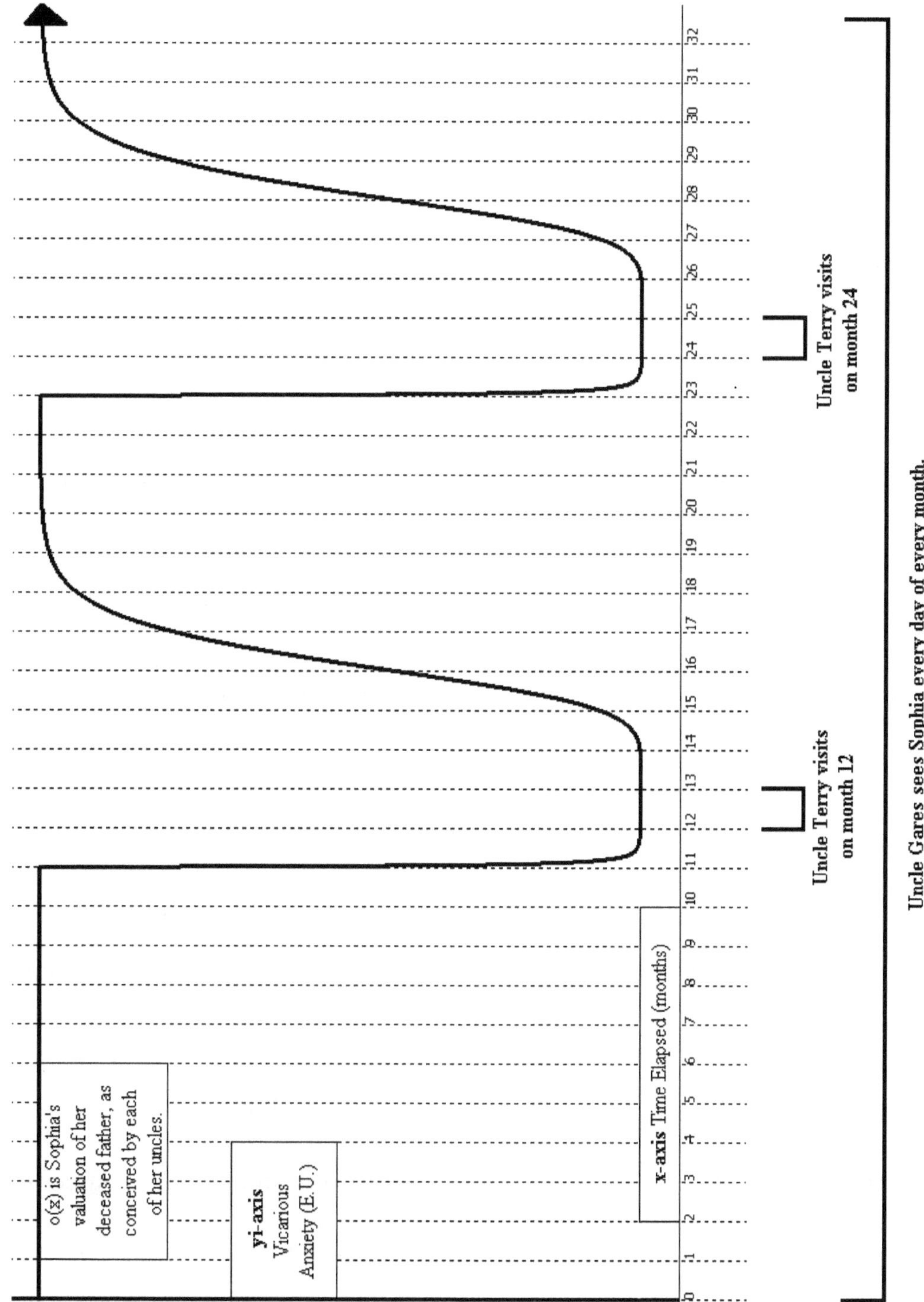

Figure 12.10 Sophia's grief for her deceased father, as gauged by her uncles.

Emotive States (Category IV Emotions) and Time

Gares, who sees Sophia every day, may realize that her happiness is only temporary and seek to find out why. Terry on the other hand, who only sees her one month out of the year and misses her bouts of grief, might have no inkling that Sophia is battling depression if he solely based his insights upon the times he has visited her. The Self-Distinction variable in both cases would be -1, as Terry and Gares each imagine themselves in Sophia's place investing anxiety in her late father and with the same level of success when they are both present.

Modeling Absolute Time Perception and a Relativistic or Cognitive Marker Based Acknowledgment of Time

Ultimately, whichever method of gauging time and the duration of events is up to the user of the equation. Each, however, will not always model the same intensity of an emotion. This can be made clear in the example of a midterm paper. For instance, Dave and Max have a midterm paper due 30 days from the first of the month and neither has completed it yet. In one scenario, Dave is thinking about the paper and its completion twenty four hours a day for the entire time. Dave's attention continuously stays on the entity as the threat of harm to the paper escalates (e.g., number of days he has not worked on the paper). All of Dave's attentional resources are devoted to it for thirty days and escalate gradually. Max, on the other hand, devotes attentional resources to it for the first eight days and then ceases to keep track of cognitive markers when he decides to go out to party for three weeks. Both reassess the paper's valuation at day twenty-nine, a day before it is due.

Increasing threat and decreasing efficacy can be modeled over thirty days by using coefficients in the exponent to represent the number of days the paper has not been worked on.

(x / 30) would be a coefficient against Threat Severity and Threat Susceptibility, and

((30 - x) / 30) would be a coefficient against Response Efficacy and Self Efficacy.

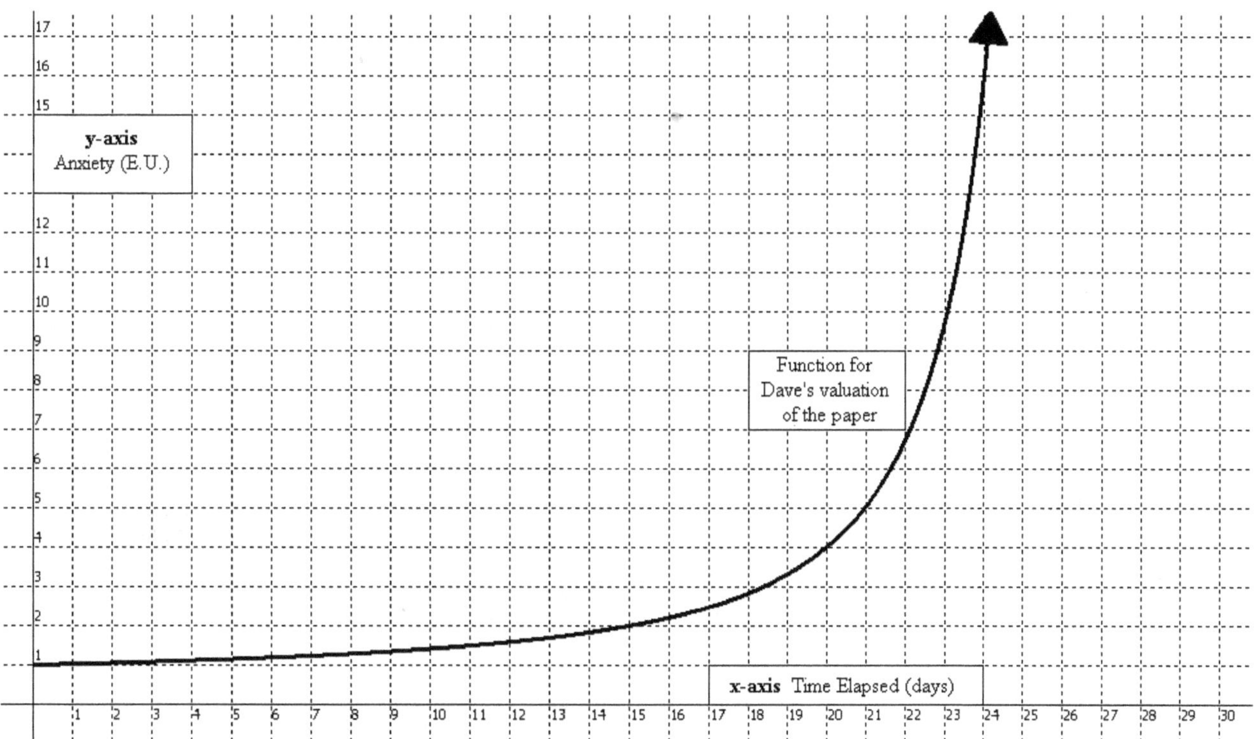

Figure 12.11 The above is a plot of the paper's value over thirty days. Dave thinks about the paper and its completion non-stop for the entire month. The paper’s value starts to elevate dramatically after day twenty and may spur Dave to act on it before too much anxiety is invested in it.

In the second scenario for Max, the value of the midterm starts out essentially the same. However, after the eighth day of the month, Max ceases to keep track of cognitive markers for twenty-one days because he goes out to party every night for three weeks. During the three weeks, and in the best case scenario, Max's valuation of the paper would keep the same trajectory it left off on at day eight. In the worst case scenario, if Max, who is initially confident that he can complete the paper, diverts Attention away from the threat components, then the valuation of the entity would diminish to its existential value of + 1. In either case, to Max it would appear as if his valuation of the paper elevated more quickly on day twenty-nine than Dave's and the graph would be more indicative of surprise for Max if neither had actually started working on it until then.

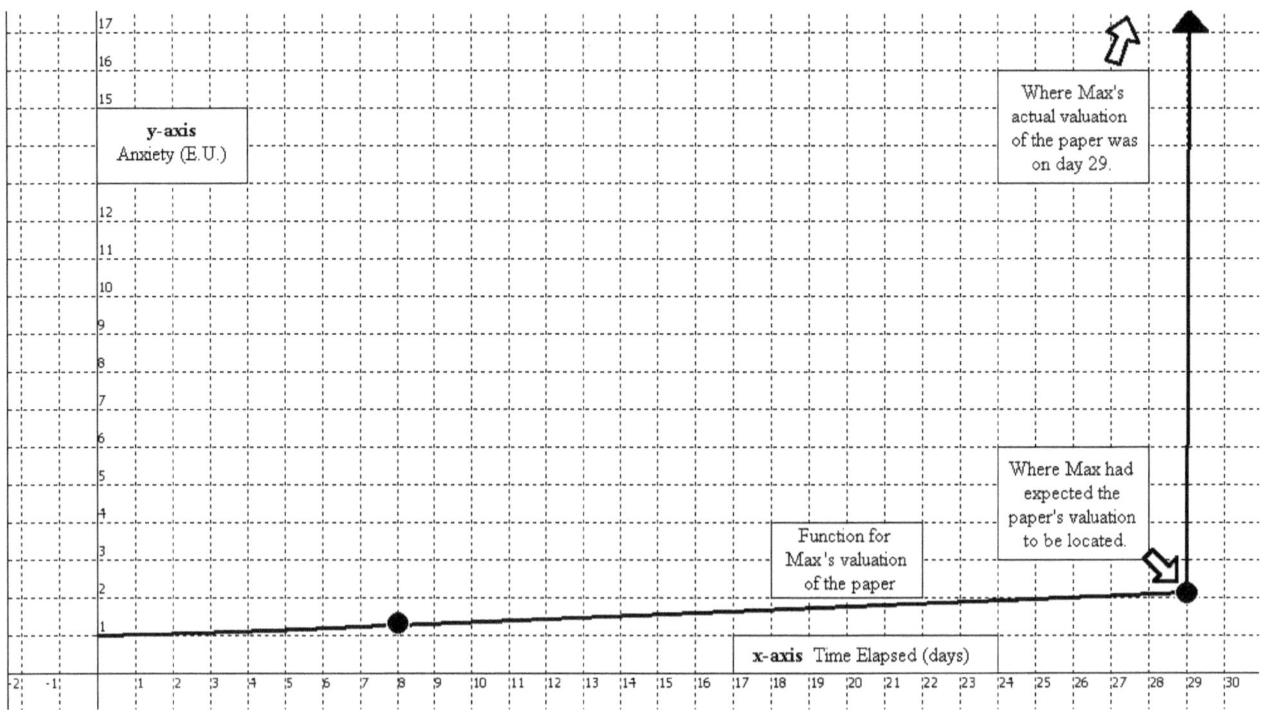

Figure 12.12 Plot of the paper's value over thirty days for Max. The graph itself is initially the same as Dave's until day eight. After the eighth day of the month, Max stops keeping track of cognitive markers and goes out every night to party for three straight weeks. The valuation of the paper, having been ignored completely may stay on the same trajectory from day eight (modeled above) if Max made a projection, or diminish to its existential value if Attention is directed away from threat components or if he temporarily alters the Sentiment variable to go party. Max's slope, when it is reassessed on day twenty-nine, appears steeper than Dave's and Max would be more likely to feel surprised, according to the model.

Emotive States (Category IV Emotions) and Time

Biological Rhythms: Infradian Rhythms vs. Ultradian Rhythms

A living organism's periodic variation in biological systems is defined as a biological rhythm, the most intensely studied of which is the circadian rhythm.[13] If these periodic variations recur in a cycle that is longer than a day or are of a "considerable duration," then they are referred to as infradian rhythms.[14] The menstrual cycle is one such example of an infradian rhythm.[15]

Alternatively, if these periodic variations recur in cycles that are shorter than a day, then they are referred to as ultradian rhythms.[16] One such example of an ultradian rhythm is the circadian rhythm of a day-light cycle, or twenty-four hours.[17] Although in technical literature "circadian time" is based on a twenty-five-hour day, as the cycle drifts from a twenty-four-hour day to twenty-five hours in most species, any cycle near this can be considered an ultradian rhythm.[18] Cortisol hormonal levels, as noted earlier, are an example of a circadian rhythm, given that they vary according to sleep-wake cycles, peaking in the morning and ebbing throughout the day. Taking into account the role of cortisol, a hormone whose activity increases blood sugar levels, this and the implications that other biological rhythms have for the equation will be considered lightly.

Fatigue

Generally, fatigue is a decline in the "ability to do work."[19] This may be due to the results of previous efforts or may be induced from a feeling of "weariness or tiredness" brought about by extended effort.[20] A number of distinctions for different types of fatigue also exist in the psychological literature: "sensory fatigue," "neural fatigue," "muscle fatigue," "emotional fatigue," and "mental fatigue."[21] The conception of fatigue in the equation incorporates emotional fatigue and may be thought of as an exhaustion of emotional units available to value. Exhaustion, in the physiological sense, is "a state in which the metabolic process has been depleted, producing fatigue, weariness and a general lack of responsiveness."[22] For the equation, it will be held that an individual who is less fatigued will have more metabolic energy, more anxiety, and more negative anxiety resources at his or her disposal to value entities, thus making entities that are being valued seemed less taxing because they occupy a smaller portion of the entire emotional units in reserve. Conversely, it will be held that an individual who is fatigued and near exhaustion will have less metabolic energy, and subsequently, less anxiety and less negative anxiety resources at his or her disposal to value entities. Entities that are being valued will seem more taxing because they will occupy a larger portion of emotion units in reserve. Because anxiety and negative anxiety are being measured by value, and value is being determined by the theoretical emotional unit, the emotional unit will need to be converted into an energy form. This would be, for example, the amount of energy required to produce one emotional unit, or joules of energy per emotional unit of value.

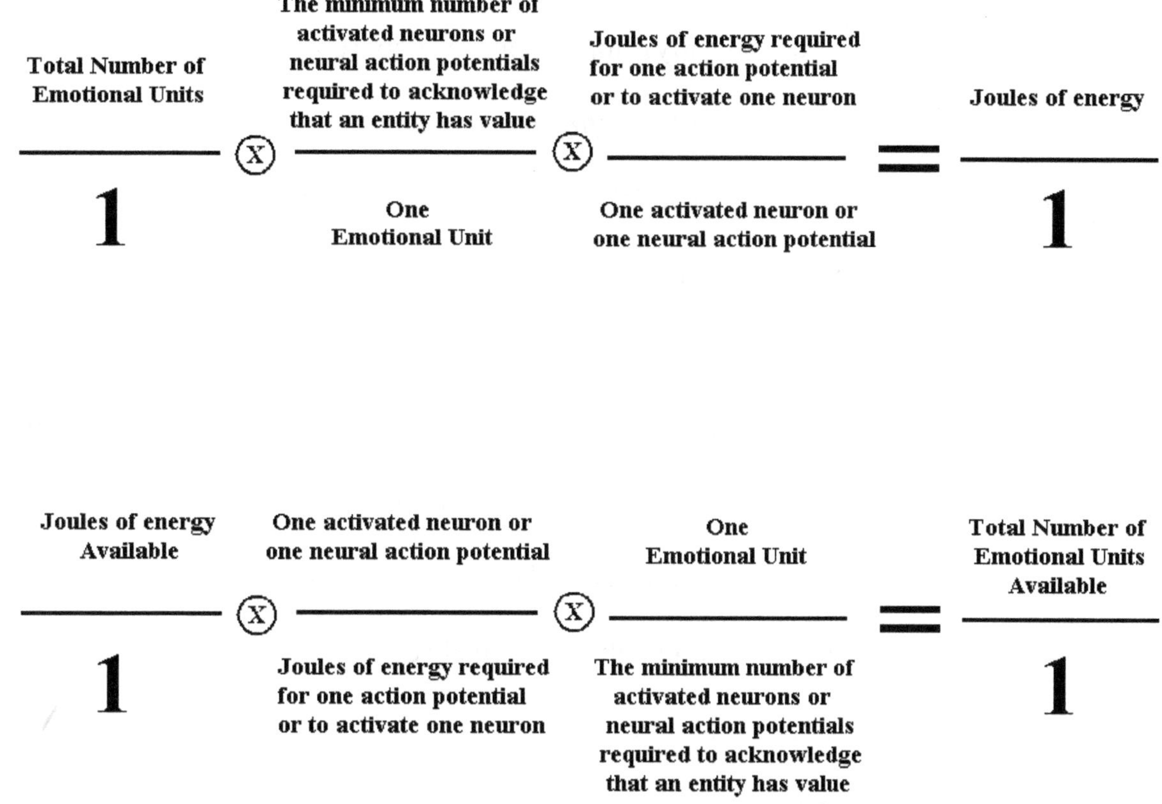

Figure 12.13 Converting emotional units used into joules of energy expended (top), and converting joules of energy available into emotional units Available (bottom)

Up until now, the equation has presumed an individual's energy level, stamina, or metabolic processes constant at 100%, with fatigue being held to zero or available energy and emotional units maximized. As biological rhythms tend to be periodic, the use of trigonometric expressions to estimate them would be ideal (e.g., sine and cosine operations). Although they are not a perfect representation of an individual's physiological state, trigonometric expressions present themselves as a good candidate for exploring this realm. The ideal location for them to be incorporated would be as a coefficient against the entire function. In this manner, multiple biological rhythms could be accounted for by the equation. For the following example, a hypothetical case will be considered where only one agent that varies periodically is influencing available energy. *The original function, if divided by the trigonometric function representing the total number of emotional units available to value entities, would change what the y-axis represents*. Instead of representing emotional units invested in an entity, the y-axis would represent emotional units invested in an entity over the total amount of emotional units available to value entities in the individual's system. This would be a fraction less than or equal to one. A biological rhythm that is on a period of 24 hours, with a maximum of 30 emotional units and a minimum of two emotional units available (i.e., amplitude of 14), where the base is two emotional units, and where the maximum and minimum emotional units made available occur at noon and midnight respectively could be in the following form:

> $15 \times \cos((x - 12) / 12\ pi) + 15 + 2$, or
>
> $a \times \cos((x + k) / (h)\ pi) + a + b$,
>
> where the cosine is measured in radians, **x** represents the number of hours elapsed since time zero, **a** represents the amplitude of the sinusoidal curve, **b** represents the emotional units or energy that cannot vary (i.e., metabolic energy on reserve for vital functions and other activities), **k** represents the horizontal shift of the sinusoidal curve 12 hours to the right so that peak energy occurs at noon, and **h** in the denominator is half the period (half of 24 hours).

In figure 12.16, the y-value will be scaled to 100 in order to see anxiety invested in an entity as a percentage of all an individual's anxiety available to value entities:

Emotive States (Category IV Emotions) and Time

$$y = Scale / (a \times \cos ((x + k) / (h) pi) + a + b)$$

$$y = 100 / (15 \times \cos ((x - 12) / 12 pi) + 15 + 2)$$

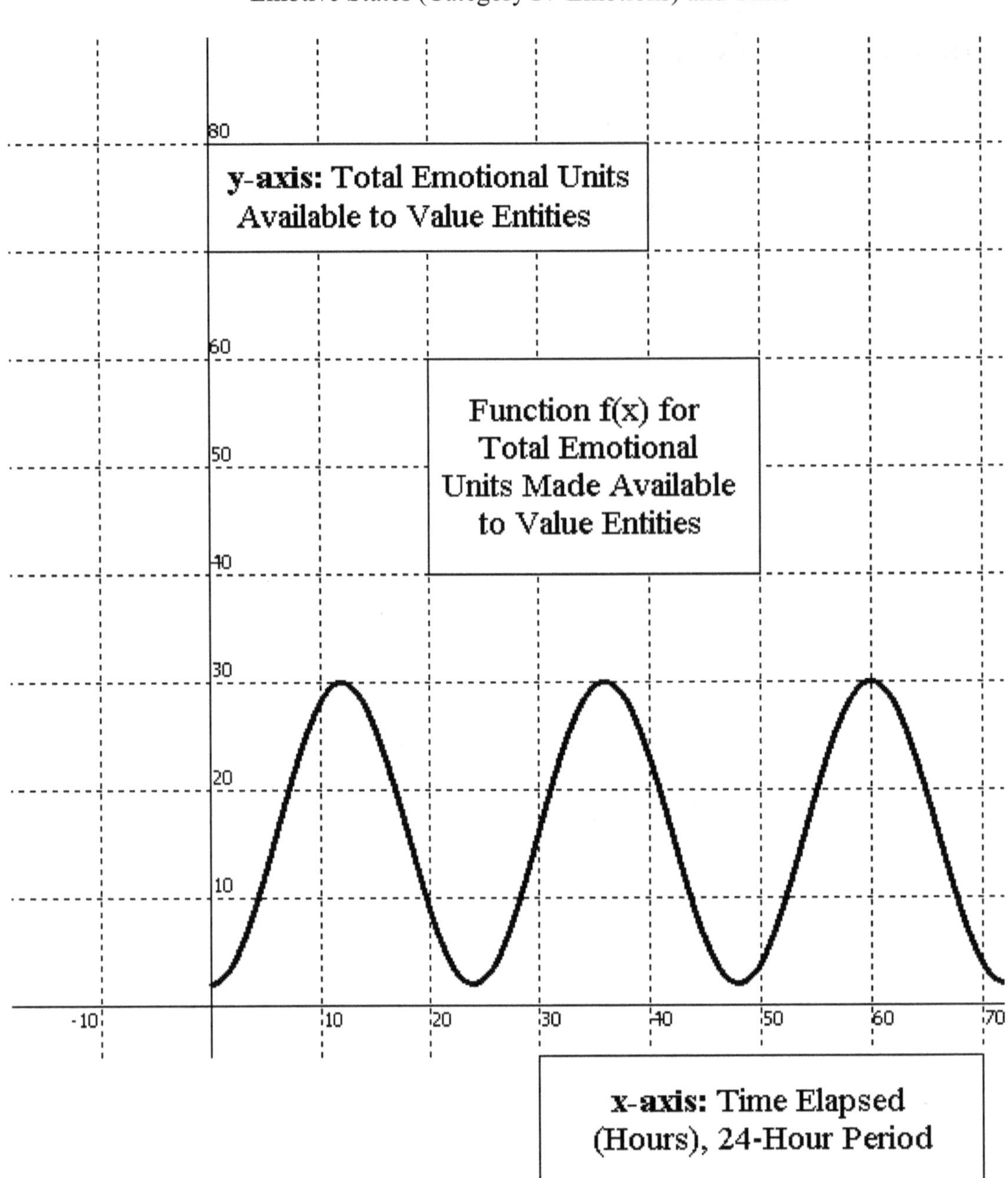

Figure 12.14 The above is a function for the total emotional units that are available to value entities. In this case, the peak amount of emotional units available (30 units) occurs at 12 hours, 36 hours, and 60 hours, or noon. The minimum amount of emotional units available (2 units) occurs at 0, 24, 48, and 72 hours or midnight.

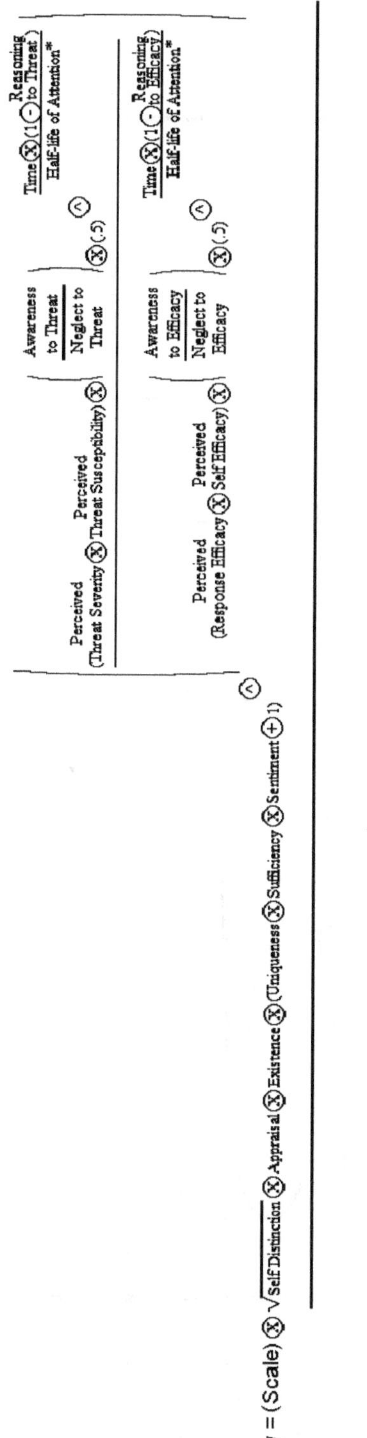

Figure 12.15 Biological Rhythms

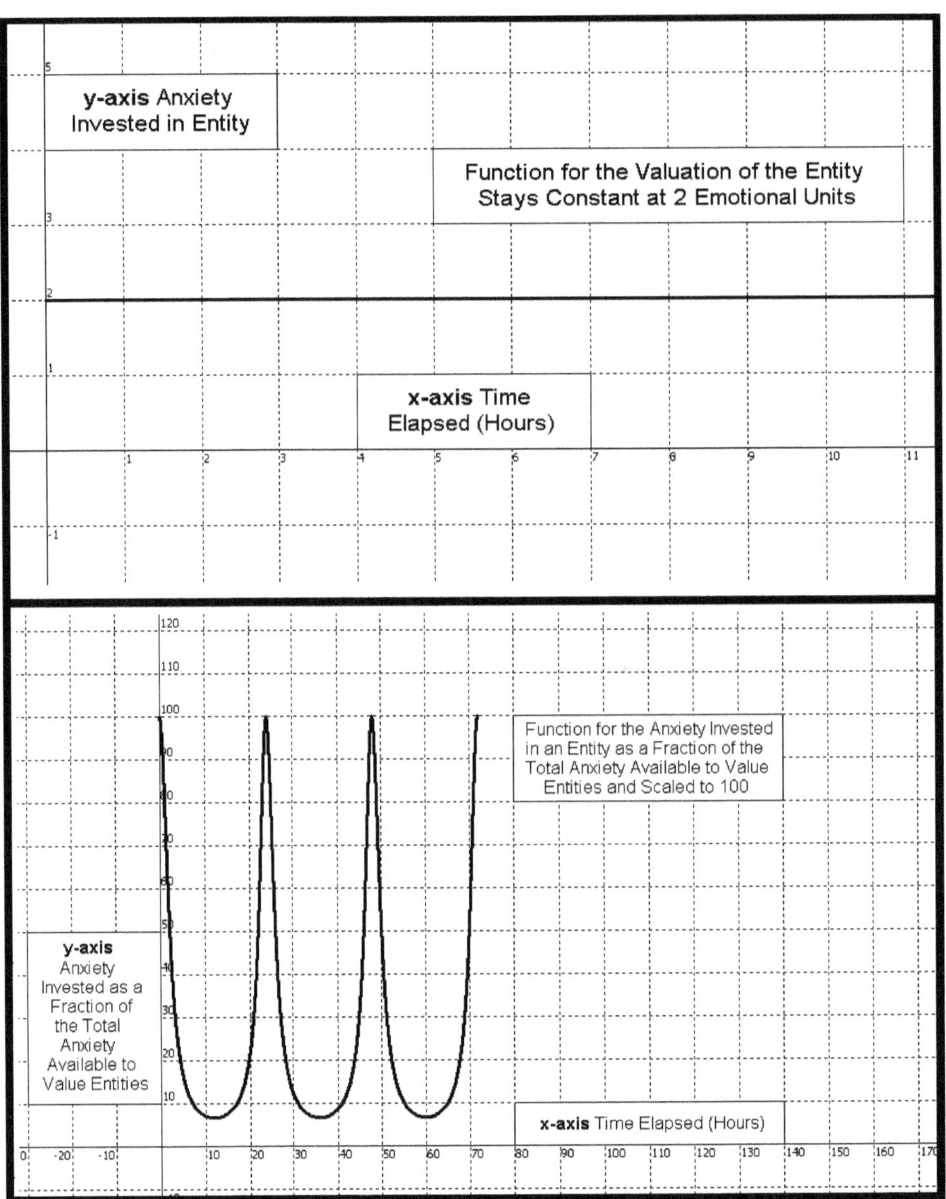

Figure 12.16 The above is a valuation of an entity (top) and graph of the function for emotional units invested in an entity as a fraction of the total emotional units available to value entities and scaled to 100 (bottom). The entity's valuation is held constant at 2 emotional units for the 72 hours that the total for the emotional units available to value entities is varying.

For example, if an individual's available energy, and subsequently available emotional units, is on a 24-hour cycle, then the denominator under the function might look like figure 12.15, where at time zero (midnight) the individual's energy available for use is at its lowest and fatigue is at its maximum (e.g., the middle of the sleep cycle). At time 12 hours (noon) the individual's energy available for use would be at its highest and fatigue would be at its minimum. If the Cosine of the function is closer to + 1, then the individual would have more emotional units available to value entities and investing two emotional units into is not terribly taxing when 30 are available. If the Cosine of the function is closer to - 1, then the individual would have fewer emotional units available to value entities and feel more fatigued. Investing two emotional units into an entity is more taxing when only four, three, or two are available. Fatigue, then, is understood to be a lack of energy or a lack of emotional units available in an individual's system to value entities.

1) A sine or a cosine expression may be used in the denominator under the entire equation so long as an appropriate shift is made horizontally.

2) Measured in radians:

   $a \times \cos ((x - 12) / 12 \text{ pi}) + a + b$,

   which reduces to

   $a \times (\cos ((x - 12) / 12 \text{ pi}) + 1) + b$.

   This would be the period for an individual's fatigue level, assuming a 24-hour cycle with peak available energy occurring at noon and the minimum available energy occurring at midnight.

   If one desired, the expression could be shifted nine hours later so that the maximum available energy occurred at 9:00 A.M. to more closely model the hormone cortisol, though this would also have the effect of shifting the minimum available energy to 9:00 P.M. as well.

This shift could be represented in a number of ways:

Scale / (a × (cos [**(x - 9)** / 12 pi] + 1) + b), or

Scale / (a × (cos [**(x + 15)** / 12 pi] + 1) + b).

If one wished to use a sine expression, then either of the following would work:

Scale / (a × (sin [**(x - 3)** / 12 pi] + 1) + b), or

Scale / (a × (sin [**(x + 21)** / 12 pi] + 1) + b).

3) The variables of ***a*** and ***b*** explained with the function itself:

a) ***a*** is the amplitude of the sinusoidal curve. It represents half of the possible range that an individual's available energy (emotional units available to value entities) can be modified by whatever agent's periodicity is being described by the trigonometric function (e.g., cortisol levels).

b) ***b*** equals the minimum amount of extra metabolic energy (measured in emotional units) that an individual must possess so that the involuntary mechanisms that keep the person alive can operate. This would be a range of available energy that must always be available and may come from other biological rhythms; this energy is vital for supporting the individual's life and also serves as an emotional unit reserve during peak fatigue. The **b** value in the equation of figure 12.16 is + 2 emotional units. This means that 2 emotional units out of a maximum possible of 30 emotional units are always available, (e.g., due to the influence of other biological rhythms). It would be present even when fatigue is at its peak for the observed rhythm (**a** = -1). The maximum percentage of energy or total emotional units in the system that can oscillate would be 93.3%. Two out of 30 E.U., or 6.6% of the individual's total energy in the system, would at least always be available from somewhere else. The ***b*** value would normally be another trigonometry function, but for the purpose of explaining can be understood as a constant value here.

c) 2a + b would represent the total emotional units available in an individual's system to value entities, or the maximum possible available energy in the individual's system to value entities. In the above scenario it is 30 emotional units.

I) If 2a + b = b, then the individual is in a state of complete rest or minimum available energy to value entities while still being alive. In the above scenario, this would be the middle of the sleep cycle at midnight.

ii) If 2a is at its maximum value, when the Cosine of ((x - 12) / 12 pi) is maximized, and b, the base available energy, is intact, then the individual is in a state of non-rest with the maximum amount of energy available. In the above scenario, this would be at noon.

iii) If 2a + b = 0, then the individual has expired (e.g., death). The equation would become undefined at this point, due to zero being in the denominator.

d) Sample (next page)

For instance, if the **b** value were set to 1% of the total energy in a person's system (e.g., 1 / 100), then for the equation, this would mean that an individual's emotional units available to value entities would only oscillate between 1% and 100% of their total number in the system. The 1% of the emotional units or the energy available corresponding to the **b** value would serve to both maintain vital functioning and act as an emotional unit reserve. The **a** value would then be approximately 99 / 2 or 49.5, meaning that his or her energy level would periodically fluctuate between 1% and 100%, or 49.5% above and below the node of 50.5%, based upon the individual's cortisol levels. In figures 12.14, 12.15, and 12.16, the individual's energy level or emotional units available to value entities, could only oscillate between 6.66% and 100%.

The Problem of Multiple Biological Rhythms

In the example above, the coefficient **a** is the maximum amplitude of the sinusoidal curve and it designates half of the possible range that an individual's available emotional units could oscillate. However, it is unlikely that only one biological rhythm, such as cortisol level, would influence an individual's energy levels on its own. A proverbial stew of biological rhythms would most likely be at play to influence one's energy level. Moreover, the potential for alterations in these periodic cycles, such as those brought on by drug use (e.g., sleep medication or caffeine), additive and synergistic effects (e.g., between hormones or chemicals) would have to be considered. Once these are determined, the mystifying realm of dreams and nightmares can be addressed with respect to biological rhythms, and modeled using valuation resilience.

Drug Use, Multiple Biological Rhythms, Alterations in Periodic Variance, and Dreams

If one wanted to account for drug use that influenced the periodicity of a biological rhythm, such as by stretching or shrinking its period, then a few things must be done. First of all, drugs must be defined; thereafter, they may be implemented in the equation.

1) Drugs are essentially non-food substances that alter the body in some way, and include antidepressants along with narcotics.
2) Drugs would be identified as an entity in the equation, as an individual would recognize that the drug has a psychological or affective value.
3) Drugs would also have a physiological value, in the sense that they change an individual's regulatory processes and biological rhythms. This physiological value would generally be beyond the capacity of the individual to control directly, but would nevertheless influence processes that concern affect. This physiological influence on the individual would have to be accounted for as well. As far as biological rhythms are concerned, analysis of the influence would entail observing a change in an item's periodicity. A reduction or change in Reasoning capacity or Attention are other areas that could be considered as well.

For instance, a drug might be shown by the equation to act by increasing the duration of a hormone's periodicity while it is in the body. In the case of cortisol, if a drug acted in such a manner as to increase the length of the period for cortisol and the time it stayed in an individual's body, then this would be represented by a change to the coefficient beside (x - 12). In the expression below, the coefficient is originally equal to one:

> a × (cos ((x - 12) / 12 pi) + 1) + b = the sinusoidal period for an individual's energy level, assuming a 24-hour cycle with peak cortisol and available energy occurring at noon, and the minimum available energy occurring at midnight.

Increasing the duration of the period by a factor of two would be accomplished by changing the coefficient to .5 instead of 1 for (x - 12):

> a × (cos (**.5** (x - 9) / 12 pi) + 1) + b, (figure 12.17).

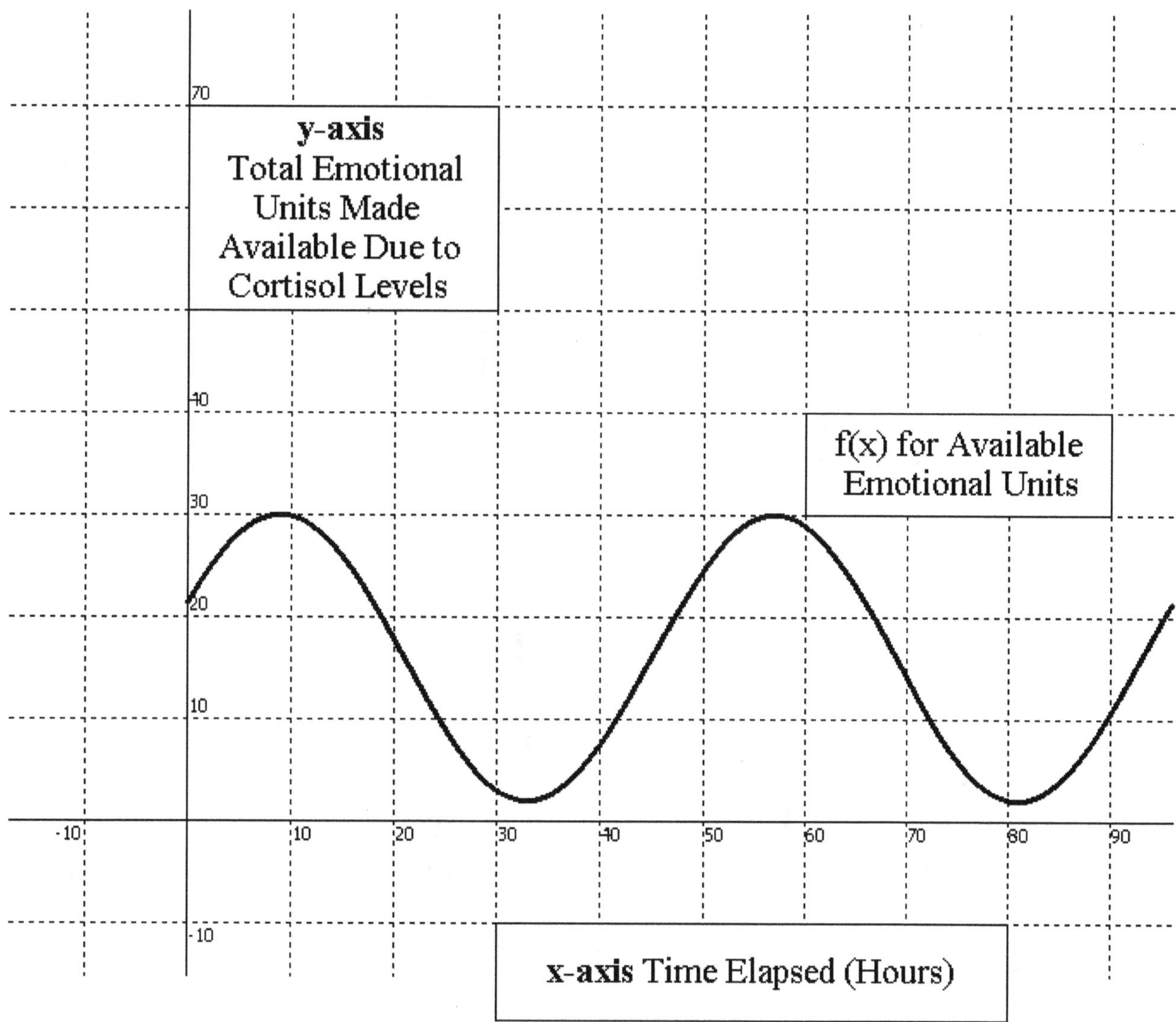

Figure 12.17 The above is a graph of cortisol's period being lengthened to 48 hours due to an increase in its presence

This would mean that an individual's cortisol levels would fluctuate as if they were on a 48-hour period instead of a 24-hour period due to the effects of a drug. The emotional units invested in an entity, as a fraction of the total energy available, would be affected accordingly (figure 12.18).

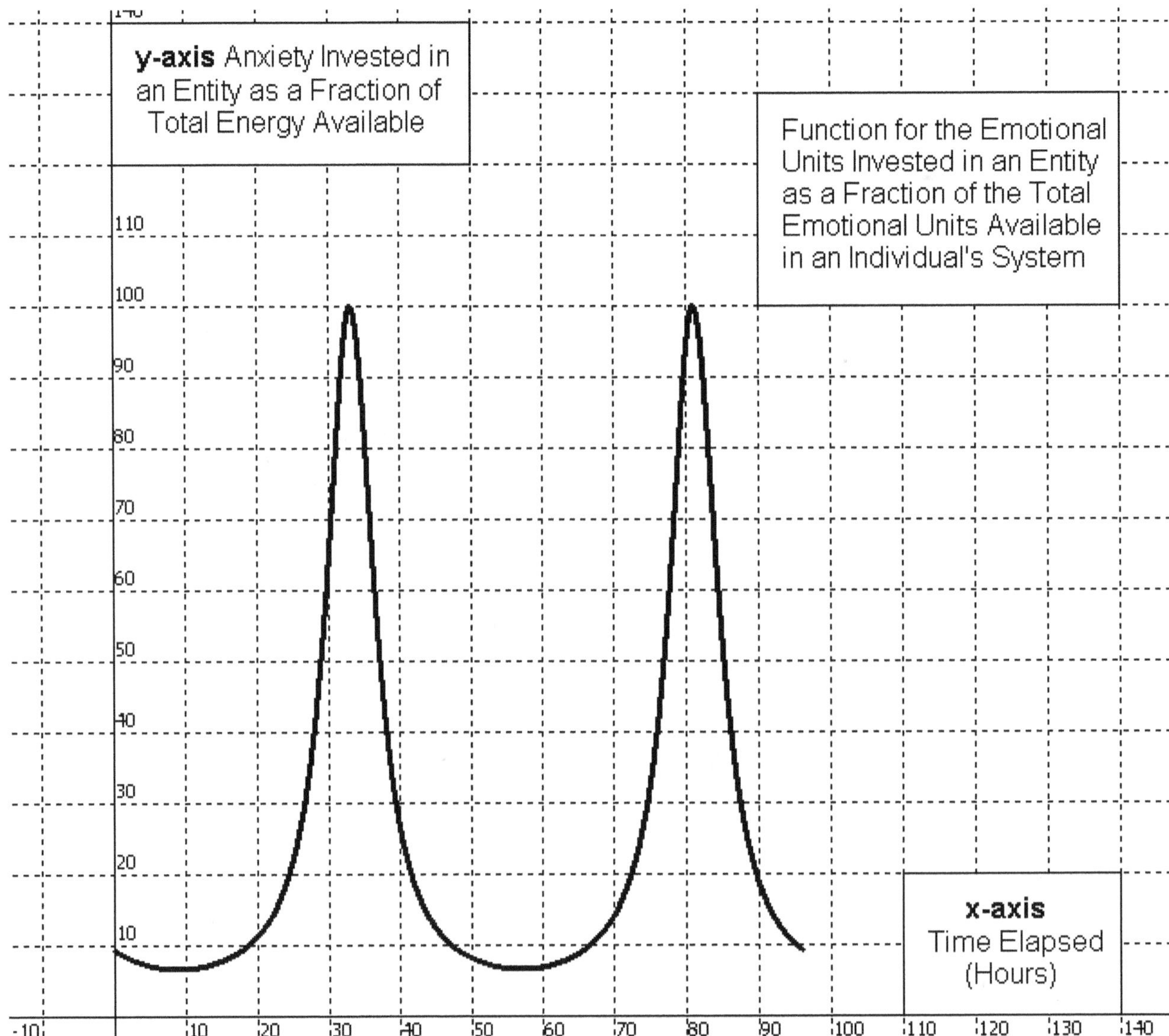

Figure 12.18 Above is a graph of an entity with a constant valuation of + 2 emotional units against the total emotional units made available due to cortisol levels, if hypothetically speaking, cortisol were the only biological rhythm that influenced metabolism and energy available to value entities and its period were doubled.

If a drug acted in such a manner as to decrease the amount of cortisol, and subsequently the emotional units available to value entities in the body, this might also be represented by a change to the coefficient beside x when it was originally one in the equation. For instance, decreasing the duration of the period by one half would be accomplished by changing the coefficient outside the parentheses of x:

a × (cos [**2** x / 12 pi] + 1) + b,

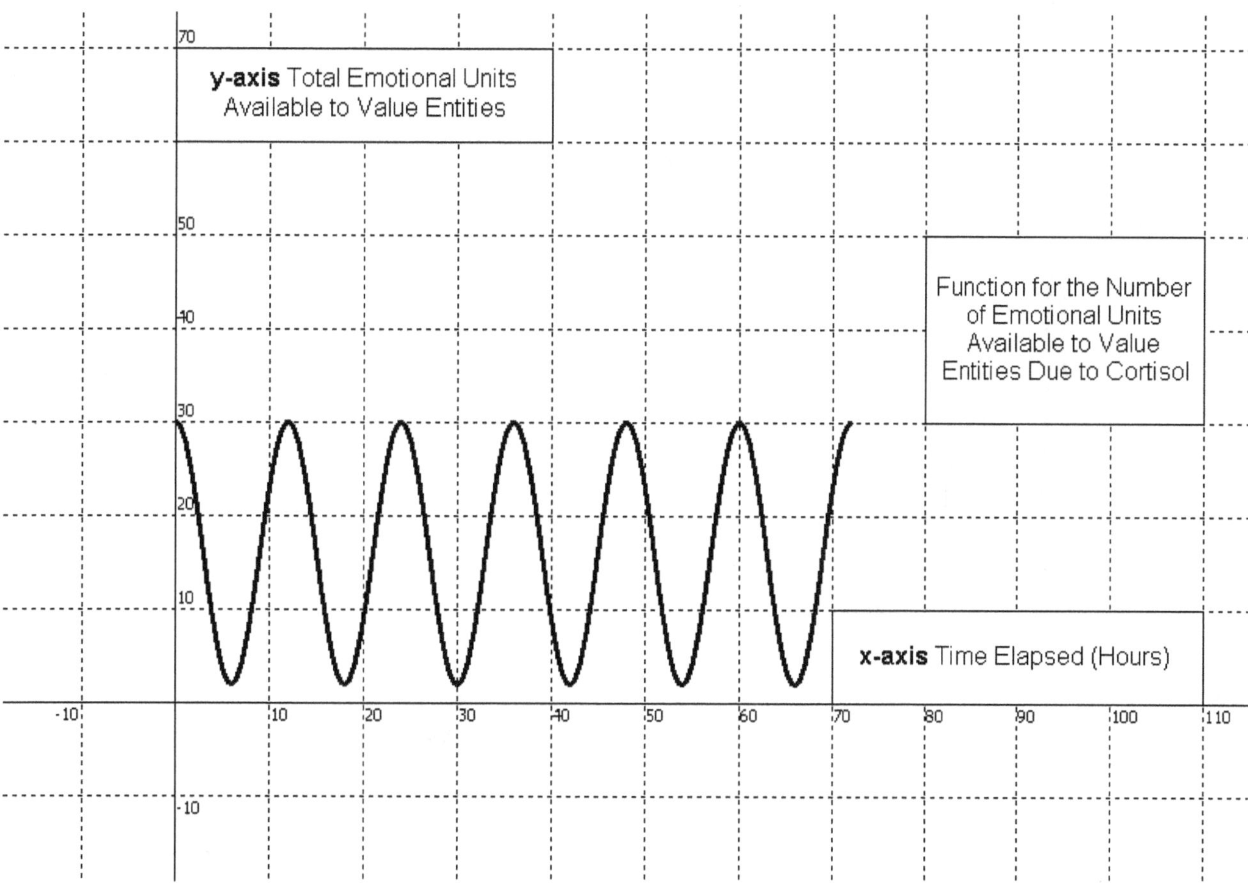

Figure 12.19 Above is a sample graph of the total emotional units made available by cortisol if the period of cortisol were doubled. This would mean that an individual's cortisol levels would fluctuate as if it were on a 12 hour cycle.

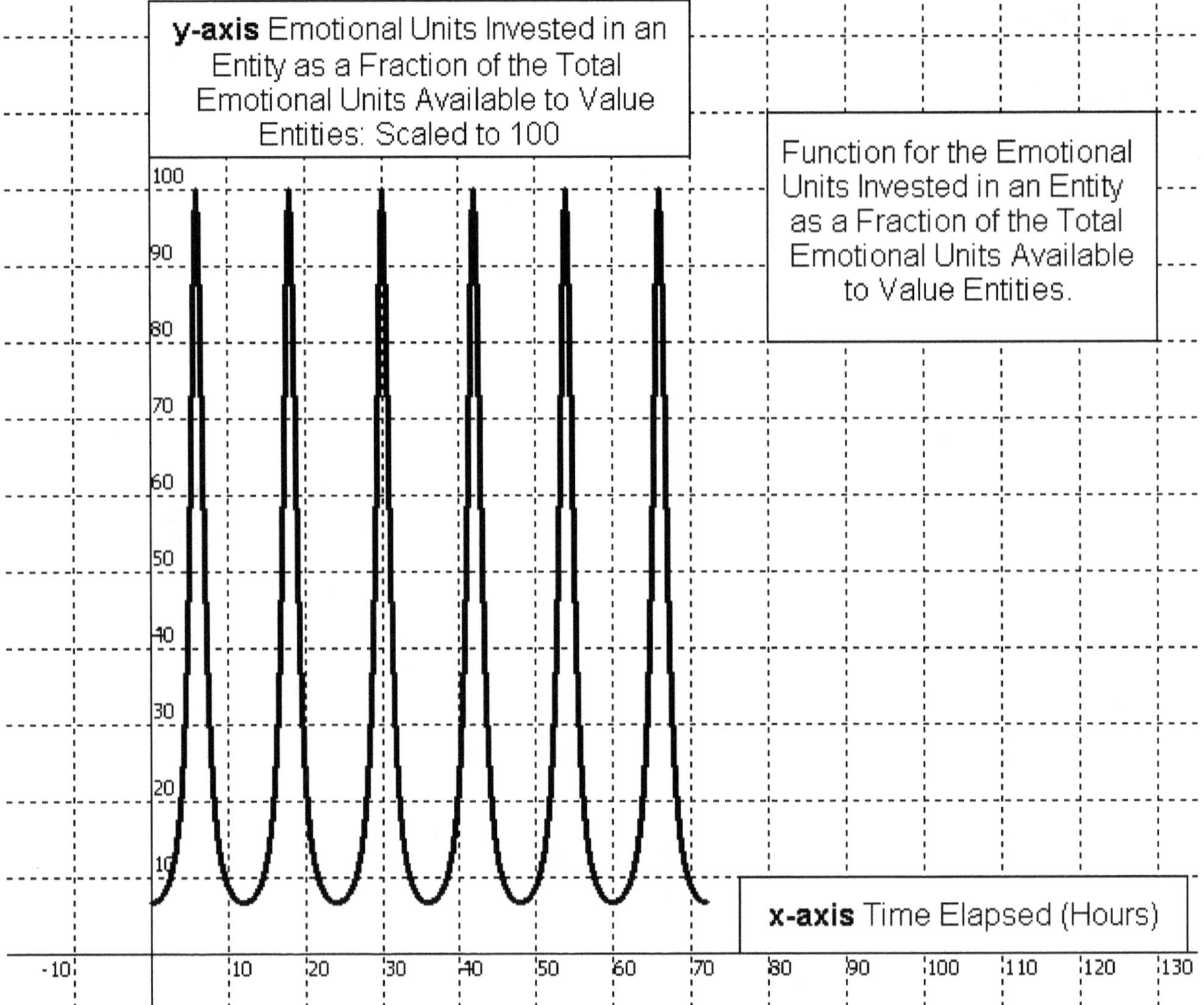

Figure 12.20 Above is a graph of an entity with a constant valuation of + 2 emotional units against the total emotional units made available due to cortisol levels, if hypothetically speaking, cortisol were the only biological rhythm that influenced metabolism and energy available to value entities and cortisol's period was doubled from 24 hours to 12 hours.

The next issue, concerning the need to account for multiple biological rhythms, can be addressed with addition; if a relationship is synergistic, then multiplication would be used. A hypothetical scenario of biological rhythms for cortisol and serotonin will now be considered.

Emotive States (Category IV Emotions) and Time

Non-synergistic Relationships

If two items have the ability to make energy available, for instance, cortisol and serotonin, if both vary on a 24-hour period, both peak at 9:00 A.M., and neither has a synergistic relationship with the other, hence, cortisol does not increase the effectiveness of serotonin and serotonin does not increase the effectiveness of cortisol, then the denominator against the equation might look like the following:

(a × (cos ((x - 9) / 12 pi) + 1)), added to

(d × (cos ((x - 9) / 12 pi) + 1)), plus

b

where ***a*** is the amplitude of the trigonometry function for cortisol, and

where ***d*** is the amplitude of the trigonometry function for serotonin.

The above would be the period for an individual's energy level, assuming a 24-hour cycle with peak available energy occurring at noon due to the effects of both cortisol and serotonin, minimum available energy occurring at midnight due to the effects of both cortisol and serotonin, and both cortisol and serotonin being presumed to have an equal influence on energy level. The physical amount of each is not exactly of concern, but rather their effect on the amount of emotional units made available to value entities based upon their amount.

1) **b** still equals the reserve energy and the minimum amount of metabolic energy that an individual must possess to operate involuntary mechanisms that keep the person alive.

Emotive States (Category IV Emotions) and Time

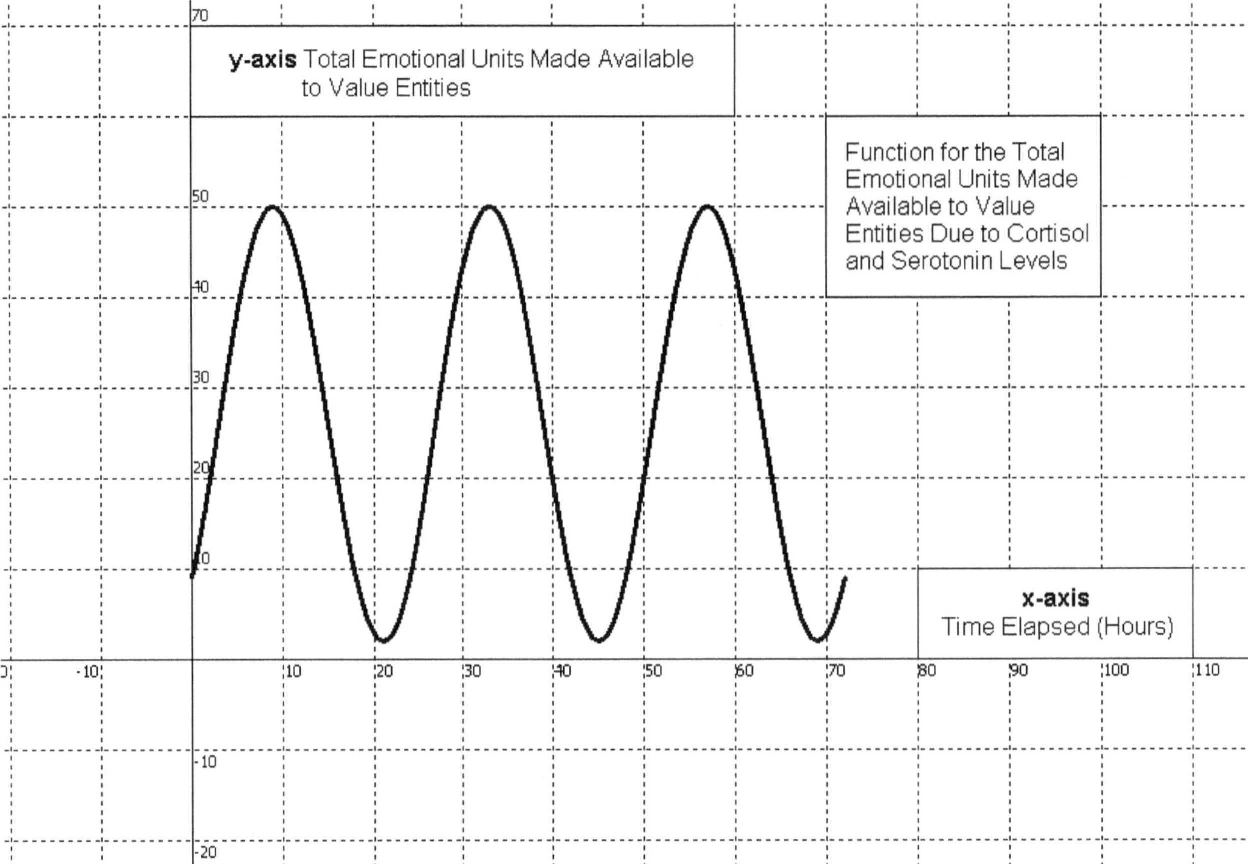

Figure 12.21 Above is a sample graph of the total emotional units made available by cortisol and serotonin if they have a non-synergistic relationship with one another. Cortisol, which is modeled to cause emotional units available to vary by 28 units (a = 14), and Serotonin, which is modeled to cause emotional units available to vary by 20 units (d = 10) are added together. Forty-eight out of a maximum fifty emotional units are varying. The two emotional units from the base value cannot vary (e.g., the energy related to them is reserved for vital life functioning).

2) Because the two biological rhythms are varying on the same period, finding the amplitude of the period is relatively straightforward. Given that the maximum value would be where the cosine of ((x - 9) / 12 pi) = 1, or when x = 9 hours after midnight, this can be placed into the expression to reveal:

$(a \times (\cos (9 - 9) + 1))$ added to $(d \times (\cos (9 - 9) + 1))$ plus b,

This would be the maximum amount of energy that could possibly be available.

This reduces to:

$a + a + d + d + b$, or

$2a + 2d + b = 2 (14) + 2 (10) + 2$.

Subtracting b from the above yields:

$2a + 2d = 2 (14) + 2 (10) = 48$ emotional units for the range, or distance between the lowest amount of energy that could possibly be available and the maximum amount of energy that could possibly be available.

In a simpler case, if **a** and **d** both equaled one, then this would mean that both cortisol and serotonin would be able to make emotional units available to the same extent (e.g., by increasing the individual's available energy to do work by valuing entities).

$1 + 1 + 1 + 1$ would equal 4, which would be the maximum range that energy levels could fluctuate, and half of this would reveal the amplitude of the sinusoidal expression.

Hence, dividing the expression in half would yield:

$(2a + 2d) \div 2 = a + d =$ Amplitude of the sinusoidal curve = two emotional units., if **a** and **d** equal one.

This would represent half of the possible range that an individual's available energy could be modified by whatever agents' periodicity is being described here, in this case cortisol and serotonin levels.

Synergistic Relationships

Alternatively, if these two items, cortisol and serotonin levels, had a synergistic relationship with one another so that one enabled the other to be more effective at making energy available but did not actually make energy available itself, then the expression would have to be changed. If it is presumed that just cortisol affects the amount of emotional units available to value entities while serotonin only makes cortisol more effective at making energy available (with ***e*** representing the level of effectiveness), then their expressions would be multiplied while leaving the 'b' value intact. For instance, this might be represented by the following:

With the influence of cortisol being represented by

(a × (cos ((x - 9) / 12 pi) + 1)),

and the synergistic influence of serotonin being represented by

(e × (cos ((x - 9) / 12 pi) + 1)),

where ***e*** represents the degree to which serotonin makes cortisol more effective, then the expression would become

(a × (cos ((x - 9) / 12 pi) + 1)) multiplied by (e × (cos ((x - 9) / 12 pi) + 1))

plus b.

The above would be the period for an individual's emotional energy level, assuming a 24-hour cycle with peak available energy occurring at noon due to the effects of both cortisol and serotonin, minimum available energy occurring at midnight due to the effects of both cortisol and serotonin. Here, cortisol is being presumed to have an influence on energy made available while serotonin is being presumed to have an influence on the effectiveness of cortisol to make energy available. This would stretch the graph vertically so that the maximum number of emotional units available, at 9:00 A.M., would be 562,

$$(a \times (\cos ((x - 9) / 12 pi) + 1)) \times (e \times (\cos ((x - 9) / 12 pi) + 1)) + b.$$

After substitutions are made,

$$(14 \times (\cos ((9 - 9) / 12 pi) + 1)) \times (10 \times (\cos ((9 - 9) / 12 pi) + 1)) + 2$$

becomes:

$(14 + 14) \times (10 + 10) + 2 = 562$ emotional units available at 9:00 A.M.

Because the two biological rhythms are varying on the same period, finding the amplitude of the period is fairly straightforward. Given that the maximum value would be where the cosine of $((x - 9) / 12 pi) = 1$, or when the expression inside the cosine's parentheses equals zero, this can be plugged into the expression and factored out to reveal:

$a (\cos 0) \times e (\cos 0) + a (\cos 0) \times e + a \times e (\cos 0) + a \times e + b =$ the maximum amount of emotional units that could possibly be available.

Subtracting ***b*** from the above yields the emotional units that are varying.

$$a (\cos 0) \times e (\cos 0) + a (\cos 0) \times e + a \times e (\cos 0) + a \times e$$

The expression above is the range or distance between the lowest amount of energy (emotional units available) that could possibly be available and the maximum amount of energy (emotional units available) that could possibly be available. Dividing this in half would yield:

$$(a (\cos 0) \times e (\cos 0) + a (\cos 0) \times e + a \times e (\cos 0) + a \times e) \div 2.$$

This would equal the amplitude of the sinusoidal curve, which would represent half of the possible range that an individual's available energy (emotional units available) could be modified by whatever agent's periodicity is being described here, in this case cortisol and serotonin levels. Thereafter, when the cos (0) coefficients are all taken out (they all equal one), then it reduces to:

$(a \times e + a \times e + a \times e + a \times e) \div 2 \ =$

$4 (a \times e) \div 2 \ =$

$2 (14 \times 10) = 280$ for the amplitude of the synergistic effect serotonin has on cortisol's ability to make emotional units available at 9:00 A.M.

If one held that cortisol levels influenced one's available emotional units on a trigonometric expression with an amplitude of three, while serotonin made cortisol twice as effective at increasing one's available energy (emotional units available), then the expression

$a (\cos 0) \times e (\cos 0) + a (\cos 0) \times e + a \times e (\cos 0) + a \times e + b$

would become

$3 (\cos 0) \times 2 (\cos 0) + 3 (\cos 0) \times 2 + 3 \times 2 (\cos 0) + 3 \times 2 + b$

reducing to

$6 + 6 + 6 + 6 + b$, or $24 + b$.

As $24 / 2 = 12$, the value 12 would represent the amplitude of the sinusoidal curve representing emotional units made available by the effects of cortisol and its synergistic relationship with serotonin.

Alternatively, if one wanted to hold that cortisol influenced emotional units available regardless of whether or not serotonin increased its effectiveness, then a cortisol trigonometry expression would be added to a modified synergistic expression between cortisol and serotonin.

Likewise, if cortisol and some other item had a synergistic effect on serotonin's ability to make emotional units available, then it would also need to be accounted for here. Exemplified:

(a × (cos ((x - 9) / 12 pi) + 1)) multiplied by (**(e - 1)** × (cos ((x - 9) / 12 pi) + 1))

plus

(a × (cos ((x - 9) / 12 pi) + 1))

plus b.

The **a**, in the top expression, would represent the amplitude of cortisol's influence on available emotional units only when it is being made more effective by serotonin, while **e** would represent the degree to which serotonin makes cortisol more effective in its synergistic relationship. The **a** in the bottom trig expression would correspond to cortisol's ability to make emotional units available; this would be independently and without serotonin's influence. Moreover, the **e** value would need to have one whole number subtracted from its effect. For instance, if serotonin, being in an equal amount to cortisol, made cortisol twice as effective, then **(e - 1)** would equal one; if serotonin made cortisol four times as effective, then **(e - 1)** would equal three. The influence of cortisol on its own would already be accounted for by the other trig expression that is added afterwards.

These reveal that sinusoidal functions, if their periods are equal and they share the same nodes, would amplify each other accordingly. This, however, begs for insight into other instances, such as those where the trigonometric expressions have different nodes and different periods.

If the expression for the periodicity of serotonin were altered to a sine, hence, shifted to the right, then with the influence of cortisol being represented by:

(a × (cos ((x - 9) / 12 pi) + 1)),

and if the synergistic effect of serotonin on cortisol is being represented by

(e × (sin ((x - 9) / 12 pi) + 1)),

and the amplitudes of the trigonometry functions are the more manageable **a** = 5, **d** = 4, and if **b** = 2, then cortisol would be able to make energy available while serotonin would only have a synergistic effect on cortisol that makes cortisol four times as effective at making energy available:

(5 × (cos ((x - 9) / 12 pi)) + 1) multiplied by (4 × (sin ((x - 9) / 12 pi)) + 1) plus b,

then figure 12.22 results.

Emotive States (Category IV Emotions) and Time

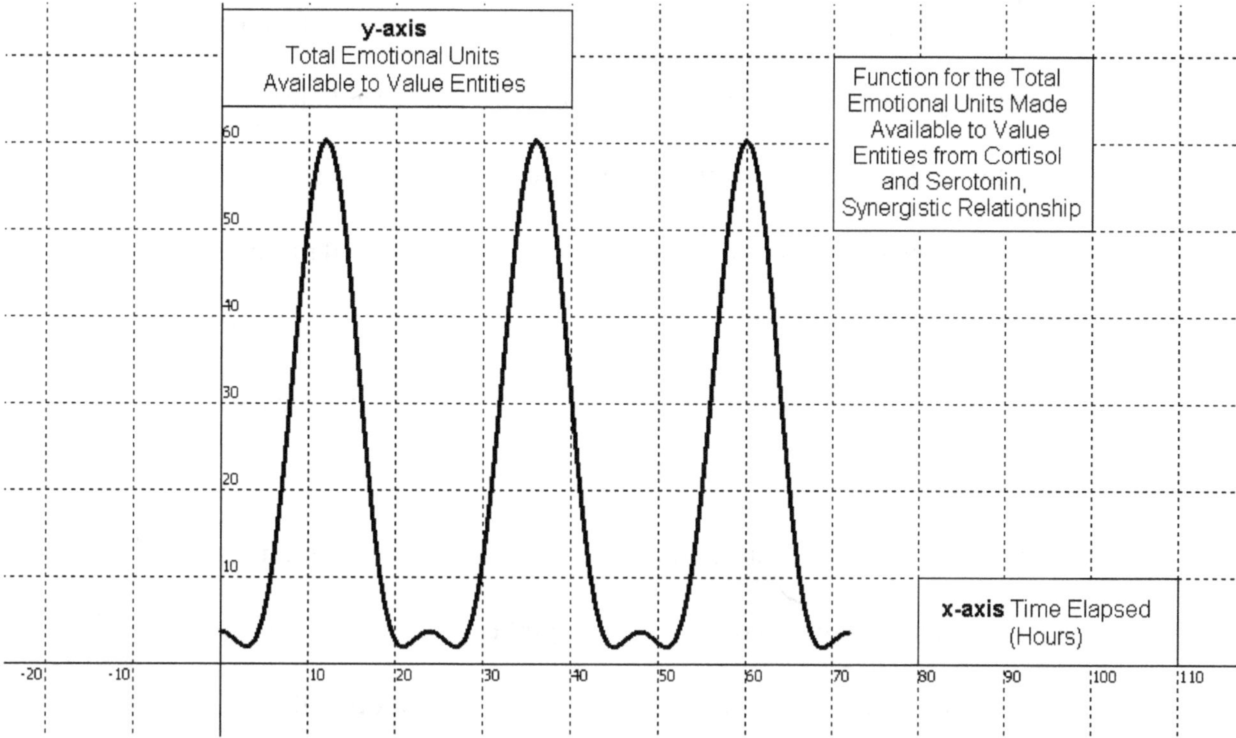

Figure 12.22 The above is a hypothetical sample graph of the total emotional units made available by cortisol and serotonin if they have a synergistic relationship and periods with different peak times. Cortisol levels, which are modeled to cause emotional units available to vary by 10 units, peak at 9:00 A.M., and serotonin, which is hypothetically modeled to make cortisol four times as effective at making energy available, peaks at 3:00 P.M. The two emotional units from the base value cannot vary (i.e., the energy related to them is reserved for vital life functioning). The maximum available energy occurs at approximately noon, halfway between, with roughly 60 emotional units being made available from serotonin's synergistic effect on cortisol. Moreover, an additional crest is created shortly before midnight when the function for the rising cortisol levels crosses with the function for decreasing serotonin levels.

Figure 12.22 depicts the period for an individual's total available energy, expressed in emotional units; it assumes a 24-hour cycle with cortisol levels peaking at 9:00 A.M. and serotonin levels peaking at 3:00 P.M. Cortisol's hypothetical influence independent of serotonin is not factored into figure 12.22, but its levels would still be at their lowest at 9:00 P.M., whereas serotonin levels would be at their lowest around 3:00 A.M.

Finding the maximum and minimum values for this range becomes more challenging as additional biological rhythms and synergies are considered. The periodicity of the function, while still evident when the nodes and periods were the same, has become less evident by shifting the period for serotonin six hours later than cortisol's period. The factoring of more and more sinusoidal expressions to represent different biological rhythms and synergies would eventually lead to an averaged period among all of the criteria concerned that would measure the emotional units made available to value entities.

Each function for separate entities would have the same expression in the denominator representing all biological rhythms concerned. Together, they would be used to assess the investment of anxiety in an entity as a percentage of the total emotional unit resources that are available in an individual's system. Prior to being scaled (e.g., to 100), the coefficient against the entire equation that represents the individual's available energy would look like the following:

(1 ÷ ("Trig expressions for extra energy available to value entities" + "Base Energy")).

The hypothetical scenario in which an individual's emotional units available to value entities is completely maximized, for instance, if all the synergies and sinusoidal expressions peaked at the same time, would be a means of gauging an upper limit of energy or emotional units. A lower limit might be found by the opposite route, hence, finding when all the synergies and sinusoidal expressions are at their lowest. The base energy still represents the energy needed to maintain vital life functions and a reserve energy, though it is not varying in this explanation. The trigonometry functions, then, would represent the ability of different biological rhythms and synergies to influence energy (e.g., emotional units) made available to do work in the system.

The ability to make emotional units or energy available for work bears a strong resemblance to the concept of entropy from physics. Entropy can be thought of as the "'shuffledness'" of a system, and it was a term originally used in order to describe heat and thermodynamic properties.[23] In thermodynamics, entropy was first conceptualized as a measure for the amount of heat not available for doing work in a system.[24] A state of maximum entropy would be "the degradation of the matter and energy in the universe to an ultimate state of inert uniformity."[5] Its use here in the equation, similarly, falls strictly along the lines of psychoanalytic theory. In psychoanalytic theory, entropy is "the degree to which psychic energy is no longer available for use, having been invested in a particular object."[25] When the trigonometry functions approach their highest values, meaning that the most amount of energy is available for work, then entropy is at its lowest value if the individual is still alive. When the trigonometry functions approach their lowest values, meaning that most of the energy is not available for work, then entropy is at its highest value. Entropy, in the equation, refers first to emotional units that are not available to value entities, for instance, due to the activation thresholds of valuing neurons being too great for metabolic processes to overcome. Secondly, entropy also refers to emotional units that are already firing and thus, are unavailable for mobilization by the individual. Additionally, energy that is being used to maintain vital life functioning is also considered to be entropy, as it is not available to do other work by valuing entities.

---

[5] By permission. From Merriam-Webster's Collegiate® Dictionary, 11th Edition ©2014 by Merriam-Webster, Inc. (www.Merriam-Webster.com).

Emotive States (Category IV Emotions) and Time

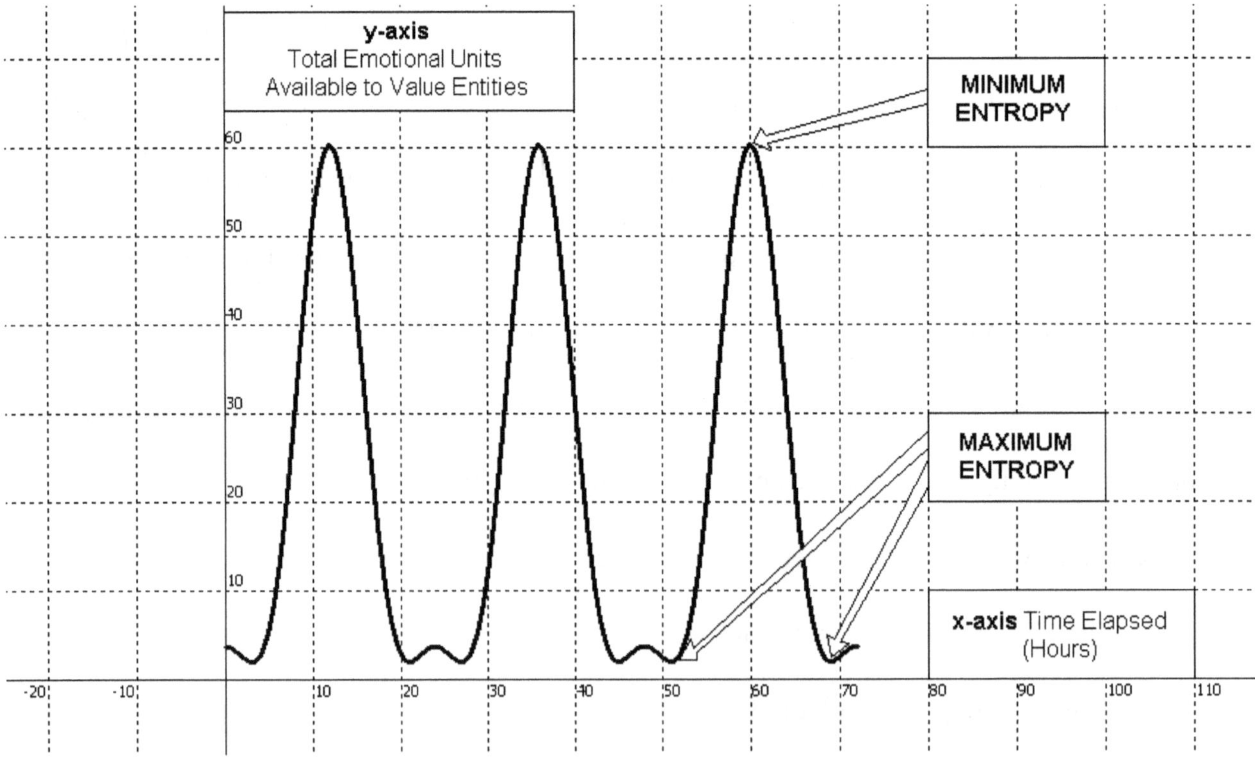

Figure 12.23 In the example from figure 12.22, maximum entropy would occur approximately at 9:00 P.M. and 3.00 A.M., while minimum entropy would occur at approximately 12:00 P.M. (noon).

In the context of the equation, work refers to the ability of the individual to allocate anxiety resources, both positive anxiety (negative affect) and negative anxiety (positive affect), in order to value entities. However, the likelihood that felt affect may be present in the absence of any association with an entity or purpose must be explained. Such an instance might occur whenever positive affect or negative affect are felt, yet they are unrelated to any specific conscious or subconscious goal harbored by the individual. If anxiety, positive or negative, is already active but not associated with any particular purpose or entity harbored by the individual, then this energy is not available for doing work and also makes up a portion of the entropy in an individual's system.

It can be likened to background noise or static on a radio frequency. Although it would have a specific physiological, and subsequently, a mathematical affective value, this value would not be directly measurable with the equation because the anxiety is unattached to a specific entity. This unusable energy could, however, be estimated by comparing it to the anxiety invested in entities that are associated with a known entity's acquisition and a purpose, and thereafter modeled using the neurological model established in chapter eight. As for how the valuing neurons became active in the first place, one explanation is that they might become activated by activity from the receptor cells that they normally target. For instance, if the self's valuing neurons measure pain and normally target the face to produce a frown, wearing a frown on one's face for no reason might activate anxiety valuing neurons in a reverse route. This would lend some credence to William James's prediction that "if one were to adopt the posture, facial expression, or other behavior associated with a particular emotion, one would come to experience that emotion."[26] Another explanation would be that the valuing neurons became hijacked by other physiological forces (e.g., drug use). The cognitive pathway established in the equation would essentially become reversed.

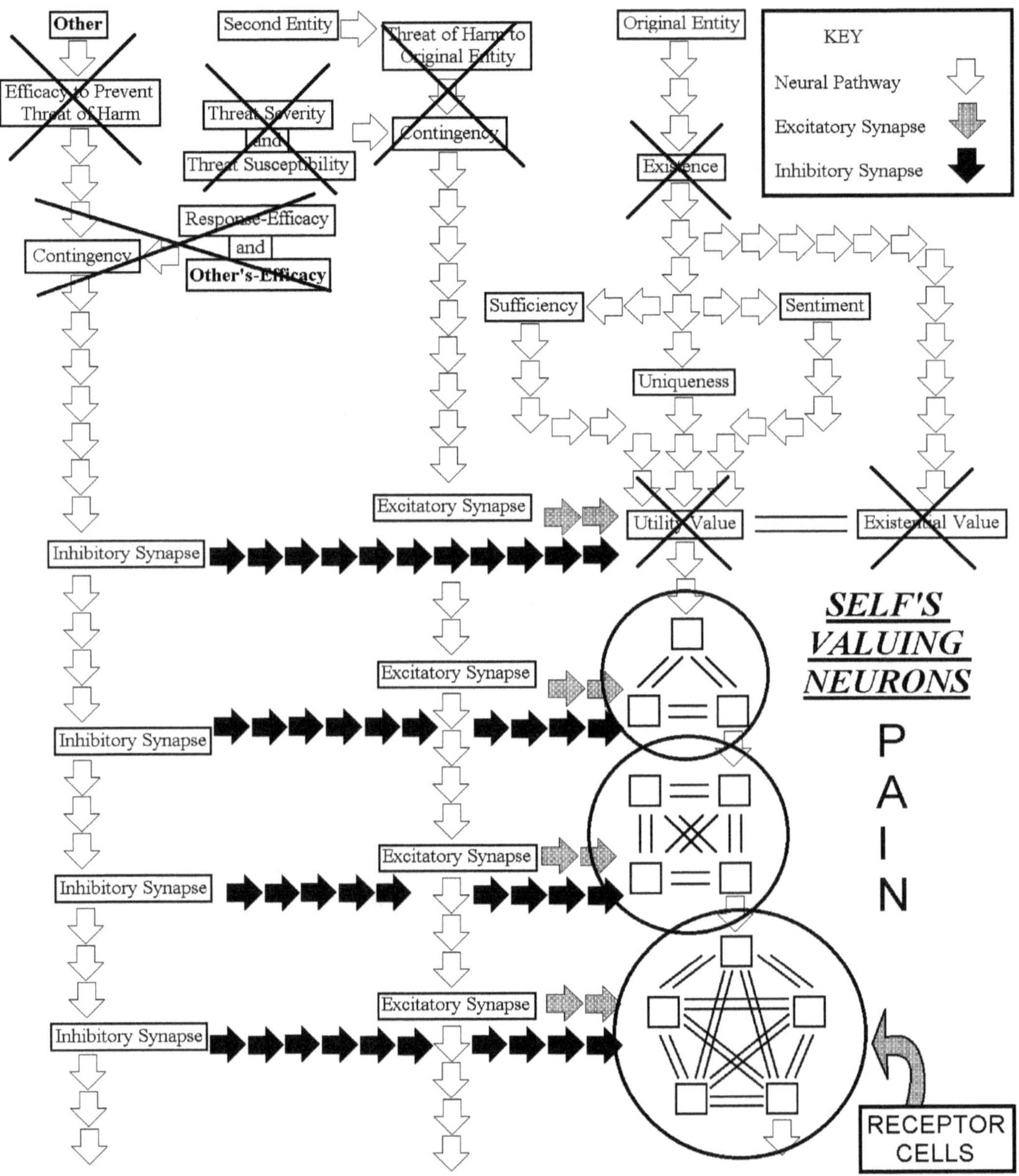

Figure 12.24 Neurological model of an individual's anxiety (O) being activated but the valuing neurons have no relationship to a cognized entity or purpose (X's). Anxiety, or alternatively negative anxiety, are present, but their activation is unrelated to a purpose or entity in the conscious or subconscious (e.g., activated via a reverse route). The anxiety and negative anxiety, having no association with an entity's acquisition or a purpose, make up a portion of the entropy in an individual's system. Their energy is not available for doing work (i.e., valuing entities) and, though unusable, they, have a physiological value comparable to valuations of known entities.

For instance, if an individual had a total of 100 emotional units in his or her system to value entities, but 10 emotional units are being activated irrespectively to any purpose or entity the individual has conceived, and the 5 emotional units making up the base value are active, then it may be said that only 85% of the individual's resources are available to do work. Those 85 emotional units would vary according to whatever influences them. The 5 emotional units making up the base value would not vary; the 10 emotional units being activated randomly would make up a portion of entropy, as they cannot be used to value entities. These would all be represented in the denominator against the entire function for every purpose in question.

$$\frac{1 \times \text{"Scale"}}{\text{"Trig expressions for energy made available to value entities"} + \text{"Base Energy"} - \text{"Energy Not Available to Do Work" (Entropy)}}$$

Figure 12.25 To assess anxiety (e.g., energy or emotional units) that are invested as a portion of the total anxiety available, the coefficient next to a function would be comprised of the number one times a scale in the numerator, with entropy being subtracted from the trigonometry expressions for energy made available to value entities and the base energy in the denominator.

The fifteen neurons that make up entropy may be thought of as the background noise or static referenced earlier. Noise, in this case, might refer to positive or negative anxiety that is fluctuating seemingly without a cause or purpose because they are unrelated to the acquisition of an entity or the fulfillment of a purpose. Alternatively, they may be unresponsive. This might be precipitated by drugs, stray hormones, loose chemicals in the body, or even physical injury for instance. While it is true that these neurons might be active and firing, their energy is unavailable to do work to value entities because the individual cannot mobilize them to direct or divert emotional units to and from entities. It should go without saying, but will be stated nevertheless, that if entropy becomes maximized and the **b** value disintegrates, then the individual will eventually expire. Lastly, it bears repeating that although the base energy includes the energy required to sustain vital functions and extra reserve energy, energy that is being used for life sustenance is inherently unavailable to value entities; entropy would remove a portion of the base energy, but never more than it. Whatever is left over would make up the individual's reserve energy, measured in emotional units. In the example that was described earlier, all of the base energy was unavailable.

Modeling Anxiety and Negative Anxiety in Dreams/Nightmares

Using the equation to model valuations that may occur during sleep is not as daunting as it may at first seem, but it does present some interesting cases. Based upon what has been established by the trigonometry expressions for fatigue and the emotional units available to value entities, it can be established that the energy used to value all entities in consideration must be less than the total energy available in the individual's system to value entities. Hence:

(Emotional units invested in entity # 1 ÷ total emotional units available), plus

(Emotional units invested in entity # 2 ÷ total emotional units available), plus

(Emotional units invested in entity # 3 ÷ total emotional units available), and so on . . .

Must be less than or equal to

(Total emotional units available ÷ Total emotional units available) = 1.

While sleeping, an individual's executive functions, modeled in the equation as Reasoning to Threat, Reasoning to Efficacy, or Reasoning to Benefit (if used), would likely be shut down or reduced. For example, if the second model of Attention was being used, certain valuations of entities might resist falling into Attentional decay, thus inhibiting sleep altogether. Alternatively, if the third model of Attention is used, hence, valuation resilience, normally suppressed valuations of entities may experience a resurgence to their true valuation once the Reasoning filters are off. If this happened in the middle of the sleep cycle and overwhelmed the available emotional units that were on standby, it might be indicative of a nightmare. The standby emotional units during sleep consist of the base or **b** value and whatever other emotional units are available from other criteria that are varying and not at their minimum level during night time sleep. This might include other hormones such as melatonin, neurotransmitters, or chemicals that may influence the energy available to value entities. In the example to follow, the **b** value makes up all of the standby emotional units and it has been raised to five emotional units from two emotional units to demonstrate how the equation might model a nightmare. Sample form III of Attention is used with value resilience taking place.

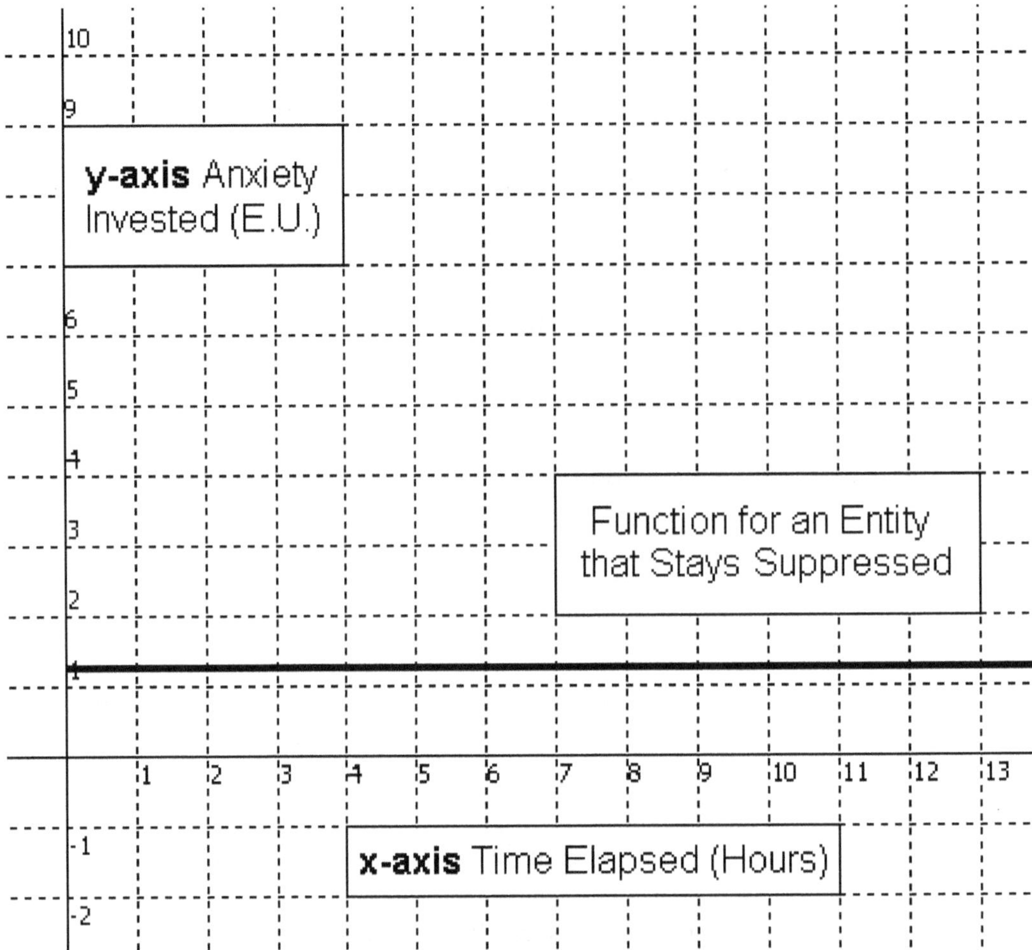

Figure 12.26 Depicted above is anxiety invested in an entity held constant just above + 1 E.U. Normally, it would surge towards a valuation of 8 emotional units (threat components are 3 times the level of efficacy components, but Attention and Reasoning are amplifying Efficacy over Threat by a factor of 9:1.

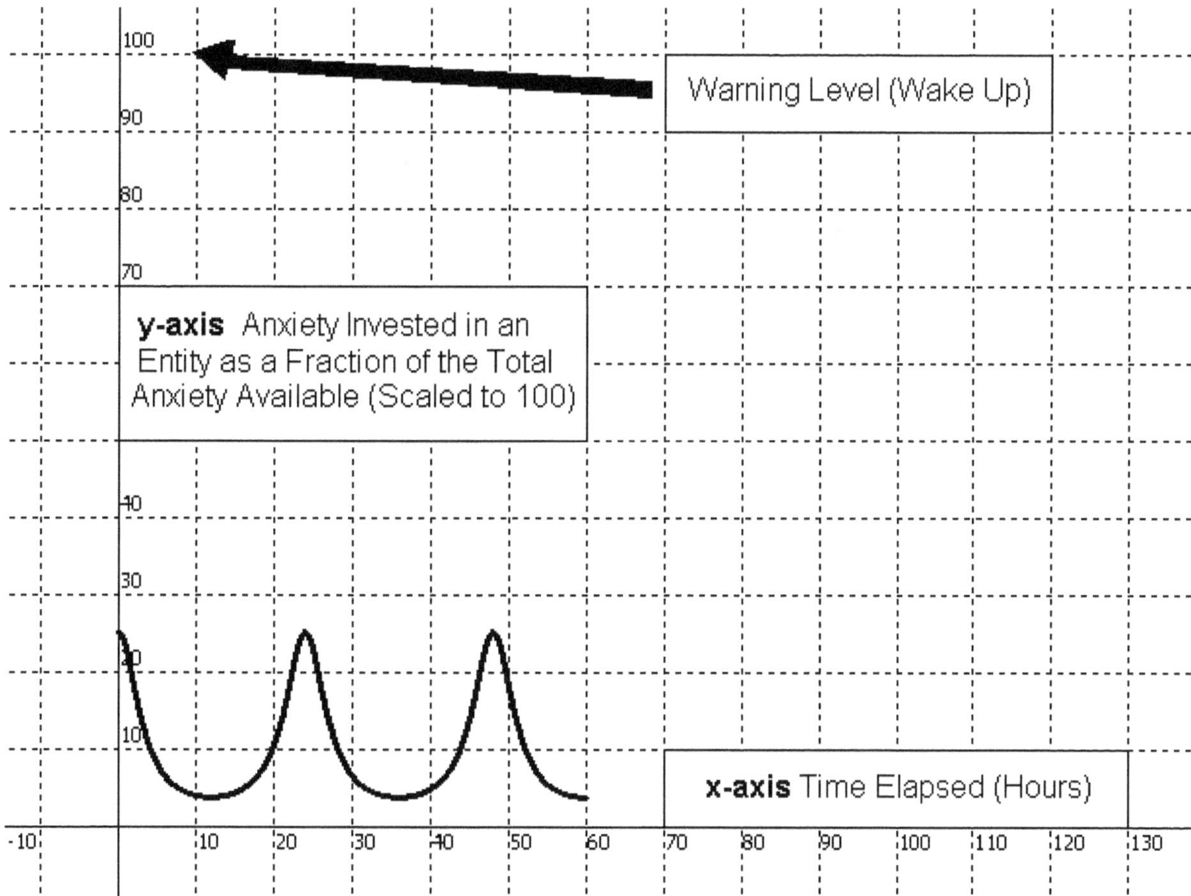

Figure 12.27 The anxiety invested in the entity in figure 12.26 is expressed as a fraction of the total energy available in an individual's system. The emotional units available to value entities are varying solely from a trigonometry function representing cortisol, with an amplitude of 14, base value of 5, and peak levels occurring at noon. It is scaled to 100:

(100 / (14 * cos ((x+12) / 12 pi) + 14 + 5))

As can be seen in figure 12.27, during sleep the individual's anxiety invested in the entity never climbs higher than 25% of the total anxiety units available to value entities. The same cannot be said for the valuation of an entity in the next example.

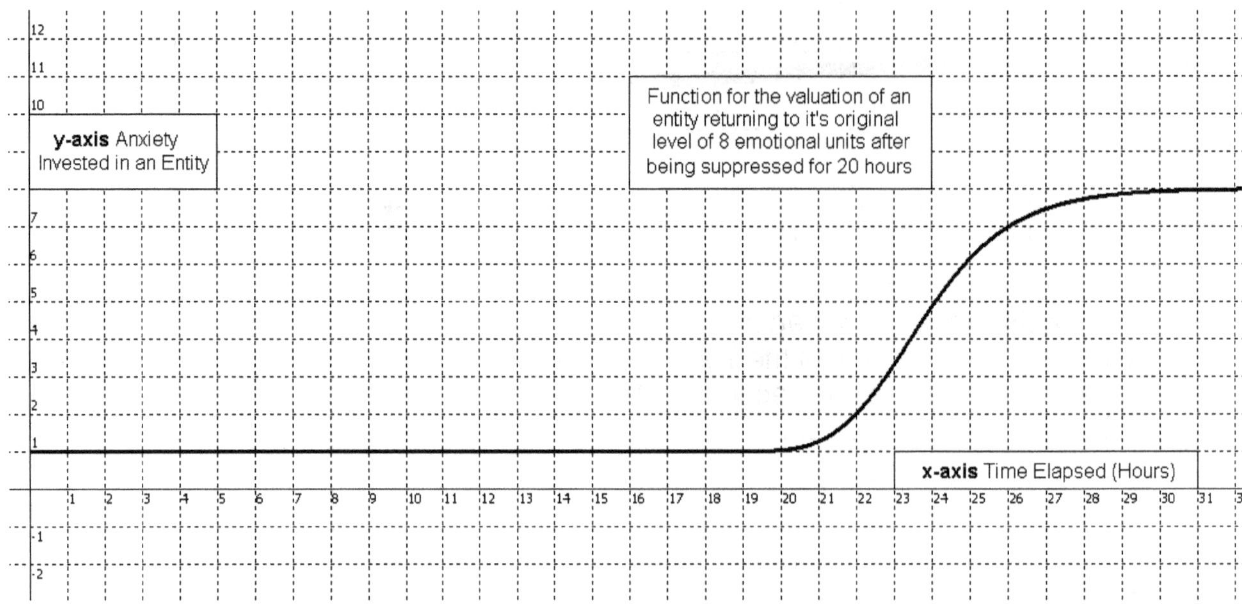

Figure 12.28 Depicted above is anxiety invested in an entity held constant just above + 1 E.U. for about 20 hours until the individual starts falling asleep. Normally, the valuation would be at 8 emotional units (threat components are 3 times the level of efficacy components, but Attention and Reasoning are amplifying Efficacy over Threat by a factor of 9:1. However, upon falling asleep, Attention reverts towards 1 in both the denominator and the numerator (Awareness : Neglect goes to .5:.5), meaning that the entity's valuation begins to return to 8. When the individual goes to sleep, Reasoning is relinquished over Attention, unlike in figure 12.26 and 12.27. Unfortunately, when cortisol levels are at their lowest, only 5 emotional units are available on standby.

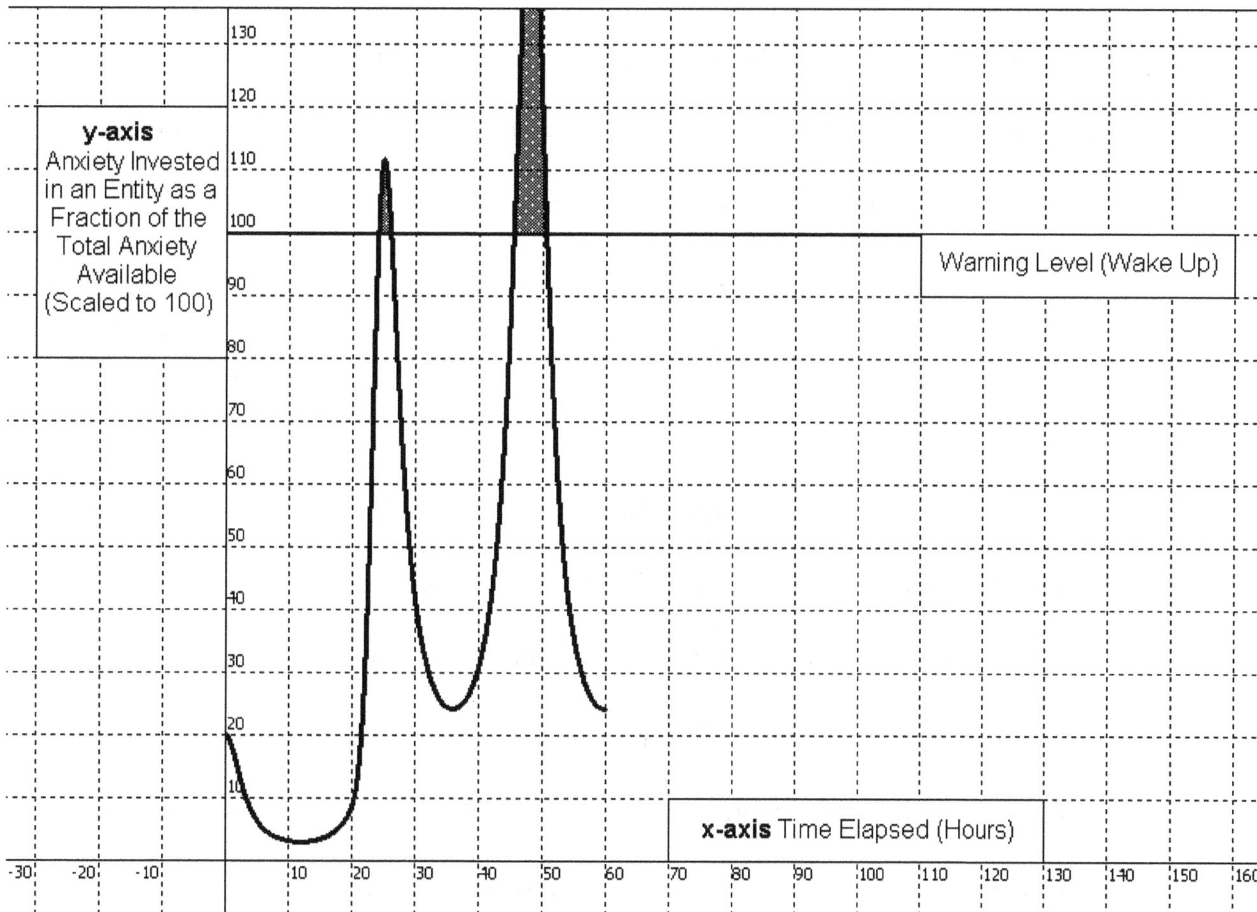

Figure 12.29 The anxiety invested in the entity in figure 12.28 is expressed as a fraction of the total energy available in an individual's system. The emotional units available to value entities are varying solely from a trigonometry function representing cortisol, with an amplitude of 14, base value of 5, and peak levels occurring at noon before being scaled to 100:

(100 / (14 * cos ((x+12) / 12 pi) + 14 + 5))

The individual's emotional units on standby are overwhelmed at midnight, when the valuation of the entity surges above 5 emotional units. The individual might then be prompted to wake up due to rising anxiety. If more emotional units became available during the sleep cycle from other synergies, hormones, neurotransmitters, or chemicals, then valuation activities and dreaming might continue without waking the individual. The shaded region corresponds to anxiety invested that is more than what would normally be available and as a percentage of the total anxiety available.

Of note, in the example above the entity's valuation is not suppressed again, and the individual may be awake for the next two hours before sufficient standby emotional units become available. If nothing changes the following day and the entity's valuation remains the same, then the anxiety invested in the entity's will overwhelm the individual's standby emotional units even sooner than before, causing him or her to wake at approximately 9:30 P.M.

Final Thoughts

The inclusion of a measure for emotional units made available (e.g., fatigue and entropy) is not essential to the understanding of the equation but it has provided interesting insights that will be useful for modeling defense mechanisms or biological processes. While it does make the equation more useful for studying an individual person, when modeling larger populations of people, such as the valuations of groups of different people concerning the same entity and purposes, it would likely prove too cumbersome for practical considerations, given that biological rhythms are specific to an individual. The next chapter will look into using the equation to create multidimensional models to model multiple goals, ethics, and the valuations of entities for multiple people. For simplicity's sake, biological rhythms will not be incorporated into the multidimensional models presented in chapter thirteen.

YOU SHOULD UNDERSTAND THIS.
CHAPTER TWELVE
CONCEPT CHECK
VOCABULARY
Biological Rhythm
Category IV Emotions:
Emotive States
Confusion
Entropy
Fatigue
Infradian Rhythm
Mood
Post-Traumatic Stress Disorder
(PTSD)
Ultradian Rhythm

Notes

1. Wilde, Oscar (2003). *The Picture of Dorian Gray.* New York: Barnes and Nobles Books. Print, p. 113. (Original work published 1890).

2. Reber, Arthur S., Rhianon Allen, and Emily S. Reber. *Penguin Dictionary of Psychology*. London. Penguin Books, 2009. Print, p. 377.

3. Reber, Arthur S., Rhianon Allen, and Emily S. Reber. *Penguin Dictionary of Psychology*. London. Penguin Books, 2009. Print, p. 198.

4. Reber, Arthur S., Rhianon Allen, and Emily S. Reber. *Penguin Dictionary of Psychology*. London. Penguin Books, 2009. Print, p. 344.

5. Reber, Arthur S., Rhianon Allen, and Emily S. Reber. *Penguin Dictionary of Psychology*. London. Penguin Books, 2009. Print, p. 344.

6. Reber, Arthur S., Rhianon Allen, and Emily S. Reber. *Penguin Dictionary of Psychology*. London. Penguin Books, 2009. Print, p. 601.

7. Reber, Arthur S., Rhianon Allen, and Emily S. Reber. *Penguin Dictionary of Psychology*. London. Penguin Books, 2009. Print, p. 818.

8. Reber, Arthur S., Rhianon Allen, and Emily S. Reber. *Penguin Dictionary of Psychology*. London. Penguin Books, 2009. Print, p. 818-819.

9. Reber, Arthur S., Rhianon Allen, and Emily S. Reber. *Penguin Dictionary of Psychology*. London. Penguin Books, 2009. Print, p. 819.

10. Reber, Arthur S., Rhianon Allen, and Emily S. Reber. *Penguin Dictionary of Psychology*. London. Penguin Books, 2009. Print, p. 819.

11. Reber, Arthur S., Rhianon Allen, and Emily S. Reber. *Penguin Dictionary of Psychology*. London. Penguin Books, 2009. Print, p. 140.

12. Reber, Arthur S., Rhianon Allen, and Emily S. Reber. *Penguin Dictionary of Psychology*. London. Penguin Books, 2009. Print, p. 140.

13. Reber, Arthur S., Rhianon Allen, and Emily S. Reber. *Penguin Dictionary of Psychology*. London. Penguin Books, 2009. Print, p. 100.

14. Reber, Arthur S., Rhianon Allen, and Emily S. Reber. *Penguin Dictionary of Psychology*. London. Penguin Books, 2009. Print, p. 381.

15. Reber, Arthur S., Rhianon Allen, and Emily S. Reber. *Penguin Dictionary of Psychology*. London. Penguin Books, 2009. Print, p. 381.

16. Reber, Arthur S., Rhianon Allen, and Emily S. Reber. *Penguin Dictionary of Psychology*. London. Penguin Books, 2009. Print, p. 841.

17. Reber, Arthur S., Rhianon Allen, and Emily S. Reber. *Penguin Dictionary of Psychology*. London. Penguin Books, 2009. Print, p. 131.

18. Reber, Arthur S., Rhianon Allen, and Emily S. Reber. *Penguin Dictionary of Psychology*. London. Penguin Books, 2009. Print, p. 131.

19. Reber, Arthur S., Rhianon Allen, and Emily S. Reber. *Penguin Dictionary of Psychology*. London. Penguin Books, 2009. Print, p. 292.

20. Reber, Arthur S., Rhianon Allen, and Emily S. Reber. *Penguin Dictionary of Psychology*. London. Penguin Books, 2009. Print, p. 292.

21. Reber, Arthur S., Rhianon Allen, and Emily S. Reber. *Penguin Dictionary of Psychology*. London. Penguin Books, 2009. Print, p. 292.

22. Reber, Arthur S., Rhianon Allen, and Emily S. Reber. *Penguin Dictionary of Psychology*. London. Penguin Books, 2009. Print, p. 276.

23. Reber, Arthur S., Rhianon Allen, and Emily S. Reber. *Penguin Dictionary of Psychology*. London. Penguin Books, 2009. Print, p. 263.

24. Reber, Arthur S., Rhianon Allen, and Emily S. Reber. *Penguin Dictionary of Psychology*. London. Penguin Books, 2009. Print, p. 263.

25. Reber, Arthur S., Rhianon Allen, and Emily S. Reber. *Penguin Dictionary of Psychology*. London. Penguin Books, 2009. Print, p. 264.

26. Cornelius, R. R. *The Science of Emotion: Research and Tradition in the Psychology of Emotion.* Upper Saddle River, NJ. Prentice-Hall, Inc. 1996. Print, p. 63.

"... to influence a person is to give him one's own soul. He does not think his natural thoughts, or burn with his natural passions. His virtues are not real to him. His sins, if there are such things as sins, are borrowed." - Oscar Wilde[1]

CHAPTER THIRTEEN

Multidimensional Models for Multiple Entities, Multiple Purposes, Multiple People, and Ethics

Front-loading: Hypercube, Nature, Non-regulatory Drives, Nurture, Primary Drives, Regulatory Drives, Secondary Drives

To this point, the values of entities have been considered in two dimensions. The x-axis has been used to represent time, while the y-axis has been used to represent the value or appraisal of an entity toward the fulfillment of a single purpose from one individual, as measured by emotional units. The potential, however, to express more information in additional dimensions has not yet been addressed. Three, four, five, or more dimensions can be used to model additional information.

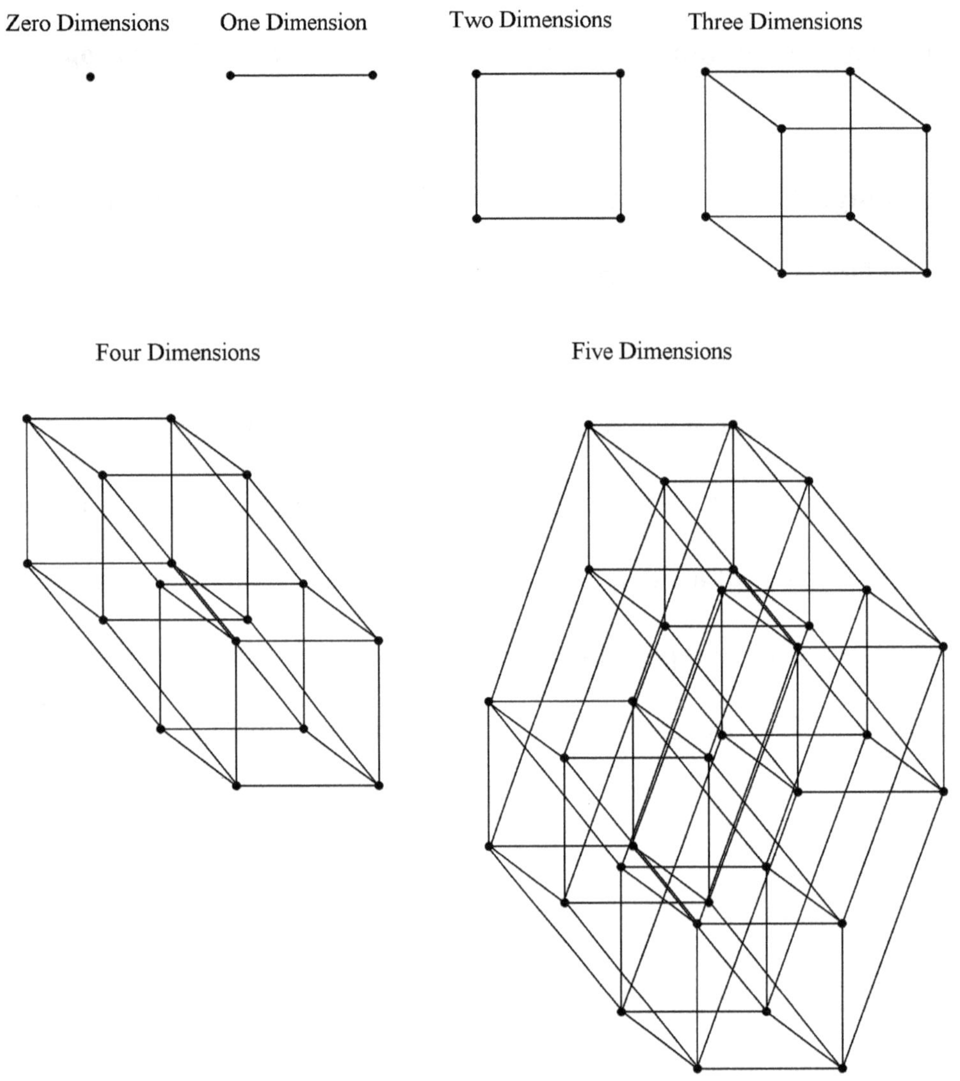

Figure 13.1 A 0-D, 1-D, 2-D, 3-D, 4-D, and 5-D object.

Multidimensional Models

The Third Dimension: Multiple Values of Entities Concerning One Purpose for One Individual

Where the x-axis still represents time, and the y-axis represents an entity's value, or collective value if it is a summation of values regarding multiple threats of harm to the entity (from chapter eight on the summation of values from multiple contingencies), the z-axis would organize different entities so that the values of multiple entities could be assessed with respect to one purpose. Valuations of entities for the complement of a specific purpose could be represented by negative *z* values and the Appraisal component would be the reverse of the original. For instance, it would be negative if the original purpose is positively appraised or positive if the original purpose is negatively appraised. However, the use of negative *z* values to organize complementary purposes is not essential; they could just as easily be organized directly alongside their corresponding purpose with positive *z* values, but across the ***x-z***-plane. Whereas on a Cartesian plane and in two-dimensional space anxiety and negative anxiety are represented by area and square emotional units invested per second, in three-dimensional space, anxiety and negative anxiety would be represented by the volume between the x-z plane, a function, and one emotional unit to one side along the z-axis. Cubic emotional units per unit of time would be the measurement. Each function for an entity's valuation would be one emotional unit away from a separate entity directly before and after it. As for deciding the order the entities might be arranged in along the z-axis (of concern if one wanted to compare different entities), entities could be conveniently arranged so that those with the highest Sufficiency and Uniqueness values are closer to the origin. All of them would have the same Sentiment ranking as they are being valued for the same purpose.

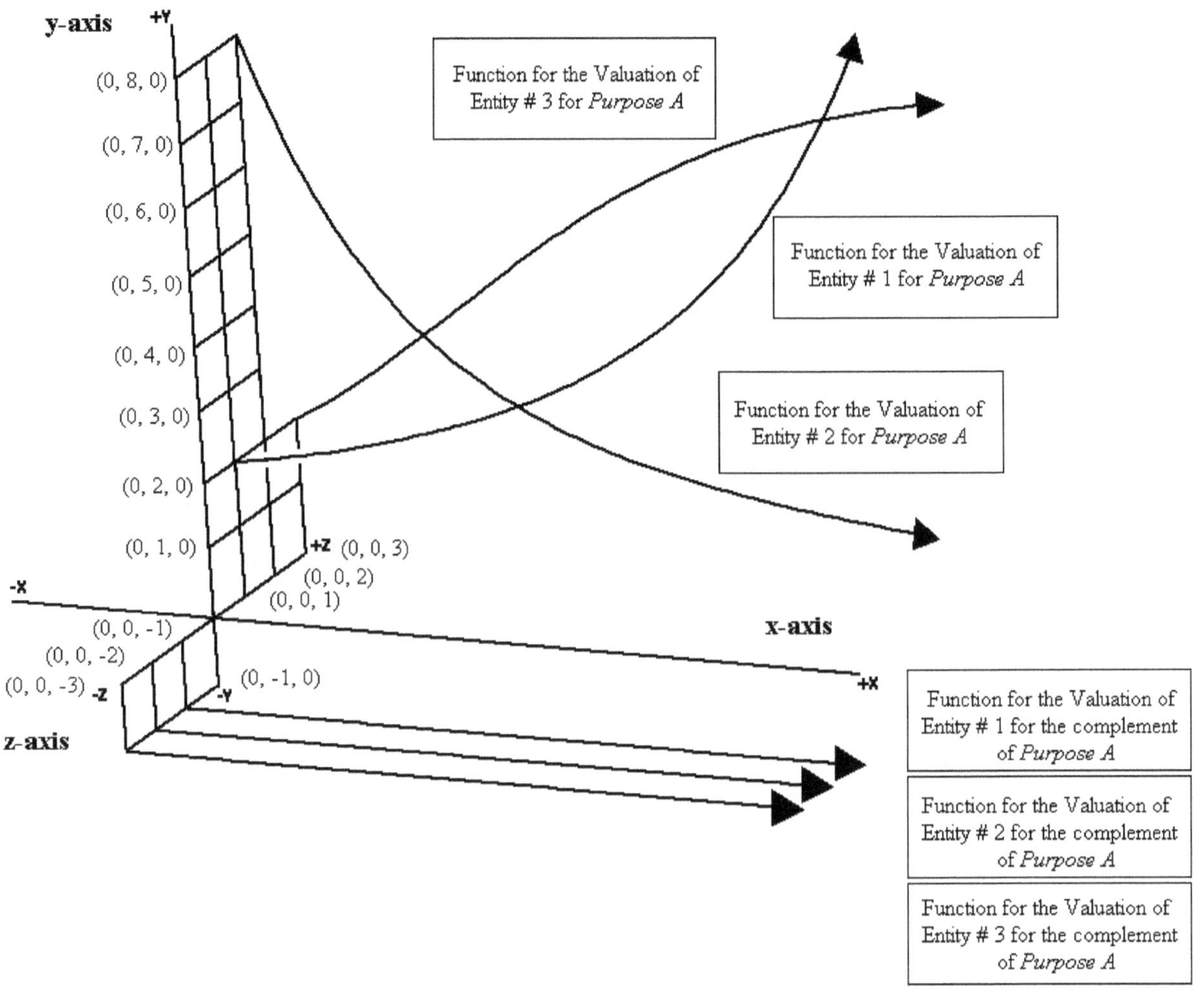

Figure 13.2 The above depicts six functions modeled on an x-y-z coordinate graph. The x-axis above measures time (e.g., seconds), and the y-axis measures an entity's valuation (i.e., emotional units invested). The z-axis organizes different entities that are being valued for the same purpose and its complementary purpose. Positive z-values are being used to indicate the valuation functions for three specific entities concerning one purpose, *A*, where anxiety is being invested. Negative z-values (optional) are being used to indicate the valuation functions for those same three entities, albeit concerning the complement to purpose *A* and where negative anxiety is being invested. If the valuation functions for the positive z-values were instead below the x-z plane (e.g., represented negative anxiety being invested), then the valuation functions for the complements and negative z-values could be positive (anxiety invested). If one desired, both purposes and complementary purposes could be represented along + **z** values and across the x-z plane.

Multidimensional Models for Multiple Entities, Multiple Purposes, Multiple People, and Ethics

The Fourth Dimension: Multiple Values of Entities Concerning Multiple Purposes and their Complementary Purposes for One Individual

Where time, an entity's value, and the values of entities concerning a single purpose and its complementary purpose are calculated along the x-axis, the y-axis, and the z-axis respectively, different drives or purposes that are being considered by the same individual would be organized along a fourth dimension. A fourth dimensional object, like a *hypercube*, could be created on paper by extending a cube away from an identical cube along a parallel set of lines.

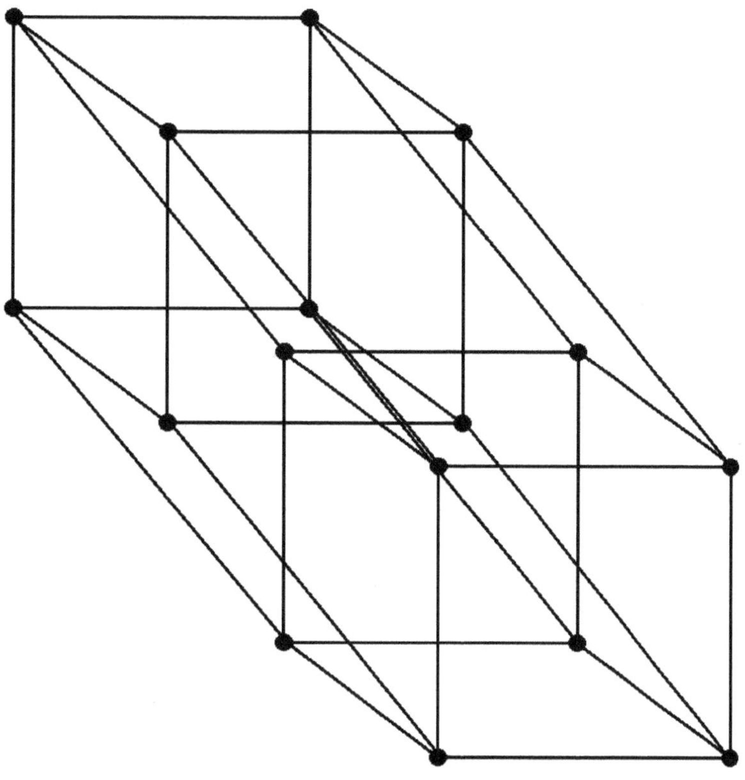

Figure 13.3 A Hypercube (Fourth Dimensional Object)

This fourth dimension will be labeled the d-axis, for drive. Drive and purpose, for all the considerations here, are synonymous with each other as they both imply a motive to acquire an entity in affect engineering. While an entity may only have one value or collective value for a specific purpose and its complementary purpose (e.g., collective if multiple contingencies considered), that entity would also have other values for different purposes. These purposes would be organized along a fourth dimension, the d-axis, and may, if one chose, be sequenced by the Sentiment ranking of the purpose. Only positive ***d*** values would be used and the origin would pertain to the purpose with the highest absolute ranking (e.g., zero, the smallest ***R*** value for Sentiment described in chapter five). For instance, a purpose with an ***R*** value of zero would have the highest priority of all purposes. In four-dimensional space, anxiety and negative anxiety would be measured with volume but in four dimensions. Like on the third plane, each purpose would be separated by an emotional unit along the fourth dimension, as each purpose is an entity.

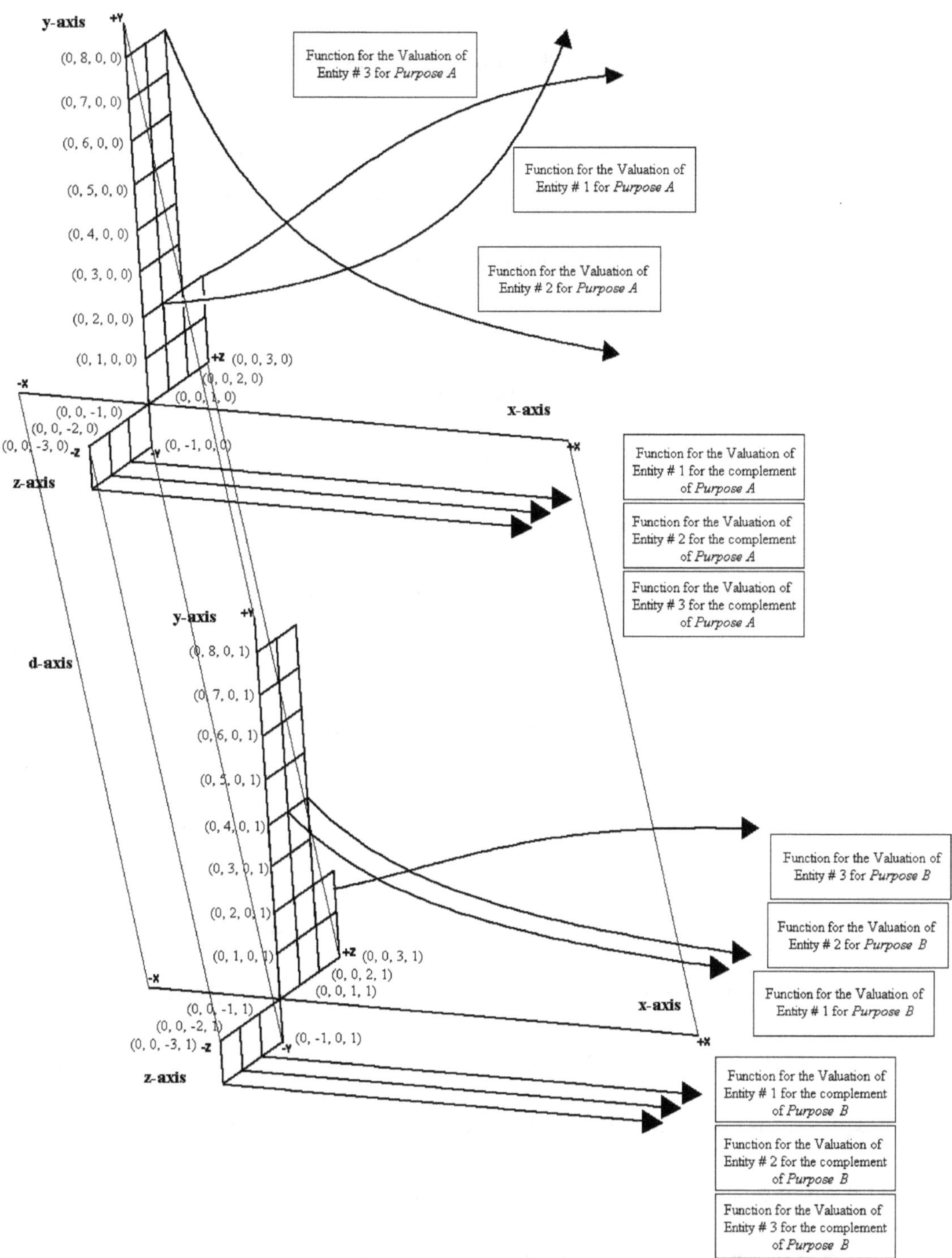

Figure 13.4 The above depicts valuations of three entities being valued for purpose *A* (Top), purpose *B* (Bottom), and their complements. The **d** variable equals zero for purpose *A*, meaning purpose *A* has the highest absolute ranking of all purposes. Purpose *B*, with a **d** value of one, has the next highest Sentiment variable, as only two purposes are being considered.

Understanding the role of the fourth dimension is crucial toward being able to model ethics with the equation. For instance, in figure 13.4, if one compares the valuation of entity # 1 for both purpose *A* and purpose *B*, it may be seen that two separate things are happening. Entity # 1's valuation begins at two emotional units then elevates for purpose *A* over time while at the same time its value for purpose *B* decreases from four emotional units to its existential level of one emotional unit. Entity # 1 is useful for both purpose *A* and purpose *B*, as its valuation is above the existential value initially. However, if entity # 1 is successfully being acquired for the fulfillment of purpose *B*, then this could be at the expense of the fulfillment of purpose *A*, particularly if entity # 1 is in limited supply. If the same entity is being valued for two non-complementary purposes and the fulfillment of one purpose, *B*, hinders the fulfillment of another purpose, *A*, then conflict erupts.

The Fifth Dimension: Multiple Values of Entities Concerning Multiple Purposes, Complementary Purposes, and Vicariously Experienced Valuations for One Individual

Where time, an entity's value, the values of entities concerning a single purpose, and different purposes are organized along the x-axis, y-axis, z-axis, and d-axis respectively, the vicarious valuations of entities would be represented along a fifth dimension, the c-axis. Complex numbers are numbers or expressions containing *i* or $\sqrt{-1}$. The c-values would correspond to *others* that the self is imagining itself to be and for whom entities are being vicariously valued. The self would be at the origin of the c-axis, or when c = zero: (x-axis, y-axis, z-axis, d-axis, c-axis).

Other's that the self is imagining itself to be would be organized along the c-axis. For instance, the self can vicariously value an entity while imagining itself to be one, two, three or more people, animals, life forms, or inanimate objects. As for how the self would organize these vicarious valuations along the c-axis, it could be done in a similar fashion to the ranking of purposes along the d-axis, except that the extent to which an individual considers an *other's* purposes as being identified with those of the self would be the means by which they would be organized. Identification is described as "the process of establishing a link between oneself and another person or group."[2]

Although it will not be modeled in the examples throughout this book, if one wished, the variable of *Identification* could be incorporated into the equation to signify the degree of vicarious experience. Its structure would be similar to that of Sentiment, Uniqueness, and Sufficiency:

$1 \div (1 + \textit{Self-Other Disparity})$

Self-Other Disparity would be restricted to positive, real numbers or zero and added to one in the denominator, while one would be by itself in the numerator. For example, if whole numbers were used, then:

A Self-Other Disparity of zero would be

$1 \div (1 + 0) = 1 = 100\%$ Identification

A Self-Other Disparity of one would be

$1 \div (1 + 1) = .5 = 50\%$ Identification

A Self-Other Disparity of two would be

$1 \div (1 + 2) = .333 = 33.3\%$ Identification

A Self-Other Disparity of three would be

$1 \div (1 + 3) = .25 = 25\%$ Identification

A Self-Other Disparity value of zero would correspond to 100% identification, meaning that the *other's* utility components in the base of the equation are treated as if they were the self's own.

Possessing an Identification level value of 50%, when the Self-Other Disparity equals one, would mean that only half of the *other's* characteristics, in this case the valuation of the utility components, are associated with the self.

Henceforth, vicarious valuations might be organized along the c-axis and fifth dimension by their Self-Other Disparity level.

An *other* with whom the *self* has a high Identification of 80% (e.g., Self-Other Disparity = .25), would have valuations that are closer to the origin along the c-axis, for instance, (0, 0, 0, 0, .25) as opposed to an *other* with whom the self has a low Identification. An Identification of 1%, for instance, would correspond to a Self-Other Disparity of 99 (0, 0, 0, 0, 99). An *other* with whom the self has a low Identification might have valuations that are further from the origin along the c-axis, for instance, (0, 0, 0, 0, 9999). The Identification variable, if implemented, would likely be best implemented as a coefficient in the Utility portion of the base of the equation alongside Uniqueness, Sufficiency, and Sentiment. As shown:

Ideal placement of the variable of Identification

Existence * (Sufficiency * Uniqueness * Sentiment * Identification + 1)

The use of the Identification variable would primarily concern instances where vicarious valuation and empathy are taking place (e.g., Category II and Category III Emotions). The influence of Identification should be nil for all valuations where no vicarious valuations occur and the Self-Distinction variable equals one. The self would be expected to associate perfectly with the characteristics and views of itself. However, if this were not the case and somehow the self did not feel as if it could associate with its own views and characteristics (evident in some psychological disorders), then this could be modeled with the equation as well by modifying the Identification variable accordingly.

For ease of comparison, one might also wish to organize the c-axis entities in such a way that the Identification variable toward *others* descends from highest to lowest moving away from the origin. This could be done in a similar manner to entities along the z-axis, and purposes along the d-axis. These extra considerations for the z-axis, d-axis, and c-axis are not essential to consider if one merely wishes to be able to conceptualize how the equation would model affect and emotion when so many factors are at play. They would, however, facilitate drawing conclusions between different valuations.

The Sixth Dimension: Multiple Values of Entities Concerning Multiple Purposes and Their Complementary Purposes for Multiple Individuals

Time elapsed, an entity's valuation, the valuations of entities concerning a single purpose along with its complement, the valuations of multiple entities for multiple purposes and their complements, and vicarious valuations of multiple entities for multiple purposes and complements are organized along the x-axis, y-axis, z-axis, d-axis, and c-axis respectively. Along a sixth dimension, multiple valuations and vicarious valuations of multiple entities, concerning multiple purposes and complementary purposes for multiple individuals, or everyone who is to be considered, can be represented along a sixth axis. This axis will be labeled the e-axis, for everyone considered, and will bring the total number of dimensions to six. The order in which people are organized along the e-axis may be decided upon at the discretion of whoever is using the equation. This dimension might prove useful if one wanted to consider the same set of purposes for multiple people. Whereas the use of negative values on the z-axis is optional, for the final three dimensions only positive valuations are used for the ***d***, ***c,*** and ***e*** values.

DIMENSION	AXIS	ORGANIZES
FIRST	x-axis	Time Elapsed
SECOND	y-axis	Valuation of a Single Entity for One Purpose
THIRD	z-axis	Multiple Valuations of Entities for One Purpose and its Complementary Purpose
FOURTH	d-axis	Multiple Valuations of Entities for Multiple Purposes and Their Complementary Purposes
FIFTH	c-axis	Multiple Valuations and Vicarious Valuations of Entities for Multiple Purposes and Their Complementary Purposes from a Single Person
SIXTH	e-axis	Multiple Valuations and Vicarious Valuations of Entities for Multiple Purposes and Their Complementary Purposes from Multiple People

Figure 13.5 The six dimensions of the equation and what each organizes are described above.

Ethics

Whenever a drive and its complement are considered in the equation, it is held that the self will aim to find a balance between the competing drives if it intends to sustain them both and continue living. As was demonstrated by the Roll Cage Theory of Drives, if the anxiety invested into a drive and into its complement are of an equal magnitude, then equilibrium would be felt halfway between them with a success ratio of one to one. If a drive is twice the magnitude of its complement, with respect to anxiety invested, then equilibrium would be felt two thirds of the way from the stronger drive toward the weaker complementary drive at a success ratio of two to one.

In the example of Flo, it was established that if the drive to hydrate by drinking water persisted all the way up to the point her body was literally made of 100% water, then she would perish. Likewise, if the drive to dehydrate oneself by not drinking water persisted until Flo's body had absolutely no water in it, or 0%, then she would perish. Flo's highest ranked drive, however, was the drive to maintain all drives and this entailed finding an equilibrium point between the drive to hydrate and the drive to dehydrate. For Flo, the ethically right thing to do would be to achieve homeostasis. Flo's deployment of anxiety and negative anxiety resources assisted her in this endeavor. Contrarily, if Flo's highest ranked drive was the drive to not maintain all drives, then the equation would maintain that holding the drive to hydrate over the drive to dehydrate, or vice versa, would be the ethically right thing to do for her.

With respect to individual purposes, if only one purpose from a complementary pair is being considered, then the self would simply aim to acquire as much of an entity as possible and the acquisition of that entity would be the morally upright thing to do. However, not only does each purpose have a complementary purpose against which it must be balanced, if the individual is alive and wishes to sustain his or her life, then these purposes have to be balanced against other purposes that the individual possesses. The question of which entities are ethically right or wrong to acquire will not simply be answered by identifying which entities facilitate the achievement of one purpose. Ultimately, the drives or purposes involved and their relationships with other drives and purposes will have to be identified in order to properly assess a code of ethics. Ethics is a "branch of philosophy concerned with that which is deemed acceptable in human behavior, with what is good or bad, right or wrong in human conduct in pursuit of goals and aims."[3]

Translating ethics into the equation, however, is a simple process and solely concerns the manner by which an individual's purposes have been ranked against the purpose with the highest priority. Henceforth, the Sentiment variable in the equation comprises the ethical code for the individual when the Self-Distinction variable is one. When the Self-Distinction variable is negative one, the Sentiment variable in the equation comprises the ethical code of an *other* as it is conceived by the self.

In psychoanalytic terms, Sentiment, paralleling the superego and its functions, houses the ego-ideal and conscience. It follows that an individual's ethical code is modeled in the equation as the manner by which an individual's purposes or drives are ranked against the purpose or drive with the utmost or highest priority. An ethical code would be assessed on a case by case basis. However, this begs consideration of the following questions:

1) If an individual's ethical code is determined by the ranking of different purposes or drives, then what determines an individual's purposes or drives in the first place? Does nature determine the ethical codes or does environment and can the equation model both?
2) What types of drives are there and can the equation model them all?
3) Can the equation also model universal drives as a precursor to a universal ethical code?

Drives Revisited: Primary vs. Secondary, and Regulatory vs. Non-regulatory

In most psychological literature, drives are defined as "motivational states produced by (a) deprivation of a needed substance such as food, a drug or a hormone, or (b) presence of a noxious stimulus such as a loud noise, excessive cold or heat, or a painful stimulus."[4] Drive, in both of the senses above, is a "hypothetical state that must be inferred from controlled operations."[5] In the equation, however, drive was first modeled as a ray, and then as having the qualities of a vector when opposing rays were overlapped. For instance, the drive to acquire food is distinguished from the drive to not acquire food and these two separate actions are not conjoined indiscriminately. They are two separate drives that must be balanced if, and only if, the individual intends to stay alive by holding the drive to maintain all drives above the drive to not maintain all drives.

Additionally, drives may be generally divided into four types: primary, acquired, regulatory, and non-regulatory:

1) Primary drives are drives that arise from an "intrinsic physiological characteristic of an organism," some of which are universal such as the drive to obtain food and water, and to reproduce, while others are specific to a species, such as nest-building and imprinting.[6]

2) Acquired or secondary drives are drives that are learned "through association with a primary drive," such as the human drive to acquire money.[7]

3) Regulatory drives are drives (such as hunger or thirst) "that function so that an organism seeks out substances that serve to maintain consistent bodily states necessary for survival."[8]

4) Non-regulatory drives are drives (such as sex) that serve "functions other than those that maintain the consistent bodily states necessary for the survival of the individual organism."[9]

Nature vs. Nurture, and the Consideration of Universal Drives as a Precursor to a Universal Code of Ethics

While nature sets the parameters for how an individual can establish an ethical code while living, based upon his or her regulatory drives, attempting to average out success ratios over large numbers of people would not reveal anything conclusive. What may be an ideal water composition for one individual might lead to the death of another; what may be an optimum caloric intake for one may starve another. Regulatory drives and their complementary drives arise out of biological necessity to keep the individual alive. In the equation, regulatory drives may be construed as the drives and their complementary drives that are necessary to sustain life. Non-regulatory drives can be construed as the drives and their complementary drives that do not necessarily support the sustenance of the individual, but make up activities that might be performed. Independently, an unchecked drive, regulatory or non-regulatory, is modeled in the equation to always lead to death.

Throughout this work, the explanation of the equation has leaned toward the nurture argument and presumed that individuals generally come into the world with a tabula rasa, "unencumbered by innate" ideas or intuitive knowledge.[10] This was chosen because it is the more difficult of the two positions for which to account. If one wanted to hold that an individual came into the world pre-loaded with knowledge, the nature argument, then this could also be represented in the equation by establishing that the cognizance of a specific entity is present at birth or sooner (e.g., holding the entity of the hunger drive to arise innately). Drives, however, and their relative force to a complementary drive, are generally assessed after experiencing them for the individual and lend themselves to the nurture argument. For instance, while an individual might acquire a drive to drink and a drive to eat based upon his or her biology, the knowledge to balance them out against their complementary drives of not drinking (e.g., dehydrating) and not eating (e.g., fasting) can only be established through the experiencing of them. The ability of the equation to account for these two very different approaches to the study of behavior while modeling emotion should not go unnoticed.

Primacy of Drives

Drives, in the equation, cannot have primacy over each other if an individual is to stay alive. Necessarily, this would appear to place the drive to maintain all drives above all others. In such a case, immortality or anything that offers a chance at eternal life would seem to be the ethical thing to pursue. However, drives are treated as a ray in the affect engineering and in their simplest form they always lead to death. Inherent in the drive to maintain all drives is the maintenance of its complement, the drive to not maintain all drives. Similarly, inherent in the drive to not maintain all drives is a check on itself, not maintaining the drive to not maintain all drives. In effect, the quest for immortality aims to shut out the drive to not maintain all drives, while the quest for death aims to shut out the drive to maintain all drives. Whichever one is held above the other would essentially become the guiding light for one's ethical code, but both are always present.

The case for holding the drive to maintain all drives above its complement is an obvious one, so attention will be turned to its opposite. For instance, if an individual's health fails to the point that the effort required to maintain life functioning inhibits the ability of the individual to pursue any other purpose or inhibits the fulfillment of vicariously felt purposes, euthanasia may become viewed as preferable to the individual. This could happen if the efficacy required to maintain regulatory drives completely and indefinitely reduces the efficacy to balance secondary drives and/or non-regulatory drives. Additionally, if the individual became entirely dependent upon an *other* for maintaining his or her own well being, then the Category I, II, and III Emotions of guilt, sadness, shame, vicarious sympathy, and jealousy might then be felt by the self. In such a scenario, the drive to maintain all drives might be pre-empted by the drive to not maintain all drives.

Similarly, the drive to maintain all drives can be pre-empted by the drive to not maintain all drives under other scenarios. These instances are not restricted to euthanasia. All of them would be modeled in the equation with the Sentiment variable and would entail a restructuring of how purposes are organized in such a manner that the drive to not maintain all drives would end up having more anxiety invested in it than the drive to maintain all drives. Some instances include the following:

1) Heroism, or apparent self sacrifice where the self will not witness the return on an investment of effort - A mother bird protecting her young at any cost to ensure the survival of her genes is one such example.
2) Some cases of terrorism or martyrdom exemplify this (i.e., a suicide bomber or a hunger strike).
3) Destitution in a time of war may lead one to opt for death over attempting to maintain a meager existence.
4) Torture might prompt the refusal to live under harsh conditions and inspire suicide.

The Role of Sentiment

Sentiment in the equation is mathematically designated in the equation as the ranking of purposes along a ratio scale with the highest ranked purpose being the origin. If Sentiment is unknown by the individual or only known to be within a certain range, then its value can be said to be either equal to one (i.e., $\boldsymbol{R}$ = zero) or undefined and said to oscillate within a range. However, if the Sentiment variable is known, then it will be a specific value as the ranking of the purpose in question against the highest ranked purpose would be known.

Psychoanalysis Components Accounted for in the Formula

1) Id: The id consists of every purpose or drive that the individual possesses. These are organized along the d-axis for drives concerning the self, and the c-axis for drives concerning an *other*. They are also the source of anxiety and negative felt as far as an individual is concerned and he or she may be consciously aware of these purposes, or unconscious and lacking in awareness of their existence.
2) Superego: Sentiment generally represents the superego and ego-ideal. Sentiment is understood in the equation as the manner by which the individual has ranked his or her own purposes or distinguished which purposes, corresponding entities, and values of entities should be weighted more heavily than others. For purposes that belong to the individual, the Self-Distinction variable equals positive one and Sentiment for these purposes may be said to form the superego. The ego-ideal is the ". . . ego's

conception of positive ideals."[11] These are ideals that the self would eventually like to live up to or become.[12] For purposes that do not belong to the individual but relate to the individual (e.g., expectations that others have of the self), Self-Distinction equals negative one and Sentiment for these purposes may be said to form the ego-ideal. The superego, in psychoanalytic theory, "is assumed to develop in response to the punishments and rewards of significant persons (usually parents)," is akin to an internalized moral code of the community, and is theorized to develop at a young age.[13] Sentiment, in affect engineering, is an idealized way for the individual to organize the priority of purposes or drives, anxiety, and negative anxiety resources (id components) of the self and of *others*.

3) Ego: The ego is represented by Reasoning and strives to reconcile the superego (Sentiment) with the demands of the id (Purposes / Drives, anxiety and negative anxiety) by mobilizing the self to act and by redirecting Attention to influence the investment of anxiety and negative anxiety. Structurally, it may be generalized that the id is primarily concerned with affect, the superego is primarily concerned with the ideal, conation, or what ought to be done, while the ego is concerned with all things relating to cognition and balancing the competing forces. The two primary ways in which the ego may mobilize resources (e.g., Reasoning) to deal with the id and superego include the following:

   a) The individual may acquire the entity (Efficacy) despite threats of harm to it (Threat severity and threat susceptibility).

   b) The individual may distort the perceived value of an entity by directing or diverting Attention away from efficacy or harm, or by altering other variables in the equation concerning the entity's value. In this manner the ego can manage the resources of the id (anxiety and negative anxiety) in order to fit them into the mold of the superego (e.g., Sentiment).

The latter of the above two will be the source point for many of the defense mechanisms modeled in chapter fourteen.

Origins of Sentiment

Sentiment is presumed to arise through experience and the efforts of the individual to organize which purposes should have priority over others. For most people, the Sentiment rankings of purposes will generally have survival (e.g., water, food, etc.) as the most important, with most other aims being subservient or less important than these. However, if an individual's needs become inverted from this norm, the equation is able to account for such scenarios, if necessary.

Preview

From the beginning of this book, the aim of *Affect Engineering* was the establishment of a foundation for a sound theory of emotion that would be capable of modeling abnormalities in normal functioning. What has resulted is not only a model that is empirically testable, but one with the capacity to expand to account for new ideas and concepts. The ultimate goal in establishing an emotional equation was to develop a new language to interpret emotions, construe emotional language, and approach psychological disorders. It was predicted that maladaptive defense mechanisms and psychological disorders could be modeled by the equation. Dysfunctional or undesirable sets of values in the equation, for instance, may help express abnormalities in normal functioning. The capacity of the equation, and its two extensions, the Roll Cage Theory of Drives and the neurological network model, will now be explored in order to determine the versatility and usefulness of the equation in practical applications.

YOU SHOULD UNDERSTAND THIS.
CHAPTER THIRTEEN
CONCEPT CHECK
VOCABULARY
Hypercube
Nature Argument
Non-regulatory Drives
Nurture Argument
Primary Drives
Regulatory Drives
Secondary Drives

Notes

1. Wilde, Oscar (2003). *The Picture of Dorian Gray.* New York: Barnes and Nobles Books. Print, p. 20. (Original work published 1890).

2. Reber, Arthur S., Rhianon Allen, and Emily S. Reber. *Penguin Dictionary of Psychology*. London. Penguin Books, 2009. Print, p. 367.

3. Reber, Arthur S., Rhianon Allen, and Emily S. Reber. *Penguin Dictionary of Psychology*. London. Penguin Books, 2009. Print, p. 272.

4. Reber, Arthur S., Rhianon Allen, and Emily S. Reber. *Penguin Dictionary of Psychology*. London. Penguin Books, 2009. Print, p. 235.

5. Reber, Arthur S., Rhianon Allen, and Emily S. Reber. *Penguin Dictionary of Psychology*. London. Penguin Books, 2009. Print, p. 235.

6. Reber, Arthur S., Rhianon Allen, and Emily S. Reber. *Penguin Dictionary of Psychology*. London. Penguin Books, 2009. Print, p. 235.

7. Reber, Arthur S., Rhianon Allen, and Emily S. Reber. *Penguin Dictionary of Psychology*. London. Penguin Books, 2009. Print, p. 235.

8. Reber, Arthur S., Rhianon Allen, and Emily S. Reber. *Penguin Dictionary of Psychology*. London. Penguin Books, 2009. Print, p. 236.

9. Reber, Arthur S., Rhianon Allen, and Emily S. Reber. *Penguin Dictionary of Psychology*. London. Penguin Books, 2009. Print, p. 235.

10. Reber, Arthur S., Rhianon Allen, and Emily S. Reber. *Penguin Dictionary of Psychology*. London. Penguin Books, 2009. Print, p. 800.

11. Reber, Arthur S., Rhianon Allen, and Emily S. Reber. *Penguin Dictionary of Psychology*. London. Penguin Books, 2009. Print, p. 249.

12. Reber, Arthur S., Rhianon Allen, and Emily S. Reber. *Penguin Dictionary of Psychology*. London. Penguin Books, 2009. Print, p. 249.

13. Reber, Arthur S., Rhianon Allen, and Emily S. Reber. *Penguin Dictionary of Psychology*. London. Penguin Books, 2009. Print, p. 789.

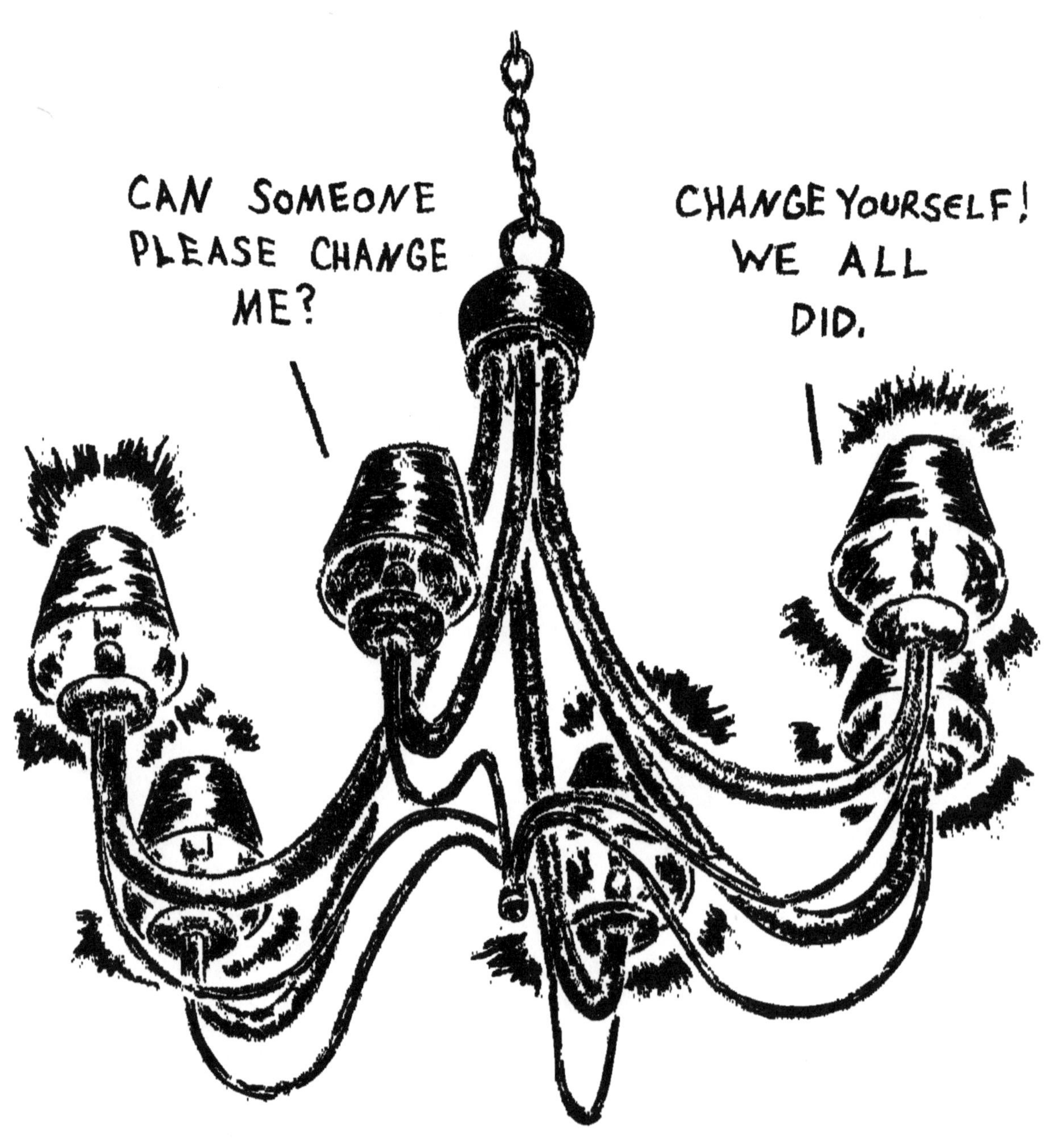

"A man who is master of himself can end a sorrow as easily as he can invent a pleasure." - Oscar Wilde[1]

CHAPTER FOURTEEN

Application to Defense Mechanisms

Front-loading: Acting Out, Altruism, Anticipation, Conversion, Delusion, Denial, Displacement, Dissociation, Distortion, Fantasy, Humor, Hypochondriasis, Idealization, Identification, Intellectualization, Introjection, Isolation, Passive Aggressive, Projection, Rationalization, Reaction Formation, Regression, Repression, Somatization, Splitting, Sublimation, Suppression, Undoing, Withdrawal

Defense mechanisms are protective behaviors that offer protection to individuals against ". . . the awareness of that which is anxiety-producing."[2] Additionally, due to their peculiarities, defense mechanisms may be seen to involve some type of irrationality. Processes that are irrational are "in violation of the rules of logic," and the term is commonly used to describe acts that "have been carried out under the pressures of emotional factors that have, somehow, overridden the logical, rational faculties."[3] Irrationality is further distinguished from non-rationality in the sense that the non-rational refers to thinking, believing, decision-making, or other actions "where the notions of reason and logic are not applicable."[4] While some of the defense mechanisms can be explained by instances where the individual may have misdirected attention away from or toward specific entities, in some cases novel approaches to the equation and the two models derived from it, the neurological network model and the Roll Cage Theory of Drives, will have to be taken in order to depict them. The following list of defense mechanisms is by no means exhaustive and it is organized in a fashion similar to George Valliant's methodology in his work *Adaptation to Life* (1977).

Application to Defense Mechanisms

Pathological Defenses

Pathologies are abnormal conditions where an individual's proper functioning is prevented.[5] Pathological defense mechanisms would be those that depart from proper functioning.

Conversion

A conversion disorder is a somatoform disorder where psychic conflict is converted into a physical manifestation.[6] Resulting symptoms, such as deafness or seizures, may appear to have an underlying physical cause, but there is usually no evidence of a physical disorder.[7] However, symptoms of a conversion disorder are "not feigned and not under conscious control."[8] Essentially, "psychic conflict" is converted into bodily sensations or "somatic form."[9]

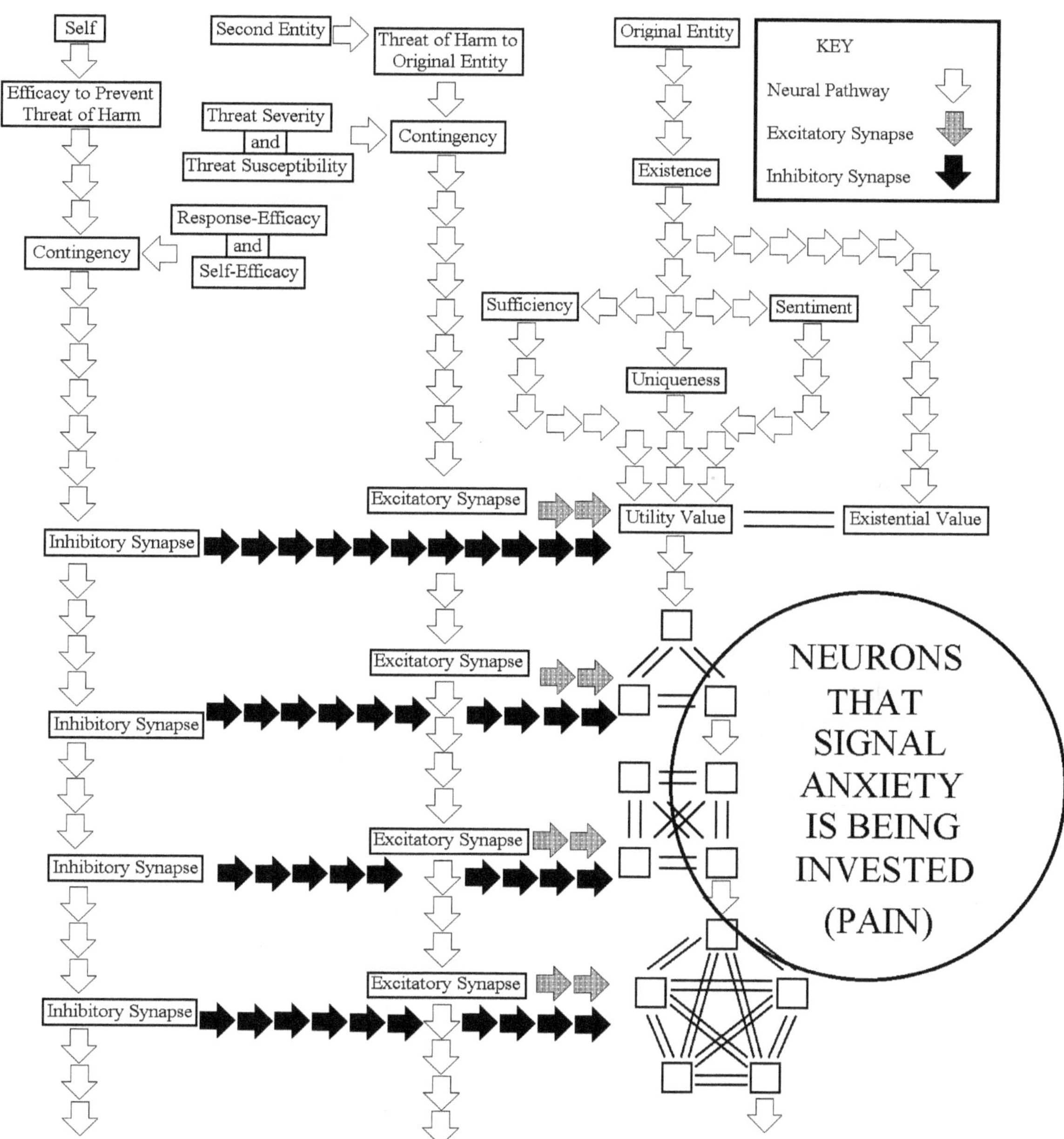

Figure 14.1 Sample Neurological Model for Conversion Disorder
The valuing neurons that normally target receptor sites relevant to the situation (e.g., targeting facial muscles to illicit a frown upon receiving bad news) may be inadvertently targeting unrelated receptor sites. This might be caused, for instance, by an overflow of anxiety produced in response to an extreme stressor or loose anxiety in an individual's system that is not available to do work (e.g., entropy, a variable in the denominator against the entire function).

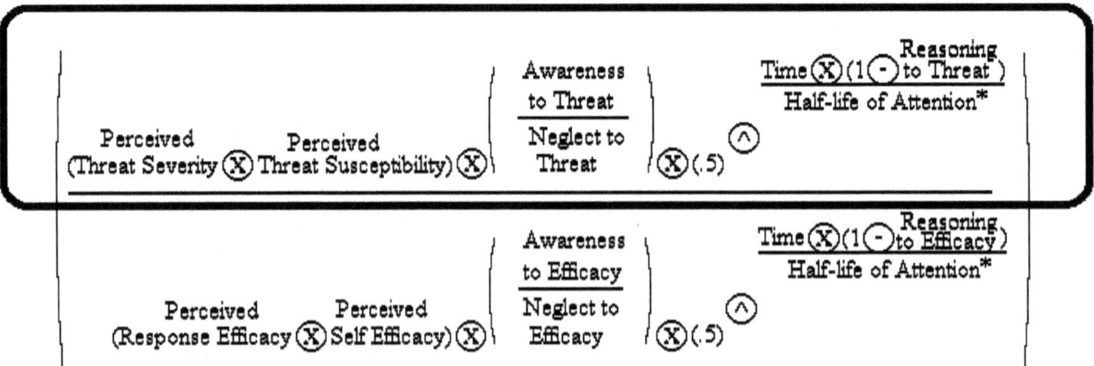

Figure 14.2 Sample Mathematical Expression of a Conversion Disorder
Using the Avoidance of Pain Formula, an elevation of the variables in the numerator (circled) over the variables in the denominator in the exponent might produce a surge of anxiety invested into an entity that creates physical symptoms in relevant and non-relevant receptor sites targeted by valuing neurons. This, in turn, could lead to an increase in overall entropy.

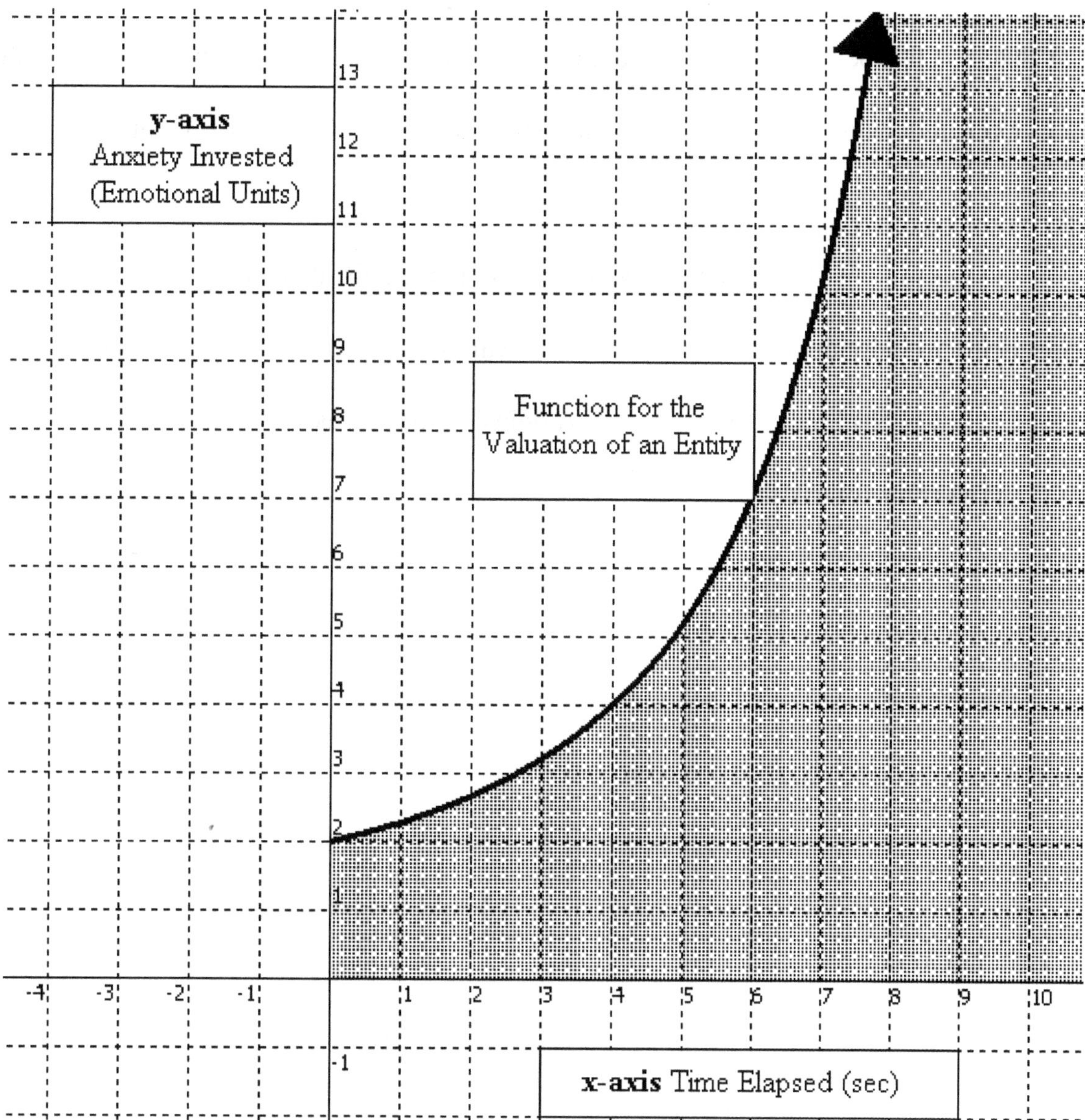

Figure 14.3 Sample Graph of Anxiety Elevating (Conversion Disorder)
An elevation of anxiety invested (shaded region) could have unintended and unexpected effects on an individual's physiology if it is in such a large quantity that it spreads beyond normally targeted receptor sites. As many disorders in psychoanalytic literature are hypothesized to result from an elevation of anxiety (e.g., stressor, trauma) and the ego's attempts to control the id, Cartesian plots will be provided if something deviates from the normal.

Application to Defense Mechanisms

Delusion

A delusion is a belief that is maintained despite evidence to the contrary which should be reasonably expected to destroy it.[10] Delusional disorders are usually one of several types, such as erotic, grandiose, persecutory, and jealous.[11]

The representation of delusions in the math equation entails identifying which relationships between entities an individual has cognized and thereafter one must determine if they differ from what is logical or would normally be expected. A delusion may be as simple as misinterpreting the Uniqueness of an entity toward the fulfillment of a purpose or as complex as believing a contingency exists where there is none, concerning threat, benefits, or efficacy.

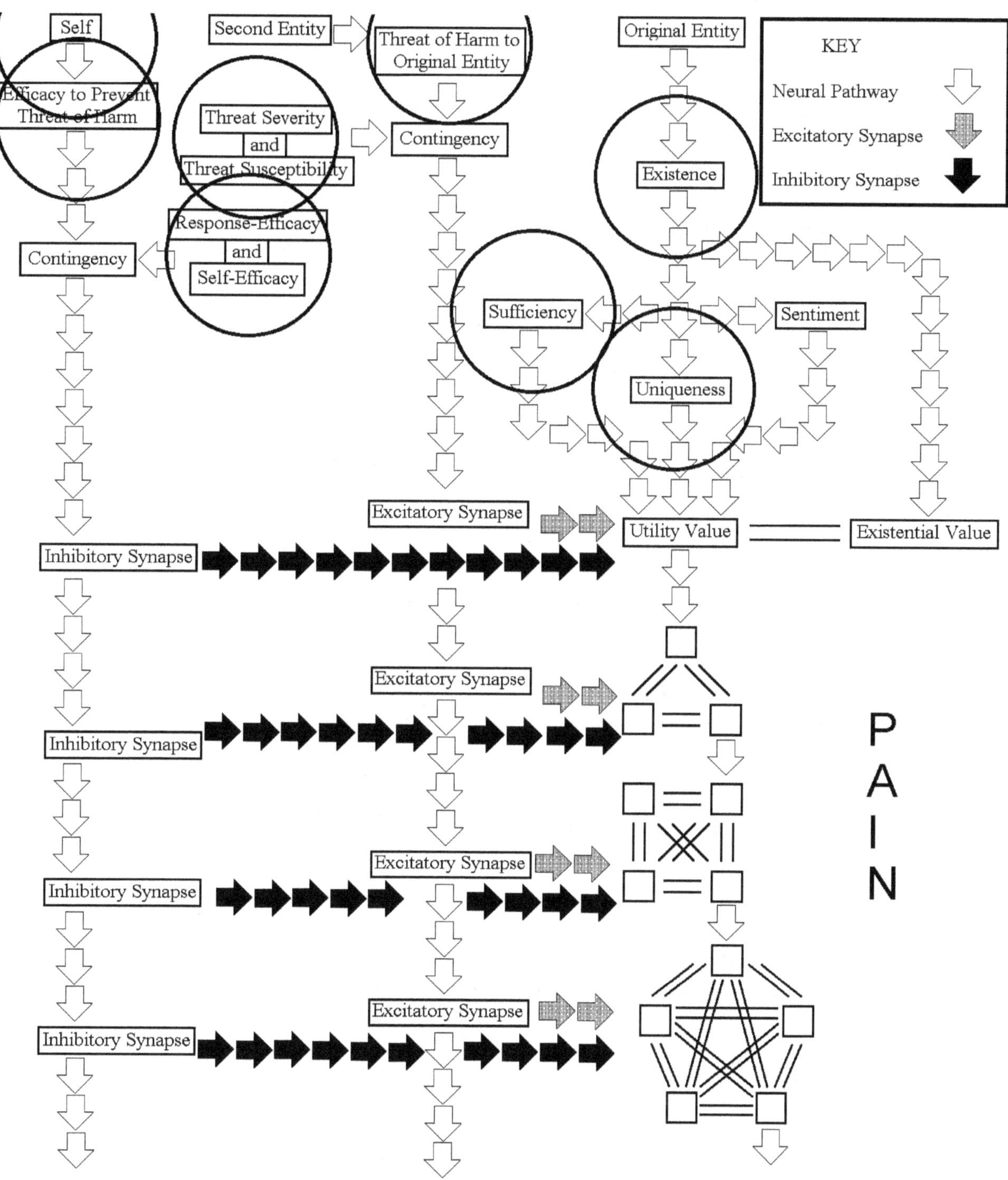

Figure 14.4 Sample Neurological Model of a Delusion The circled variables in the neurological model are ones that might be involved (e.g., at fault or in error) when an entity is being given a valuation and a delusion is occurring.

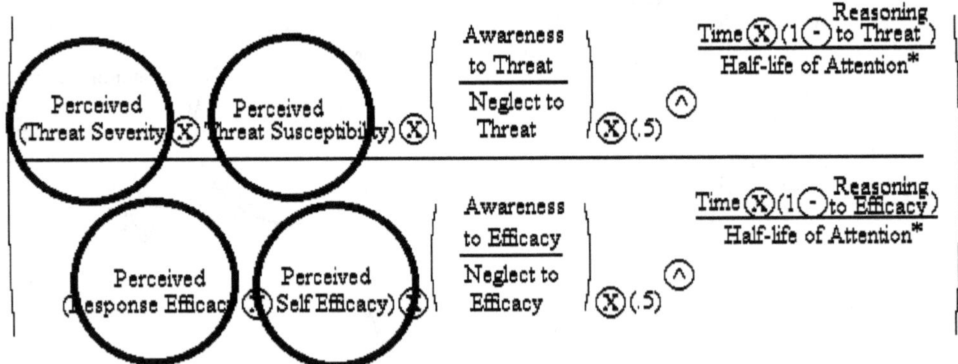

EXPONENT OF THE EQUATION

BASE OF THE EQUATION

Figure 14.5 The circled mathematical variables above are likely to be found at fault or in error when an individual is under a delusion.

Denial

Denial is a defense mechanism where "thoughts, feelings, wishes or needs that cause anxiety" are simply disavowed.[12] Sample Form III of attention, with valuation resilience, would best model denial based upon the description above. Thoughts, feelings, wishes, or events that are of true importance to an individual and are assessed accordingly can still have their valuations drastically reduced if Attention is directed elsewhere, for instance, to something less important. If the anxiety related to an entity is merely pushed into the subconscious or unconscious, then this could be represented in either the Roll Cage Theory (e.g., anxiety-mass attached below the disc) or in the neurological model where Reasoning processes that influence Attention cannot be brought under the focus of one's Attention.

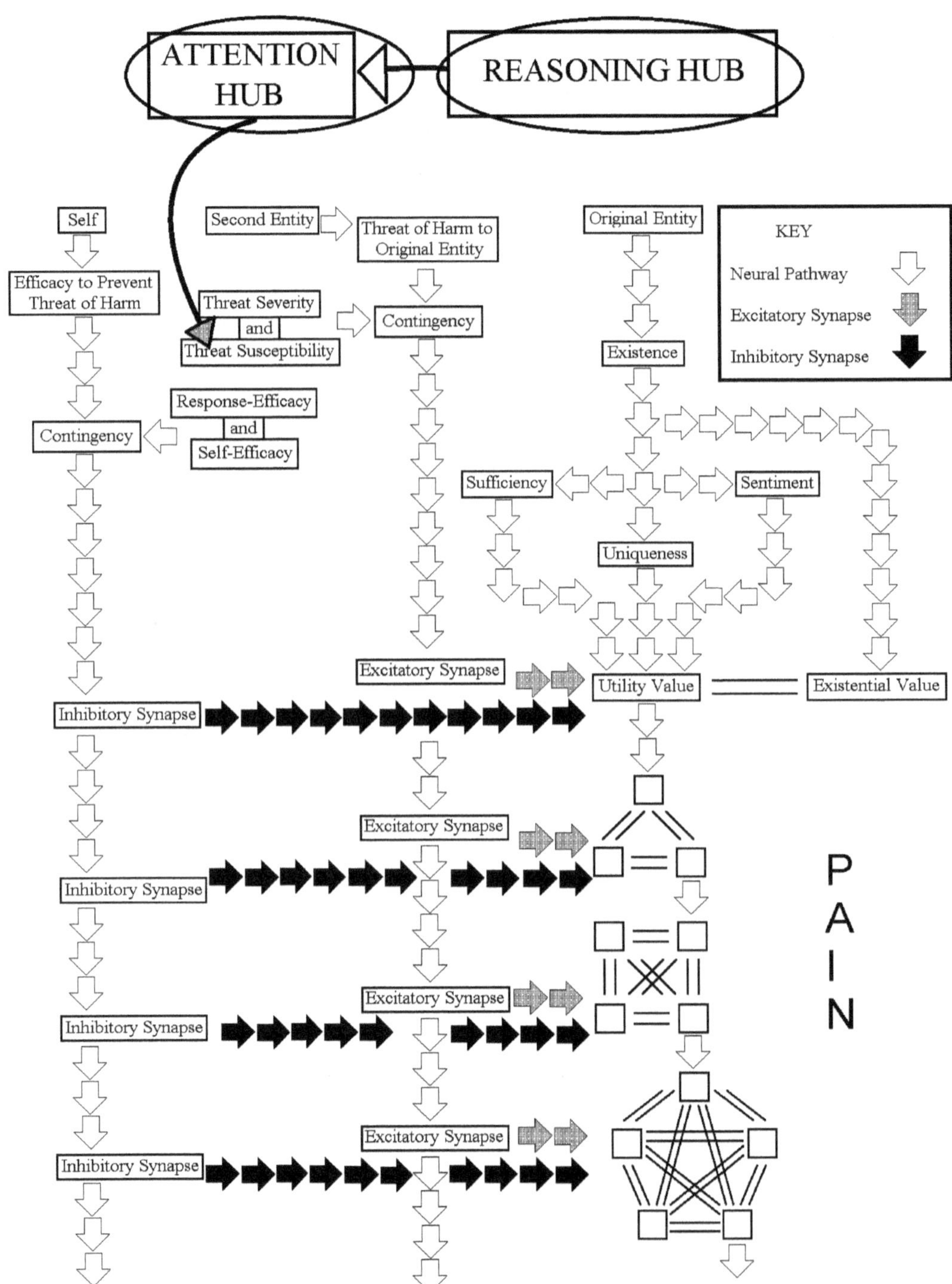

Figure 14.6 Neurological Model of Denial Taking Place
The circled variables, Attention and Reasoning, would be the key players.

Application to Defense Mechanisms

Conscious, subconscious, and unconscious activity can be represented in the neurological model by interaction between Attention and Reasoning. For instance, Reasoning that the individual is conscious of would have a portion of Attention redirected back unto Reasoning. Activity taking place in the subconscious, sometimes called the preconscious, refers to information that lies momentarily out of awareness, however, it may "easily be brought into consciousness."[13] Contrarily, activity taking place in the unconscious would correspond to mental activity occurring largely outside of an individual's awareness.[14] In depth psychology, these processes are theorized to be largely inaccessible, for example, on account of being too anxiety-provoking.[15]

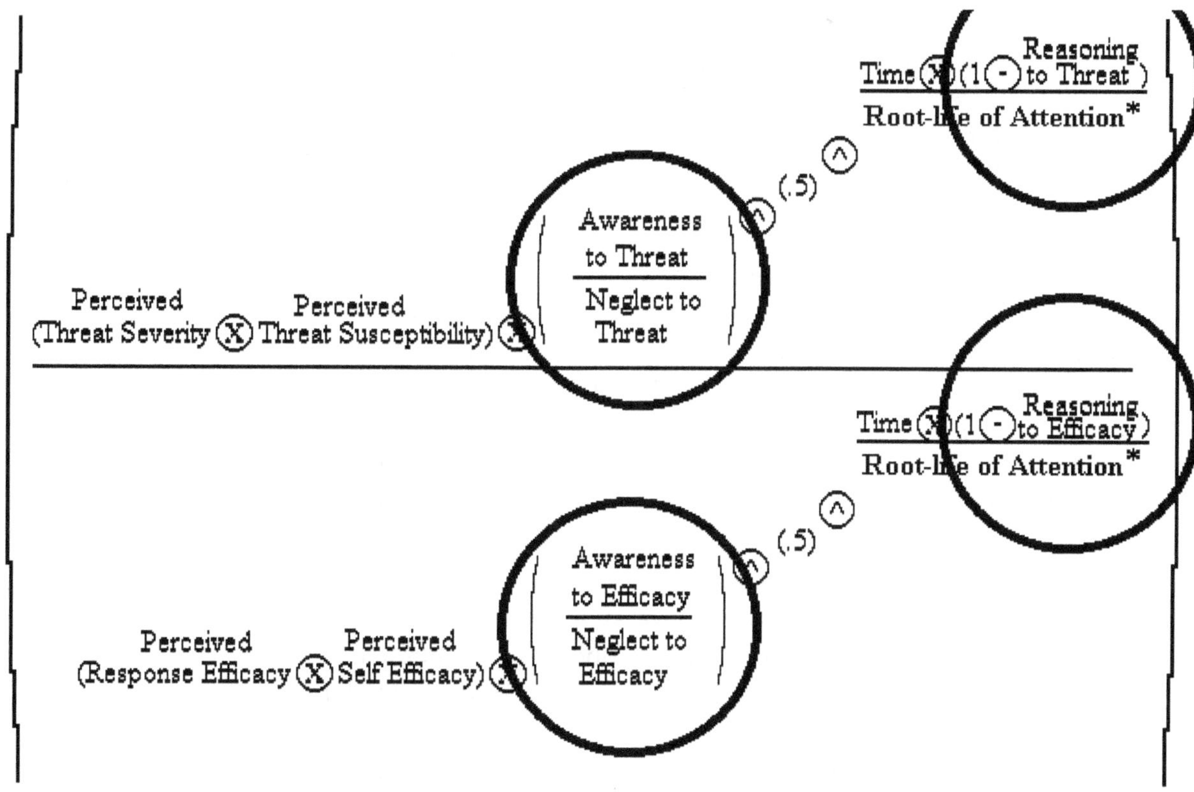

Figure 14.7 Sample Mathematical Expression of Denial Taking Place
Denial would likely involve an amplification of threat over efficacy or efficacy over threat (alternatively, benefit over efficacy or vice versa if benefit is being used). Reasoning would direct Attention toward or away from one or the other unevenly. This may be done either consciously, if a portion of Attention is directed to the reasoning processes going on, or unconsciously if otherwise.

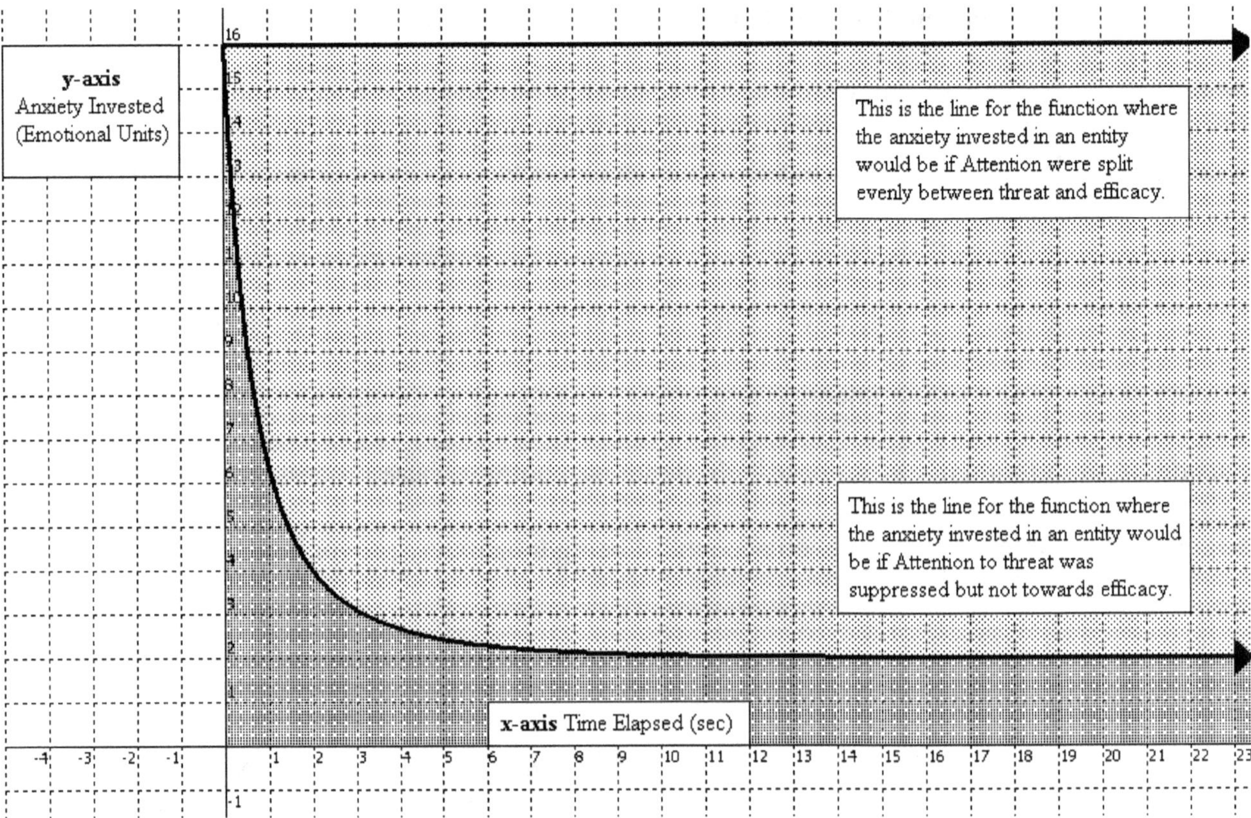

Figure 14.8 Sample Graph of Denial Taking Place
Using form three of attention, with valuation resilience, the same function was modeled in two cases, with one difference. The threat to efficacy ratio was 4:1 in both cases (Threat Severity * Threat Susceptibility : Response Efficacy * Self Efficacy = 4:1), and the ratio of Awareness to Neglect equaled 1:4 for each (e.g. .2 / .8). However, in the first case (top), neither the threat nor efficacy components had Attention directed away from them and the anxiety invested remained at 16 emotional units, where it should be based on the threat to efficacy ratio. In the second case (bottom) Reasoning to Threat stayed suppressed while Reasoning to Efficacy was permitted to surge toward one. The self effectively denies the legitimacy of the threat by overestimating efficacy. The darkly shaded region (bottom) depicts the anxiety being invested in the entity when the threat components are ignored and the efficacy components exaggerated. The lightly shaded region represents the disparity in anxiety investment, roughly 14 emotional units after 14 seconds.

Distortion

In psychoanalysis, distortion is a "defense mechanism that (presumably) functions to alter or 'disguise' dream content that would be unacceptable in non-distorted form."[16] If a desire is taboo or too disturbing, it may be distorted or masked as something benign. For instance, a disturbing entity that an individual wants to acquire may have its valuations placed on or disguised as an entirely different entity, hence, one more acceptable to the person.

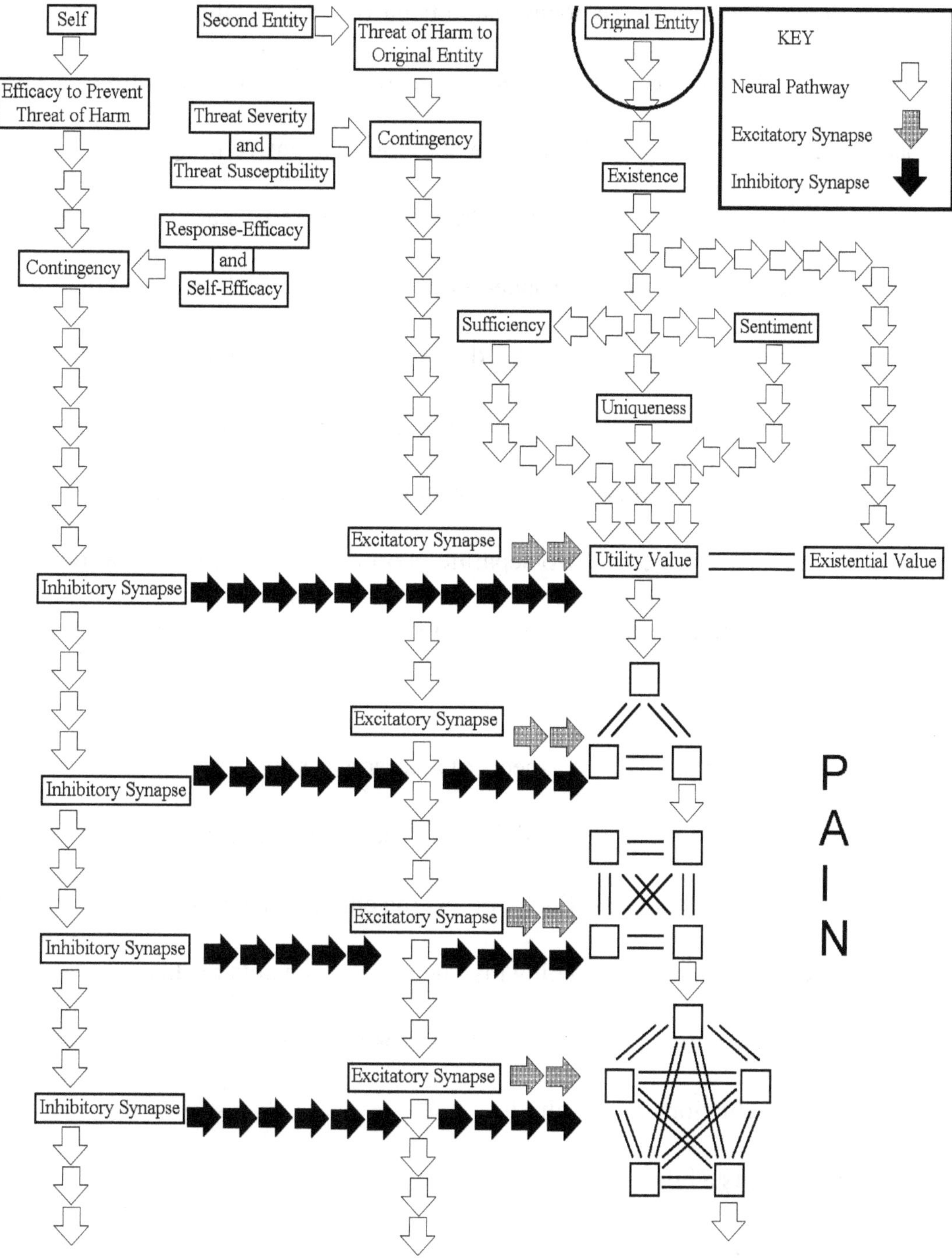

Figure 14.9 Above is a sample neurological model for distortion where the circled item, the entity itself, becomes ambiguous.

Application to Defense Mechanisms

Splitting

Splitting is a "defense mechanism whereby one deals with conflict and stress by compartmentalizing (i.e., splitting) the positive and negative aspects of oneself or others."[17] "Use of the mechanism is marked by a tendency to view oneself and others in alternating polar opposites, switching back and forth between highly positive and highly negative images."[18]

The modeling of splitting with the use of the equation is remarkably straightforward, as the equation necessitates dividing the different aspects of a person (e.g., self or *other*) in order to adhere to the 1:1:1:1 ratio, for instance, by imparting a different value on an entity for different purposes. The problem with splitting, according to the definition offered above and concerning psychoanalytic theory, is that an individual using this defense may view a single object, such as a parent, as both a good parent and an evil parent, hence, two separate people. However, the equation sidesteps the complicated process of dividing a whole entity into two separate ones by not even assuming it was whole to begin with in the mind of the individual.

Rather than viewing the different aspects of a person as split off from a whole, the equation models splitting as a failure to integrate the different aspects of an *other* or the self into a cohesive whole (i.e., into an associative percept). For example, the entity of a parent will be considered in the case where a child wants a hug after scraping his knee and also wants to stay up late. Adhering to the 1:1:1:1 ratio, it soon becomes evident that four purposes are at play here for which a parent will be given four separate valuations by the child:

1) Child wants a hug.
2) Child does not want a hug.
3) Child wants to go to sleep.

4) Child does not want to go to sleep.

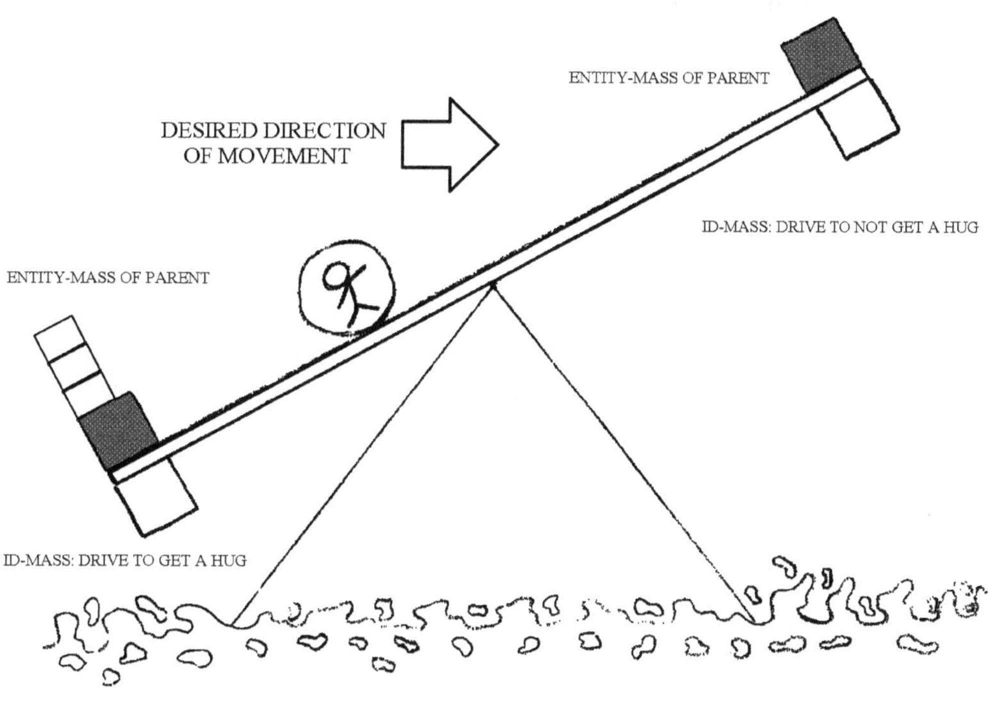

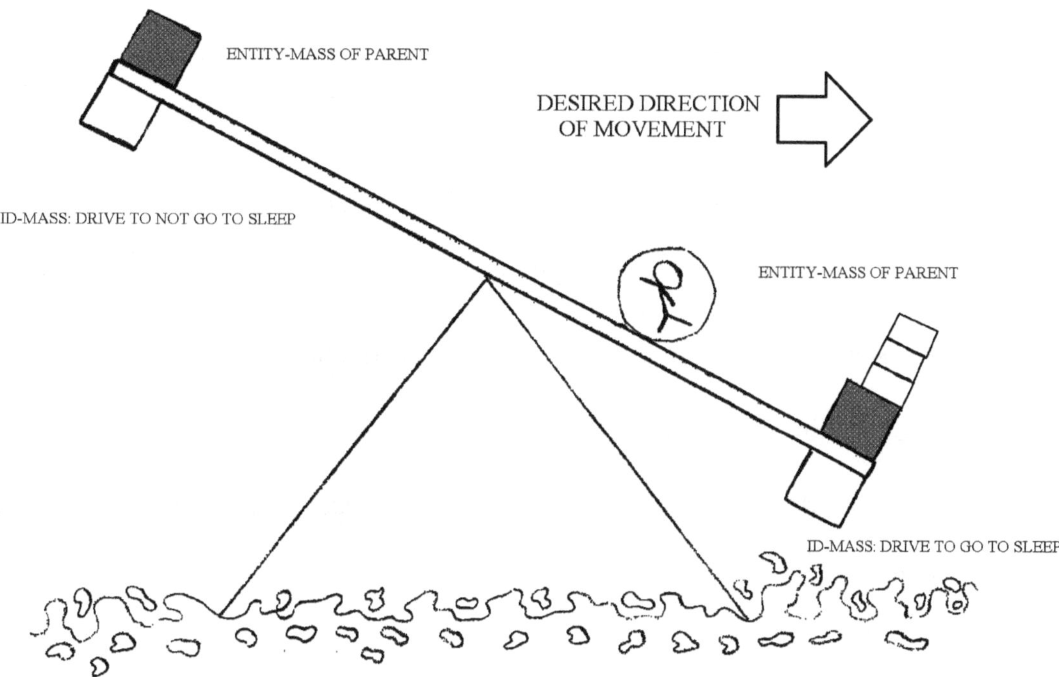

Figure 14.10 Roll Cage Theory of Drives Depicting Splitting
The shaded entities above are unintegrated into a whole.

Application to Defense Mechanisms

Splitting, modeled in the Roll Cage Theory of Drives, is depicted as a failure by the individual to integrate the four valuations of an entity (e.g., a parent) from four separate purposes into one entity via bottom-up processing. The valuations stay separate. Henceforth, the notion of a good parent and an evil parent do not come together to form the associative percept of just parent.

Immature Defenses

Immature can refer to anything that is "not fully developed" in the neutral sense, or to things "displaying less-well-developed traits and characteristics than the norm for one's age," which carries a distinctively "negative connotation."[19] Immature defense mechanisms would be those that either are not fully developed or are inappropriate and deviate from the norm for one's age.

Acting Out

Acting out is a "coping style in which the individual deals with conflict or stress through actions rather than through reflections or feelings."[20] For instance, unconscious psychic conflict within an individual may manifest itself as a behavior that eases the latent anxiety. The key element distinguishing acting out from misbehaving is that the individual does not realize the nature of the conflict (e.g., unconscious), but feels that performing a particular behavior mitigates some of the anxiety and emotional pain felt. In short, an individual engages in a specific behavior to alleviate anxiety resulting from hidden psychic conflict that may or may not be related to the expressed behavior. The Roll Cage Theory of Drives can best depict the defense mechanism of acting out.

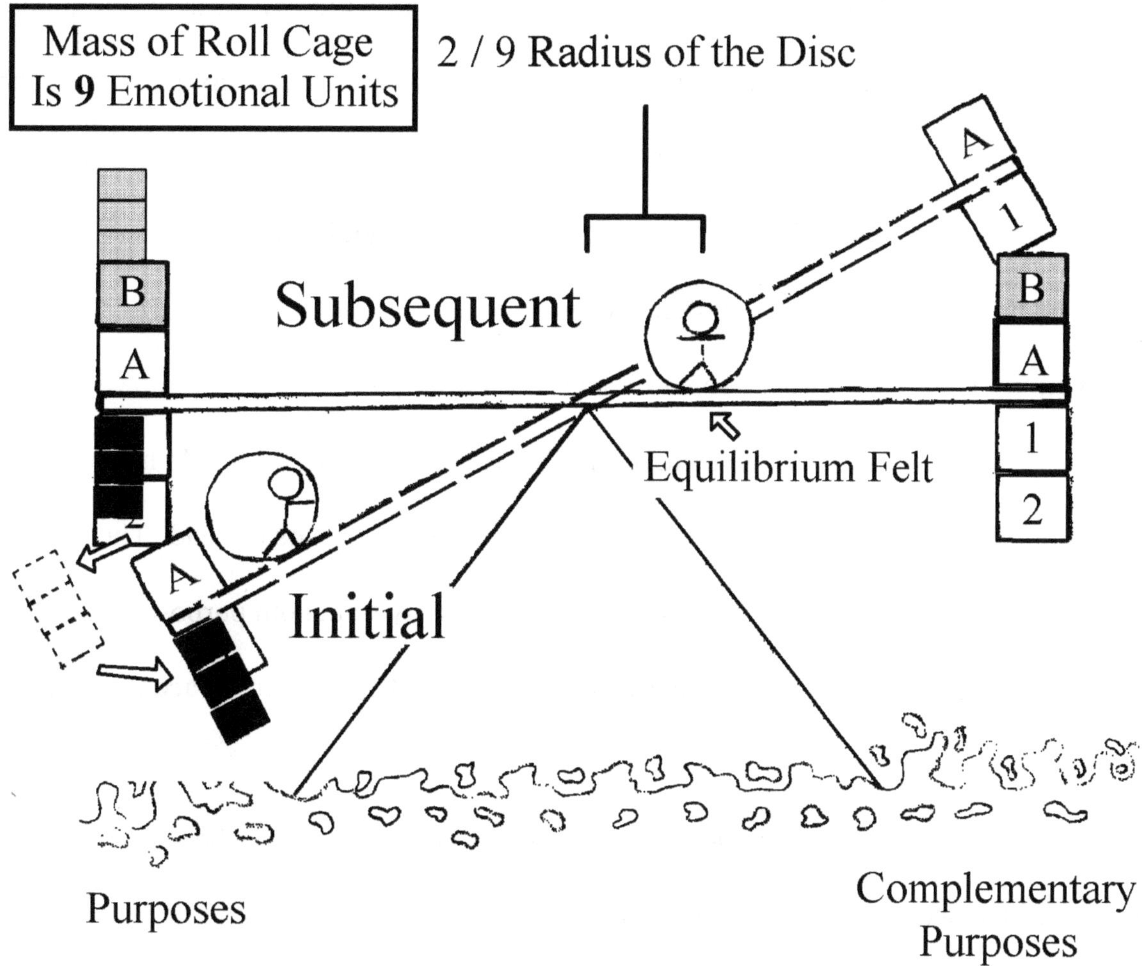

Figure 14.11 The above is a Roll Cage Model depicting acting-out (the ego underestimates the roll cage's anxiety-mass). The utility-mass from entity *A* has been pushed off by the ego, but it reattaches in the unconscious. The ego does not account for the hidden anxiety that disrupts the balance of the disc and later begins to associate purpose # 1 with purpose # 2 because they occur together (e.g., the id-masses move closer together on the circumference of the disc apparatus). Entity *B*, which initially only alleviated anxiety for purpose # 2, subsequently alleviates anxiety for both purpose # 1 and # 2 when it is acquired.

The ego, in figure 14.11, wants to acquire entity *A* to fulfill the drive represented by id-mass # 1. However, entity *A*, for one reason or another, is also a taboo or is forbidden to the individual. In a desperate attempt to rid itself of excess anxiety-mass influencing the disc, entity *A*, valuable to the fulfillment of the id-mass and purpose on the left, has had its utility-mass unceremoniously pushed off the disc by the ego in an attempt to eliminate it. However, the utility-mass reattaches to the disc apparatus on the underside and remains in the unconscious. The ego, now unable to notice the hidden anxiety-mass disturbing the balance of the disc, does not invest enough anxiety into the purpose of maintaining equilibrium between the two drives. Instead of investing 10 units of anxiety mass onto the roll cage, only nine units are invested. Over time, purpose # 2, one with which the ego has no reservations, became associated with purpose # 1. For example, this could happen if action is taken to fulfill purpose # 2 anytime that the thought of purpose # 1 rears its head. The acquisition of entity *B*, thereafter, performs double-duty, alleviating anxiety from both purpose # 2 (conscious anxiety) and from purpose # 1 (unconscious anxiety), as the individual cannot acquire entity *A*.

Fantasy

Fantasy, if it comes to dominates a person's mental activities, can become maladaptive. Dreams were modeled using Sample Form III of attention in chapter twelve. If the suppression of an entity's valuation is released (e.g., Reasoning suppressing Attention to an entity's valuation), then an entity's valuation can undergo a resurgence. Such a resurgence could result in an entity's valuation swinging toward a value closer to positive infinity (anxiety and pain) or toward a value closer to negative infinity (negative anxiety and pleasure). Additionally, fantasy might also occur as a result of an individual mistakenly believing the self to be in the place of an *other*.

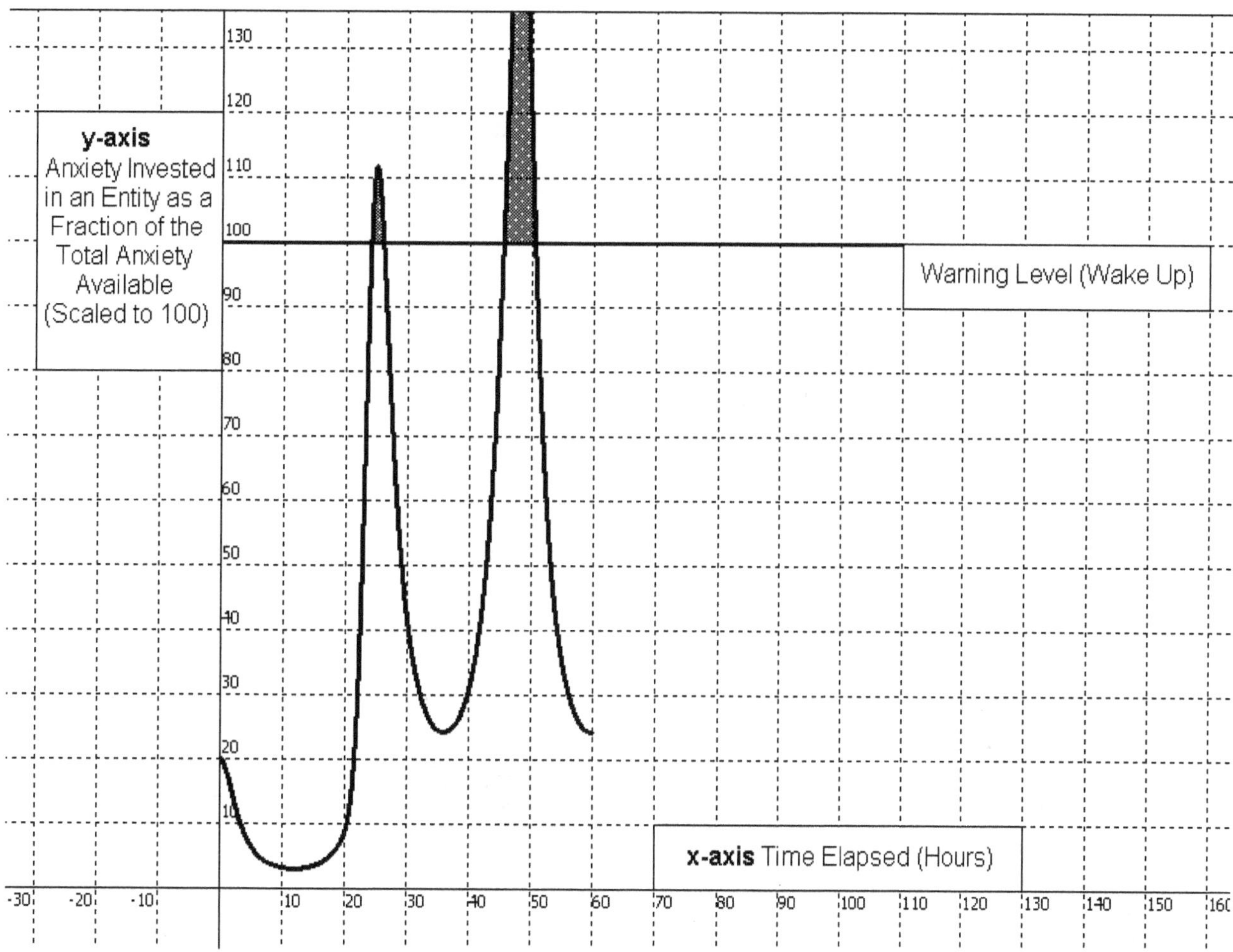

Figure 14.12 Dream modeling from **figure 12.29** (Fantasy)
It was seen that when Reasoning to Efficacy and Threat are both released, an entity's valuation can rebound back to where it would be without any interference from Reasoning processes. This could happen during sleep (above) just as easily as it might happen during the day when someone is awake.

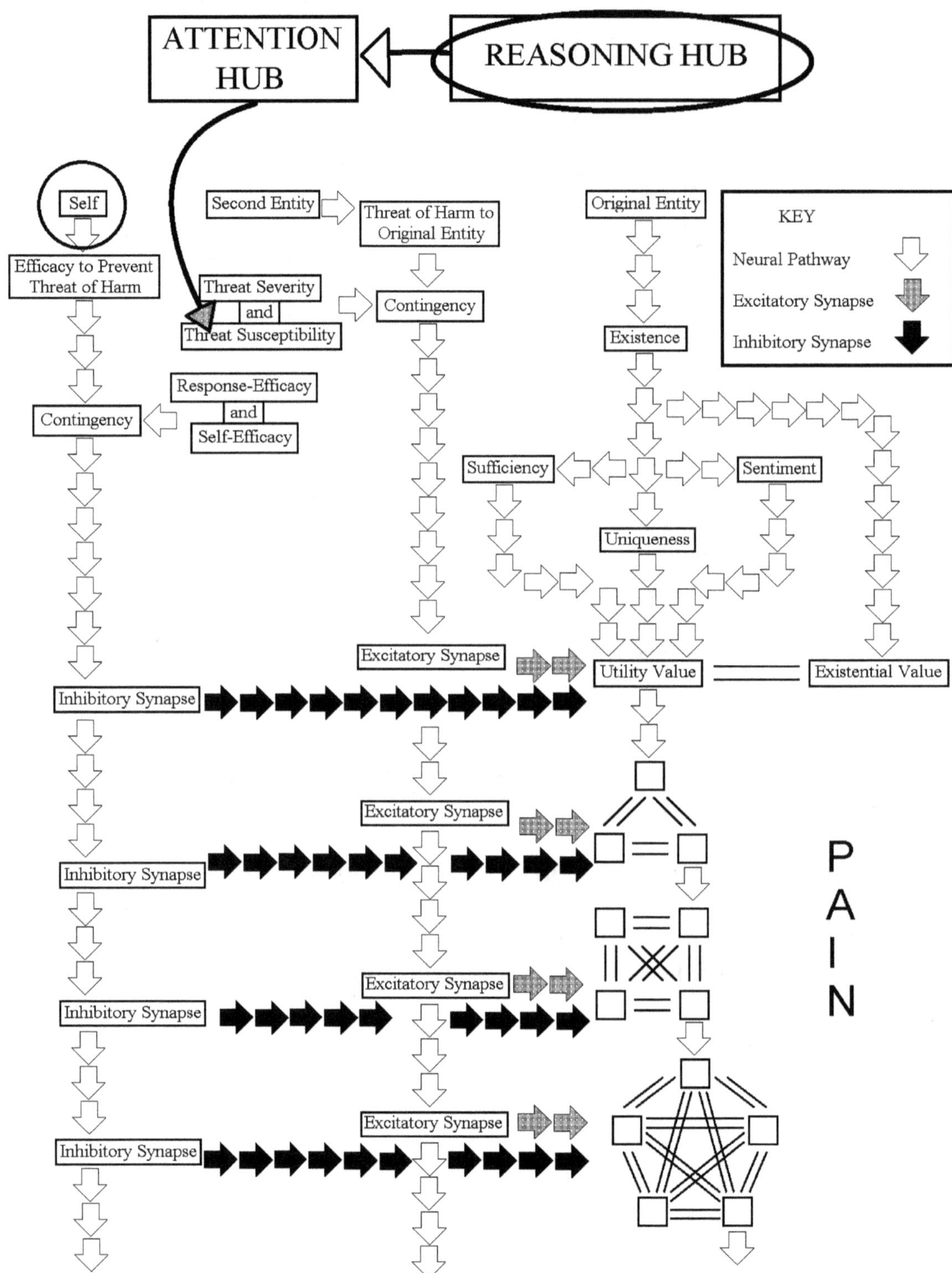

Figure 14.13 Neurological Model: Reasoning, and its relinquishment of control over Attention, would be a factor in assessing fantasy as a defense mechanism, as would the ability of the self to distinguish itself from an *other*.

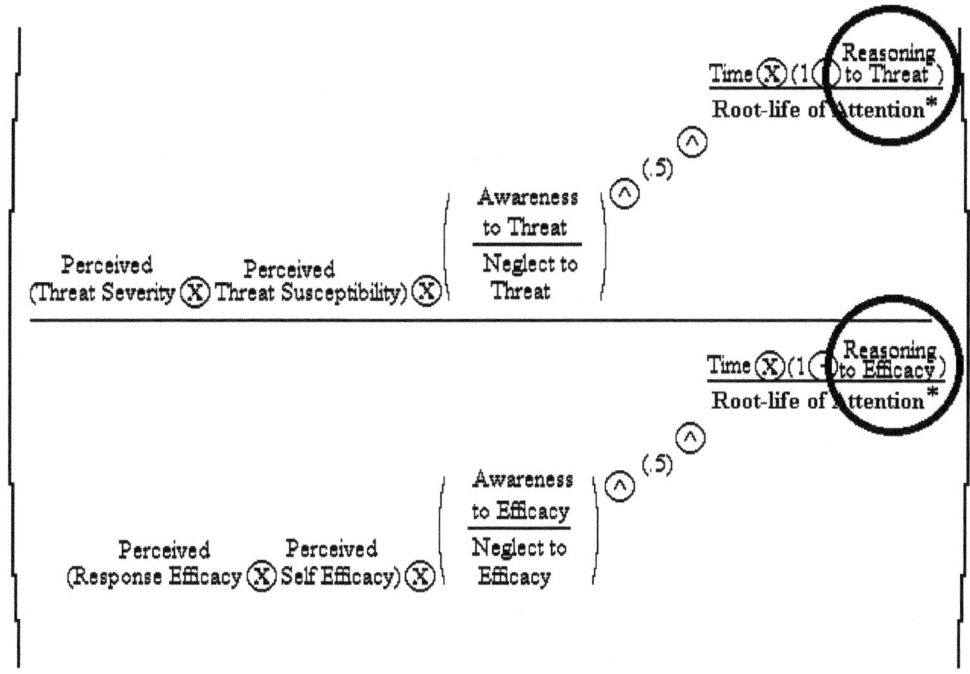

EXPONENT OF THE EQUATION

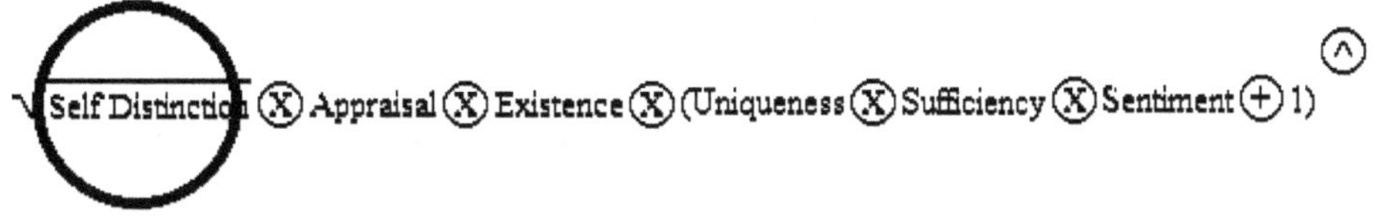

BASE OF THE EQUATION

Figure 14.14 Variables in the equation that would likely contribute to the use of the defense mechanism of Fantasy include Self-Distinction and Reasoning.

Idealization

In psychoanalytic theory, idealization is "a defense mechanism in which an object about which one is ambivalent is split into two conceptual representations, one wholly bad and one completely, ideally good."[21] The counterpart of idealization would be devaluation, “a defense mechanism where one attributes exaggerated and negative qualities to oneself or others.”[22]

Structurally, the representation of idealization and devaluation would be similar to splitting, mentioned earlier in figure 14.10. Foremost, the different valuations of an entity for different purposes are not integrated. Idealization would then occur whenever the valuations pertaining to a purpose preferable to the self are given more importance than other valuations. Devaluation would occur whenever the valuations pertaining to a purpose not preferable to the self are given more importance than other valuations.

Passive Aggression

Passive-aggressive personality disorder is a personality disorder "marked by a pattern of passive resistance to request for appropriate social and/or occupational performance."[23] "An individual with the condition typically procrastinates, becomes sulky and irritable when asked to do things he or she does not wish to do, works slowly on jobs (in a seemingly deliberate manner) and avoids obligations or responsibilities."[24] Passive-aggressive behavior refers to "behavior in which aggressiveness is displayed in a passive rather than in an active manner."[25] "Overt aggressiveness" would likely "lead to reprisal" if there is a power disparity between the two individuals.[26]

The transformation of passive-aggressiveness into the equation will be similar to that of acting out, save that the individual might be conscious of the psychic conflict and anger most assuredly would be involved. At least two associated purposes and two complementary purposes would be involved as well. The conditions that might foster passive-aggressiveness are exemplified in William Shakespeare's tragedy of Hamlet, prince of Denmark. For example, Hamlet's purpose of acquiring the entity of revenge (e.g., purpose *A*) against King Claudius is initially threatened by his doubts of his uncle's guilt, and then by his desire to ensure that he does not send Claudius to heaven upon exacting vengeance. Hamlet's efficacy to acquire entity *A*, vengeance, is questioned all throughout the play until the fatal blow is delivered in the final act. The passive-aggressiveness, however, takes place before this.

Heretofore, a second purpose, the acquisition of the entity of his uncle's vilification (e.g., purpose *B*), is safer for Hamlet than outright retribution. Hamlet's extensive mourning for his murdered father can be construed as a condemnation of both the insincerity of the grief exhibited by King Claudius, Queen Gertrude, and of their hasty marriage thereafter. While acquiring entity *B* does not mitigate as much anxiety as would the acquisition of entity *A*, it is all that the individual, Hamlet, can afford to do for the time being.

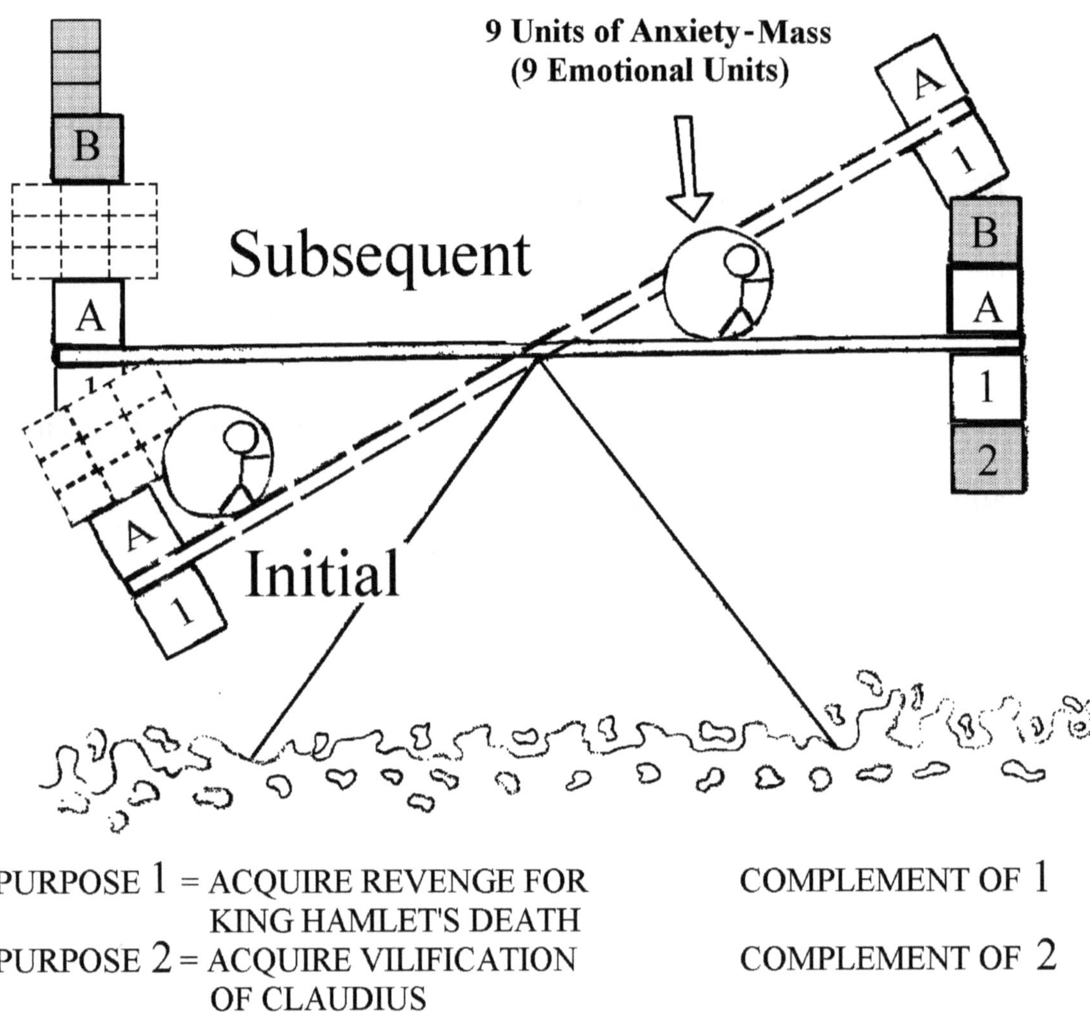

Figure 14.15 Roll Cage Theory Depicting Passive Aggression
Although acquiring entity *A* would alleviate more anxiety (unshaded-dashed mass), it is riskier and the efficacy of the individual to acquire it is in doubt. However, a purpose associated with acquiring revenge for his father's murder, hence, acquiring vilification of Claudius, carries less of a risk of reprisal and is initially within Hamlet's capabilities (high efficacy) as it is not overt accusation. Although the anxiety stemming from the desire to fulfill purpose *A* is substantial and in the conscious, the ego does not attach enough anxiety-mass to the roll cage. Instead of having 12 units of anxiety-mass, it only has 9, as the utility-mass in the dashed outline is not matched by the ego. Entity *B*'s acquisition must make up the difference, as entity *A*'s utility mass (dashed-line) still influences the disc.

Henceforth, by deciding to not invest additional anxiety-mass in the roll cage, the ego or self inside the roll cage has made it clear that attempting to maintain equilibrium between the drive to acquire revenge and the drive to not acquire revenge is less important. If only purpose # 1 were being considered, then that would suggest the drive to maintain all drives had lost ground against the drive to not maintain all drives. The extra anxiety, nevertheless, unsettles the disc, so another means by which to defuse the anxiety is sought and found. After purpose # 2 (acquiring the entity of his uncle's vilification) becomes associated with purpose # 1 (avenging his father's murder), the acquisition of entity *B* can help to mitigate the anxiety Hamlet feels for both purpose #1 and purpose # 2, but must be acquired in a larger amount than otherwise if only purpose # 2 were considered.

Projection

Projection is the "process by which one ascribes one's own traits, emotions, dispositions, etc. to another."[27] When used, one's own faults and shortcomings may be attributed to another, and this typically includes the stipulation that it is done unwittingly.[28]

Projection, when translated into the equation, would include cases where a valuation made by the self is mistakenly assigned as a valuation made by an *other*. The self literally replaces the *other's* valuation with his or her own. Actual valuations made by the *other*, as assessed by the self and concerning threat, benefit, efficacy, or utility components, would likely be reduced if not eliminated outright.

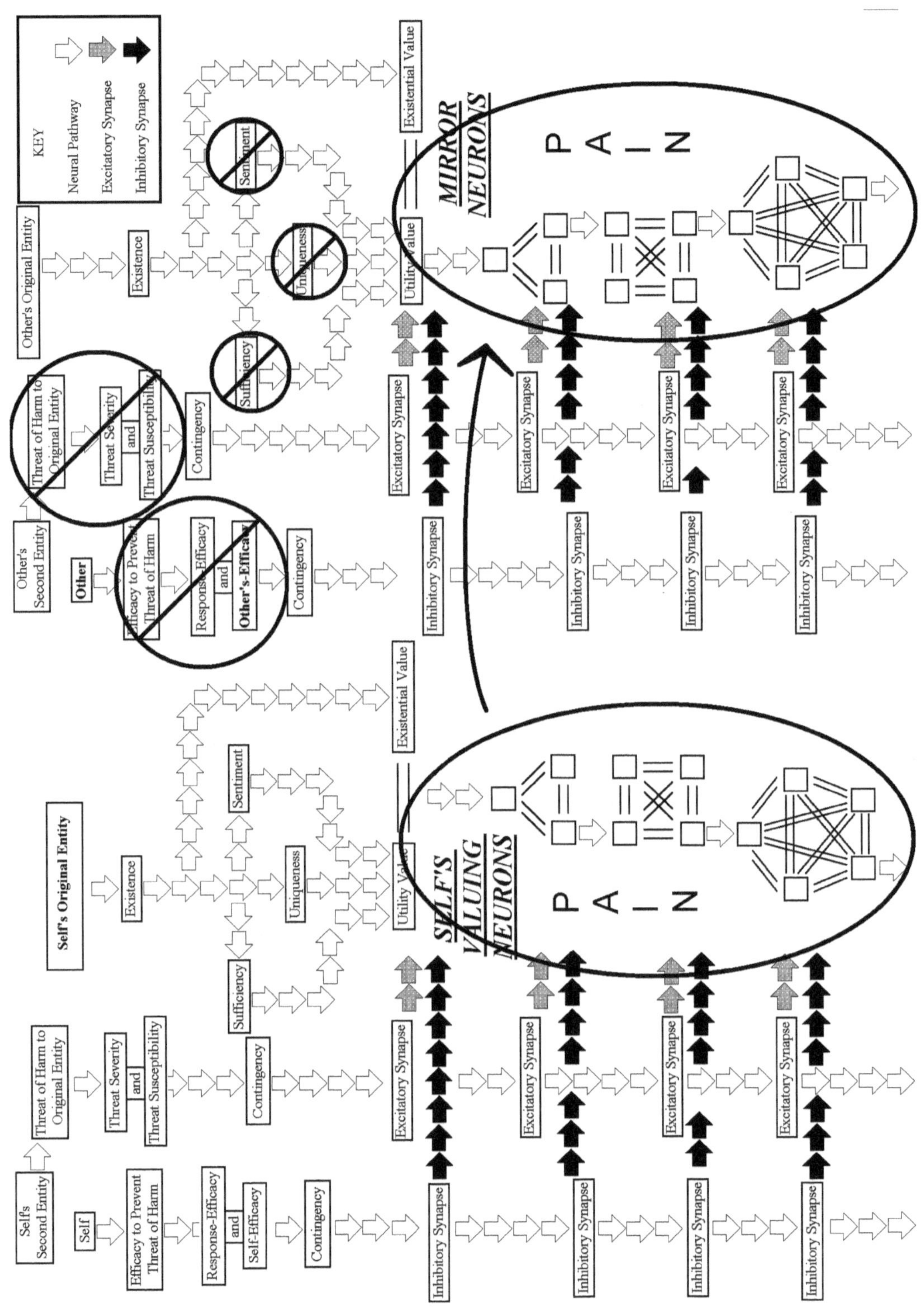

Figure 14.16 Sample Neurological Model for the Defense Mechanism of Projection

Projective Identification

Projective identification is similar to projection with the exception that the individual "does not fully disavow what is projected."[29] "The person remains aware of his or her own feelings or impulses, but misattributes them and regards them as being justifiable reactions to the behavior of the other persons involved."[30]

The representation of projective identification in the equation would be similar to that of projection. The difference, however, would be that the self finds evidence to justify what is projected onto the *other*. Additionally, the self might take action to elicit a response from the *other* that confirms what is projected onto the *other*. The circled variables in figure 14.18 (e.g., threat components, efficacy components, and utility components) are ones that would likely be involved when projective identification is occurring; moreover, the self might use the interpretation of them (circled components in figure 14.18) to justify a valuation of an entity that is being projected onto an *other*.

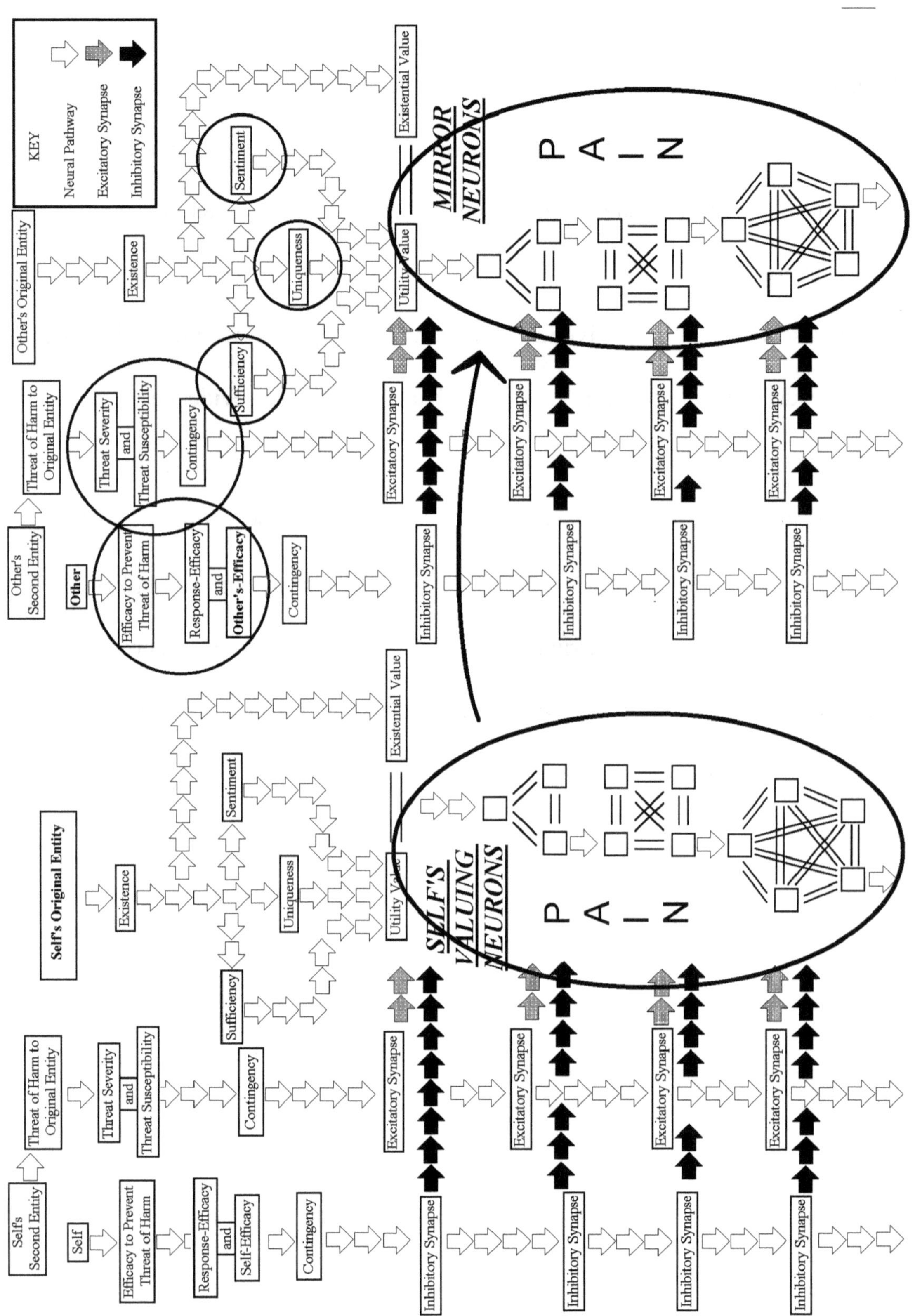

Figure 14.17 Sample Neurological Model for Projective Identification

Somatization

A somatization disorder is a "somatoform disorder characterized by a history of recurrent and multiple physical symptoms for which there are no apparent physical causes."[31] Vague pains, allergies, and gastrointestinal problems, psychosexual symptoms are some of the complaints involved in somatization disorder.[32] Moreover, "no detectable organic damage or neurophysiological dysfunction" can explain somatoform disorders, which has lead to the "strong presumption that they are linked to psychological factors."[33]

The representation of somatization disorder and other somatoform disorders would be similar to that of conversion mentioned earlier. An overflow of unusable anxiety in an individual's system (e.g., entropy or anxiety that cannot be mobilized to value entities) would be a likely culprit for a somatization disorder. If an intense overflow of anxiety felt for a specific entity is found to be the source, then the neurological model for conversion disorder (figure 14.1) or even a graphical representation of valuations for entities in four dimensions (i.e., along the x, y, z, and d-axises) may prove to be more helpful in modeling somatization disorder and entropy. If a somatization disorder is found to have an underlying psychological factor or psychic conflict that lies deep within the unconscious, such as in severe trauma, then it might be best to represent somatization disorder more like the defense mechanism of acting-out and the Roll Cage Theory of Drives (figure 14.11) may be more helpful.

Application to Defense Mechanisms

Neurotic Defenses

The term neurotic is used to indicate items that pertain to or characterize "specific behaviors that are actually displayed by a person diagnosed as having a neurosis."[34] The term neurosis is used to indicate "a personality or mental disturbance not due to any known neurological or organic dysfunction."[35] In the "five factor theory," neuroticism is a "broad personality disposition that characterizes the degree to which one is chronically emotionally unstable," "prone to anxiety" and to "psychological distress."[36]

Displacement

Displacement, as it relates to behavior, occurs when one response is substituted for another, for instance, in the case where the original response is blocked or thwarted.[37] For instance, anger or fear for one item may become discharged onto another, neutral item. Modeling displaced anger in the equation involves identifying two original entities (e.g., *A* and *B*) the self wants and two contingencies with a threatening entity (*C* to *A*) and perhaps a nonthreatening entity (*D* to *B*). For instance, entity *A* might have a relatively high amount of anxiety invested in it while entity *B* might not. Anxiety invested in the first entity might also be rising beyond a level that the individual feels he or she can control (e.g., high threat components and low efficacy components), and the self does not feel capable of preventing entity *C* from threatening entity *A*. Entity *B*, meanwhile, may have less anxiety invested in it because entity *D* poses little or no threat to entity *B*. Hence, entity *D's* threat to entity *C* would be infinitesimal, if not zero, and the efficacy of the self to prevent it high. It might even be the case that entity *D* benefits entity *C* and has a positive contingency with it (explained in chapter eight), if the exponent for benefit were being used in the Avoidance of Pain formula.

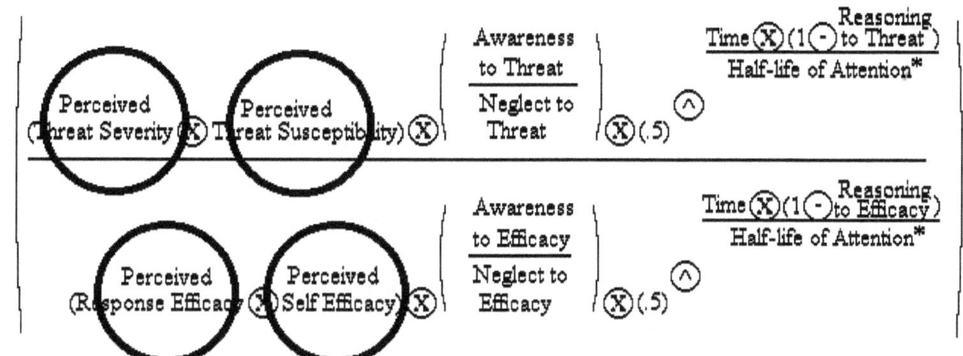

EXPONENT OF THE EQUATION

BASE OF THE EQUATION

Figure 14.18 Displacement
The circled variables above are variables that would be involved in the case of displaced anger. The efficacy variables of a second purpose might be switched with those of the first purpose, as if a recommended response for preventing a contingency in the second purpose influenced the first purpose.

In a fit of frustration at his or her inability to prevent entity *C* (e.g., a lack of efficacy), the individual might lash out against entity *D* by preventing it, as if it were entity *C*. Though lashing out provides a temporary easing of mounting anxiety invested in entity *A*, once the individual realizes the error in the calculation, anxiety will likely resurface for entity *A* and would be expected to elevate for entity *B*, especially if entity *B* was positively contingent to entity *D*.

In figure 14.19, at least two purposes would be under consideration. For one purpose, anxiety invested in an entity would be mounting and very little or nothing could be done to prevent it. The second purpose would be a case where anxiety is not mounting and the valued entity's contingency to another entity is likely low or may even be positive. The numerical value of the efficacy to prevent a small threat or quite possibly a benefit in the second purpose (high efficacy value) might be switched with the efficacy to prevent an incredibly severe and probable threat for which the self has no satisfactory response in the first purpose (low efficacy value) and then acted upon with the impression that efficacy has elevated. A less pleasant scenario might occur if the individual displaced the efficacy in the first purpose with the efficacy concerning an equally severe threat and low efficacy of another, third purpose, as it would likely prove itself futile or fatal.

Dissociation and Dissociative Disorders

Dissociation generally describes processes where a "coordinated set of activities, thoughts, attitudes or emotions becomes separated from the rest of a persons' personality and function independently."[38] Dissociative disorders are "characterized by a breakdown in the usual integrated functions of consciousness, perception of self, and sensory/motor behavior."[39] "Multiple personality disorder, depersonalization, and some forms of amnesia and fugue states" would be classified as dissociative disorders.[40]

The breaking point of the ego from the Roll Cage Theory of Drives is an ideal place to begin an attempt to model dissociation and dissociative disorders. If, after reaching its breaking point (i.e., the maximum measurement of gravitational force that the ego can handle at angle p) the self loses its integrity, then the roll cage may theoretically split into two new selves. These two new selves, having a smaller total amount of anxiety invested in them, would be easier to manage than the integrated self precisely because the anxiety invested in each would be less, like in a two for one stock split. The downside of this, however, is that likely only one could be successfully managed at a time.

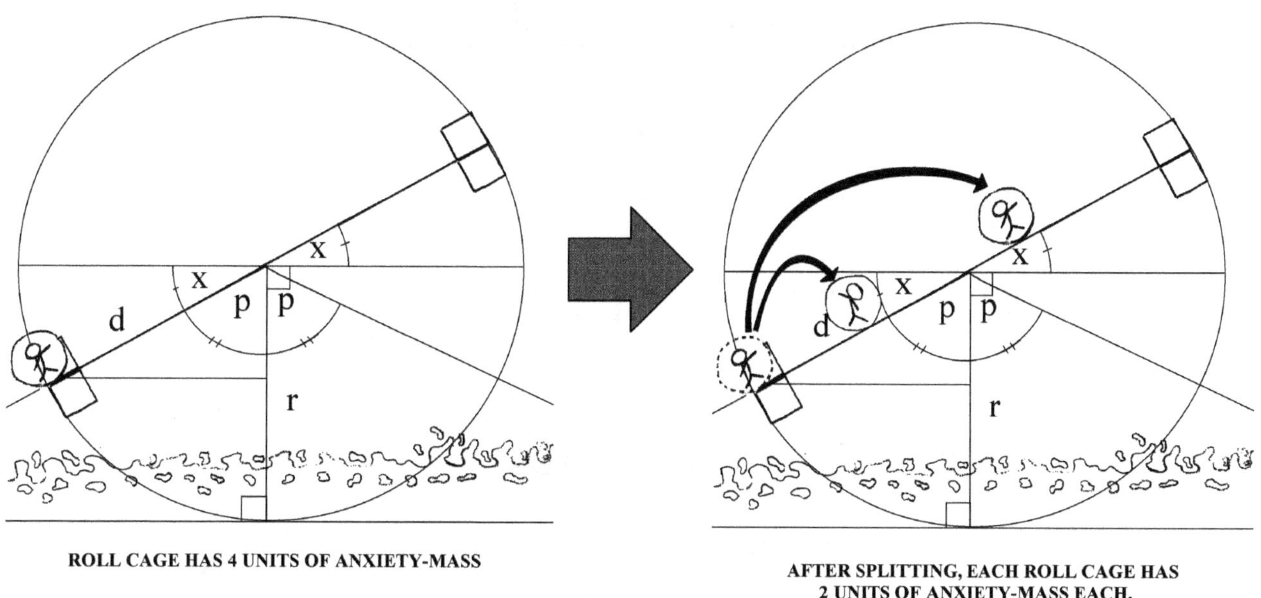

Figure 14.19 The Roll Cage model above depicts dissociation into multiple selves, as in the case of multiple personality disorder or some fugue and amnesia states. In the case of depersonalization, the split self may view its other parts as someone or something else (e.g. an *other*) with whom it can barely relate to vicariously, if at all.

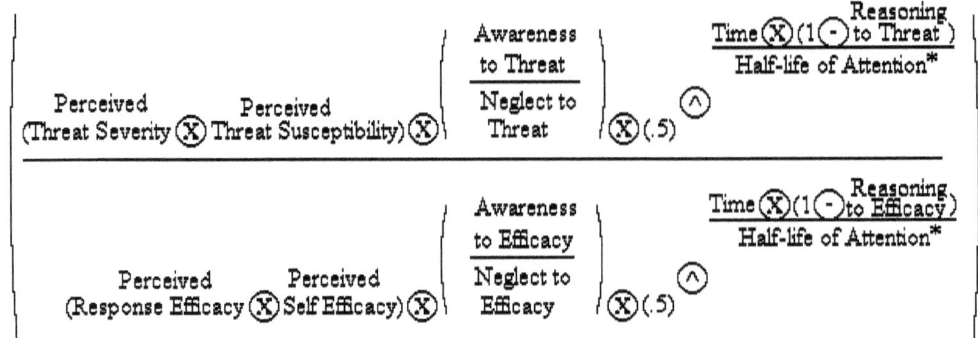

EXPONENT OF THE EQUATION

Figure 14.20 In the case of a dissociative disorder and a split self, the Self-Distinction variable might alternate back and forth between positive one and negative one when in reality it should just be positive one to indicate the self.

Hypochondriasis

Hypochondriasis is characterized by persisting fears that an individual has a serious physical disease or ailment. The representation of hypochondriasis in the equation would be similar to that of somatization disorder or conversion disorder, where an abnormally high level of anxiety invested in an entity might be triggering physical symptoms. If the valued entity is not directly accessible to one's awareness (e.g., lies deep in the unconscious from subtle reasoning processes) and Attention cannot be brought to the forces behind it, then a model similar to acting out may be the most accurate. If the physiological symptoms are real, then the lost anxiety energy should be considered entropy in the equation. The individual would be unable to mobilize the anxiety to value entities or stop valuing entities; the unusable anxiety may be provoking physical symptoms, mimicking an illness.

Intellectualization

Intellectualization is "a defense mechanism whereby problems are analyzed in remote, intellectual terms while emotion, affect, and feeling are ignored."[41]

For instance, when faced with mounting anxiety, an individual may choose to focus on the facts of a situation or information, rather than the feeling itself. The defense mechanism of intellectualization, when modeled in the equation, would be represented by a redirection of one's Attention from an entity that has a lot of anxiety invested in it and toward an entity that has very little anxiety invested in it (e.g., statistics and figures) but is still somewhat related to the first entity.

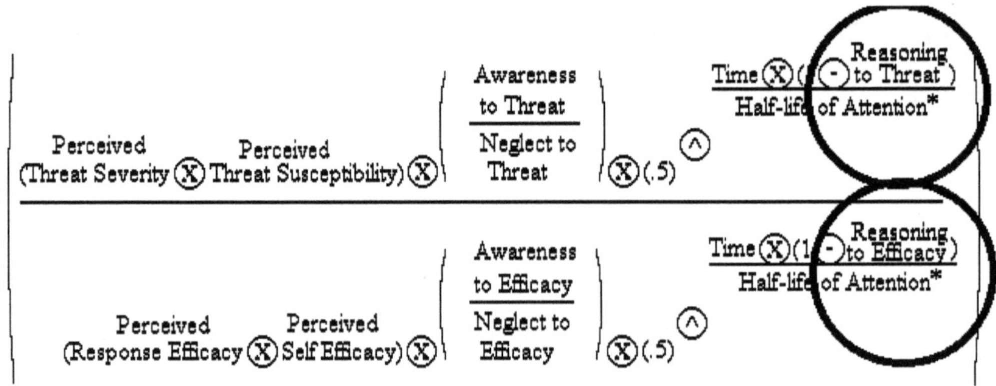

EXPONENT OF THE EQUATION

BASE OF THE EQUATION

Figure 14.21 In the defense mechanism of Intellectualization, Attention is directed away from an entity evoking strong feeling due to having substantial anxiety invested in it, and toward an entity that is comparatively benign, such as information surrounding the original entity that evoked a strong feeling. The ignored entity either has its valuation fall into attentional decay or it is repressed, depending on which form of attention is being used.

Isolation

Isolation, in psychoanalytic theory, is a defense mechanism assumed to function by severing ties between an unacceptable impulse or act and its original memory source.[42] The impulse, act, or memory essentially becomes isolated in the individual's mind.

An Isolated Purpose and Complementary Purpose

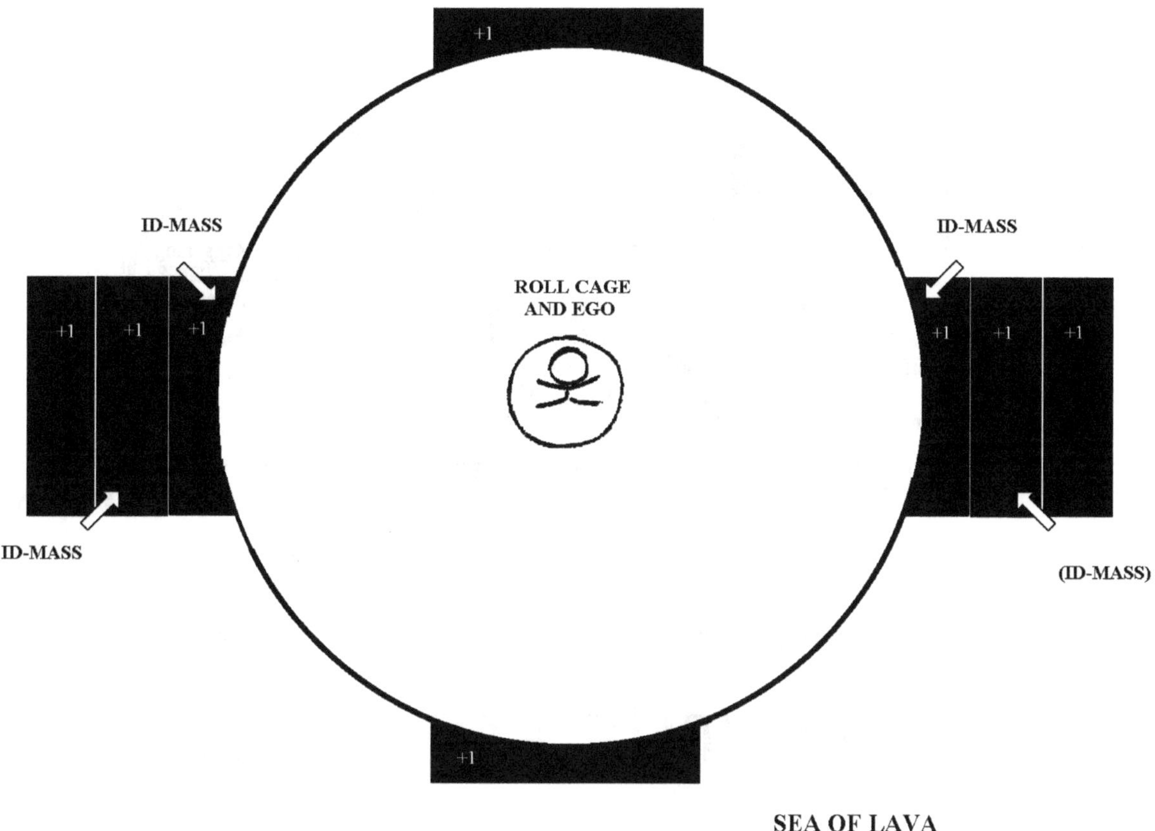

Figure 14.22 The above is a bird's eye view of the Roll Cage Theory of Drives exemplifying isolation. The isolated id-masses (top and bottom) are prevented from forming associative links with the other six id-masses representing purposes and complementary purposes, that are on the left and right. The individual is less likely to consider their influence when fulfilling other purposes to restore equilibrium.

Rationalization

Rationalization, as a defense mechanism, generally serves to make things that are confusing more clear, but has the added connotation of serving to conceal the true motivations for one's actions.[43] Rationalization, when modeled in the equation, might be indicated by variables that are skewed so as to present the self in a better light. For example, guilt has already been established in the equation, and a number of actions might be taken to minimize, or change completely, an instance of guilt to make it seem justifiable.

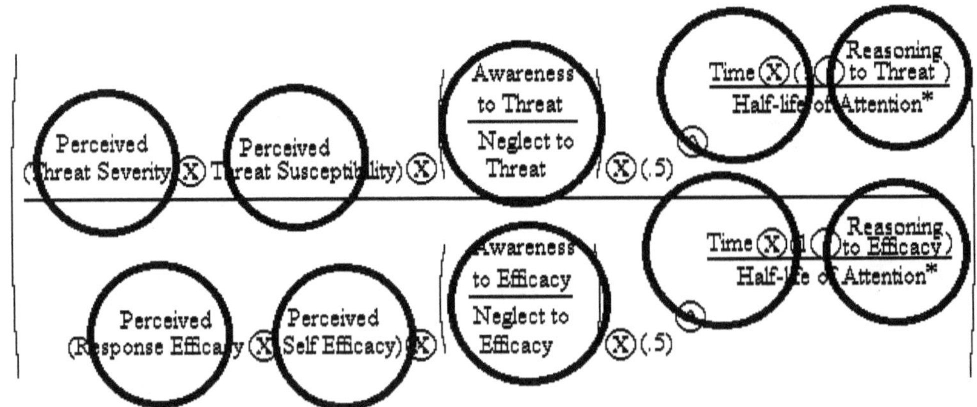

EXPONENT OF THE EQUATION

BASE OF THE EQUATION

Figure 14.23 These variables in the equation could potentially be in error or at fault if an individual is using Rationalization as a defense mechanism. Upon retrospection, an individual might rewind time (e.g., replay a scenario) to change an estimation of initial threat components to make them seem higher than they were, or the efficacy components to make them seem lower than they were. Likewise, Sentiment may be altered after the fact (e.g., by lowering it) in an attempt to mask disappointment felt at not acquiring a specific entity.

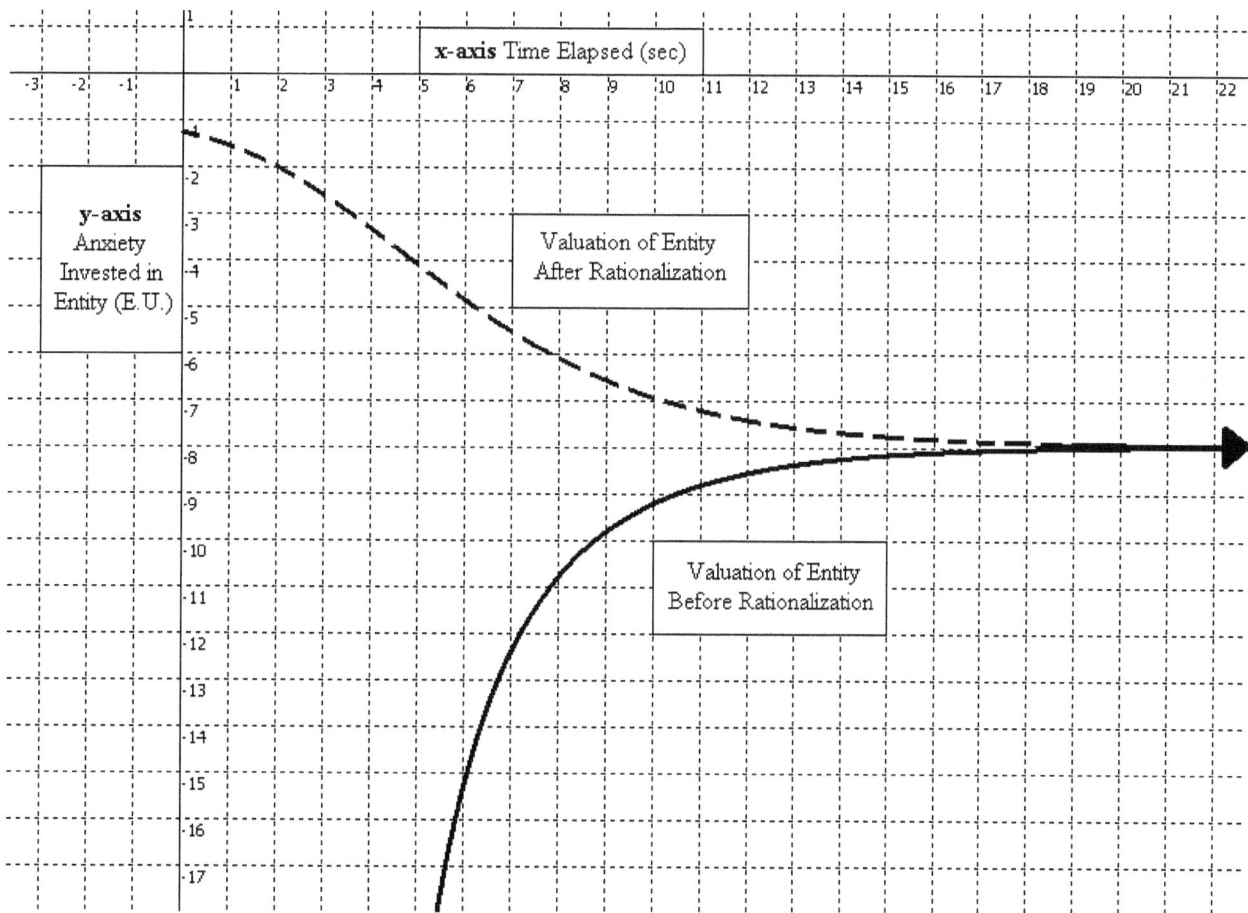

Figure 14.24 Sample Graph Depicting the Defense Mechanism of Rationalization
The solid line represents the valuation of an entity prior to the defense mechanism of rationalization being used; the dashed line represents the valuation of an entity after the individual reassesses the event by going back chronologically(e.g., rewinding time) and changing the initial valuation by overestimating the threat and underestimating efficacy relative to the first valuation (solid line). Initially, guilt is clearly occurring, as negative anxiety invested is decreasing. Sample Form III of attention is used with valuation resilience. Attention, in this case the Neglect to Awareness ratio, is allowed to bounce back to .5 : .5 or 1:1 for both the numerator and the denominator and the end valuation is the same prior to Rationalization occurring and afterwards. The equations are identical except the Neglect to Awareness ratio equaled 3:1 and 1:3 for Threat and Efficacy, respectively, prior to rationalization being used (solid line function). After rationalization was used (dashed line function), these values were switched so that the Neglect to Awareness ratio became 1:3 for Threat and 3:1 for Efficacy. The individual replayed an event in time, elevated the initial estimate of threat and lowered the initial estimate of efficacy in order to create the sensation that he or she made the best of a raw deal, instead of feeling guilty for a missed opportunity which was actually the case.

Application to Defense Mechanisms

Reaction Formation

In psychoanalytic theory, a reaction formation is a process where "unacceptable feelings or impulses are controlled by establishing behavior patterns directly opposed to them."[44] For instance, an individual who is prejudiced and using the defense mechanism of reaction formation to avoid psychic conflict might engage in behavior that seems to contradict this.

The defense mechanism of reaction formation can be illustrated using the Roll Cage Theory of Drives. If an individual were prejudiced toward tall people, for instance, but wished to push these unacceptable impulses out, then they might be sent to the unconscious. The black anxiety-mass beneath the disc in figure 14.26 represents unconscious anxiety invested into entity *A*, aversion of tall people, which is useful for fulfilling purpose # 1. To conceal this unconscious prejudice toward tall people, the individual may have seeking out tall friends as a purpose, purpose # 2 and entity *B* respectively in figure 14.26. Purpose # 1 and purpose # 2 become associated but are set up against one another. Like other unconscious anxiety-mass, the ego would fail to account for the unconscious anxiety mass beneath the disc when assessing the anxiety to invest in the roll cage. Instead of having an anxiety-mass of 10 emotional units, the roll cage in figure 14.26 only has nine emotional units invested in it. Although true equilibrium is felt when the roll cage is at the fulcrum, this will conflict with the information that the ego sees above the disc. Based upon what the ego sees above the disc, it will be lead to mistakenly believe that it must acquire tall friends (i.e., move to the left) in order to restore equilibrium, but this actually disrupts balance between the competing drives.

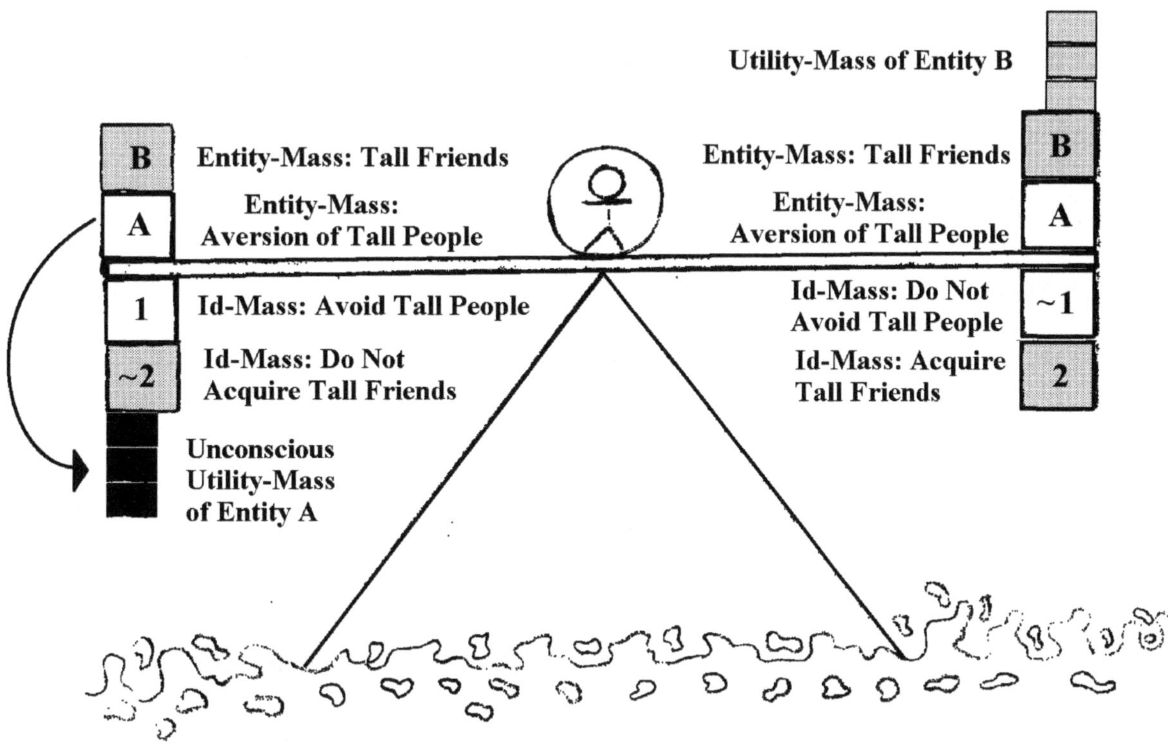

Figure 14.25 The above is the Roll Cage Theory Depicting the Defense mechanism of Reaction Formation. The lightly shaded gray boxes are entities and purposes resulting from the reaction formation that is in response to the unconscious anxiety (black boxes) from entity *A* and purpose # 1 (unshaded boxes). True equilibrium is already felt by the roll cage, but what it sees based upon the anxiety-mass above the disc can lead it to believe that equilibrium is somewhere to the left, as the unconscious utility-mass is not accounted for by the ego despite its ability to influence the disc.

Regression

Regression is the opposite of progression, and generally means going backward.[45] Regression might be indicated by instances where an older individual (e.g., adolescent or adult) reverts to childlike behavior, such as thumb-sucking.[46] Regression would concern instances where an individual resorts to the use of less mature defense mechanisms, for instance, pathological or immature ones, when the anxiety invested in entities becomes more than the ego can handle. The individual, in essence, backslides from the use of more appropriate defense mechanisms (e.g., mature ones) to less mature ones such as acting out or projection. If one models two distinctly different defense mechanisms for an individual, and the latter one is found to be a less mature one than the former, regression may have occurred.

Repression

In psychoanalytic theory and other forms of depth psychology, repression is a defense mechanism that operates on the unconscious level to block or prevent anxiety producing impulses or ideas from ever reaching consciousness.[47]

Sample Form III of attention would be ideal for representing the defense mechanism of repression, as this model of attention forcibly alters the valuation of any entity in unattended and attended streams alike. Unlike Sample Form II of attention where the anxiety invested can only increase if Reasoning loosens its grip selectively, the valuation of an entity may surge if Reasoning to Attention is loosened on both threat and efficacy components under Sample Form III of attention.

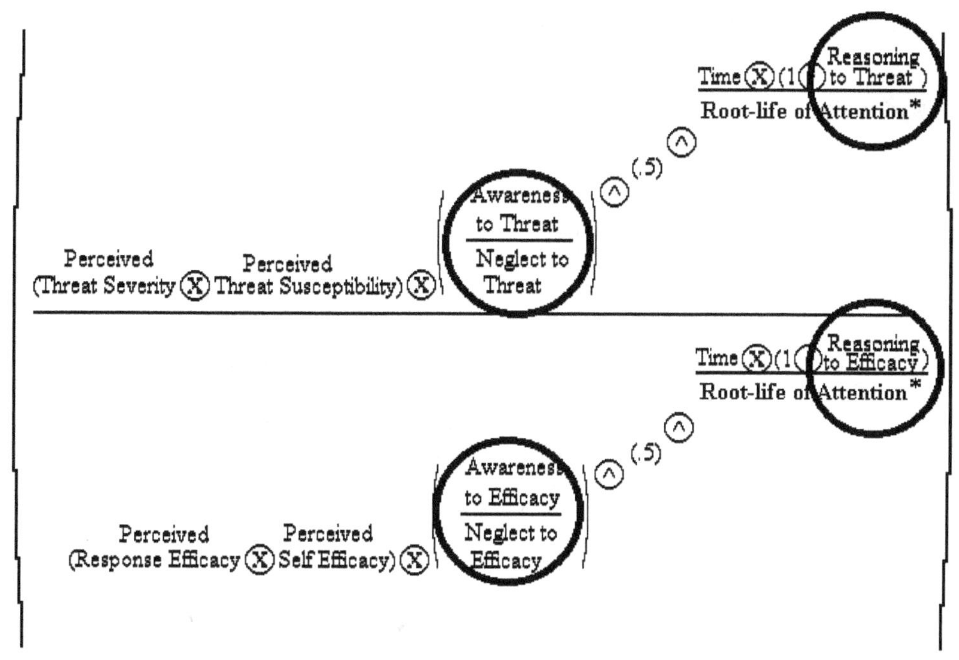

EXPONENT OF THE EQUATION

BASE OF THE EQUATION

Figure 14.26 Sample Form III of attention best illustrates the defense mechanism of repression. Reasoning to Threat (Alternatively Reasoning to Benefit), Reasoning to Efficacy, and Attention (ratio of Neglect to Awareness) are the variables that would be involved.

The problem, however, is that these repressed valuation may not stay repressed, and might resurface as a result of reasoning processes, or a lack there of, that permit them to do so. Moreover, these reasoning processes, being so deep in the unconscious, would not have Awareness and subsequently, Attention, redirected to them by the self. Reasoning processes that are strongly resistant to being brought under scrutiny may even be directing Attention away from themselves or onto something innocuous (Figure 14.28).

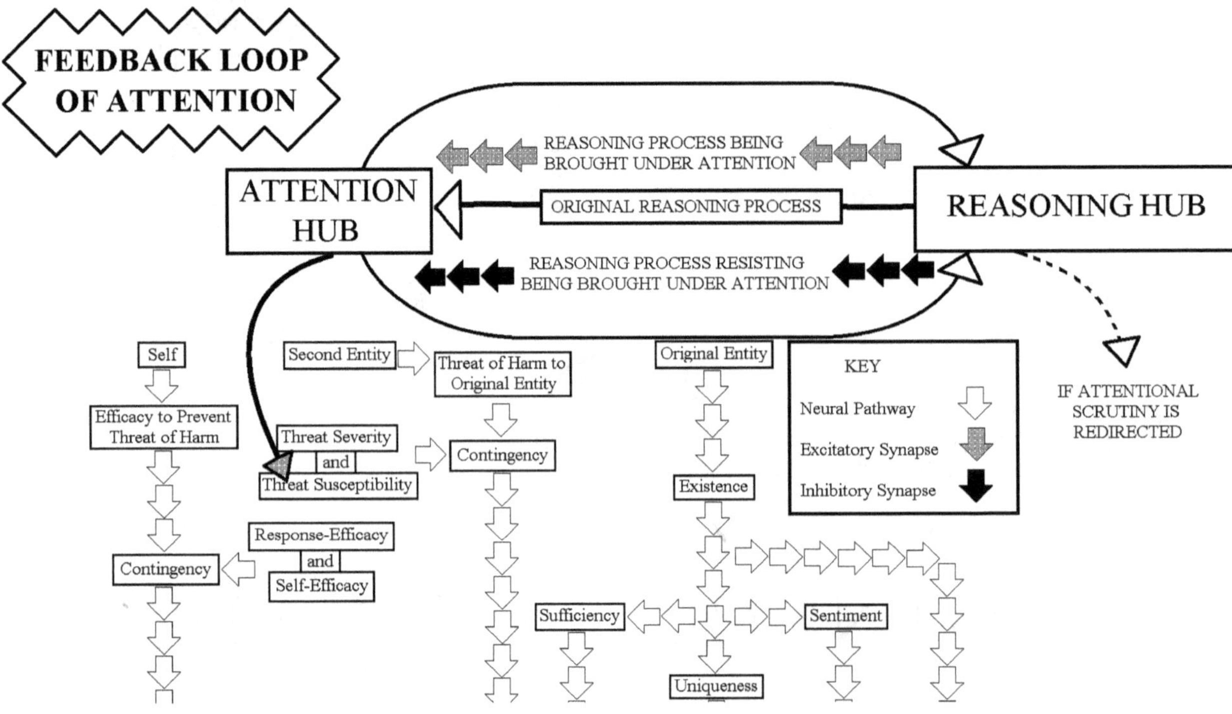

Figure 14.27 Above is a sample attention feedback loop with reasoning processes. In one case (top) the reasoning processes are brought under attentional focus and the loop perpetuates, firing more frequently. In a second case (bottom) the reasoning processes resist being brought under attentional focus and the activation threshold for the feedback loop is raised, firing less frequently. Alternatively, the reasoning hub may redirect signals from an attention hub elsewhere, thus staying cloaked from scrutiny.

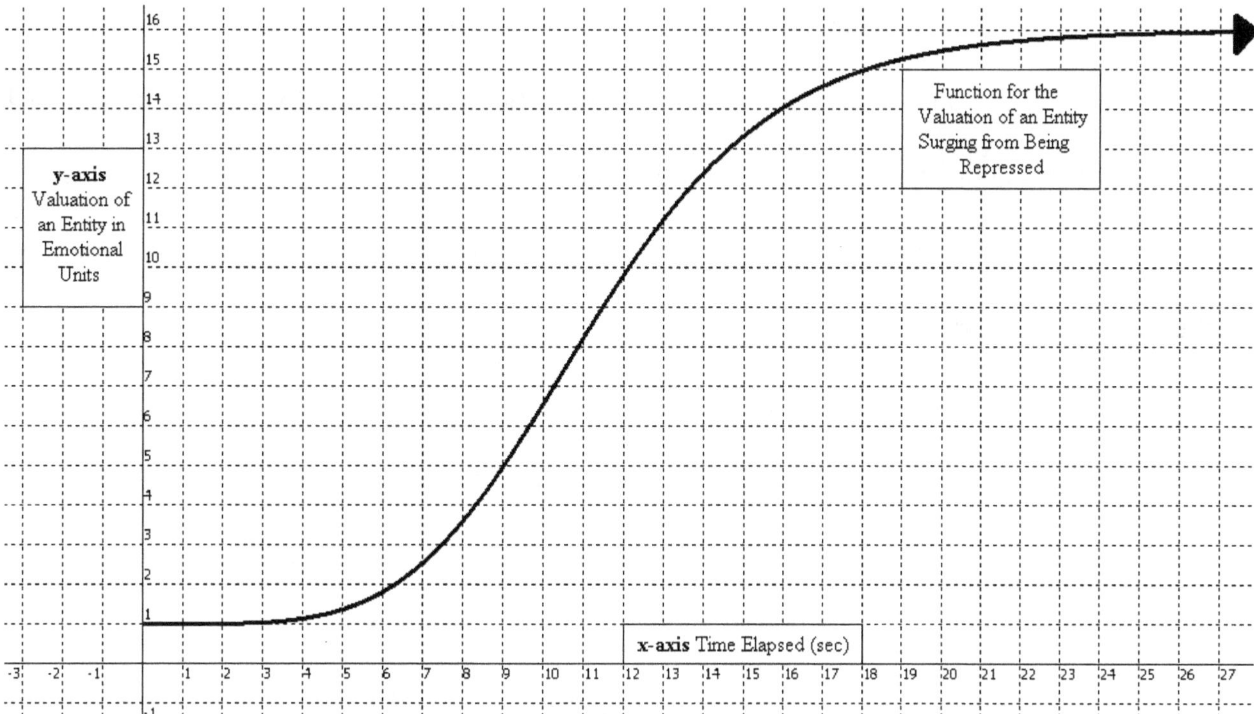

Figure 14.28 Sample Form III of attention above depicts an entity's valuation that was repressed and then experienced a resurgence after the Reasoning resources being used to repress the valuation were allocated elsewhere. Without Reasoning to Threat and Reasoning to Efficacy to keep the entity's valuation artificially lower than where it would normally be, the valuation of and anxiety invested in the entity elevates back to its original level.

Figure 14.29 above depicts an entity's valuation and the anxiety invested in it rebounding from being repressed after Reasoning processes ceased subduing it. The entity's valuation, had it not been repressed, would have been 16 emotional units. The threat components were maximized after being factored, and the efficacy components equaled .25 after being factored. Awareness over Neglect equaled 1 / 3 in the numerator for Threat (.25 : .75) and 3 / 1 in the denominator for Efficacy (.75 : .25). Reasoning to Threat and Reasoning to Efficacy were both zero after shifting the function to the right five seconds.

Although an Awareness level of 75% for one specific entity does not qualify it for being in the unconscious or even the subconscious, the same effect in the graph could be realized with a much smaller percentage of the total attention. A one to nine proportion of Attention to Threat vs. Attention to Efficacy would accomplish this aim. If the ratio of Awareness to Neglect equaled .001 and .009 in the numerator and the denominator respectively (.1% and .9%), then the plot of the function in figure 14.29 would be similar if Reasoning resources were retracted and the Attention level between the two reached parity at .5%.

Undoing

Undoing is a defense mechanism "in which a person attempts to annul (i.e., undo) the unpleasant outcome of some act by mentally replaying or in some cases ritualistically re-enacting the sequence of events but with a different, more acceptable ending."[48]

The defense mechanism of undoing principally concerns the variable of time and likely efficacy or threat/benefit. The self may replay an event, for instance by resetting the valuation of entities to an earlier x-value in the equation, and attempt to change an aspect of the situation (e.g., raising the efficacy components or lowering the threat components) in order to make the outcome more different.

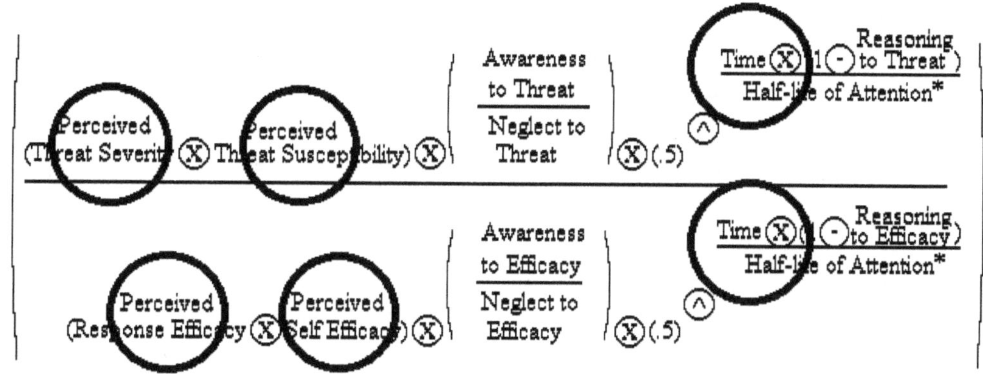

EXPONENT OF THE EQUATION

BASE OF THE EQUATION

Figure 14.29 If the defense mechanism of undoing is being used, the self may rewind the variable of time and reconstruct a scenario before attempting to alter variables. The threat components or efficacy components may be altered in such a way as to reduce the amount of anxiety invested in an entity, for instance by elevating efficacy or reducing threat in hypothetical scenarios.

Withdrawal

Withdrawal can refer to a "pattern of behavior characterized by a person removing him- or herself form normal day-to-day functioning and all of its attendant frustrations, tensions, and disappointments."[49] In this sense of withdrawal's meaning, it refers to "a neurotic removal of the self from normal social discourse, accompanied by uncooperativeness, irresponsibility, and often a reliance on drugs and alcohol to facilitate this social remoteness."[50]

The defense mechanism of withdrawal necessarily entails a retraction of most or all of the anxiety that an individual has invested in entities. As its name implies, the individual literally withdraws from anything and everything in the world that may provoke anxiety being invested. This can be depicted in a number of ways in the equation, but the quickest and most extreme method by which individuals can remove themselves from the tension and frustration of day to day functioning is to corrupt the variables at the base of the equation. In other words, if the assessment of the existential value or the utility variables of Sufficiency, Uniqueness, and Sentiment become impaired, then all value and valuations might potentially be removed from his or her world. A sense of apparent aimlessness might result as the individual would have no means of distinguishing between purposes and entities, assuming these things would even be conceptualized at that point.

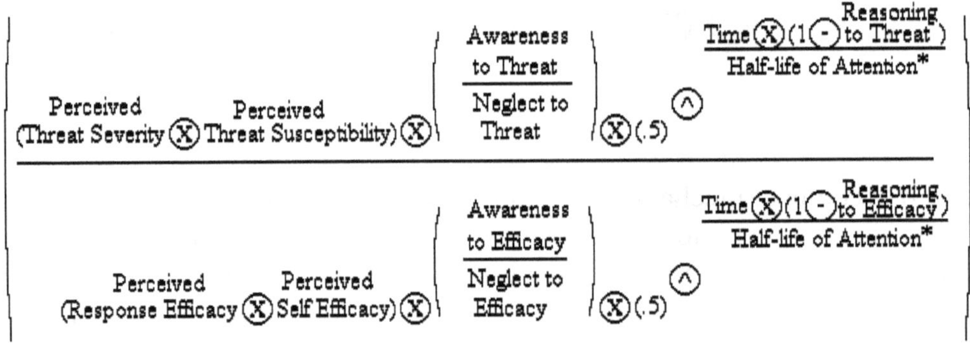

EXPONENT OF THE EQUATION

Figure 14.30 Variables that may become corrupted, if the defense mechanism of withdrawal is being used, include Existence, Uniqueness, Sufficiency, and Sentiment..

Application to Defense Mechanisms

Mature Defense

Maturity "generally entails value judgments made of persons to reflect how successfully they correspond to socially and culturally accepted norms."[51] Mature defenses would be those that are viewed as healthy, normal, and adaptive.

Altruism

Altruism consists in the "elevation of the welfare, happiness, interests, or even the survival of others above one's own."[52] For instance, in "kin-selection altruism one acts in a manner that jeopardizes one's safety but protects or promotes that of one's own kin, hence the behavior is arguably in one's interests in that it increases the likelihood of the survival of one's own genes."[53] The employment of altruism as a defense mechanism is not very far off then, if it mitigates anxiety.

The closest parallel to altruism is the Category III Emotion of benevolence. In benevolence, an individual was said to pursue a course of action designed to help an other achieve a purpose so that the self could vicariously experience the other's success. Benevolence entailed either a decrease in anxiety invested if it were in the protective form, or an increase in negative anxiety invested if it were in the indulgent form. This net lowering of anxiety might be used to offset an increase in anxiety elsewhere concerning a completely different purpose. The use of altruism, then, as a defense mechanism, may be construed as an instance where the self engages in an action in order to produce a feeling of benevolence and this benevolence might likely be used to offset a rise in anxiety or a decrease in negative anxiety elsewhere.

Anticipation

Anticipation is a "preparatory mental set in which one is primed for the perception of a particular stimulus."[54] As a defense mechanism, this would consist in the ability of an individual to prepare for anxiety that is likely to occur in the future.

Similar to undoing and rationalization, the defense mechanism of anticipation would involve a manipulation of the variable of Time. However, instead of traveling backward in time, the self would travel forward in time to predict what may happen in the future and how anxiety would be invested if events continued to unfold as they are; pre-emptive action might thereafter be taken to ensure or prevent the scenario. For example, in the case from chapter twelve where Dave and Max were valuing the paper due at the end of the month, if they each used the defense mechanism of anticipation at different instances and over different lengths of time, then it may be seen why one might procrastinate and the other might avoid postponing the work on it.

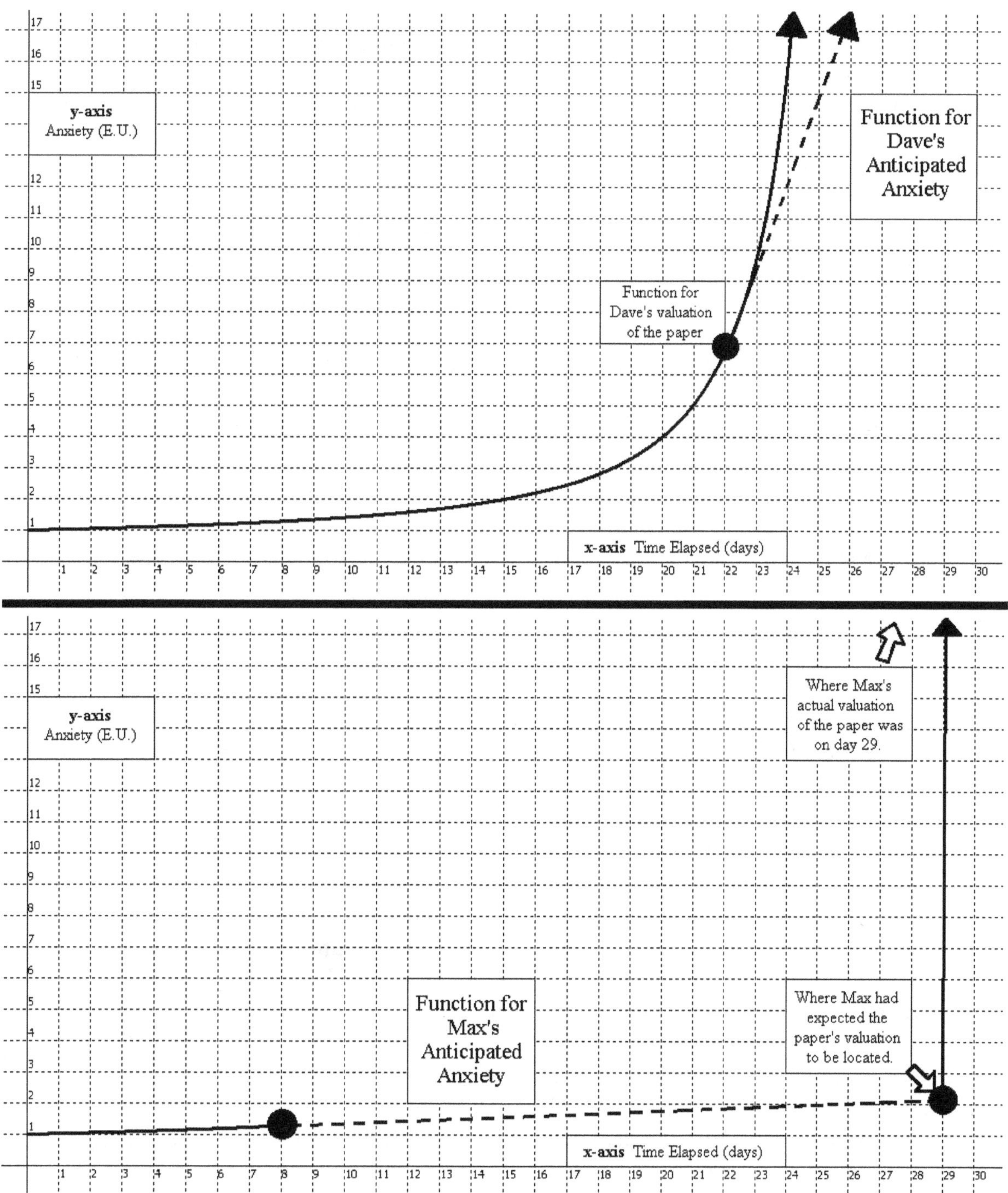

Figure 14.31 If Dave used a 7 day forecast beginning on day 22, compared to Max's 21 day forecast beginning on day 8, the two arrive at vastly different anticipated valuations for the test based on the number of days remaining. In Dave's case, anticipation spurred him to action to resolve it. However, in Max's case, he used it too soon and for too long; it lulled him into a false sense of security.

Application to Defense Mechanisms

Humor

Humor is generally the quality of "being pleasant, sympathetic, amusing, or funny."[55] As a defense mechanism, and given the general elusive nature of humor, describing humor as a process where something amusing or funny is found to relieve tension in an anxiety inducing situation may be the only option for now. This description provides perhaps the best conception of humor as a defense mechanism that can be translated into the equation. If an entity is initially seen as serious or painful (e.g., has a substantial amount of anxiety invested in it), discourse that enables an individual to alter this valuation so that it becomes assessed as pleasurable, thus reducing the anxiety invested in it, is perhaps the least inaccurate way to model humor in the equation. It would follow that any variable in the equation might be subject to modification if the defense mechanism of humor is at play.

Identification

Identification can be thought of as a "mental operation whereby one attributes to oneself, either consciously or unconsciously, the characteristics of another person or group," thus establishing a "link between oneself and another person or group."[56] As a defense mechanism, identification describes a process where an individual comes incorporate features of another person or group in order to alleviate psychic tension.

The use of the defense mechanism of identification can be accounted for along the c-axis and fifth dimension that were mentioned in chapter thirteen with the variable of Identification. If there is no disparity between how the self values an entity with respect to a purpose and how the self perceives an *other* has valued that same entity for the same purpose, then the Identification level would be 100% if that was the only purpose and entity under consideration. The process used by the self to identify with an *other* would consist in the steps taken to make the self's valuation of an entity for a particular purpose similar to an *other's* valuation of that entity for that purpose. Identification, if initially thought of as a percentage, would be translated into the equation with the notion of Self-Other Disparity and be in the base of the equation.

Identification = 1 ÷ (1 + Self-Other Disparity)

... Existence * (Sufficiency * Uniqueness * Sentiment * Identification + 1) ^ . . .

As noted in chapter thirteen, an Identification level of 100% corresponded to a Self-Other Disparity value of zero, 50% to one, 33.3% to two, 25% to three and so on. Whole numbers do not have to be used for Self-Other Disparity, as fractions are viable. Zero, or any positive real number would suffice.

Introjection

Introjection is the "process by which aspects of the external world are absorbed into or incorporated within the self."[57] An instance where the defense mechanism of introjection is used would entail the self absorbing an aspect, trait, or characteristic of an *other* to make one's own. If an *other's* valuation of an entity became the self's own, along with the purpose this valuation was ascribed to, then introjection could be modeled using the neurological model. In figure 14.33 the group of neurons corresponding to an *other* are transformed into neurons corresponding to the self. The *other* literally becomes an extension of the self, and a direct link between the two would be established (circled variables in figure 14.33).

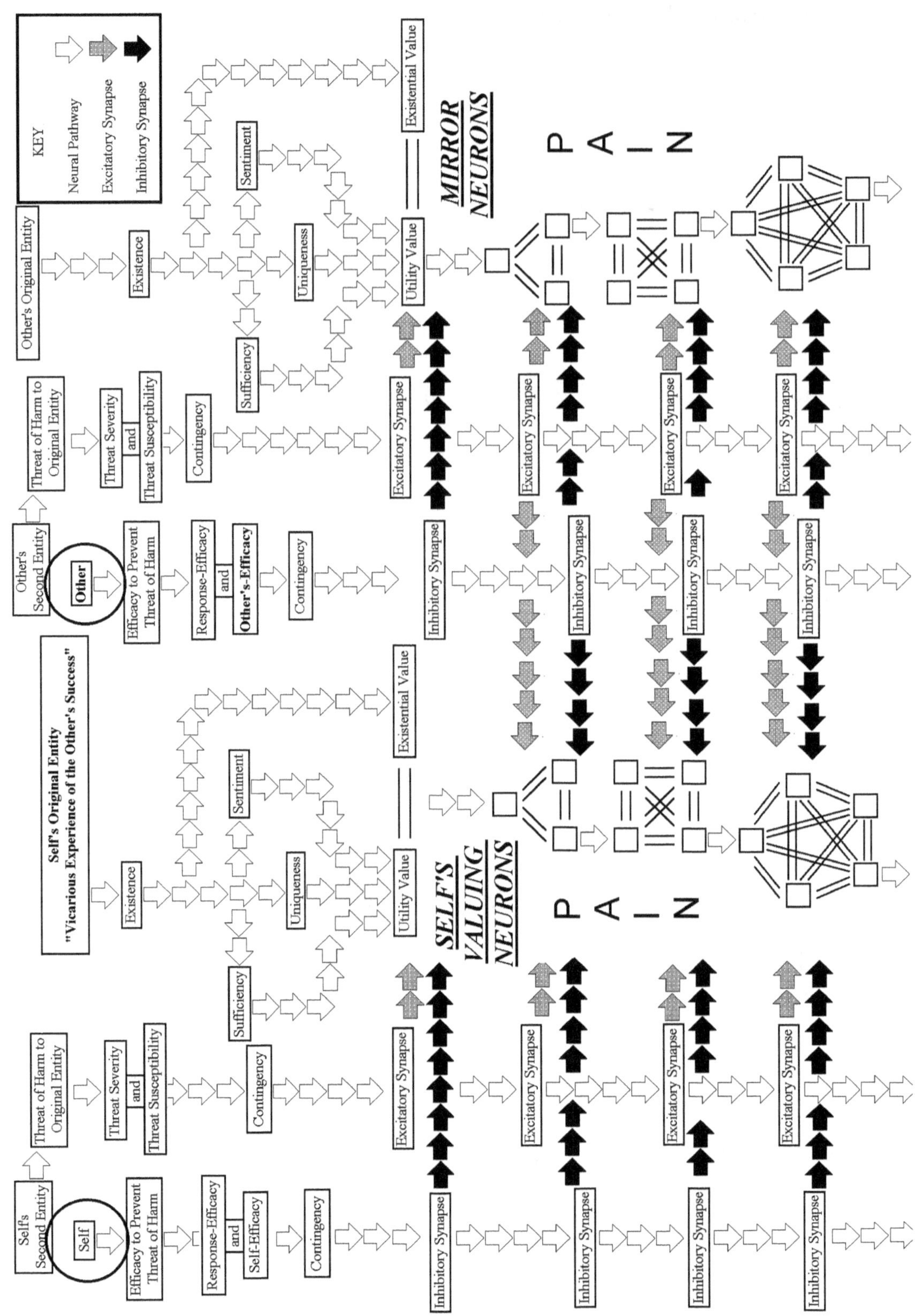

Figure 14.32 Neurological Model Depicting Introjection and Key Variables

Sublimation

In psychoanalytic theory, sublimation is the "process whereby primitive, libidinous impulses are redirected and refined into new . . . behaviors" that are socially acceptable.[58]

The modeling of sublimation would be similar to the defense mechanism of acting-out in figure 14.11, save that the mode the unacceptable drive is channeled into is a socially acceptable one. Contrarily, in the case of acting-out the unacceptable drive or conflict was channeled into an inappropriate behavior.

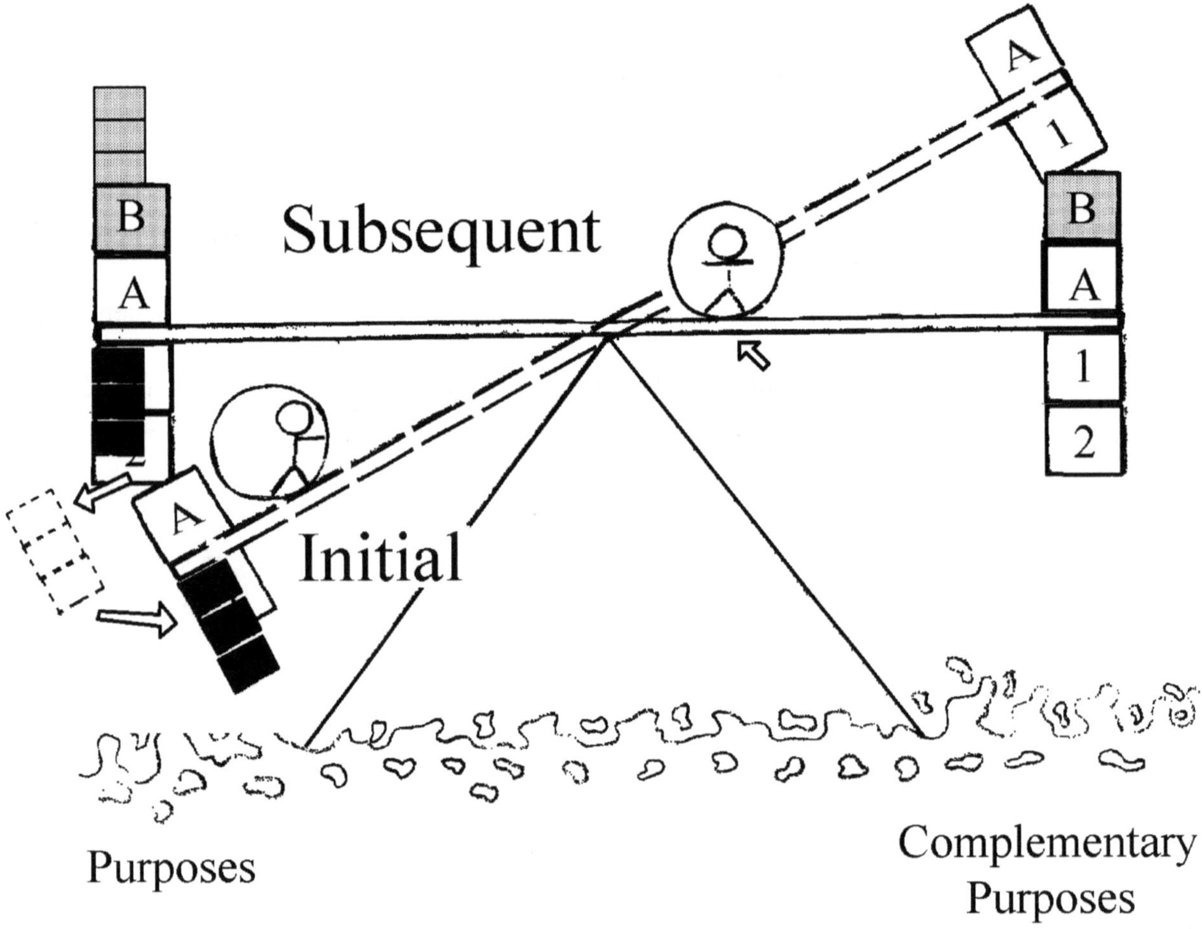

Purpose 1 and Entity A = Socially unacceptable drive and corresponding anxiety invested in it (Unconscious)

Purpose 2 and Entity B = Socially acceptable drive and corresponding anxiety invested in it (Conscious)

Figure 14.33 Roll Cage Theory of Drives Depicting Defense Mechanism of Sublimation
The primary difference between sublimation and acting-out is that the acquisition of entity *B* corresponds to a socially acceptable drive (e.g., purpose # 2). In the case of acting out, neither drive # 1 nor drive # 2 are socially acceptable.

Suppression

Suppression, in psychoanalytic theory, is the "conscious exclusion of impulses, thoughts and desires that are felt to be unacceptable to the individual."[59] It is distinguished from repression in the sense that repression is an unconscious process while suppression is a conscious one.[60]

The primary distinction between suppression and repression concerns the level of consciousness (e.g., Attention and Awareness level) that is present when it occurs. If the entity's valuation in figure 14.35 is inhibited at a fairly high level of Attention, (e.g., 1 / 3 in the Numerator vs. 3 / 1 in the denominator, or 25% Awareness and 75% respectively), then suppression would be said to have occurred. Alternatively, if the entity's valuation is inhibited at a fairly low level of Attention, (e.g., 9 / 1000 in the numerator and 1 / 1000 in the denominator, approximately .8920% Awareness and .0999% respectively.), then repression would be said to have occurred.

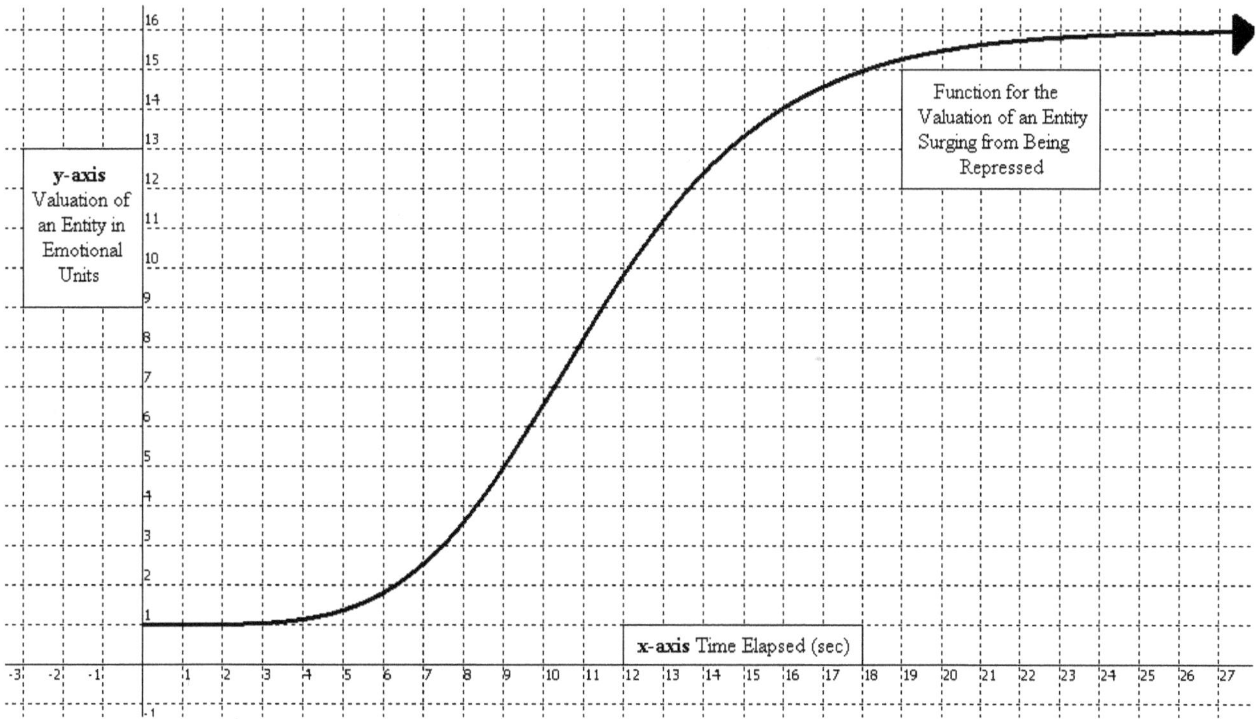

Figure 14.34 From **figure 14.29**. An Attention level ratio of 3 vs. 1 / 3 favoring the numerator over the denominator would be more indicative of suppression. Contrarily, an Attention level ratio of .009 vs .001 favoring the numerator over the denominator would be more indicative of repression.

Application to Defense Mechanisms

Preview

While this list of defense mechanisms is not a complete list of all the ones out there, it provides a glimpse at how the functions and the two extensions from it might be utilized for endeavors less specifically related to affect psychology and emotions, such as psychodynamic psychology. But where does the equation go to from here? Why does the subtitle of this book read A Unified Field Theory of Emotions if fields have not been mentioned at all? Answers await.

YOU SHOULD UNDERSTAND THIS.
CHAPTER FOURTEEN
CONCEPT CHECK
VOCABULARY
Pathological Defenses
Immature Defenses
Neurotic Defenses
Mature Defenses

Notes

1. Wilde, Oscar (2003). *The Picture of Dorian Gray.* New York: Barnes and Nobles Books. Print, p. 112. (Original work published 1890).

2. Reber, Arthur S., Rhianon Allen, and Emily S. Reber. *Penguin Dictionary of Psychology*. London. Penguin Books, 2009. Print, p. 194.

3. Reber, Arthur S., Rhianon Allen, and Emily S. Reber. *Penguin Dictionary of Psychology*. London. Penguin Books, 2009. Print, p. 402.

4. Reber, Arthur S., Rhianon Allen, and Emily S. Reber. *Penguin Dictionary of Psychology*. London. Penguin Books, 2009. Print, p. 402. p. 514.

5. Reber, Arthur S., Rhianon Allen, and Emily S. Reber. *Penguin Dictionary of Psychology*. London. Penguin Books, 2009. Print, p. 561.

6. Reber, Arthur S., Rhianon Allen, and Emily S. Reber. *Penguin Dictionary of Psychology*. London. Penguin Books, 2009. Print, p. 169.

7. Reber, Arthur S., Rhianon Allen, and Emily S. Reber. *Penguin Dictionary of Psychology*. London. Penguin Books, 2009. Print, p. 169.

8. Reber, Arthur S., Rhianon Allen, and Emily S. Reber. *Penguin Dictionary of Psychology*. London. Penguin Books, 2009. Print, p. 170.

9. Reber, Arthur S., Rhianon Allen, and Emily S. Reber. *Penguin Dictionary of Psychology*. London. Penguin Books, 2009. Print, p. 402. p. 169.

10. Reber, Arthur S., Rhianon Allen, and Emily S. Reber. *Penguin Dictionary of Psychology*. London. Penguin Books, 2009. Print, p. 199.

11. Reber, Arthur S., Rhianon Allen, and Emily S. Reber. *Penguin Dictionary of Psychology*. London. Penguin Books, 2009. Print, p. 199.

12. Reber, Arthur S., Rhianon Allen, and Emily S. Reber. *Penguin Dictionary of Psychology*. London. Penguin Books, 2009. Print, p. 203.

13. Reber, Arthur S., Rhianon Allen, and Emily S. Reber. *Penguin Dictionary of Psychology*. London. Penguin Books, 2009. Print, p. 782.

14. Reber, Arthur S., Rhianon Allen, and Emily S. Reber. *Penguin Dictionary of Psychology*. London. Penguin Books, 2009. Print, p. 843.

15. Reber, Arthur S., Rhianon Allen, and Emily S. Reber. *Penguin Dictionary of Psychology*. London. Penguin Books, 2009. Print, p. 843.

16. Reber, Arthur S., Rhianon Allen, and Emily S. Reber. *Penguin Dictionary of Psychology*. London. Penguin Books, 2009. Print, p. 227.

17. Reber, Arthur S., Rhianon Allen, and Emily S. Reber. *Penguin Dictionary of Psychology*. London. Penguin Books, 2009. Print, p. 765.

18. Reber, Arthur S., Rhianon Allen, and Emily S. Reber. *Penguin Dictionary of Psychology*. London. Penguin Books, 2009. Print, p. 765.

19. Reber, Arthur S., Rhianon Allen, and Emily S. Reber. *Penguin Dictionary of Psychology*. London. Penguin Books, 2009. Print, p. 371.

20. Reber, Arthur S., Rhianon Allen, and Emily S. Reber. *Penguin Dictionary of Psychology*. London. Penguin Books, 2009. Print, p. 10.

21. Reber, Arthur S., Rhianon Allen, and Emily S. Reber. *Penguin Dictionary of Psychology*. London. Penguin Books, 2009. Print, p. 366.

22. Reber, Arthur S., Rhianon Allen, and Emily S. Reber. *Penguin Dictionary of Psychology*. London. Penguin Books, 2009. Print, p. 210.

23. Reber, Arthur S., Rhianon Allen, and Emily S. Reber. *Penguin Dictionary of Psychology*. London. Penguin Books, 2009. Print, p. 560.

24. Reber, Arthur S., Rhianon Allen, and Emily S. Reber. *Penguin Dictionary of Psychology*. London. Penguin Books, 2009. Print, p. 560.

25. Reber, Arthur S., Rhianon Allen, and Emily S. Reber. *Penguin Dictionary of Psychology*. London. Penguin Books, 2009. Print, p. 560.

26. Reber, Arthur S., Rhianon Allen, and Emily S. Reber. *Penguin Dictionary of Psychology*. London. Penguin Books, 2009. Print, p. 560.

27. Reber, Arthur S., Rhianon Allen, and Emily S. Reber. *Penguin Dictionary of Psychology*. London. Penguin Books, 2009. Print, p. 622.

28. Reber, Arthur S., Rhianon Allen, and Emily S. Reber. *Penguin Dictionary of Psychology*. London. Penguin Books, 2009. Print, p. 622.

29. Reber, Arthur S., Rhianon Allen, and Emily S. Reber. *Penguin Dictionary of Psychology*. London. Penguin Books, 2009. Print, p. 622.

30. Reber, Arthur S., Rhianon Allen, and Emily S. Reber. *Penguin Dictionary of Psychology*. London. Penguin Books, 2009. Print, p. 624.

31. Reber, Arthur S., Rhianon Allen, and Emily S. Reber. *Penguin Dictionary of Psychology*. London. Penguin Books, 2009. Print, p. 756.

32. Reber, Arthur S., Rhianon Allen, and Emily S. Reber. *Penguin Dictionary of Psychology*. London. Penguin Books, 2009. Print, p. 756.

33. Reber, Arthur S., Rhianon Allen, and Emily S. Reber. *Penguin Dictionary of Psychology*. London. Penguin Books, 2009. Print, p. 756.

34. Reber, Arthur S., Rhianon Allen, and Emily S. Reber. *Penguin Dictionary of Psychology*. London. Penguin Books, 2009. Print, p. 508.

35. Reber, Arthur S., Rhianon Allen, and Emily S. Reber. *Penguin Dictionary of Psychology*. London. Penguin Books, 2009. Print, p. 507.

36. Reber, Arthur S., Rhianon Allen, and Emily S. Reber. *Penguin Dictionary of Psychology*. London. Penguin Books, 2009. Print, p. 508.

37. Reber, Arthur S., Rhianon Allen, and Emily S. Reber. *Penguin Dictionary of Psychology*. London. Penguin Books, 2009. Print, p. 225.

38. Reber, Arthur S., Rhianon Allen, and Emily S. Reber. *Penguin Dictionary of Psychology*. London. Penguin Books, 2009. Print, p. 226.

39. Reber, Arthur S., Rhianon Allen, and Emily S. Reber. *Penguin Dictionary of Psychology*. London. Penguin Books, 2009. Print, p. 226.

40. Reber, Arthur S., Rhianon Allen, and Emily S. Reber. *Penguin Dictionary of Psychology*. London. Penguin Books, 2009. Print, p. 226.

41. Reber, Arthur S., Rhianon Allen, and Emily S. Reber. *Penguin Dictionary of Psychology*. London. Penguin Books, 2009. Print, p. 390.

42. Reber, Arthur S., Rhianon Allen, and Emily S. Reber. *Penguin Dictionary of Psychology*. London. Penguin Books, 2009. Print, p. 403.

43. Reber, Arthur S., Rhianon Allen, and Emily S. Reber. *Penguin Dictionary of Psychology*. London. Penguin Books, 2009. Print, p. 653.

44. Reber, Arthur S., Rhianon Allen, and Emily S. Reber. *Penguin Dictionary of Psychology*. London. Penguin Books, 2009. Print, p. 654.

45. Reber, Arthur S., Rhianon Allen, and Emily S. Reber. *Penguin Dictionary of Psychology*. London. Penguin Books, 2009. Print, p. 666.

46. Reber, Arthur S., Rhianon Allen, and Emily S. Reber. *Penguin Dictionary of Psychology*. London. Penguin Books, 2009. Print, p. 666.

47. Reber, Arthur S., Rhianon Allen, and Emily S. Reber. *Penguin Dictionary of Psychology*. London. Penguin Books, 2009. Print, p. 692.

48. Reber, Arthur S., Rhianon Allen, and Emily S. Reber. *Penguin Dictionary of Psychology*. London. Penguin Books, 2009. Print, p. 844.

49. Reber, Arthur S., Rhianon Allen, and Emily S. Reber. *Penguin Dictionary of Psychology*. London. Penguin Books, 2009. Print, p. 876.

50. Reber, Arthur S., Rhianon Allen, and Emily S. Reber. *Penguin Dictionary of Psychology*. London. Penguin Books, 2009. Print, p. 876.

51. Reber, Arthur S., Rhianon Allen, and Emily S. Reber. *Penguin Dictionary of Psychology*. London. Penguin Books, 2009. Print, p. 450.

52. Reber, Arthur S., Rhianon Allen, and Emily S. Reber. *Penguin Dictionary of Psychology*. London. Penguin Books, 2009. Print, p. 29.

53. Reber, Arthur S., Rhianon Allen, and Emily S. Reber. *Penguin Dictionary of Psychology*. London. Penguin Books, 2009. Print, p. 29.

54. Reber, Arthur S., Rhianon Allen, and Emily S. Reber. *Penguin Dictionary of Psychology*. London. Penguin Books, 2009. Print, p. 46.

55. Reber, Arthur S., Rhianon Allen, and Emily S. Reber. *Penguin Dictionary of Psychology*. London. Penguin Books, 2009. Print, p. 357.

56. Reber, Arthur S., Rhianon Allen, and Emily S. Reber. *Penguin Dictionary of Psychology*. London. Penguin Books, 2009. Print, p. 367.

57. Reber, Arthur S., Rhianon Allen, and Emily S. Reber. *Penguin Dictionary of Psychology*. London. Penguin Books, 2009. Print, p. 399.

58. Reber, Arthur S., Rhianon Allen, and Emily S. Reber. *Penguin Dictionary of Psychology*. London. Penguin Books, 2009. Print, p. 784.

59. Reber, Arthur S., Rhianon Allen, and Emily S. Reber. *Penguin Dictionary of Psychology*. London. Penguin Books, 2009. Print, p. 791.

60. Reber, Arthur S., Rhianon Allen, and Emily S. Reber. *Penguin Dictionary of Psychology*. London. Penguin Books, 2009. Print, p. 791.

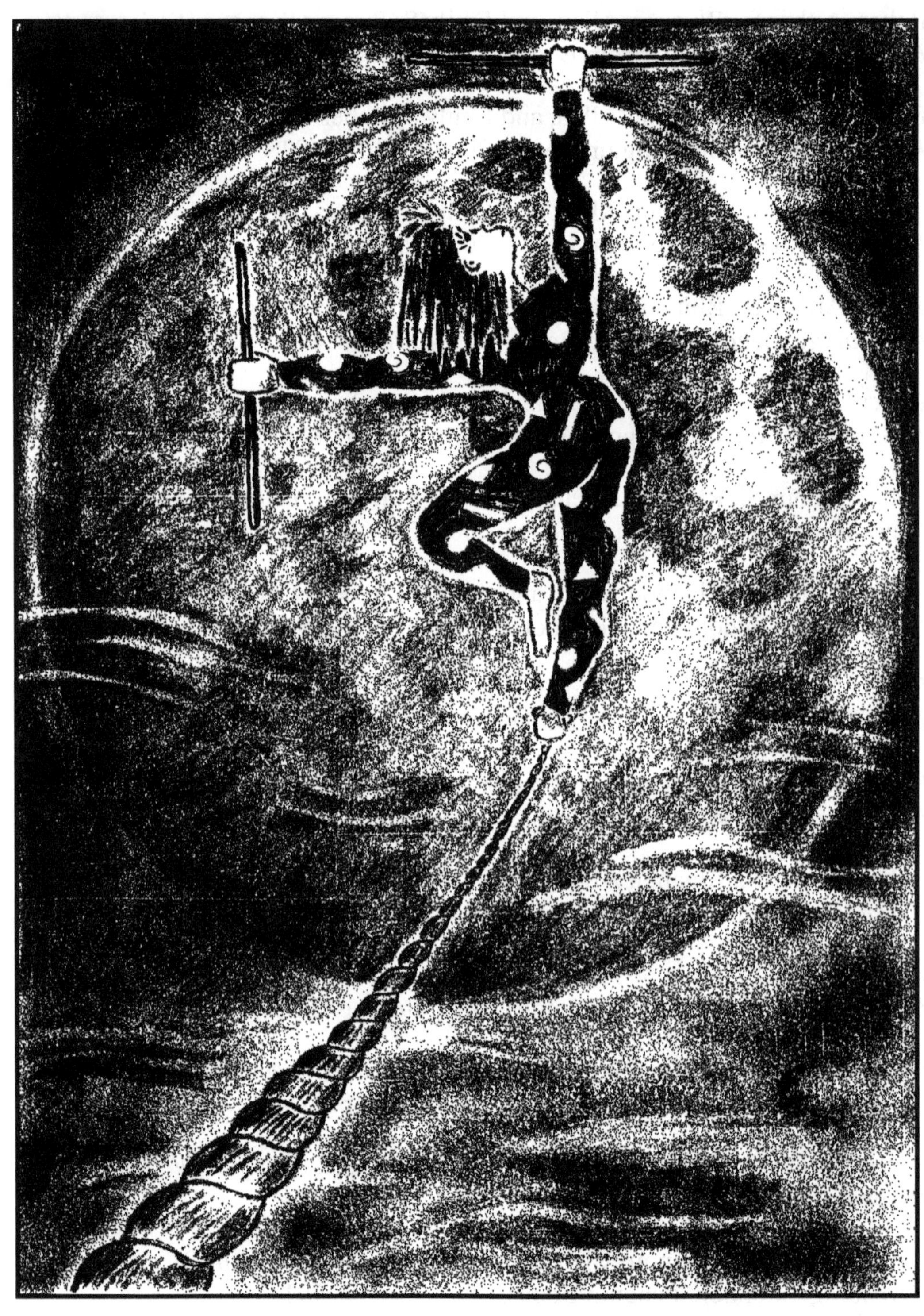

"... the way of paradoxes is the way of truth. To test Reality we must see it on the tight-rope. When the Verities become acrobats we can judge them." - Oscar Wilde[1]

CHAPTER FIFTEEN

Conclusion and Future

Front-loading: Cognitive Tradition, Darwinian Tradition, Jamesian Tradition, Natural Selection, Social Constructivist Tradition

The title of this book, *Affect Engineering: Toward a Unified Field Theory of Emotion*, certainly implores the question, "When will fields ever come into consideration?" In a vector field, points in 2-D or 3-D space are assigned vectors that measure specific quantities, such as the speed and direction of wind or of fluids. With the functions, a vector field would model the amount of anxiety or negative anxiety invested in an entity for a purpose as it moves through time.

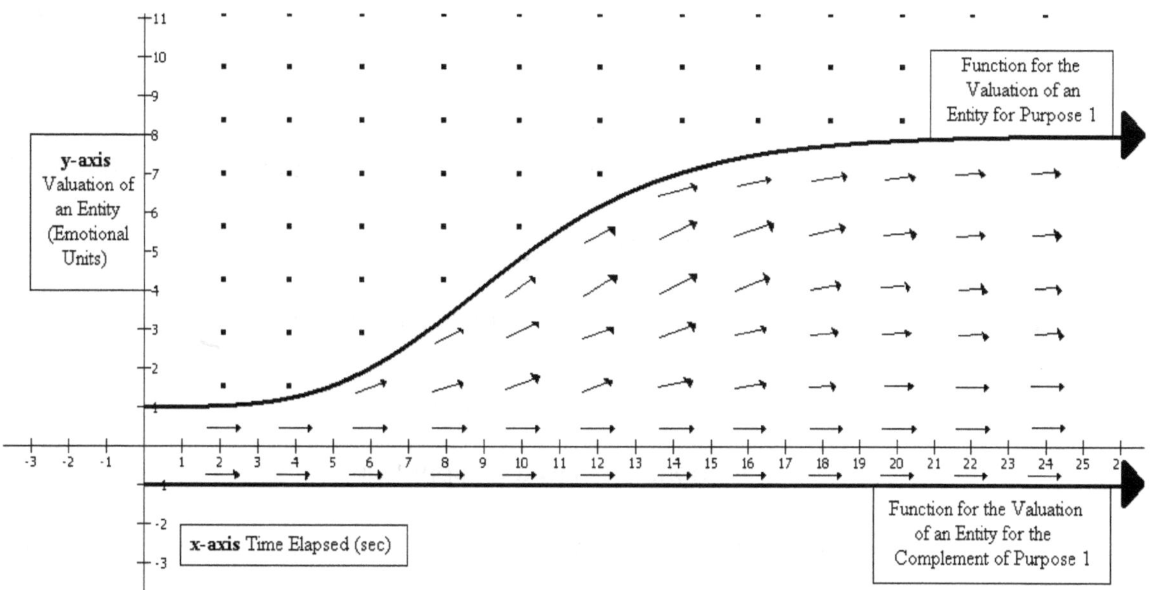

Figure 15.1 Above is a portion of a sample vector plot of anxiety and negative anxiety being invested in an entity for a purpose (top) and its complement (bottom) as they move through time to the right and as an entity's valuation rebounds after being suppressed for the main purpose. Sample Form III of attention is used.

Conclusion and Future

The employment of vector fields in two or three dimensions would permit someone using the equation to compare the valuation of an entity for a single purpose and complement to valuations of entities for multiple purposes, vicarious valuations, and the valuations or vicarious valuations from multiple people over time. This would enable the comparison of a much greater number of valuations on a large scale as all six dimensions of the equation could be taken into account if one desired. In this context, affect engineering would have countless applications too expansive to go into detail here.

A Look at Four Traditions of Emotion Perspectives

Relatively under-addressed thus far has been an investigation of facial expressions and body language as they relate to emotions. They do, however, have a place in relationship to the equation and the study of emotion. Concerning the display of emotions and their nature, Cornelius identified four theoretical traditions of research concerning the study of emotion in psychology: the "Darwinian, Jamesian, the cognitive, and the social constructivist perspectives."[2]

1) Darwinian tradition: "Emotions have adaptive functions" and "are universal."[3]
2) Jamesian tradition: Emotions are "bodily responses."[4]
3) Cognitive tradition: "Emotions are based on appraisals."[5]
4) Social Constructivist tradition: "Emotions are social constructions, serve social purposes."[6]

For the most part, the equation has emphasized the cognitive tradition and has kept in accordance with the observation that "most, if not all, contemporary cognitive theories of emotion," such as those of Lazarus and Smith, "hold that specific emotions are preceded by specific patterns of Appraisal."[7] Specific valuations and the manner by which an entity's valuation is appraised, changes, or does not change, has lead to the classification of the emotions in figure 15.2 into four categories. A total of ten Category I Emotions, twenty Category II Emotions (of five types), eight Category III Emotions (of four types), and at least five Category IV Emotions were modeled and distinguished by the appraised valuation of entities, with room to grow for miscellaneous emotions. For instance, isolated or culturally specific emotions could eventually be mapped out and incorporated under Category IV emotions if the patterns of an entity's appraised valuations do not resemble any of the first three categories.

CATEGORY I EMOTION					
AVOIDANCE OF PAIN	Grief	Sadness	Fear	Anger	Content
PURSUIT OF PLEASURE	Euphoria	Happiness	Courage	Guilt	Content

CATEGORY II EMOTION	SELF IS PASSIVE		SELF IS ACTIVE	
	VICARIOUS EMOTION	SELF	VICARIOUS EMOTION	SELF
Loving Pride	Vicarious Pride	Love	Vicarious Love	Pride
Sympathetic Shame	Vicarious Shame	Sympathy	Vicarious Sympathy	Shame
Hateful Humiliation	Vicarious Humiliation	Hatred	Vicarious Hatred	Humiliation
Antipathetic Mercy	Vicarious Mercy	Antipathy	Vicarious Antipathy	Mercy
Indifference	Vicarious Loneliness	Neutrality	Vicarious Neutrality	Loneliness

CATEGORY III EMOTION	INDULGENT TYPE		PROTECTIVE TYPE	
	CAT. I EMOTION +	CAT. II EMOTION	CAT. I EMOTION +	CAT. II EMOTION
BENEVOLENCE	Happiness	Loving Pride	Anger	Loving Pride
JEALOUSY	Guilt	Sympathetic Shame	Sadness	Sympathetic Shame
MALEVOLENCE	Happiness	Hateful Humiliation	Anger	Hateful Humiliation
ENVY	Guilt	Antipathetic Mercy	Sadness	Antipathetic Mercy

CATEGORY IV EMOTIONS			
SURPRISE	JOYFULNESS	RESTLESSNESS	HELPLESSNESS
LIMERENCE	CONFUSION	GREED	ETC.

Figure 15.2 The Four Categories of Emotions

Conclusion and Future

Affect (i.e., positive anxiety and negative anxiety in the equation) that is felt but divorced from the cognitive processes and variables in the equation falls under the classification of entropy. Entropy, in the equation, is emotional energy (e.g., an emotional unit) that is not available to do work by cognitively valuing entities. Where entropy exists, one of the following must be true:

1) The valuing neurons are in a state of permanent rest (i.e., expired).
2) The activation threshold of the valuing neurons threshold is too high for the individual's cognitive processes to overcome (i.e., neural fatigue).
3) The valuing neurons are already active and cannot be brought under control by cognitive processes and awareness. In other words, the deactivation threshold of the valuing neurons is too low for the individual's cognitive processes to overcome (e.g., seizure, unconscious processes, etc.).

Effectively, the counter-argument from Zajonc's position that affective reactions can take place "without the participation of cognitive processes" can be reconciled with Lazarus' Cognitive-Motivational-Relational Theory when viewed through these lenses.[8] Affective reactions that take place without intentional cognitive activity, such as valuing neurons firing on their own or as a result of some non-cognitive stimulus, make up entropy in the equation. Their energy, while exhibiting an effect on the body, would temporarily be unavailable for use to cognitively value entities and become background noise against which the individual must operate.

For example, it would be implausible to say that an individual who, after shattering a leg in four places, only felt psychological pain after cognitively assessing that his or her leg was valuable to the purpose of going mountain climbing next weekend and that breaking them has threatened this purpose. A more plausible scenario is that shooting pain was felt from the broken appendage and thereafter the individual lay on the ground crippled in searing agony. The normal neurological pathway would then be reversed, with the valuing neurons firing at a high rate due to stimulation being received from receptor cites normally targeted (i.e., damaged muscle tissue).

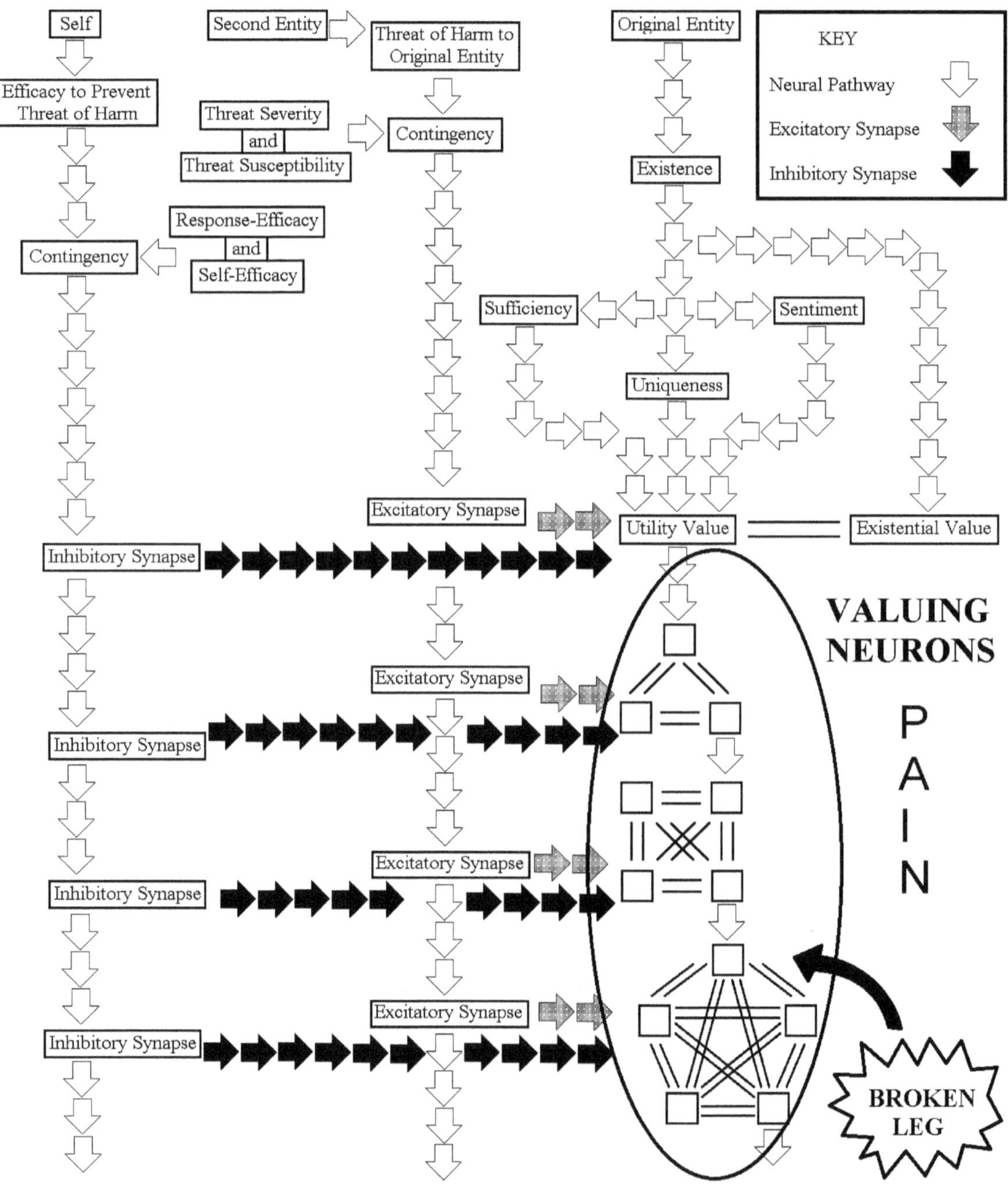

Figure 15.3 Valuing neurons that fire intensely in response to stimulation that is not initiated cognitively would likely become entropy unless brought under control.

However, if the individual who has just broken the leg is particularly steadfast and stoic, it may be the case that he or she is not overwhelmed by the accident. If the valuing neurons and the anxiety concerning them is completely subdued by the individual (e.g., through mental fortitude), then the previously unusable emotional energy could be harnessed and become usable again.

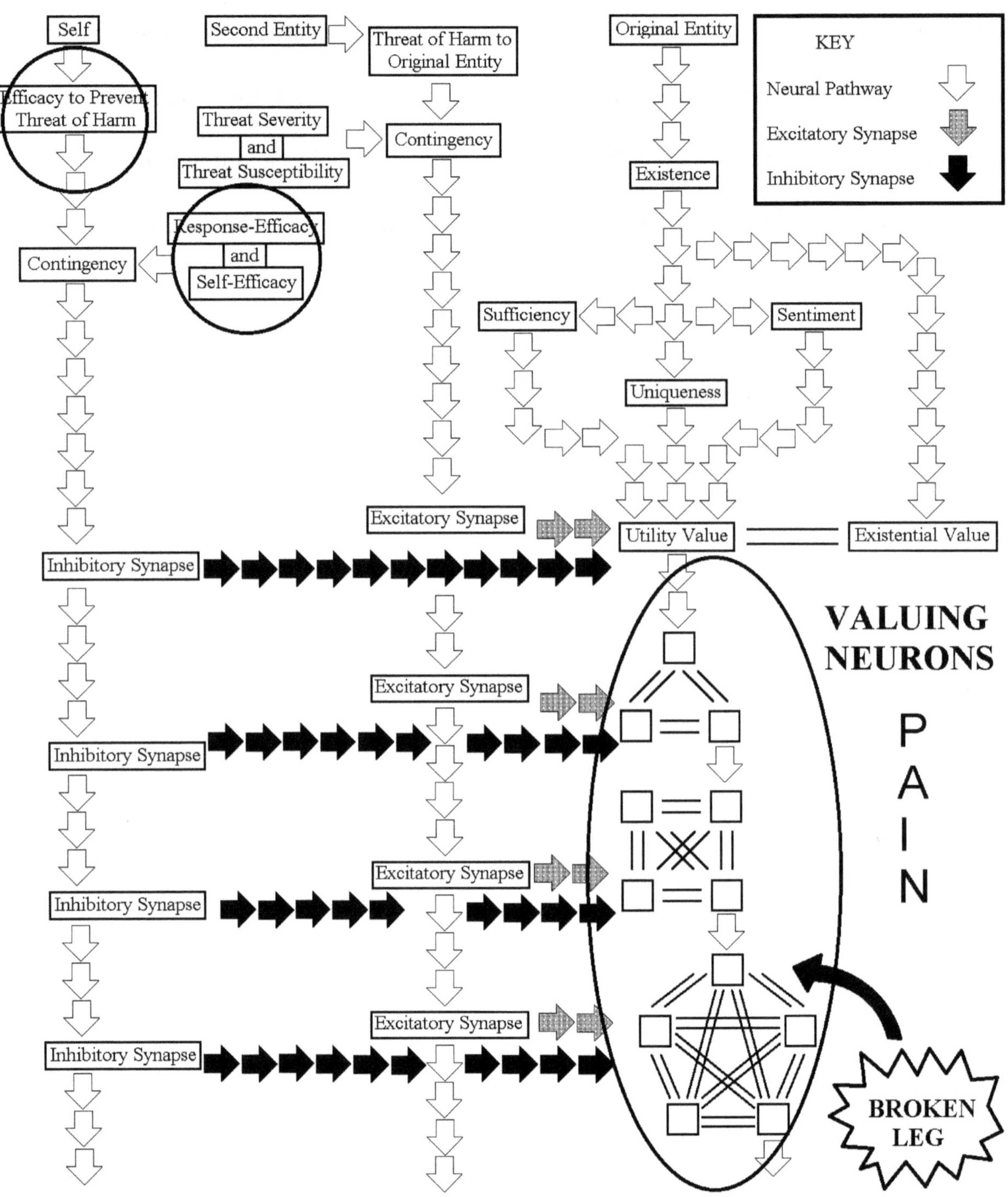

Figure 15.4 To minimize the entropy in an individual's system following the broken leg, an individual may direct all of his or her attentional energy to response and self efficacy (i.e., mind-over-matter).

The example in figure 15.3 and 15.4 draws attention to the Jamesian perspective of emotion.

William James held that "common sense got the sequence of emotional experience all wrong."[9] Rather than first perceiving an "emotion-eliciting stimulus," experiencing an emotion, and expressing it, James argued that one must first "experience the bodily changes that have been initiated by the emotion-eliciting stimulus."[10]

1) Common sense order: Perception; Emotion; Bodily Response.[11]

2) William James and Carl Lange: Perception; Bodily Response, Emotion[12]

Cornelius gives a compelling example of the difference in his example of an individual walking through the woods and being accosted by a "rabid woodchuck."[13] A common sense argument would hold that the individual perceives the stimulus, the woodchuck, becomes afraid, and then feels a bodily response to it such as a "pounding heart" or "sweaty hands."[14] Under the James-Lange model, an individual would perceive the woodchuck, feel a bodily response such as the "pounding heart" and "sweaty hands," and then become afraid.[15] Also, among William James' claims is the notion that if one were to merely "adopt the posture, facial expression, or other behavior associated with a particular emotion," then "one would come to experience that emotion."[16] In other words, if one were feeling miserable and desired to experience a particular emotion, such as happiness, James would have held that one merely had to take on the posture or facial expressions associated with the emotion.[17]

Although James' use of the ambiguous term "bodily changes" opened himself up to a barrage of criticism, the idea he was attempting to convey, that "one must first have bodily changes" in order to experience an emotion, can be accounted for in the neurological model and the calculus function as entropy, or conceived as a type of nonretractable anxiety-mass in the Roll Cage Theory of Drives.[18] If affect and bodily changes are divorced from cognition, then the emotional energy associated with it (e.g., positive anxiety, negative anxiety, emotional units) is understood as entropy in the equation, from chapter twelve. Unless the valuing neurons corresponding to specific bodily processes can be brought under the influence of cognitive processes, then they will remain unusable for cognitive work.

Emotions, in the equation, are modeled as changes in the valuation of an entity over time (e.g., affect felt for or anxiety invested in) concerning a specific purpose. Affect that is divorced from cognition, from an entity, and from a purpose, but still directs behavior, would be indicative of impulsive behavior. Actions that seem to be divorced from affect would be indicative of flat behavior.

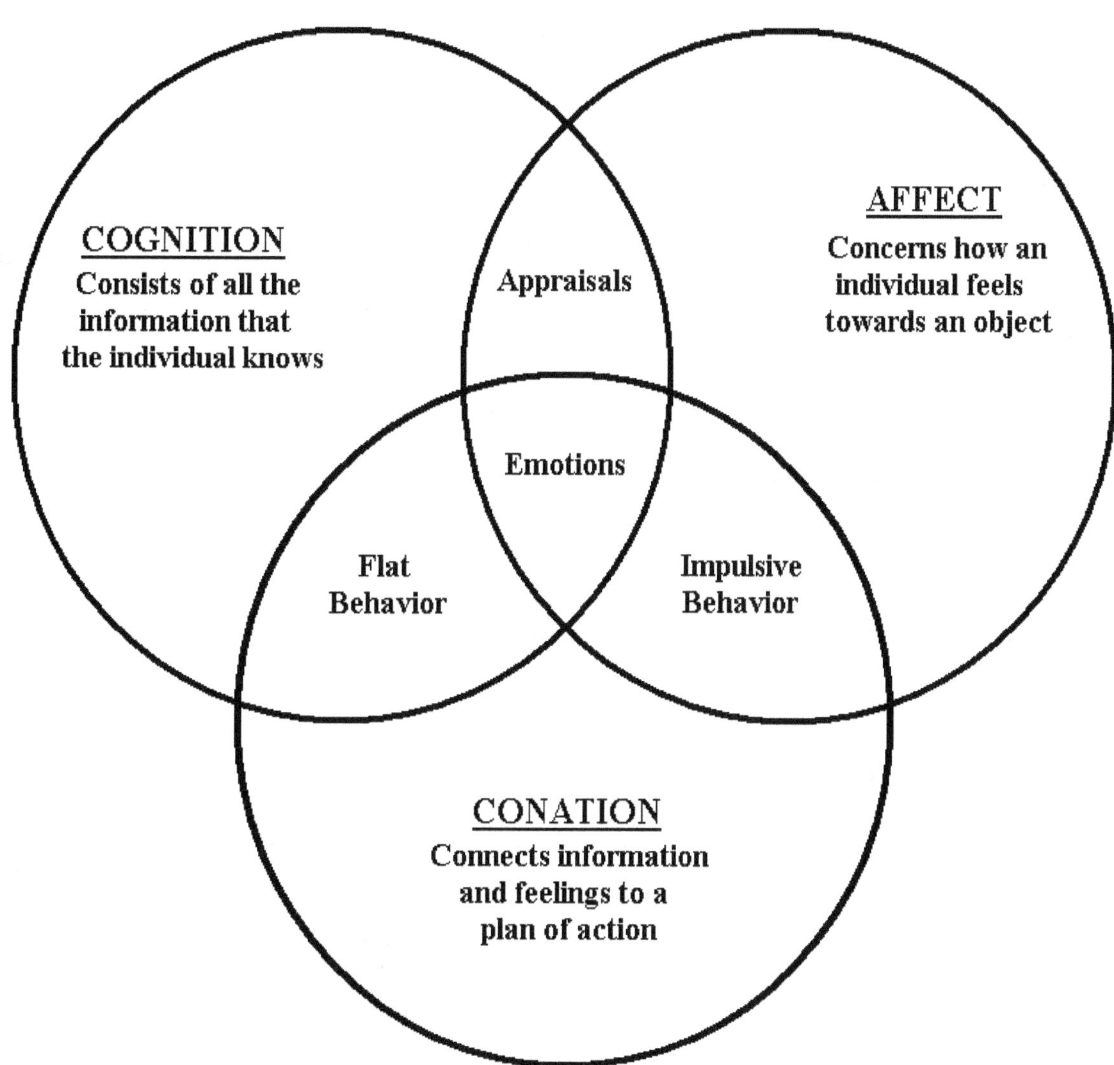

Figure 15.5 From chapter three, affect that motivates behavior independently of cognition (e.g., entropy) would be indicative of impulsive behavior. Similarly, cognition that motivates behavior independently of affect would be indicative of flat behavior.

Subsequently, these changes in valuation, measured in emotional units, would correspond to neurological activity and subsequently, the physiology of an individual. Henceforth, the sensation of an entity's valuation changing or remaining constant, and the bodily processes associated with this (e.g., elevation of neural activity), is emotion in the equation. In the equation, the mere presence of bodily changes without a subsequent change in an entity's cognitive valuation does not constitute an emotion. Likewise, the presence of cognitive processes without an appropriate level of affect involved does not constitute an emotion either.

The social constructivist viewpoint, bearing much in common with social constructionism, can be accounted for by assessing the manners by which a single idea may be construed as two, three, or more separate purposes without any loss of meaning between them (e.g., comparing the effectiveness of exponents of the equation in figure 8.18 to 8.20, or figure 8.19 to 8.21). Different constructions of the same idea, for instance, might be processed by an individual more or less readily depending on whether threat is more salient to an individual, benefit, the efficacy to prevent something or the efficacy to ensure something when contingencies and relationships between entities are being considered. The eight different manners by which the equation can be arranged (e.g., set up of the threat/benefit and efficacy components in the exponent and whether they will be ensured or prevented) should account for a number of differences in semantics and social constructions.

Finally, in consideration of the Darwinian perspective, the research of Paul Ekman and Carroll Izard, both pioneers in the Darwinian tradition who aimed to demonstrate the "cross-cultural universality of facial expressions" in order to validate Darwin's view of emotion and natural selection, must be considered.[19] Natural selection holds that traits beneficial to an organism's survival are more likely to be passed on to offspring. The incorporation of facial expressions into the equation is a fairly straightforward one and is nearly identical to the instance of the broken leg in figure 15.3 and figure 15.4. Specific patterns of valuations for entities might be expected to correspond to specific facial expressions. For instance, happiness, modeled as the valuation of an entity decreasing so that larger amounts of negative anxiety become present, might be expected to

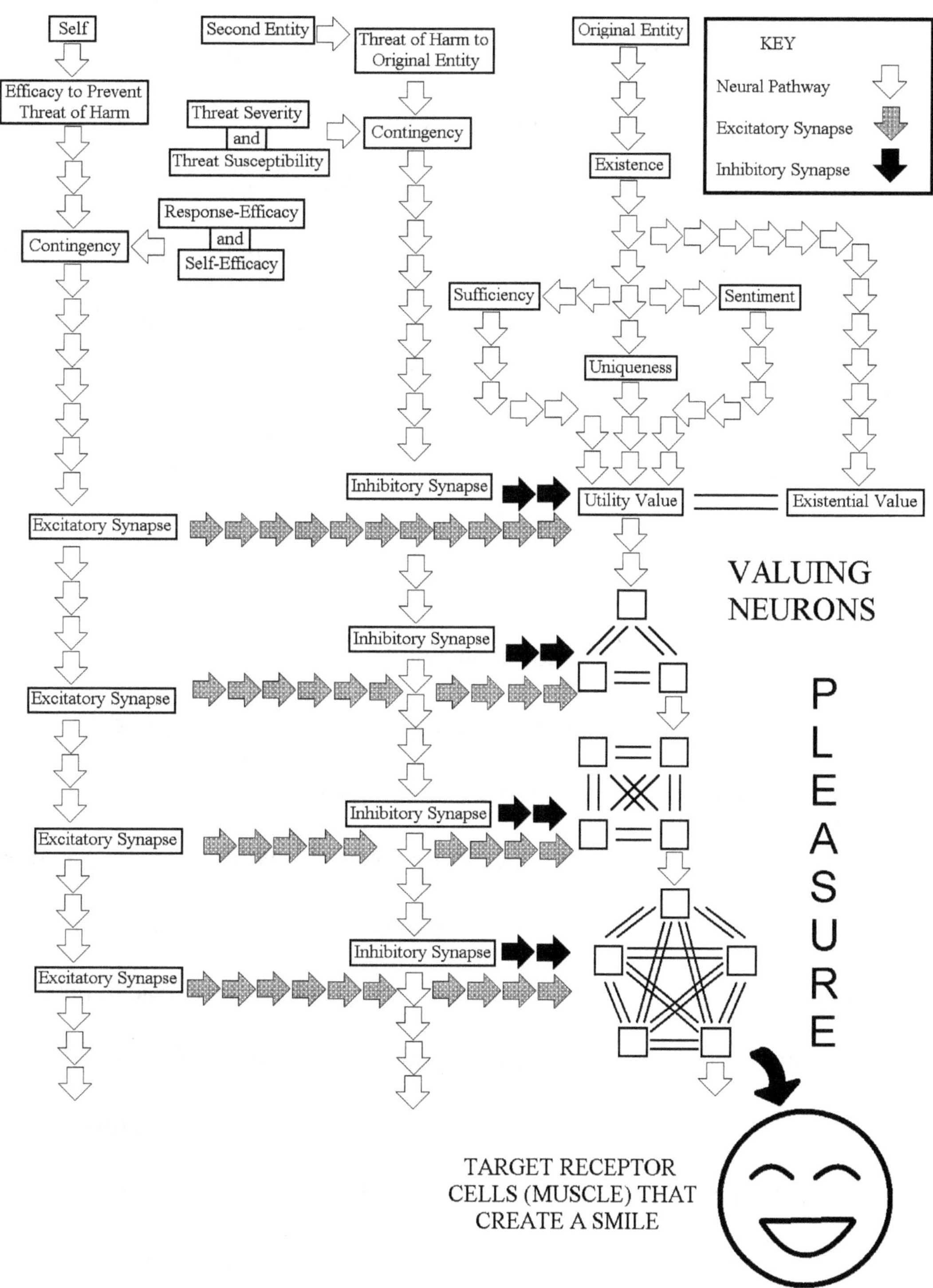

Figure 15.6 Valuing neurons that invest negative anxiety into an entity could, after targeting specific receptor cells for facial muscles, create a smile.

target facial muscles creating a smile (figure 15.6). Like the broken leg in figure 15.3 and 15.4, the possibility is left open that simply smiling may, through a reverse route, create pleasure and entropy, (of the more preferable variety, but entropy nonetheless). As facial expressions are entities themselves, determining whether or not they are inherited from genes or learned is a question the equation does not have to answer; it only has to be able to model both likelihoods, which it can do. The choice then, is up to the users of the functions to decide for themselves what they wish to emphasize and hold as true.

Concluding Thoughts

There are of course many more theories and models that could have been incorporated into the equation and were not included within the confines of this already large book. The calculus functions, the Roll Cage Theory of Drives, and the neurological network model were three tools for modeling affect and emotion that were outlined in this book. Providing someone a compass and demonstrating its use, however, does not mean that it must be used in the suggested manner. If one is fortunate enough, not using it in any of the manners specified may even lead to the discovery of new frontiers and novel ways to approach and solve problems. Yet, just as any lifesaving device can become a weapon of destruction, so too can the ideas presented in this work be utilized for aims that can either promote or inhibit the well-being of oneself and others. To say the least, affect engineering is a discipline that has the potential to bring out both the best and the worst in humanity. It should be held in high esteem and treated accordingly.

What would Protagoras have said?

Notes

1. Wilde, Oscar (2003). *The Picture of Dorian Gray.* New York: Barnes and Nobles Books. Print, p. 43. (Original work published 1890).

2. Cornelius, R. R. *The Science of Emotion: Research and Tradition in the Psychology of Emotion.* Upper Saddle River, NJ. Prentice-Hall, Inc. 1996. Print, p. 11.

3. Cornelius, R. R. *The Science of Emotion: Research and Tradition in the Psychology of Emotion.* Upper Saddle River, NJ. Prentice-Hall, Inc. 1996. Print, p. 12.

4. Cornelius, R. R. *The Science of Emotion: Research and Tradition in the Psychology of Emotion.* Upper Saddle River, NJ. Prentice-Hall, Inc. 1996. Print, p. 12.

5. Cornelius, R. R. *The Science of Emotion: Research and Tradition in the Psychology of Emotion.* Upper Saddle River, NJ. Prentice-Hall, Inc. 1996. Print, p. 12.

6. Cornelius, R. R. *The Science of Emotion: Research and Tradition in the Psychology of Emotion.* Upper Saddle River, NJ. Prentice-Hall, Inc. 1996. Print, p. 12.

7. Cornelius, R. R. *The Science of Emotion: Research and Tradition in the Psychology of Emotion.* Upper Saddle River, NJ. Prentice-Hall, Inc. 1996. Print, p. 124.

8. Cornelius, R. R. *The Science of Emotion: Research and Tradition in the Psychology of Emotion.* Upper Saddle River, NJ. Prentice-Hall, Inc. 1996. Print, p. 128.

9. Cornelius, R. R. *The Science of Emotion: Research and Tradition in the Psychology of Emotion.* Upper Saddle River, NJ. Prentice-Hall, Inc. 1996. Print, p. 60.

10. Cornelius, R. R. *The Science of Emotion: Research and Tradition in the Psychology of Emotion.* Upper Saddle River, NJ. Prentice-Hall, Inc. 1996. Print, p. 60.

11. Cornelius, R. R. *The Science of Emotion: Research and Tradition in the Psychology of Emotion.* Upper Saddle River, NJ. Prentice-Hall, Inc. 1996. Print, p. 60-62.

12. Cornelius, R. R. *The Science of Emotion: Research and Tradition in the Psychology of Emotion.* Upper Saddle River, NJ. Prentice-Hall, Inc. 1996. Print, p. 60-62.

13. Cornelius, R. R. *The Science of Emotion: Research and Tradition in the Psychology of Emotion.* Upper Saddle River, NJ. Prentice-Hall, Inc. 1996. Print, p. 61.

14. Cornelius, R. R. *The Science of Emotion: Research and Tradition in the Psychology of Emotion.* Upper Saddle River, NJ. Prentice-Hall, Inc. 1996. Print, p. 62.

15. Cornelius, R. R. *The Science of Emotion: Research and Tradition in the Psychology of Emotion.* Upper Saddle River, NJ. Prentice-Hall, Inc. 1996. Print, p. 62.

16. Cornelius, R. R. *The Science of Emotion: Research and Tradition in the Psychology of Emotion.* Upper Saddle River, NJ. Prentice-Hall, Inc. 1996. Print, p. 63.

17. Cornelius, R. R. *The Science of Emotion: Research and Tradition in the Psychology of Emotion.* Upper Saddle River, NJ. Prentice-Hall, Inc. 1996. Print, p. 63.

18. Cornelius, R. R. *The Science of Emotion: Research and Tradition in the Psychology of Emotion.* Upper Saddle River, NJ. Prentice-Hall, Inc. 1996. Print, p. 61.

19. Cornelius, R. R. *The Science of Emotion: Research and Tradition in the Psychology of Emotion.* Upper Saddle River, NJ. Prentice-Hall, Inc. 1996. Print, p. 31.

LIST OF FIGURES

List of Figures

List of Figures

INDEX

www.ingramcontent.com/pod-product-compliance
Lightning Source LLC
Chambersburg PA
CBHW052128020426
42334CB00023B/2639

* 9 7 8 0 9 9 6 0 4 9 3 1 3 *